HAESE MATHEMATICS

Advanced Mathematics

for A Level

2

Michael Haese   *Mark Humphries*   *Chris Sangwin*   *Ngoc Vo*

ADVANCED MATHEMATICS 2 FOR A LEVEL

Michael Haese   B.Sc.(Hons.), Ph.D.
Mark Humphries   B.Sc.(Hons.)
Chris Sangwin   M.A., M.Sc., Ph.D.
Ngoc Vo   B.Ma.Sc.

Published by Haese Mathematics
152 Richmond Road, Marleston, SA 5033, AUSTRALIA
Telephone: +61 8 8210 4666, Fax: +61 8 8354 1238
Email: info@haesemathematics.com.au
Web: www.haesemathematics.com.au

National Library of Australia Card Number & ISBN 978-1-925489-32-3

© Haese & Harris Publications 2017

    First Edition   2017

Cartoon artwork by John Martin and Rebecca Huang.

Cover art by Jennifer Court.

Photo of Fraser Corsan on page 377 by Jarno Cordia.

Artwork by Brian Houston and Bronson Mathews.

Computer software by Huda Kharrufa, Brett Laishley, Bronson Mathews, Linden May, Joshua Douglass-Molloy, Jonathan Petrinolis, and Nicole Szymanczyk.

Production work by Sandra Haese, Bradley Steventon, and Rebecca Huang.

Typeset in Australia by Deanne Gallasch and Charlotte Frost. Typeset in Times Roman $10\frac{1}{2}$.

Printed in China by Prolong Press Limited.

FOREWORD

This book is written for the revised GCE Advanced level (A level) Mathematics specifications for first teaching in 2017.

The book is designed to complete the A level Mathematics course in conjunction with **Advanced Mathematics 1 for AS and A Level**.

Since the content for A level Mathematics is now prescribed in the syllabus, this book is suitable for any of the awarding organisations.

To reflect the principles on which the course is based, we have attempted to produce a book that embraces understanding and problem solving in order to give students different learning experiences.

The textbook and interactive online features provide an engaging and structured package, allowing students to explore and develop their confidence in mathematics. The material is presented in a clear, easy-to-follow style, free from unnecessary distractions, while effort has been made to contextualise questions so that students can relate concepts to everyday use.

Each chapter begins with an Opening Problem, offering an insight into the application of the mathematics that will be studied in the chapter. Important information and key notes are highlighted, while worked examples provide step-by-step instructions with concise and relevant explanations. Discussions, Activities, Investigations, Puzzles, and Research exercises are used throughout the chapters to develop understanding, problem solving, and reasoning, within an interactive environment.

The interactive online features include our SELF TUTOR software (see p. 6), links to graphing software, statistics software, demonstrations, calculator instructions, and a range of printable worksheets, tables, and diagrams, allowing teachers to demonstrate concepts and students to experiment for themselves.

A chapter summary for teachers is available on our website.

We welcome your feedback.   Email: info@haesemathematics.com.au
                              Web: www.haesemathematics.com.au

PMH, MAH, CS, NV

ABOUT THE AUTHORS

Michael Haese completed a BSc at the University of Adelaide, majoring in Infection and Immunity, and Applied Mathematics. He completed Honours in Applied Mathematics, and a PhD in high speed fluid flows. Michael has a keen interest in education and a desire to see mathematics come alive in the classroom through its history and relationship with other subject areas. He is passionate about girls' education and ensuring they have the same access and opportunities that boys do. His other interests are wide-ranging, including show jumping, cycling, and agriculture. He has been the principal editor for Haese Mathematics since 2008.

Mark Humphries completed a degree in Mathematical and Computer Science, and an Economics degree at the University of Adelaide. He then completed an Honours degree in Pure Mathematics. His mathematical interests include public key cryptography, elliptic curves, and number theory. Mark enjoys the challenge of piquing students' curiosity in mathematics, and encouraging students to think about mathematics in different ways. He has been working at Haese Mathematics since 2006, and is currently the writing manager.

Chris Sangwin completed a BA in Mathematics at the University of Oxford, and an MSc and PhD in Mathematics at the University of Bath. He spent thirteen years in the Mathematics Department at the University of Birmingham, and from 2000 - 2011 was seconded half time to the UK Higher Education Academy "Maths Stats and OR Network" to promote learning and teaching of university mathematics. He was awarded a National Teaching Fellowship in 2006, and is now Professor of Technology Enhanced Science Education at the University of Edinburgh.

His research interests focus on technology and mathematics education and include automatic assessment of mathematics using computer algebra, and problem solving using the Moore method and similar student-centred approaches.

Ngoc Vo completed a BMaSc at the University of Adelaide, majoring in Statistics and Applied Mathematics. Her Mathematical interests include regression analysis, Bayesian statistics, and statistical computing. Ngoc has been working at Haese Mathematics as a proof reader and writer since 2016.

Dedicated to Prof. John Blake.

Mathematician, teacher, mentor, and friend.

ONLINE FEATURES

With the purchase of a new textbook, you will gain 27 months subscription to our online product. This subscription can be renewed for a small fee.

Access is granted through **SNOWFLAKE**, our book viewing software that can be used in your web browser or may be installed to your tablet or computer.

Students can revisit concepts taught in class and undertake their own revision and practice online.

COMPATIBILITY

For iPads, tablets, and other mobile devices, some of the interactive features may not work. However, the digital version of the textbook can be viewed online using any of these devices.

REGISTERING

You will need to register to access the online features of this textbook.

Visit www.haesemathematics.com.au/register and follow the instructions. Once registered, you can:
- activate your digital textbook
- use your account to make additional purchases.

To activate your digital textbook, contact Haese Mathematics. On providing proof of purchase, your digital textbook will be activated. **It is important that you keep your receipt as proof of purchase.**

For general queries about registering and subscriptions:
- Visit our **SNOWFLAKE** help page: http://snowflake.haesemathematics.com.au/help
- Contact Haese Mathematics: info@haesemathematics.com.au

ONLINE VERSION OF THE TEXTBOOK

The entire text of the book can be viewed online, allowing you to leave your textbook at school.

SELF TUTOR

Self tutor is an exciting feature of this book.

The ◀) **Self Tutor** icon on each worked example denotes an active online link.

> Simply 'click' on the ◀) **Self Tutor** (or anywhere in the example box) to access the worked example, with a teacher's voice explaining each step necessary to reach the answer.
>
> Play any line as often as you like. See how the basic processes come alive using movement and colour on the screen.

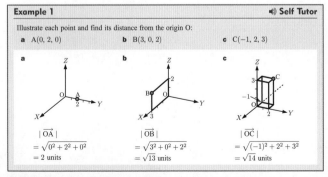

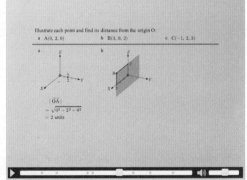

See **Chapter 14, Vectors,** p. 316

INTERACTIVE LINKS

Throughout your digital textbook, you will find interactive links to:

- Graphing software
- Statistics software
- Games
- Demonstrations
- Printable pages

CLICK ON THESE ICONS ONLINE

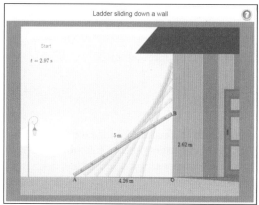

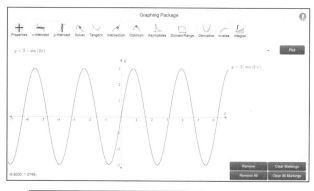

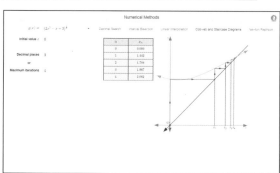

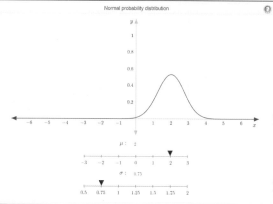

GRAPHICS CALCULATOR INSTRUCTIONS

Printable graphics calculator instruction booklets are available for the **Casio fx-9860G PLUS**, **Casio fx-CG20**, and the **TI-84 Plus CE**. Click on the relevant icon below.

CASIO fx-9860G PLUS

CASIO fx-CG20

TI-84 Plus CE

When additional calculator help may be needed, specific instructions can be printed from icons within the text.

GRAPHICS CALCULATOR INSTRUCTIONS

TABLE OF CONTENTS

SYMBOLS AND NOTATION USED IN THIS BOOK

$\in$	is an element of
$\notin$	is not an element of
$\varnothing$	the empty set
$\mathcal{E}$	the universal set
$\mathbb{N}$	the set of natural numbers, $\{1, 2, 3, ...\}$
$\mathbb{Z}$	the set of integers, $\{0, \pm 1, \pm 2, \pm 3, ...\}$
$\mathbb{Z}^+$	the set of positive integers, $\{1, 2, 3, ...\}$
$\mathbb{Z}_0^+$	the set of non-negative integers, $\{0, 1, 2, 3, ...\}$
$\mathbb{R}$	the set of real numbers
$\mathbb{Q}$	the set of rational numbers, $\{\frac{p}{q} : p \in \mathbb{Z}, \ q \in \mathbb{Z}^+\}$
$[a, b]$	the closed interval $\{x \in \mathbb{R} : a \leqslant x \leqslant b\}$
$[a, b)$	the interval $\{x \in \mathbb{R} : a \leqslant x < b\}$
$(a, b]$	the interval $\{x \in \mathbb{R} : a < x \leqslant b\}$
(a, b)	the open interval $\{x \in \mathbb{R} : a < x < b\}$
$\approx$	is approximately equal to
$\equiv$	is identical to or is congruent to
∞	infinity
$\propto$	is proportional to
$\therefore$	therefore
$<$	is less than
$\leqslant, \leq$	is less than or equal to, is not greater than
$>$	is greater than
$\geqslant, \geq$	is greater than or equal to, is not less than
$p \Rightarrow q$	p implies q (if p then q)
$p \Leftrightarrow q$	p implies and is implied by q (p is equivalent to q)
$\sum\limits_{i=1}^{n} a_i$	$a_1 + a_2 + ... + a_n$
d	common difference for an arithmetic sequence

r	common ratio for a geometric sequence
S_n	sum to n terms of a sequence
$\sqrt{a}$	the non-negative square root of a
$|a|$	the modulus of a
$n!$	n factorial: $n! = n \times (n-1) \times .... \times 2 \times 1$, $n \in \mathbb{N}; \ 0! = 1$
$\binom{n}{r}$	the binomial coefficient $\dfrac{n!}{r!(n-r)!}$ for $n, r \in \mathbb{Z}_0^+$, $r \leqslant n$ or $\dfrac{n(n-1)....(n-r+1)}{r!}$ for $n \in \mathbb{Q}, \ r \in \mathbb{Z}_0^+$
$f(x)$	the value of the function f at x
$f : x \mapsto y$	the function f maps the element x to the element y
$\lim\limits_{x \to a} f(x)$	the limit of $f(x)$ as x tends to a
$\Delta x, \delta x$	an increment of x
$\dfrac{dy}{dx}, \dfrac{d^2 y}{dx^2}, ..., \dfrac{d^n y}{dx^n}$	the first, second, ..., nth derivatives of y with respect to x
$f'(x), f''(x), ..., f^{(n)}(x)$	the first, second, ..., nth derivatives of $f(x)$ with respect to x
$\int y \, dx$	the indefinite integral of y with respect to x
$\int_a^b y \, dx$	the definite integral of y with respect to x between $x = a$ and $x = b$
e	base of natural logarithms
$e^x, \exp x$	exponential function of x
$\log_a x$	logarithm to the base a of x
$\ln x, \log_e x$	natural logarithm of x
sin, cos, tan, cosec, sec, cot	the trigonometric functions
$\sin^{-1}, \cos^{-1}, \tan^{-1}$, arcsin, arccos, arctan	the inverse trigonometric functions
$^\circ$	degrees
rad	radians

$\mathbf{a}$	the vector $\mathbf{a}$
$\overrightarrow{AB}$	the vector represented by the directed line segment AB
$\mathbf{i}, \mathbf{j}$	unit vectors in the directions of the Cartesian coordinate axes
$|\mathbf{a}|$	the magnitude of $\mathbf{a}$
$|\overrightarrow{AB}|$, AB	the magnitude of $\overrightarrow{AB}$
$\binom{a}{b}$, $a\mathbf{i} + b\mathbf{j}$	column vector and corresponding unit vector notation
$A \cup B$	union of the events A and B
$A \cap B$	intersection of the events A and B
$P(A)$	probability of the event A
A'	complement of the event A
$P(A \mid B)$	probability of event A conditional on the event B
X, Y, R, etc.	random variables
x, y, r, etc.	values of the random variables X, Y, R, etc.
$x_1, x_2, \ldots$	values of observations
$f_1, f_2, \ldots$	frequencies with which the observations $x_1, x_2, \ldots$ occur
$p(x)$, $P(X = x)$	probability function of the discrete random variable X
$p_1, p_2, \ldots$	probabilities of the values $x_1, x_2, \ldots$ of the discrete random variable X
$\sim$	has the distribution
$B(n, p)$	binomial distribution with parameters n and p, where n is the number of trials and p is the probability of success in a trial

$N(\mu, \sigma^2)$	normal distribution with mean μ and variance σ^2
$Z \sim N(0, 1)$	standard normal distribution
μ	population mean
σ^2	population variance
σ	population standard deviation
$\bar{x}$	sample mean
s^2	sample variance
s	sample standard deviation
H_0	Null hypothesis
H_1	Alternative hypothesis
r	product moment correlation coefficient
ρ	product moment correlation coefficient for a population
t	time
s	displacement
u	initial velocity
v	velocity or final velocity
a	acceleration
g	acceleration due to gravity
$\mathbf{F}$	force or resultant force
N	Newton
N m	Newton metre (moment of a force)
μ	coefficient of friction

LARGE DATA SETS

A Level Mathematics specifications require students to become familiar with one or more specified large data sets.

It is important that students explore the data set(s) during the course of their study, so that they are aware of the terminology and contexts relating to the data.

In the final examination, students will be required to answer questions based on selected data or summary statistics from the large data set(s). Students who are familiar with the context and structure of the large data set(s) will be more able to engage in realistic interpretation in an examination setting.

Sample questions associated with each examination board's large data sets are provided in the icons below.

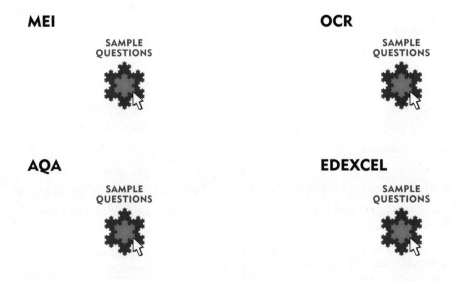

MEI

SAMPLE QUESTIONS

OCR

SAMPLE QUESTIONS

AQA

SAMPLE QUESTIONS

EDEXCEL

SAMPLE QUESTIONS

1

Sequences and series

Contents:

Opening problem   The legend of Sissa ibn Dahir

Around 1260 AD, the Kurdish historian Ibn Khallikān recorded the following story about Sissa ibn Dahir and a chess game against the Indian King Shihram.

King Shihram was a tyrant king, and his subject Sissa ibn Dahir wanted to teach him how important all of his people were. He invented the game of chess for the king, and the king was greatly impressed. He insisted on Sissa ibn Dahir naming his reward, and the wise man asked for one grain of wheat for the first square, two grains of wheat for the second square, four grains of wheat for the third square, and so on, doubling the wheat on each successive square on the board.

The king laughed at first and agreed, for there was so little grain on the first few squares. By halfway he was surprised at the amount of grain being paid, and soon he realised his great error: that he owed more grain than there was in the world.

Things to think about:

a How can we describe the number of grains of wheat for each square?

b What expression gives the number of grains of wheat for the nth square?

c Find the total number of grains of wheat that the king owed.

To help understand problems like the **Opening Problem**, we need to study **sequences** and their sums which are called **series**.

A   NUMBER SEQUENCES

Consider the illustrated tower of bricks:

We let the top row be the first row.

The first row has 3 bricks.
The second row has 4 bricks.
The third row has 5 bricks.
The fourth row has 6 bricks.

1st row ⟶
2nd row ⟶
3rd row ⟶

If we let t_n represent the number of bricks in the nth row, then $t_1 = 3$, $t_2 = 4$, $t_3 = 5$, and $t_4 = 6$.

The pattern could be continued forever, generating the **sequence** of numbers:
3, 4, 5, 6, 7, 8, 9, ...

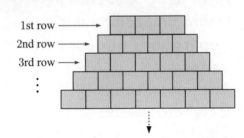

The string of dots indicates that the pattern continues forever.

A **number sequence** is an ordered list of numbers defined by a rule.

The numbers in a sequence are called the **terms** of the sequence.

We will look at three ways of describing a number sequence.

(1) Using **words**.
The sequence for the tower of bricks can be described as "starting at 3, and increasing by 1 each time".

(2) Using an **explicit formula**, which gives the nth term of the sequence t_n in terms of n.
The explicit formula for the tower of bricks is $t_n = n + 2$. t_n is called the **nth term** or the **general term**.
We can use this formula to find, for example, the 20th term of the sequence, which is $t_{20} = 20 + 2 = 22$.

(3) Using a **recursive formula**, which gives the nth term of the sequence in terms of one or more of the preceding terms.
The recursive formula for the tower of bricks is $t_1 = 3$, $t_{n+1} = t_n + 1$ for $n \geqslant 1$.
So, we have $t_1 = 3$
$$t_2 = t_1 + 1 = 3 + 1 = 4$$
$$t_3 = t_2 + 1 = 4 + 1 = 5, \quad \text{and so on.}$$

In a recursive formula, at least one initial term must be defined.

EXERCISE 1A

1 Write down the first five terms of the sequence:
  a starting at 7, and increasing by 6 each time
  b starting at 23, and decreasing by 4 each time
  c whose nth term is the nth square number.

2 The sequence of prime numbers is 2, 3, 5, 7, 11, 13, 17, 19, ... Write down the value of:
  a t_2 **b** t_5 **c** t_{10}.

3 Consider the sequence 4, 7, 10, 13, 16, ...
  a Describe the sequence in words.
  b Write down the values of t_1 and t_4.
  c Assuming the pattern continues, find the value of t_8.

4 Find the first four terms of the sequence defined by the explicit formula $t_n = 2n + 5$.

5 A sequence is defined by the explicit formula $t_n = 3n - 2$. Find:
  a t_1 **b** t_5 **c** t_{27}.

6 Consider the number sequence $-9, -6, -1, 6, 15,$ ....
  a Which of these is the correct explicit formula for this sequence?
    A $t_n = n - 10$ **B** $t_n = n^2 - 10$ **C** $t_n = n^3 - 10$
  b Use the correct formula to find the 20th term of the sequence.

7   Find the first four terms of the sequence with recursive formula:

   a $t_1 = 5$, $t_{n+1} = t_n + 6$, $n \geqslant 1$
      b $t_1 = 8$, $t_{n+1} = t_n - 5$, $n \geqslant 1$

   c $t_1 = 4$, $t_{n+1} = 3 \times t_n$, $n \geqslant 1$
      d $t_1 = 80$, $t_{n+1} = \dfrac{t_n}{2}$, $n \geqslant 1$

   e $t_1 = 3$, $t_2 = 7$, $t_{n+2} = t_{n+1} - t_n$, $n \geqslant 1$

8   Match each explicit formula with the corresponding recursive formula.

   a $t_n = 2n + 5$
      A $t_1 = 5$, $t_{n+1} = t_n - 2$

   b $t_n = 2^n$
      B $t_1 = 7$, $t_{n+1} = t_n + 2$

   c $t_n = 7 - 2n$
      C $t_1 = 2$, $t_{n+1} = 2 \times t_n$

9   A sequence is defined by the recursive formula $t_1 = 7$, $t_{n+1} = 10 - t_n$.

   a Show that $t_{n+2} = t_n$.
      b Hence, find t_{200} and t_{375}.

10 A sequence is defined by $t_1 = 1$, $t_{n+1} = \dfrac{1}{1 + t_n}$, $n \geqslant 1$.

   a Find the next four terms of the sequence.

   b Display the first 5 terms on a graph.

   c Assuming that the terms of the sequence tend to the constant value t, show that $t^2 + t - 1 = 0$.

   d Hence show that $\dfrac{1}{1 + \dfrac{1}{1 + \dfrac{1}{1 + \frac{1}{1 + ...}}}} = \dfrac{-1 + \sqrt{5}}{2}$.

11 A sequence is defined by $t_1 = 1$, $t_{n+1} = \sqrt{\dfrac{1}{1 + t_n}}$, $n \geqslant 1$.

   a Find the next four terms of the sequence, writing your answers correct to 5 decimal places.

   b Assuming that the sequence terms tend to the constant value t, show that $t^3 + t^2 - 1 = 0$.

   c Hence find $\sqrt{\dfrac{1}{1 + \sqrt{\dfrac{1}{1 + \sqrt{\frac{1}{1 + \sqrt{....}}}}}}}$ correct to 5 decimal places.

   d Check your answer using technology.

GRAPHICS
CALCULATOR
INSTRUCTIONS

12 Find, correct to 5 decimal places: $\dfrac{1}{1 + \left(\dfrac{1}{1 + \left(\dfrac{1}{1 + \left(\frac{1}{1 + (....)}\right)^2}\right)^2}\right)^2}$

Research   The Fibonacci sequence

Leonardo Pisano Bigollo, known commonly as Fibonacci, was born in Pisa around 1170 AD. He is best known for the **Fibonacci sequence** 1, 1, 2, 3, 5, 8, 13, 21, ... which starts with 1 and 1, and then each subsequent member of the sequence is the sum of the preceding two members.

   1 How would you write a recursive formula for the Fibonacci sequence?

   2 Is there an explicit formula for the Fibonacci sequence?

   3 Where do we see the Fibonacci sequence in nature?

B ARITHMETIC SEQUENCES

An **arithmetic sequence** is a sequence in which each term differs from the previous one by the same fixed number. We call this number the **common difference** d.

A sequence is arithmetic $\Leftrightarrow t_{n+1} - t_n = d$ for all $n \in \mathbb{Z}^+$.

For example:

- the tower of bricks in the previous Section forms an arithmetic sequence with common difference 1
- 2, 5, 8, 11, 14, ... is arithmetic with common difference 3 since
$$5 - 2 = 3$$
$$8 - 5 = 3$$
$$11 - 8 = 3, \quad \text{and so on.}$$

  The recursive formula for this sequence is $t_1 = 2$, $t_{n+1} = t_n + 3$, $n \geqslant 1$.

- 30, 25, 20, 15, 10, ... is arithmetic with common difference -5 since
$$25 - 30 = -5$$
$$20 - 25 = -5$$
$$15 - 20 = -5, \quad \text{and so on.}$$

  The recursive formula for this sequence is $t_1 = 30$, $t_{n+1} = t_n - 5$, $n \geqslant 1$.

THE GENERAL TERM FORMULA

If we know that a sequence is arithmetic, we can use a formula to find the value of any term of the sequence.

Suppose the first term of an arithmetic sequence is t_1 and the common difference is d.

Then $t_2 = t_1 + d$, $t_3 = t_1 + 2d$, $t_4 = t_1 + 3d$, and so on.

Hence $t_n = t_1 + \underbrace{(n-1)}\,d$

term number

the coefficient of d is one less than the term number

> For an **arithmetic sequence** with **first term t_1** and **common difference d**, the **general term** or **nth term** is $t_n = t_1 + (n-1)d$.

Example 1   ◆) Self Tutor

Consider the sequence   2, 9, 16, 23, 30, ...

a Show that the sequence is arithmetic.

b Find a formula for the general term t_n.

c Find the 100th term of the sequence.

d Is   **i** 828   **ii** 2341   a term of the sequence?

a $9 - 2 = 7$ The difference between successive terms is constant.
$16 - 9 = 7$ $\therefore$ the sequence is arithmetic with $t_1 = 2$ and $d = 7$.
$23 - 16 = 7$
$30 - 23 = 7$

b $t_n = t_1 + (n-1)d$
∴ $t_n = 2 + 7(n-1)$
∴ $t_n = 7n - 5$

c $t_{100} = 7(100) - 5$
$= 695$

d   **i**   Let $t_n = 828$
∴ $7n - 5 = 828$
∴ $7n = 833$
∴ $n = 119$

∴   828 is a term of the sequence,
and in fact is the 119th term.

ii   Let $t_n = 2341$
∴ $7n - 5 = 2341$
∴ $7n = 2346$
∴ $n = 335\frac{1}{7}$

But n must be an integer, so 2341
is not a member of the sequence.

EXERCISE 1B

1 Decide whether these sequences are arithmetic:

    a   7, 15, 23, 31, 39, ...

    b   10, 14, 18, 20, 24, ...

    c   41, 35, 29, 23, 17, ...

    d   6, 1, -6, -11, -16, ...

2 State the first term t_1 and common difference d for these arithmetic sequences:

    a   5, 9, 13, 17, 21, ...

    b -4, 3, 10, 17, 24, ...

    c   23, 18, 13, 8, 3, ...

    d -6, -15, -24, -33, ...

3 For each of these arithmetic sequences:

    i   State t_1 and d.
    ii   Find the formula for the general term t_n.
    iii   Find the 15th term of the sequence.

    a   19, 25, 31, 37, ...

    b   101, 97, 93, 89, ...

    c   8, $9\frac{1}{2}$, 11, $12\frac{1}{2}$, ...

    d   31, 36, 41, 46, ...

    e   5, -3, -11, -19, ...

    f a, $a + d$, $a + 2d$, $a + 3d$, ...

4 Consider the sequence   6, 17, 28, 39, 50, ...

    a Show that the sequence is arithmetic.

    b Find the formula for its general term.

    c Find the 50th term.

    d Is 325 a member?

    e Is 761 a member?

5 Consider the sequence   87, 83, 79, 75, 71, ...

    a Show that the sequence is arithmetic.

    b Find the formula for its general term.

    c Find the 40th term.

    d Which term of the sequence is -297?

6 A sequence is defined by $t_n = 3n - 2$.

    a Prove that the sequence is arithmetic.   **Hint:** Find $t_{n+1} - t_n$.

    b Find t_1 and d.

    c Find the 57th term.

    d What is the largest term of the sequence that is smaller than 450? Which term is this?

7 A sequence is defined by $t_n = \dfrac{71 - 7n}{2}$.

    a Prove that the sequence is arithmetic.   **b** Find t_1 and d.   **c** Find t_{75}.

    d For what values of n are the terms of the sequence less than -200?

8   A sequence is defined by the recursive formula $t_1 = -12$, $t_{n+1} = t_n + 7$, $n \geqslant 1$.

    a   Prove that the sequence is arithmetic.   **b**   Find the 200th term of the sequence.

    c   Is 1000 a member of the sequence?

Example 2   ◀) **Self Tutor**

Find k given that $3k + 1$, k,   and -3 are consecutive terms of an arithmetic sequence.

Since the terms are consecutive, $k - (3k + 1) = -3 - k$ {equating differences}

$$\therefore \quad k - 3k - 1 = -3 - k$$
$$\therefore \quad -2k - 1 = -3 - k$$
$$\therefore \quad -1 + 3 = -k + 2k$$
$$\therefore \quad k = 2$$

---

9   Find k given the consecutive arithmetic terms:

    a $32, k, 3$ **b** $k, 7, 10$ **c** $k + 1, 2k + 1, 13$

    d $k - 1, 2k + 3, 7 - k$ **e** $k, k^2, k^2 + 6$ **f** $5, k, k^2 - 8$

Example 3   ◀) **Self Tutor**

Find the general term t_n for an arithmetic sequence with $t_3 = 8$ and $t_8 = -17$.

$t_3 = 8$ $\therefore \quad t_1 + 2d = 8$ .... (1)   {using $t_n = t_1 + (n - 1)d$}

$t_8 = -17$ $\therefore \quad t_1 + 7d = -17$ .... (2)

We now solve (1) and (2) simultaneously:

$$-t_1 - 2d = -8 \qquad \text{\{multiplying both sides of (1) by } -1\}$$
$$\underline{t_1 + 7d = -17}$$
$$\therefore \quad 5d = -25 \qquad \text{\{adding the equations\}}$$
$$\therefore \quad d = -5$$

So, in (1): $t_1 + 2(-5) = 8$ *Check*:

$$\therefore \quad t_1 - 10 = 8 \qquad\qquad t_3 = 23 - 5(3)$$
$$\therefore \quad t_1 = 18 \qquad\qquad\quad = 23 - 15$$
$$\text{Now} \quad t_n = t_1 + (n - 1)d \qquad = 8 \ \checkmark$$
$$\therefore \quad t_n = 18 - 5(n - 1) \qquad t_8 = 23 - 5(8)$$
$$\therefore \quad t_n = 18 - 5n + 5 \qquad\quad = 23 - 40$$
$$\therefore \quad t_n = 23 - 5n \qquad\qquad = -17 \ \checkmark$$

---

10   Find the general term t_n for an arithmetic sequence with:

    a $t_7 = 41$ and $t_{13} = 77$

    b $t_5 = -2$ and $t_{12} = -12\frac{1}{2}$

    c   seventh term 1 and fifteenth term -39

    d   eleventh and eighth terms being -16 and $-11\frac{1}{2}$ respectively.

Example 4   ◀)) Self Tutor

Insert four numbers between 3 and 12 so that all six numbers are in arithmetic sequence.

Suppose the common difference is d.

$\therefore$ the numbers are   3, $3+d$, $3+2d$, $3+3d$, $3+4d$, and 12

$$\therefore \ 3+5d = 12$$
$$\therefore \ 5d = 9$$
$$\therefore \ d = \tfrac{9}{5} = 1.8$$

So, the sequence is   3, 4.8, 6.6, 8.4, 10.2, 12.

11 Insert three numbers between 5 and 10 so that all five numbers are in arithmetic sequence.

12 Insert six numbers between -1 and 32 so that all eight numbers are in arithmetic sequence.

13   **a** Insert three numbers between 50 and 44 so that all five numbers are in arithmetic sequence.

   b Assuming the sequence continues, find the first negative term of the sequence.

Example 5   ◀)) Self Tutor

Ryan is a cartoonist. His comic strip has just been bought by a newspaper, so he sends them the 28 comic strips he has drawn so far. Each week after the first he sends 3 more comic strips to the newspaper.

a Find the total number of comic strips sent after 1, 2, 3, and 4 weeks.

b Show that the total number of comic strips sent after n weeks forms an arithmetic sequence.

c Find the number of comic strips sent after 15 weeks.

d When does Ryan send his 120th comic strip?

a *Week 1:*   28 comic strips
   Week 2: $28 + 3 = 31$ comic strips
   Week 3: $31 + 3 = 34$ comic strips
   Week 4: $34 + 3 = 37$ comic strips

b Every week, Ryan sends 3 comic strips, so the difference between successive weeks is always 3. We have an arithmetic sequence with $t_1 = 28$ and common difference $d = 3$.

c $t_n = t_1 + (n-1)d$
   $= 28 + (n-1) \times 3$ $\therefore \ t_{15} = 25 + 3 \times 15$
   $= 25 + 3n$ $= 70$

After 15 weeks Ryan has sent 70 comic strips.

d We want to find n such that $t_n = 120$
                    $\therefore \ 25 + 3n = 120$
                    $\therefore \ 3n = 95$
                    $\therefore \ n = 31\tfrac{2}{3}$

Ryan sends the 120th comic strip in the 32nd week.

14 A luxury car manufacturer sets up a factory for a new model vehicle. In the first month only 5 cars are made. After this, 13 cars are made every month.

    a List the total number of cars that have been made by the end of each of the first six months.

    b Explain why the total number of cars made after n months forms an arithmetic sequence.

    c How many cars are made in the first year?

    d How long is it until the 250th car is made?

15 Valéria joins a social networking website. After 1 week she has 34 online friends. After 2 weeks she has 41 friends, after 3 weeks she has 48 friends, and after 4 weeks she has 55 friends.

    a Show that Valéria's number of friends forms an arithmetic sequence.

    b Assuming the pattern continues, find the number of online friends Valéria will have after 12 weeks.

    c After how many weeks will Valéria have 150 online friends?

16 A farmer feeds hay to his cattle herd every day in July. The amount of hay in his barn at the end of day n is given by the arithmetic sequence $t_n = 100 - 2.7n$ tonnes.

    a Write down the amount of hay in the barn on the first three days of July.

    b Find and interpret the common difference.

    c Find and interpret t_{25}.

    d How much hay is in the barn at the beginning of August?

Discussion

- What can you say about t_1 and d for an arithmetic sequence if:
  - all of the terms are even
  - all of the terms are odd?

- It is known that there are infinitely many prime numbers. Is it possible to construct an infinite arithmetic sequence such that all of the terms are prime?

C   GEOMETRIC SEQUENCES

A **geometric sequence** is a sequence in which each term can be obtained from the previous one by multiplying by the same non-zero number. We call this number the **common ratio** r.

A sequence is geometric $\Leftrightarrow \dfrac{t_{n+1}}{t_n} = r$ for all $n \in \mathbb{Z}^+$.

For example:

- 2, 10, 50, 250, ... is a geometric sequence as each term can be obtained by multiplying the previous term by 5.

  The common ratio is 5 since $\dfrac{10}{2} = \dfrac{50}{10} = \dfrac{250}{50} = 5$.

- 2, −10, 50, −250, ... is a geometric sequence with common ratio −5.

  The recursive formula for this sequence is $t_1 = 2$, $t_{n+1} = -5 \times t_n$, $n \geqslant 1$.

THE GENERAL TERM FORMULA

Suppose the first term of a geometric sequence is t_1 and the common ratio is r.

Then $t_2 = t_1 r$, $t_3 = t_1 r^2$, $t_4 = t_1 r^3$,   and so on.

Hence $t_n = t_1 r^{n-1}$

term number   The power of r is one less than the term number.

> For a **geometric sequence** with **first term t_1** and **common ratio r**,
> the **general term** or **nth term** is $t_n = t_1 r^{n-1}$.

Example 6   ◀) **Self Tutor**

Consider the sequence $8, 4, 2, 1, \frac{1}{2}, ...$

  a Show that the sequence is geometric.   **b** Find the general term t_n.

  c Hence, find the 12th term as a fraction.

a $\dfrac{4}{8} = \dfrac{1}{2}$ $\dfrac{2}{4} = \dfrac{1}{2}$ $\dfrac{1}{2} = \dfrac{1}{2}$ $\dfrac{\frac{1}{2}}{1} = \dfrac{1}{2}$

Consecutive terms have a common ratio of $\frac{1}{2}$.

∴ the sequence is geometric with $t_1 = 8$ and $r = \frac{1}{2}$.

b $t_n = t_1 r^{n-1}$

∴ $t_n = 8\left(\frac{1}{2}\right)^{n-1}$ *or* $t_n = 2^3 \times (2^{-1})^{n-1}$

$= 2^3 \times 2^{-n+1}$

$= 2^{3+(-n+1)}$

$= 2^{4-n}$

c $t_{12} = 8 \times \left(\frac{1}{2}\right)^{11}$

$= \frac{1}{256}$

EXERCISE 1C

1 State the first term t_1 and common ratio r for these geometric sequences:

  a $5, 15, 45, 135, ...$ **b** $72, 36, 18, 9, ...$

  c $2, -8, 32, -128, ...$ **d** $6, -2, \frac{2}{3}, -\frac{2}{9}, ...$

2 For the geometric sequence with first two terms given, find b and c:

  a $2, 6, b, c, ...$ **b** $10, 5, b, c, ...$ **c** $12, -6, b, c, ...$

3 For each of these geometric sequences:

    i State t_1 and r.

   ii Find the formula for the general term t_n.

  iii Find the 9th term of the sequence.

  a $3, 6, 12, 24, ...$ **b** $2, 10, 50, ...$ **c** $512, 256, 128, ...$

  d $1, 3, 9, 27, ...$ **e** $12, 18, 27, ...$ **f** $\frac{1}{16}, -\frac{1}{8}, \frac{1}{4}, -\frac{1}{2}, ...$

4 **a** Show that the sequence $5, 10, 20, 40, ...$ is geometric.

    b Find t_n, and hence find the 15th term.

5 a Show that the sequence $12,\ -6,\ 3,\ -\tfrac{3}{2},\ ....$ is geometric.

  b Find t_n,   and hence write the 13th term as a rational number.

6 Show that the sequence $8,\ -6,\ 4.5,\ -3.375,\ ....$ is geometric. Hence find the 10th term as a decimal.

7 Show that the sequence $8,\ 4\sqrt{2},\ 4,\ 2\sqrt{2},\ ....$ is geometric. Hence show that the general term of the sequence is $t_n = 2^{\frac{7}{2} - \frac{1}{2}n}$.

8 A geometric sequence is defined by the recursive formula $t_1 = \tfrac{4}{27},\ \ t_{n+1} = 3 \times t_n,\ \ n \geqslant 1$.

  a Find the common ratio.   **b** Find the 10th term of the sequence.

  c How many terms of the sequence are not integers?

Example 7   ◀) **Self Tutor**

$k - 1,\ 2k,$ and $21 - k$ are consecutive terms of a geometric sequence. Find k.

Since the terms are geometric, $\dfrac{2k}{k-1} = \dfrac{21-k}{2k}$ {equating the common ratio r}

$$\therefore\ \ 4k^2 = (21 - k)(k - 1)$$
$$\therefore\ \ 4k^2 = 21k - 21 - k^2 + k$$
$$\therefore\ \ 5k^2 - 22k + 21 = 0$$
$$\therefore\ \ (5k - 7)(k - 3) = 0$$
$$\therefore\ \ k = \tfrac{7}{5} \text{ or } 3$$

Check: If $k = \tfrac{7}{5}$ the terms are: $\tfrac{2}{5},\ \tfrac{14}{5},\ \tfrac{98}{5}.$ ✓ $\{r = 7\}$

    If $k = 3$ the terms are: $2,\ 6,\ 18.$ ✓ $\{r = 3\}$

---

9 Find k given that the following are consecutive terms of a geometric sequence:

  a $7,\ k,\ 28$ **b** $k,\ 3k,\ 20 - k$ **c** $k,\ k + 8,\ 9k$

10 The first three terms of a geometric sequence are $k - 1,\ 6,$ and $3k$.

  a Find the possible values of k.

  b For each value of k, find the next term in the sequence.

Example 8   ◀) **Self Tutor**

A geometric sequence has $t_2 = -6$ and $t_5 = 162$.   Find its general term.

$$t_2 = t_1 r = -6 \ \ .... \ (1)$$
$$\text{and}\ \ t_5 = t_1 r^4 = 162 \ \ .... \ (2)$$
$$\text{Now}\ \ \frac{t_1 r^4}{t_1 r} = \frac{162}{-6} \ \ \ \{(2) \div (1)\}$$
$$\therefore\ \ r^3 = -27$$
$$\therefore\ \ r = \sqrt[3]{-27}$$
$$\therefore\ \ r = -3$$
$$\text{Using (1),}\ \ \ t_1(-3) = -6$$
$$\therefore\ \ t_1 = 2$$
$$\text{Thus}\ \ t_n = 2 \times (-3)^{n-1}$$

11   Find the general term t_n of the geometric sequence which has:

  a $t_4 = 24$ and $t_7 = 192$ **b** $t_3 = 8$ and $t_6 = -1$

  c $t_7 = 24$ and $t_{15} = 384$ **d** $t_3 = 5$ and $t_7 = \frac{5}{4}$

Example 9   ◀୬ **Self Tutor**

Find the first term of the sequence $6, 6\sqrt{2}, 12, 12\sqrt{2}, ...$ which exceeds 1400.

The sequence is geometric with $t_1 = 6$ and $r = \sqrt{2}$
$$\therefore \quad t_n = 6 \times (\sqrt{2})^{n-1}$$

We need to find n such that $t_n > 1400$.

Using a graphics calculator with $Y_1 = 6 \times (\sqrt{2})\wedge(X - 1)$,   we view a table of values:

Casio fx-9860G PLUS	**Casio fx-CG20**	**TI-84 Plus CE**
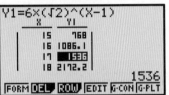	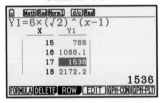	

The first term to exceed 1400 is $t_{17} = 1536$.

---

12   **a**   Find the first term of the sequence $2, 6, 18, 54, ...$ which exceeds $10\,000$.

  b   Find the first term of the sequence $4, 4\sqrt{3}, 12, 12\sqrt{3}, ...$ which exceeds 4800.

  c   Find the first term of the sequence $12, 6, 3, 1.5, ...$ which is less than 0.0001.

D   APPLICATIONS OF GEOMETRIC SEQUENCES

Geometric sequences are observed when a quantity increases or decreases by a fixed percentage of its size each time period.

For most real-world situations, we have an **initial condition** corresponding to time zero. This may be an initial investment or an initial population. We therefore allow a "zeroeth" term t_0 to begin the sequence.

For example, suppose you invest £1000 in the bank. The account pays an interest rate of 10% per annum (p.a.). The interest is added to your investment each year, so at the end of each year you will have $100\% + 10\% = 110\%$ of the value at its start. This corresponds to a *multiplier* of 1.1.

per annum means each year.

After one year your investment is worth $£1000 \times 1.1 = £1100$.

After two years it is worth   After three years it is worth
$\qquad £1100 \times 1.1$ $\qquad £1210 \times 1.1$
$\quad = £1000 \times 1.1 \times 1.1$ $\quad = £1000 \times (1.1)^2 \times 1.1$
$\quad = £1000 \times (1.1)^2 = £1210$ $\quad = £1000 \times (1.1)^3 = £1331$

Observe that:

$t_0 = £1000$ $\quad = $ initial investment

$t_1 = t_0 \times 1.1$ $\quad = $ amount after 1 year

$t_2 = t_0 \times (1.1)^2$ $\quad = $ amount after 2 years

$t_3 = t_0 \times (1.1)^3$ $\quad = $ amount after 3 years

$\vdots$

$t_n = t_0 \times (1.1)^n$ $\quad = $ amount after n years

The amount in the account after each year forms a geometric sequence!

For **compound interest**, and other situations where there is **growth** or **decay** in a geometric sequence:

$$t_n = t_0 \times r^n$$

where $\quad t_0 = $ initial amount

$r = $ growth multiplier for each time period

$n = $ number of time periods

$t_n = $ amount after n time periods.

Example 10 🔊 Self Tutor

£5000 is invested for 4 years at 3% p.a. interest compounded quarterly. Find the value of the investment at the end of this period.

There are $4 \times 4 = 16$ time periods.

Each time period the investment increases by $\dfrac{3\%}{4} = 0.75\%$.

$\therefore \quad r = 1.0075$

$\therefore \quad$ the amount after 4 years is $\quad t_{16} = t_0 \times r^{16}$

$= 5000 \times (1.0075)^{16}$

≈ 5634.96

The investment will amount to £5634.96.

EXERCISE 1D

1 Lucy invested £7000 at 6% p.a. interest compounded annually. Find the value of this investment after 5 years.

2 What will an investment of £3000 at 4.8% p.a. interest compounded annually amount to after 3 years?

3 £20 000 is invested at 4.8% p.a. interest compounded quarterly. Find the value of this investment after:

    a 1 year   **b** 3 years.

4 **a** What will an investment of £30 000 at 5.6% p.a. interest compounded annually amount to after 4 years?

    b How much of this is interest?

5 How much interest is earned by investing £80 000 at 4.4% p.a. for a 3 year period with interest compounded quarterly?

Example 11   ◀) Self Tutor

How much does Ivana need to invest now, to get a maturing value of £10 000 in 4 years' time, given interest at 8% p.a. compounded twice annually? Give your answer to the nearest pound.

The initial investment t_0 is unknown.

The investment is compounded twice annually, so the multiplier $r = 1 + \dfrac{0.08}{2} = 1.04$.

There are $4 \times 2 = 8$ compounding periods, so $n = 8$.

Now $t_8 = t_0 \times r^8$ {using $t_n = t_0 \times r^n$}

$\therefore \quad 10\,000 = t_0 \times (1.04)^8$

$\therefore \quad t_0 = \dfrac{10\,000}{(1.04)^8}$

$\therefore \quad t_0 \approx 7306.90$

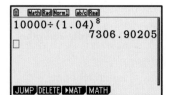

Ivana needs to invest £7307 now.

6   How much does Habib need to invest now, to get a maturing value of £20 000 in 4 years' time, given the money can be invested at a fixed rate of 7.5% p.a. compounded annually? Round your answer up to the next pound.

7   What initial investment is required to produce a maturing amount of £15 000 in 60 months' time given a guaranteed fixed interest rate of 5.5% p.a. compounded annually? Round your answer up to the next pound.

8   How much should I invest now to yield £25 000 in 3 years' time, if the money can be invested at a fixed rate of 4.2% p.a. compounded quarterly?

9   What initial investment will yield £40 000 in 8 years' time if your money can be invested at 3.6% p.a. compounded monthly?

Example 12   ◀) Self Tutor

The initial population of rabbits on a farm was 50.
The population increased by 7% each week.

a   How many rabbits were present after:

    i   15 weeks   **ii**   30 weeks?

b   How long will it take for the population to reach 500?

There is a fixed percentage increase each week, so the population forms a geometric sequence.

$t_0 = 50$ and $r = 1.07$

$\therefore$ the population after n weeks is $t_n = 50 \times 1.07^n$.

a   **i** $t_{15} = 50 \times (1.07)^{15} \approx 137.95$ **ii** $t_{30} = 50 \times (1.07)^{30} \approx 380.61$

       There were 138 rabbits.   There were 381 rabbits.

b We need to solve $50 \times (1.07)^n = 500$

$$\therefore \quad (1.07)^n = 10$$

$$\therefore \quad n = \frac{\ln 10}{\ln(1.07)} \approx 34.03$$

```
50×(1.07)³⁴
              498.9056769
50×(1.07)³⁵
              533.8290742
□
JUMP DEL ▸MAT MATH
```

The population will reach 500 early in the 35th week.

10 A nest of ants initially contains 500 individuals.
The population is increasing by 12% each week.

  a How many ants will there be after:

   i 10 weeks   **ii** 20 weeks?

  b How many weeks will it take for the ant population to reach 2000?

11 The animal *Eraticus* is endangered. Since 2005 there has only been one colony remaining, and in 2005 the population of that colony was 555. The population has been steadily decreasing by 4.5% per year.

  a Estimate the population in the year 2020.

  b In what year do we expect the population to have declined to 50?

12

A herd of 32 deer is to be left unchecked in a new sanctuary. It is estimated that the size of the herd will increase each year by 18%.

  a Estimate the size of the herd after:

   i 5 years   **ii** 10 years.

  b How long will it take for the herd size to reach 5000?

13 An endangered species of marsupials has a population of 178. However, with a successful breeding program it is expected to increase by 32% each year.

  a Find the expected population size after: **i** 10 years   **ii** 25 years.

  b Estimate how long it will take for the population to reach 10 000.

E SERIES

There are many situations where we are interested in finding the sum of the terms of a number sequence.

A **series** is the sum of the terms of a sequence.

For a **finite** sequence with n terms, the corresponding series is $t_1 + t_2 + t_3 + ... + t_n$.

The sum of this series is $S_n = t_1 + t_2 + t_3 + ... + t_n$ and this will always be a finite real number.

For an **infinite** sequence the corresponding series is $t_1 + t_2 + t_3 + ... + t_n + ...$

In many cases, the sum of an infinite series cannot be calculated. In some cases, however, it does **converge** to a finite number.

SIGMA NOTATION

$t_1 + t_2 + t_3 + t_4 + ... + t_n$ can be written more compactly using **sigma notation** or **summation notation**.

The symbol $\sum$ is called **sigma**. It is the equivalent of capital S in the Greek alphabet.

We write $t_1 + t_2 + t_3 + t_4 + ... + t_n$ as $\displaystyle\sum_{k=1}^{n} t_k$.

$\displaystyle\sum_{k=1}^{n} t_k$ reads "the sum of all numbers of the form t_k where $k = 1, 2, 3, ...,$ up to n".

Example 13
◀)) **Self Tutor**

Consider the sequence $1, 4, 9, 16, 25, ...$

a Write down an expression for S_n.   **b** Find S_n for $n = 1, 2, 3, 4,$ and 5.

a $S_n = 1^2 + 2^2 + 3^2 + 4^2 + ... + n^2$
{all terms are squares}

$= \displaystyle\sum_{k=1}^{n} k^2$

b $S_1 = 1$
$S_2 = 1 + 4 = 5$
$S_3 = 1 + 4 + 9 = 14$
$S_4 = 1 + 4 + 9 + 16 = 30$
$S_5 = 1 + 4 + 9 + 16 + 25 = 55$

Example 14
◀)) **Self Tutor**

Expand and evaluate: **a** $\displaystyle\sum_{k=1}^{7} (k+1)$ **b** $\displaystyle\sum_{k=1}^{5} \dfrac{1}{2^k}$

a $\displaystyle\sum_{k=1}^{7} (k+1)$

$= 2 + 3 + 4 + 5 + 6 + 7 + 8$

$= 35$

b $\displaystyle\sum_{k=1}^{5} \dfrac{1}{2^k}$

$= \frac{1}{2} + \frac{1}{4} + \frac{1}{8} + \frac{1}{16} + \frac{1}{32}$

$= \frac{31}{32}$

You can also use technology to evaluate the sum of a series in sigma notation.
Click on the icon for instructions.

GRAPHICS CALCULATOR INSTRUCTIONS

PROPERTIES OF SIGMA NOTATION

$$\sum_{k=1}^{n} (a_k + b_k) = \sum_{k=1}^{n} a_k + \sum_{k=1}^{n} b_k$$

If c is a constant, $\displaystyle\sum_{k=1}^{n} ca_k = c \sum_{k=1}^{n} a_k$ and $\displaystyle\sum_{k=1}^{n} c = cn$.

EXERCISE 1E

1 Consider the sequence of composite numbers   4, 6, 8, 9, 10, 12, 14, 15, 16, ...
Find:

    a S_3 **b** S_5 **c** S_{12}

2 Suppose a sequence has $S_4 = 13$ and $S_5 = 20$. Find the value of t_5.

3 For each of the following sequences:

    i Write down an expression for S_n.   **ii** Find S_5.

    a 3, 11, 19, 27, ...   **b** 42, 37, 32, 27, ...   **c** 12, 6, 3, $1\frac{1}{2}$, ...

    d 2, 3, $4\frac{1}{2}$, $6\frac{3}{4}$, ...   **e** 1, $\frac{1}{2}$, $\frac{1}{4}$, $\frac{1}{8}$, ...   **f** 1, 8, 27, 64, ...

4 Expand and evaluate:

    a $\displaystyle\sum_{k=1}^{3} 4k$ **b** $\displaystyle\sum_{k=1}^{6} (k+1)$ **c** $\displaystyle\sum_{k=1}^{4} (3k-5)$

    d $\displaystyle\sum_{k=1}^{5} (11-2k)$ **e** $\displaystyle\sum_{k=1}^{7} k(k+1)$ **f** $\displaystyle\sum_{k=1}^{5} 10 \times 2^{k-1}$

5 For $t_n = 3n - 1$, write $t_1 + t_2 + t_3 + ... + t_{20}$ using sigma notation and evaluate the sum.

6 Show that:

    a $\displaystyle\sum_{k=1}^{n} c = cn$ **b** $\displaystyle\sum_{k=1}^{n} ca_k = c \sum_{k=1}^{n} a_k$ **c** $\displaystyle\sum_{k=1}^{n} (a_k + b_k) = \sum_{k=1}^{n} a_k + \sum_{k=1}^{n} b_k$

Example 15   ◀》 Self Tutor

a Expand $\displaystyle\sum_{k=1}^{n} 2k$ then write the expansion again underneath but with the terms in the reverse order.

b Add the terms vertically, and hence write an expression for the sum S_n of the first n even integers.

a $\displaystyle\sum_{k=1}^{n} 2k = 2 + \quad 4 \quad + \quad 6 \quad + ... + (2n-2) + 2n$

        or $2n + (2n-2) + (2n-4) + ... + \quad 4 \quad + 2$

b $2\displaystyle\sum_{k=1}^{n} 2k = (2n+2) + (2n+2) + (2n+2) + ... + (2n+2) + (2n+2)$

           $= n(2n+2)$

           $= 2n(n+1)$

$\therefore \displaystyle\sum_{k=1}^{n} 2k = n(n+1)$

---

7 **a** Expand $\displaystyle\sum_{k=1}^{n} k$ then write the expansion again underneath with the terms in the reverse order.

    b Add the terms vertically, and hence write an expression for the sum S_n of the first n integers.

    c Hence find a and b such that $\displaystyle\sum_{k=1}^{n} (ak + b) = 8n^2 + 11n$ for all positive integers n.

8 Write an expression for the sum of the first n positive odd integers.

9 Given that $\displaystyle\sum_{k=1}^{n} k = \frac{n(n+1)}{2}$ and $\displaystyle\sum_{k=1}^{n} k^2 = \frac{n(n+1)(2n+1)}{6}$, write $\displaystyle\sum_{k=1}^{n} (k+1)(k+2)$ in simplest form.

Check your answer in the case when $n = 10$.

F   ARITHMETIC SERIES

> An **arithmetic series** is the sum of the terms of an arithmetic sequence.

For example: 21, 23, 25, 27, ..., 49   is a finite arithmetic sequence.

$21 + 23 + 25 + 27 + ... + 49$ is the corresponding arithmetic series.

SUM OF A FINITE ARITHMETIC SERIES

Rather than adding all the terms individually, we can use a formula to find the sum of a finite arithmetic series.

If the first term is t_1, the final term is t_n, and the common difference is d, the terms are
t_1, $t_1 + d$, $t_1 + 2d$, ..., $(t_n - 2d)$, $(t_n - d)$, t_n.

$$\therefore \quad S_n = t_1 + (t_1 + d) + (t_1 + 2d) + ... + (t_n - 2d) + (t_n - d) + t_n$$

But $S_n = t_n + (t_n - d) + (t_n - 2d) + ... + (t_1 + 2d) + (t_1 + d) + t_1$ {reversing them}

Adding these two equations vertically, we get:

$$2S_n = \underbrace{(t_1 + t_n) + (t_1 + t_n) + (t_1 + t_n) + ... + (t_1 + t_n) + (t_1 + t_n) + (t_1 + t_n)}_{n \text{ of these}}$$

$$\therefore \quad 2S_n = n(t_1 + t_n)$$

$$\therefore \quad S_n = \frac{n}{2}(t_1 + t_n) \quad \text{where} \quad t_n = t_1 + (n-1)d$$

> The sum of a finite arithmetic series with first term t_1, common difference d, and last term t_n, is
>
> $$S_n = \frac{n}{2}(t_1 + t_n) \quad or \quad S_n = \frac{n}{2}(2t_1 + (n-1)d).$$

Example 16   ◀◈ **Self Tutor**

Find the sum of $4 + 7 + 10 + 13 + ...$ to 50 terms.

The series is arithmetic with $t_1 = 4$, $d = 3$, and $n = 50$.

Now $S_n = \dfrac{n}{2}(2t_1 + (n-1)d)$

$\therefore \quad S_{50} = \frac{50}{2}(2 \times 4 + 49 \times 3)$

$\qquad\qquad = 3875$

Example 17

◀ッ **Self Tutor**

Find the sum of $-6 + 1 + 8 + 15 + ... + 141$.

The series is arithmetic with $t_1 = -6$, $d = 7$, and $t_n = 141$.

First we need to find n.

Now $t_n = 141$

$\therefore \quad t_1 + (n-1)d = 141$

$\therefore \quad -6 + 7(n-1) = 141$

$\therefore \quad 7(n-1) = 147$

$\therefore \quad n - 1 = 21$

$\therefore \quad n = 22$

Using $S_n = \dfrac{n}{2}(t_1 + t_n)$,

$S_{22} = \dfrac{22}{2}(-6 + 141)$

$= 11 \times 135$

$= 1485$

You can also use technology to evaluate series, although for some calculator models this is tedious.

GRAPHICS
CALCULATOR
INSTRUCTIONS

EXERCISE 1F

1 Find the sum of the arithmetic series $2 + 6 + 10 + 14 + 18 + 22 + 26 + 30$:

    a by direct addition

    b using $S_n = \dfrac{n}{2}(t_1 + t_n)$

    c using $S_n = \dfrac{n}{2}(2t_1 + (n-1)d)$.

2 Find the sum of:

    a $3 + 7 + 11 + 15 + ...$ to 20 terms

    b $\frac{1}{2} + 3 + 5\frac{1}{2} + 8 + ...$ to 50 terms

    c $100 + 93 + 86 + 79 + ...$ to 40 terms

    d $50 + 48\frac{1}{2} + 47 + 45\frac{1}{2} + ...$ to 80 terms.

3 Find the sum of:

    a $5 + 8 + 11 + 14 + ... + 101$

    b $37 + 33 + 29 + 25 + ... + 9$

    c $50 + 49\frac{1}{2} + 49 + 48\frac{1}{2} + ... + (-20)$

    d $8 + 10\frac{1}{2} + 13 + 15\frac{1}{2} + ... + 83$

4 Evaluate these arithmetic series:

    a $\displaystyle\sum_{k=1}^{10}(2k + 5)$

    b $\displaystyle\sum_{k=1}^{15}(k - 50)$

    c $\displaystyle\sum_{k=1}^{20}\left(\dfrac{k+3}{2}\right)$

Check your answers using technology.

5 An arithmetic series has seven terms. The first term is 5 and the last term is 53. Find the sum of the series.

6 An arithmetic series has eleven terms. The first term is 6 and the last term is -27. Find the sum of the series.

7

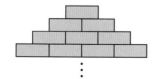

A bricklayer builds a triangular wall with layers of bricks as shown. If the bricklayer uses 171 bricks, how many layers did he build?

8 A soccer stadium has 25 sections of seating. Each section has 44 rows of seats, with 22 seats in the first row, 23 in the second row, 24 in the third row, and so on. How many seats are there in:

    a row 44 of one section   **b** each section   **c** the whole stadium?

9 Find the sum of:

    a the first 50 multiples of 11   **b** the multiples of 7 between 0 and 1000

    c the integers from 1 to 100 which are not divisible by 3.

10 The sixth term of an arithmetic sequence is 21, and the sum of the first seventeen terms is 0. Find the first two terms of the sequence.

11 Three consecutive terms of an arithmetic sequence have a sum of 12 and a product of -80. Find the terms.

    Hint: Let the terms be $x - d$, x, and $x + d$.

12 The sum of the first 15 terms of an arithmetic sequence is 480. Find the 8th term of the sequence.

13 Five consecutive terms of an arithmetic sequence have a sum of 40. The product of the first, middle, and last terms is 224. Find the terms of the sequence.

14 The sum of the first n terms of an arithmetic sequence is $\dfrac{n(3n + 11)}{2}$.

    a Find its first two terms.

    b Find the twentieth term of the sequence.

15 Find $3 - 5 + 7 - 9 + 11 - 13 + 15 - \ldots$ to 80 terms.

16 Let $t_n = 3 + 2n$.

    a For $n = 1, \ldots, 4$, plot the points (n, t_n) on a graph, and draw rectangles with vertices (n, t_n), $(n + 1, t_n)$, $(n, 0)$, and $(n + 1, 0)$.

    b Explain how S_n relates to the areas of the rectangles.

    c Using your sketch, explain why:

        i $t_{n+1} = t_n + 2$ **ii** $S_{n+1} = S_n + t_{n+1}$.

Activity 1   Stadium seating

A circular stadium consists of sections as illustrated, with aisles in between. The diagram shows the 13 tiers of concrete steps for the final section, Section K. Seats are placed along every concrete step, with each seat 0.45 m wide. The arc AB at the front of the first row is 14.4 m long, while the arc CD at the back of the back row is 20.25 m long.

1 How wide is each concrete step?

2 What is the length of the arc of the back of Row 1, Row 2, Row 3, and so on?

3 How many seats are there in Row 1, Row 2, Row 3, ..., Row 13?

4 How many sections are there in the stadium?

5 What is the total seating capacity of the stadium?

6 What is the radius r of the 'playing surface'?

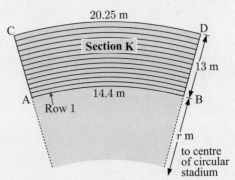

G | GEOMETRIC SERIES

A **geometric series** is the sum of the terms of a geometric sequence.

For example: 1, 2, 4, 8, 16, ..., 1024   is a finite geometric sequence.

$1 + 2 + 4 + 8 + 16 + ... + 1024$ is the corresponding finite geometric series.

If we are adding the first n terms of an infinite geometric sequence, we are then calculating a finite geometric series called the **nth partial sum** of the corresponding infinite series.

If we are adding all of the terms in an infinite geometric sequence, we have an **infinite geometric series**.

SUM OF A FINITE GEOMETRIC SERIES

If the first term is t_1 and the common ratio is r, then the terms are: $t_1, t_1r, t_1r^2, t_1r^3, ..., t_1r^{n-1}$.

So, $S_n = t_1 + \underset{\underset{t_2}{\uparrow}}{t_1r} + \underset{\underset{t_3}{\uparrow}}{t_1r^2} + \underset{\underset{t_4}{\uparrow}}{t_1r^3} + ... + \underset{\underset{t_{n-1}}{\uparrow}}{t_1r^{n-2}} + \underset{\underset{t_n}{\uparrow}}{t_1r^{n-1}}$

For a finite geometric series with $r \neq 1$,

$$S_n = \frac{t_1(r^n - 1)}{r - 1} \quad or \quad S_n = \frac{t_1(1 - r^n)}{1 - r}.$$

Proof:

If $\;S_n = t_1 + t_1r + t_1r^2 + t_1r^3 + ... + t_1r^{n-2} + t_1r^{n-1}$ $(*)$

then $\;rS_n = (t_1r + t_1r^2 + t_1r^3 + t_1r^4 + ... + t_1r^{n-1}) + t_1r^n$

$\therefore \;rS_n = (S_n - t_1) + t_1r^n \quad \{$from $(*)\}$

$\therefore \;rS_n - S_n = t_1r^n - t_1$

$\therefore \;S_n(r - 1) = t_1(r^n - 1)$

$\therefore \;S_n = \dfrac{t_1(r^n - 1)}{r - 1} \quad$ or $\quad \dfrac{t_1(1 - r^n)}{1 - r} \quad$ provided $r \neq 1$.

In the case $r = 1$ we have a sequence in which all terms are the same. The sequence is also arithmetic (with $d = 0$), and $S_n = t_1 n$.

Example 18   ◀) Self Tutor

Find the sum of $\;2 + 6 + 18 + 54 + ...\;$ to 12 terms.

The series is geometric with $t_1 = 2$, $r = 3$, and $n = 12$.

$S_n = \dfrac{t_1(r^n - 1)}{r - 1}$

$\therefore \;S_{12} = \dfrac{2(3^{12} - 1)}{3 - 1} = 531\,440$

Example 19

◀)) **Self Tutor**

Find a formula for S_n, the sum of the first n terms of the series $9 - 3 + 1 - \frac{1}{3} + \ldots$

This answer cannot be simplified as we do not know if n is odd or even.

The series is geometric with $t_1 = 9$ and $r = -\frac{1}{3}$

$$S_n = \frac{t_1(1 - r^n)}{1 - r} = \frac{9(1 - (-\frac{1}{3})^n)}{\frac{4}{3}}$$

$$\therefore \quad S_n = \frac{27}{4}(1 - (-\frac{1}{3})^n)$$

EXERCISE 1G.1

1 Find the sum of the following series:

   a $12 + 6 + 3 + 1.5 + \ldots$ to 10 terms

   b $\sqrt{7} + 7 + 7\sqrt{7} + 49 + \ldots$ to 12 terms

   c $6 - 3 + 1\frac{1}{2} - \frac{3}{4} + \ldots$ to 15 terms

   d $1 - \frac{1}{\sqrt{2}} + \frac{1}{2} - \frac{1}{2\sqrt{2}} + \ldots$ to 20 terms

2 Find a formula for S_n, the sum of the first n terms of the series:

   a $\sqrt{3} + 3 + 3\sqrt{3} + 9 + \ldots$

   b $12 + 6 + 3 + 1\frac{1}{2} + \ldots$

   c $0.9 + 0.09 + 0.009 + 0.0009 + \ldots$

   d $20 - 10 + 5 - 2\frac{1}{2} + \ldots$

3 A geometric sequence has partial sums $S_1 = 3$ and $S_2 = 4$.

   a State the first term t_1.

   b Calculate the common ratio r.

   c Calculate the fifth term t_5 of the series.

   d Find the value of S_5.

4 Evaluate these geometric series:

   a $\displaystyle\sum_{k=1}^{10} 3 \times 2^{k-1}$

   b $\displaystyle\sum_{k=1}^{12} \left(\frac{1}{2}\right)^{k-2}$

   c $\displaystyle\sum_{k=1}^{25} 6 \times (-2)^k$

5 At the end of each year, a salesperson is paid a bonus of £2000 which is banked into the same account. It earns a fixed rate of interest of 6% p.a. with interest being paid annually. The total amount in the account at the end of each year will be:

$$A_1 = 2000$$
$$A_2 = A_1 \times 1.06 + 2000$$
$$A_3 = A_2 \times 1.06 + 2000 \quad \text{and so on.}$$

   a Show that $A_3 = 2000 + 2000 \times 1.06 + 2000 \times (1.06)^2$.

   b Show that $A_4 = 2000[1 + 1.06 + (1.06)^2 + (1.06)^3]$.

   c Find the total bank balance after 10 years, assuming there are no fees or withdrawals.

6 Paula has started renting an apartment. She paid £5000 rent in the first year, and the rent increased by 5% each year.

   a Find, to the nearest £10, the rent paid by Paula in the 4th year.

   b Write an expression for the total rent paid by Paula during the first n years.

   c How much rent did Paula pay during the first 7 years? Give your answer to the nearest £10.

7 Consider $S_n = \frac{1}{2} + \frac{1}{4} + \frac{1}{8} + \frac{1}{16} + \ldots + \frac{1}{2^n}$.

a Find S_1, S_2, S_3, S_4, and S_5 in fractional form.

b Hence guess the formula for S_n.

c Find S_n using $S_n = \dfrac{t_1(1 - r^n)}{1 - r}$.

d Comment on S_n as n gets very large.

e Explain the relationship between the given diagram and **d**.

8 A geometric series has second term 6. The sum of its first three terms is -14. Find its fourth term.

9 Find n given that $\displaystyle\sum_{k=1}^{n} 2 \times 3^{k-1} = 177\,146$.

10 Suppose $t_1, t_2, \ldots, t_n$ is a geometric sequence with common ratio r. Show that

$$(t_1 + t_2)^2 + (t_2 + t_3)^2 + (t_3 + t_4)^2 + \ldots + (t_{n-1} + t_n)^2 = \frac{2t_1^{\,2}(r^{2n-1} - 1)}{r - 1} - (t_1^2 + t_n^2).$$

11 £8000 is borrowed over a 2-year period at a rate of 12% p.a. compounded quarterly. Quarterly repayments are made and the interest is adjusted each quarter, which means that at the end of each quarter, interest is charged on the previous balance and then the balance is reduced by the amount repaid.

There are $2 \times 4 = 8$ repayments and the interest per quarter is $\dfrac{12\%}{4} = 3\%$.

At the end of the first quarter, the amount owed is given by $A_1 = \text{£}8000 \times 1.03 - R$, where R is the amount of each repayment.

At the end of the second quarter, the amount owed is given by:

$$\begin{aligned} A_2 &= A_1 \times 1.03 - R \\ &= (\text{£}8000 \times 1.03 - R) \times 1.03 - R \\ &= \text{£}8000 \times (1.03)^2 - 1.03R - R \end{aligned}$$

a Write an expression for the amount owed at the end of the third quarter, A_3.

b Write an expression for the amount owed at the end of the eighth quarter, A_8.

c Given that $A_8 = 0$ for the loan to be fully repaid, deduce the value of R.

d Now suppose the amount borrowed was £P, the interest rate was $r\%$ per repayment interval, and

there were m repayments. Show that each repayment would be $R = \text{£}\dfrac{P(1 + \frac{r}{100})^m \times \frac{r}{100}}{(1 + \frac{r}{100})^m - 1}$.

SUM OF AN INFINITE GEOMETRIC SERIES

To examine the sum of all the terms of an infinite geometric sequence, we need to consider $S_n = \dfrac{t_1(1 - r^n)}{1 - r}$ when n gets very large.

If $|r| > 1$, the series is said to be **divergent** and the sum becomes infinitely large.

For example, when $r = 2$, $1 + 2 + 4 + 8 + 16 + \ldots$ is infinitely large.

$|r|$ is the *size* of r.
If $|r| > 1$ then $r < -1$ or $r > 1$.

If $|r| < 1$, or in other words $-1 < r < 1$, then as n becomes very large, r^n approaches 0.

This means that S_n will get closer and closer to $\dfrac{t_1}{1 - r}$.

If $|r| < 1$, an infinite geometric series of the form $t_1 + t_1 r + t_1 r^2 + ... = \sum\limits_{k=1}^{\infty} t_1 r^{k-1}$ will **converge** to the **limiting sum** $S = \dfrac{t_1}{1-r}$.

Proof: If the first term is t_1 and the common ratio is r, the terms are $t_1, t_1 r, t_1 r^2, t_1 r^3, ...$

Suppose the sum of the corresponding infinite series is

$S = t_1 + t_1 r + t_1 r^2 + t_1 r^3 + ...$ $(*)$

$\therefore \quad rS = t_1 r + t_1 r^2 + t_1 r^3 + t_1 r^4 + ...$

$\therefore \quad rS = S - t_1$ {comparing with $(*)$}

$\therefore \quad S(r-1) = -t_1$

$\therefore \quad S = \dfrac{t_1}{1-r}$ {provided $r \neq 1$}

This result can be used to find the value of recurring decimals.

Example 20   ◀) **Self Tutor**

Write $0.\overline{7}$ as a rational number.

$0.\overline{7} = \frac{7}{10} + \frac{7}{100} + \frac{7}{1000} + \frac{7}{10\,000} + ...$ is an infinite geometric

series with $t_1 = \frac{7}{10}$ and $r = \frac{1}{10}$.

$\therefore \quad S = \dfrac{t_1}{1-r} = \dfrac{\frac{7}{10}}{1 - \frac{1}{10}} = \dfrac{7}{9}$

$\therefore \quad 0.\overline{7} = \frac{7}{9}$

EXERCISE 1G.2

1 a Explain why $0.\overline{3} = \frac{3}{10} + \frac{3}{100} + \frac{3}{1000} + ...$ is an infinite geometric series.

  b Hence show that $0.\overline{3} = \frac{1}{3}$.

2 Write as a rational number:

  a $0.\overline{4}$ **b** $0.\overline{16}$ **c** $0.\overline{312}$

3 Use $S = \dfrac{t_1}{1-r}$ to check your answer to **Exercise 1G.1** question **7 d**.

4 Find the sum of each of the following infinite geometric series:

  a $18 + 12 + 8 + \frac{16}{3} + ...$ **b** $18.9 - 6.3 + 2.1 - 0.7 + ...$

5 Find:

  a $\sum\limits_{k=1}^{\infty} \dfrac{3}{4^k}$ **b** $\sum\limits_{k=0}^{\infty} 6\left(-\frac{2}{5}\right)^k$

6 The sum of the first three terms of a convergent infinite geometric series is 19. The sum of the series is 27. Find the first term and the common ratio.

7 The second term of a convergent infinite geometric series is $\frac{8}{5}$. The sum of the series is 10. Show that there are two possible series, and find the first term and the common ratio in each case.

8 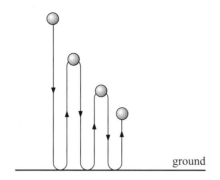 When dropped, a ball takes 1 second to hit the ground. It then takes 90% of this time to rebound to its new height, and this continues until the ball comes to rest.

    a Show that the total time of motion is given by
$$1 + 2(0.9) + 2(0.9)^2 + 2(0.9)^3 + \dots.$$

    b Find S_n for the series in **a**.

    c How long does it take for the ball to come to rest?

ground

9 When a ball is dropped, it rebounds 75% of its height after each bounce. If the ball travels a total distance of 490 cm, from what height was the ball dropped?

10   **a** Explain why $0.\overline{9} = 1$ exactly.

    b Show that if $t_n = \dfrac{9}{10^n}$, then $S_n = 1 - \dfrac{1}{10^n}$.

    c On a graph, plot the points (n, t_n) and (n, S_n) for $n = 1, 2, \dots., 10$. Connect each set of points with a smooth curve.

11 Find x if $\displaystyle\sum_{k=1}^{\infty} \left(\frac{3x}{2}\right)^{k-1} = 4$.

Discussion

- Can we explain through intuition how a sum of non-zero terms, which goes on and on for ever and ever, could actually be a finite number?

- What happens to the infinite "geometric" series $\displaystyle\sum_{k=1}^{\infty} t_1 r^{k-1}$ in the case $r = 1$?

- In the case $r = -1$, the infinite geometric series $\displaystyle\sum_{k=1}^{\infty} t_1 r^{k-1}$ is $t_1 - t_1 + t_1 - t_1 + \dots.$

  If we take partial sums of the series, the answer is always t_1 or 0.

  Is $\displaystyle\sum_{k=1}^{\infty} t_1 r^{k-1}$ defined for $r = -1$?

Activity 2

Click on the icon to run a card game for sequences and series.

CARD GAME

Investigation 1   von Koch's snowflake curve

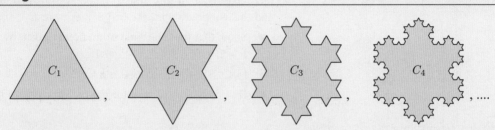

In this Investigation we consider a **limit curve** named after the Swedish mathematician Niels Fabian Helge von Koch (1870 - 1924).

To draw **von Koch's snowflake curve** we:

- start with an equilateral triangle, C_1
- divide each side into 3 equal parts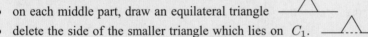
- on each middle part, draw an equilateral triangle
- delete the side of the smaller triangle which lies on C_1.

DEMO

The resulting curve is C_2. By repeating this process on every edge of C_2, we generate curve C_3.

We hence obtain a sequence of special curves C_1, C_2, C_3, C_4, ... and von Koch's curve is the limiting case when n is infinitely large.

Your task is to investigate the perimeter and area of von Koch's curve.

What to do:

1 **a** Suppose C_1 has perimeter 3 units. Find the perimeter of C_2, C_3, C_4, and C_5.

   Hint: _____ becomes ⟋⟍ so 3 parts become 4 parts.

   b Remembering that von Koch's curve is C_n, where n is infinitely large, find the perimeter of von Koch's curve.

2 Suppose the area of C_1 is 1 unit^2.

   a Explain why the areas of C_2, C_3, C_4, and C_5 are:

   $A_2 = 1 + \frac{1}{3}$ units^2 $A_3 = 1 + \frac{1}{3}[1 + \frac{4}{9}]$ units^2

   $A_4 = 1 + \frac{1}{3}[1 + \frac{4}{9} + (\frac{4}{9})^2]$ units^2 $A_5 = 1 + \frac{1}{3}[1 + \frac{4}{9} + (\frac{4}{9})^2 + (\frac{4}{9})^3]$ units^2.

   b Use your calculator to find A_n where $n = 1, 2, 3, 4, 5, 6,$ and 7, giving answers which are as accurate as your calculator permits.

   c What do you think will be the area within von Koch's snowflake curve?

3 Is there anything remarkable about your answers to **1** and **2**?

4 Investigate the sequence of curves obtained by adding squares on successive curves from the middle third of each side. These are the curves C_1, C_2, C_3, ... shown below.

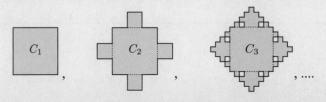

H   THE BINOMIAL EXPANSION

The sum $a + b$ is called a **binomial** as it contains two terms.

Any expression of the form $(a + b)^n$ is called a **power of a binomial**.

In this Section we consider how the power of a binomial can be expanded as a series without brackets.

FACTORIAL NOTATION

For $n \geqslant 1$, $n!$ is the product of the first n positive integers.

$$n! = n(n-1)(n-2).... \times 2 \times 1$$

We also define $0! = 1$.

THE BINOMIAL EXPANSION FOR $n \in \mathbb{Z}^+$

You should have previously seen that:

For $n \in \mathbb{Z}^+$,

$$(a + b)^n = a^n + \binom{n}{1}a^{n-1}b + ... + \binom{n}{r}a^{n-r}b^r + ... + b^n$$

$$= \sum_{r=0}^{n} \binom{n}{r}a^{n-r}b^r$$

where $\binom{n}{r} = \dfrac{n!}{r!\,(n-r)!}$ is the **binomial coefficient** for $n, r \in \mathbb{Z}_0^+$, $r \leqslant n$.

The **general term** or $(r+1)$**th term** in the binomial expansion is $T_{r+1} = \binom{n}{r}a^{n-r}b^r$.

Example 21
◀) **Self Tutor**

In the expansion of $\left(x - \dfrac{2}{x^2}\right)^9$, find:

a the constant term

b the coefficient of x^3.

$a = (x)$, $b = \left(\dfrac{-2}{x^2}\right)$, and $n = 9$

$\therefore$ the general term $T_{r+1} = \binom{9}{r}(x)^{9-r}\left(\dfrac{-2}{x^2}\right)^r$

$$= \binom{9}{r}x^{9-r} \times \dfrac{(-2)^r}{x^{2r}}$$

$$= \binom{9}{r}(-2)^r x^{9-3r}$$

a If $9 - 3r = 0$

then $3r = 9$

$\therefore$ $r = 3$

$\therefore$ $T_4 = \binom{9}{3}(-2)^3 x^0$

$\therefore$ the constant term is $\binom{9}{3}(-2)^3 = -672$

b If $9 - 3r = 3$

then $3r = 6$

$\therefore$ $r = 2$

$\therefore$ $T_3 = \binom{9}{2}(-2)^2 x^3$

$\therefore$ the coefficient of x^3 is $\binom{9}{2}(-2)^2 = 144$

EXERCISE 1H.1

1   Expand and simplify:

    a $(x-2)^3$
    b $(2x+1)^4$
    c $(m+n)^5$

    d $(a^2-1)^3$
    e $\left(x-\dfrac{2}{x}\right)^4$
    f $(x+\sqrt{x})^4$

2   Without simplifying, write down:

    a   the 5th term of $(3x-2)^8$
    b   the 4th term of $(x+y^2)^{10}$

    c   the 6th term of $\left(2x-\dfrac{1}{x}\right)^{11}$
    d   the 7th term of $(\sqrt{x}+1)^{16}$

3   Consider the expansion of $\left(\frac{1}{2}x+\frac{1}{3}\right)^{12}$.

    a   Write down the general term of the expansion.

    b   Find the coefficient of:

        i x^3
        ii x^5
        iii x^9

4   Find the constant term in the expansion of:

    a $\left(\dfrac{1}{x^2}-x\right)^6$
    b $\left(x^2+\dfrac{5}{x}\right)^{15}$

5   Find the value of a given:

    a   the coefficient of x^6 in the expansion of $(2x+a)^8$ is 448

    b   the coefficient of $\dfrac{y^2}{x^6}$ in the expansion of $\left(y-\dfrac{a}{x}\right)^8$ is 1792

    c   the fourth term in the expansion of $\left(ax+\sqrt{2}\right)^{10}$ is $30x^7$

6   Find the coefficient of:

    a x^4 in the expansion of $(x-3)(2x+1)^7$

    b x in the expansion of $(2+x)\left(\dfrac{1}{x}-2x\right)^4$

    c x^3 in the expansion of $(x-3)(2x+1)^5$

THE BINOMIAL EXPANSION FOR $n \in \mathbb{Q}$

We have seen that for $n \in \mathbb{Z}^+$,

$$(a+b)^n = a^n + \binom{n}{1}a^{n-1}b + \ldots + \binom{n}{r}a^{n-r}b^r + \ldots + b^n$$

$$= a^n + na^{n-1}b + \frac{n(n-1)}{2!}a^{n-2}b^2 + \frac{n(n-1)(n-2)}{3!}a^{n-3}b^3 + \ldots + b^n$$

Observe how with each term:

- the power of a decreases by 1
- the power of b increases by 1

- the coefficient is multiplied by $\dfrac{n-(r-1)}{r}$ where $r+1$ is the term number.

Sir Isaac Newton generalised this to expand a binomial raised to any rational power using an infinite series.

For $n \in \mathbb{Q}$, $(a + b)^n = \displaystyle\sum_{r=0}^{\infty} \binom{n}{r} a^{n-r} b^r$

where $\binom{n}{r} = \dfrac{n(n-1)(n-2)....(n-r+1)}{r!}$ is the **binomial coefficient** for $n \in \mathbb{Q}$, $r \in \mathbb{Z}_0^+$.

In the case where $n \in \mathbb{Z}^+$, the series terminates after $n + 1$ terms. This is because with the $(n+2)$th term, $r = n + 1$, and so the coefficient is multiplied by $\dfrac{n - (n+1-1)}{n+1} = 0$. Every subsequent term will also have coefficient 0.

Having observed that an infinite geometric series will only converge under particular conditions, we recognise that this infinite series expansion for $(a+b)^n$, $n \in \mathbb{Q}$ will also only converge under particular conditions.

THE BINOMIAL EXPANSION OF $(a + bx)^n$ FOR $n \in \mathbb{Q}$

In this course we consider values of x for which the binomial expansion of $(a + bx)^n$, $n \in \mathbb{Q}$ converges.

We first notice that $(a + bx)^n = a^n \left(1 + \dfrac{b}{a}x\right)^n$

For $n \in \mathbb{Q}$, $(a + bx)^n = a^n \displaystyle\sum_{r=0}^{\infty} \binom{n}{r} \left(\dfrac{bx}{a}\right)^r$

where $\binom{n}{r} = \dfrac{n(n-1)(n-2)....(n-r+1)}{r!}$ is the **binomial**

coefficient for $n \in \mathbb{Q}$, $r \in \mathbb{Z}_0^+$.

The expansion **converges** for x such that $\left|\dfrac{bx}{a}\right| < 1$.

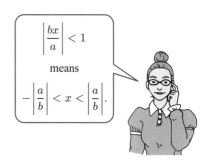

$\left|\dfrac{bx}{a}\right| < 1$

means

$-\left|\dfrac{a}{b}\right| < x < \left|\dfrac{a}{b}\right|$.

The proof of this result is beyond the scope of the course. However, we may think of it intuitively as follows:

$(a + bx)^n = a^n \displaystyle\sum_{r=0}^{\infty} \binom{n}{r} \left(\dfrac{bx}{a}\right)^r$

$= a^n \displaystyle\sum_{k=1}^{\infty} \binom{n}{k-1} \left(\dfrac{bx}{a}\right)^{k-1}$ {letting $k = r + 1$}

where the series $\displaystyle\sum_{k=1}^{\infty} \binom{n}{k-1} \left(\dfrac{bx}{a}\right)^{k-1}$ is similar to the infinite geometric sequence $\displaystyle\sum_{k=1}^{\infty} t_1 r^{k-1}$ with $r = \dfrac{bx}{a}$.

The difference is that rather than the constant t_1 in the infinite geometric sequence, we now have the binomial coefficient $\binom{n}{k-1}$ which varies with each term. However, just as the infinite geometric sequence converges for $|r| < 1$, the binomial expansion converges for $\left|\dfrac{bx}{a}\right| < 1$.

Most importantly, for the series to converge, the terms must become smaller in size as k increases. This means that we can approximate the binomial expansion by truncating it at a finite number of terms.

Example 22   ◄⑴ **Self Tutor**

Consider the binomial expansion of $\dfrac{1}{\sqrt{2+x}}$.

  a Write down the first 4 terms of the expansion.

  b For what values of x does the expansion converge?

  c Use the expansion to estimate $\dfrac{1}{\sqrt{2.1}}$. Check your answer by direct calculation.

a $\dfrac{1}{\sqrt{2+x}} = (2+x)^{-\frac{1}{2}}$

$\qquad\qquad = 2^{-\frac{1}{2}}\left(1+\dfrac{x}{2}\right)^{-\frac{1}{2}}$

$\qquad\qquad = \dfrac{1}{\sqrt{2}}\left(1+\dfrac{x}{2}\right)^{-\frac{1}{2}}$

$\qquad\qquad = \dfrac{1}{\sqrt{2}}\displaystyle\sum_{r=0}^{\infty}\binom{-\frac{1}{2}}{r}\left(\dfrac{x}{2}\right)^{r}$

$\qquad\qquad = \dfrac{1}{\sqrt{2}}\left(1 + \left(-\tfrac{1}{2}\right)\left(\tfrac{x}{2}\right) + \dfrac{\left(-\frac{1}{2}\right)\left(-\frac{3}{2}\right)}{2!}\left(\tfrac{x}{2}\right)^{2} + \dfrac{\left(-\frac{1}{2}\right)\left(-\frac{3}{2}\right)\left(-\frac{5}{2}\right)}{3!}\left(\tfrac{x}{2}\right)^{3} + ...\right)$

$\qquad\qquad = \dfrac{1}{\sqrt{2}}\left(1 - \tfrac{1}{4}x + \tfrac{3}{32}x^{2} - \tfrac{5}{128}x^{3} + ...\right)$

b The series converges provided $\left|\dfrac{x}{2}\right| < 1$

$\qquad\therefore\quad -2 < x < 2$

c Letting $x = 0.1$, $\quad\dfrac{1}{\sqrt{2.1}} \approx \dfrac{1}{\sqrt{2}}\left(1 - \tfrac{1}{4}(0.1) + \tfrac{3}{32}(0.1)^{2} - \tfrac{5}{128}(0.1)^{3}\right)$

$\qquad\qquad\qquad\qquad\qquad\quad \approx 0.690\,064$

   Using technology, $\quad\dfrac{1}{\sqrt{2.1}} \approx 0.690\,066$.

EXERCISE 1H.2

1 Consider the binomial expansion of $\dfrac{1}{1+x}$.

  a Write down the first 4 terms of the expansion.

  b For what values of x does the expansion converge?

  c Use the expansion to estimate $\dfrac{1}{1.1}$. Check your answer by direct calculation.

2 Consider the binomial expansion of $\dfrac{1}{\sqrt{4-x}}$.

  a Write down the first 4 terms of the expansion.

  b For what values of x does the expansion converge?

  c Use the expansion to estimate $\dfrac{1}{\sqrt{3.6}}$. Check your answer by direct calculation.

3 Expand $\dfrac{1}{(1-2x)^2}$ up to the term in x^3 and state the values of x for which the expansion is valid.

Hence estimate $\dfrac{1}{(0.96)^2}$.

4 **a** Expand $\sqrt{1+x}$ up to the term in x^3 and state the values of x for which the expansion is valid.

b To estimate $\sqrt{37}$, explain why you cannot substitute $x = 36$.

c Estimate $\sqrt{37}$ by first letting $\sqrt{37} = \sqrt{36} \times \sqrt{\frac{37}{36}} = \sqrt{36} \times \sqrt{1 + \frac{1}{36}}$.

5 **a** Expand $\dfrac{1}{(1+2x)^2}$ up to the term in x^3.

b Hence find the coefficient of x^2 in the expansion of $\dfrac{(1-3x)^3}{(1+2x)^2}$.

Investigation 2 — Finite differences

Given a sequence of numbers t_n, we can create a sequence of *first differences* Δ_n by finding the difference between successive terms of the sequence:

$$\Delta_n = t_{n+1} - t_n$$

For example, for the sequence   1,   3,   6,   10,   15,   21, ...,

the sequence of first differences is   2,   3,   4,   5,   6, ...

$$3-1 \quad 6-3 \quad 10-6 \quad 15-10 \quad 21-15$$

RECURRENCE RELATIONS

Rearranging $\Delta_n = t_{n+1} - t_n$, we have

$$t_{n+1} = t_n + \Delta_n.$$

This is a **recurrence relationship** connecting consecutive terms of the sequence.

The next two terms in the sequence of first differences above are $\Delta_6 = 7$ and $\Delta_7 = 8$. So, the next two terms in the original sequence are

$$t_7 = t_6 + \Delta_6 = 21 + 7 = 28 \quad \text{and} \quad t_8 = t_7 + \Delta_7 = 28 + 8 = 36.$$

FORMULAE FOR FINITE DIFFERENCES

We may be able to find a formula for the sequence Δ_n directly from the sequence. Alternatively, if we have a formula for t_n, it can be used to calculate the formula for Δ_n.

For example, the original sequence 1, 3, 6, 10, 15, ... has formula

$$t_n = \tfrac{1}{2}n(n+1) = \tfrac{1}{2}(n^2 + n).$$

Therefore the sequence of first differences has formula

$$\Delta_n = t_{n+1} - t_n = \tfrac{1}{2}\big((n+1)^2 + (n+1)\big) - \tfrac{1}{2}(n^2 + n) = n+1.$$

What to do:

1 For each of the following sequences:

    i Calculate the sequence of first differences, and try to find a pattern.

    ii Continue the original sequence for the next three terms.

    iii If possible, write down formulae for the original sequence t_n and the sequence of first differences Δ_n.

  a 3, 5, 7, 9, 11, 13, 15, 17, ...

  b 6, 11, 16, 21, 26, 31, 36, 41, ...

  c 1, 4, 9, 16, 25, 36, 49, 64, ...

  d −3, 3, 13, 27, 45, 67, 93, 123, ...

  e 2, 4, 8, 16, 32, 64, 128, 256, ...

  f −1, 1, −1, 1, −1, 1, −1, 1, ...

  g 1, 1, 2, 3, 5, 8, 13, 21, 34, 55, ...

  h 2, 1, 3, 4, 7, 11, 18, 29, 47, ...

  i 2, 3, 5, 7, 11, 13, 17, 19, 23, 29, 31, 37, 41, 43, ...

2 Find a formula for Δ_n for a sequence with general term:

  a $t_n = n^2$

  b $t_n = 3n + 2$

  c $t_n = an^2 + bn + c$

3 Prove that if the terms in t_n are defined by a polynomial of degree m, then the sequence of first differences Δ_n is a polynomial of degree $m - 1$.

4 Find a formula for Δ_n for a sequence with general term:

  a $t_n = 5$

  b $t_n = \dfrac{1}{n}$

  c $t_n = \dfrac{1}{n^2}$

5 **a** Construct a table with four columns as shown below. Use your results from **2** and **4** to complete the table.

t_n	Δ_n	$f(x)$	$f'(x)$
n^2	$2n + 1$	x^2	$2x$
$3n + 2$		$3x + 2$	
$\vdots$	$\vdots$	$\vdots$	$\vdots$

  b Can you find any pattern between finite differences for sequences, and derivatives for functions?

  c Prove that for any sequence, $\Delta_1 + \Delta_2 + \Delta_3 + \dots + \Delta_{n-1} = t_n - t_1$.

  d Describe the similarity between this result and $\displaystyle\int_a^b f'(x)\,dx = f(b) - f(a)$.

6 The *difference operator* Δ takes a sequence and returns another sequence. We can apply the difference operator multiple times. For example: $t_n = 1, 3, 6, 10, 15, 21, 28, 36, ...$

$$\Delta_n^{(1)} = 2, 3, 4, 5, 6, 7, 8, ...$$

$$\Delta_n^{(2)} = 1, 1, 1, 1, 1, 1, 1, ...$$

The first sequence of finite differences Δ_n is $\Delta_n^{(1)}$. The second sequence of finite differences $\Delta_n^{(2)}$ is the difference between elements in $\Delta_n^{(1)}$.

For each of the sequences in **1**, calculate the second and third sequences of finite differences. Describe any patterns that you notice.

Review set 1A

1 Consider the number sequence 5, 9, 11, 12, 15, 19. Find:

   a t_2 **b** t_6 **c** S_4.

2 Identify the following sequences as arithmetic, geometric, or neither:

   a $7, -1, -9, -17, \ldots.$ **b** $4, -2, 1, -\frac{1}{2}, \ldots.$ **c** $1, 1, 2, 3, 5, 8, \ldots.$

3 Find k if $3k$, $k-2$, and $k+7$ are consecutive terms of an arithmetic sequence.

4 A sequence is defined by $t_n = 6\left(\frac{1}{2}\right)^{n-1}$.

   a Prove that the sequence is geometric.   **b** Find t_1 and r.

   c Find the 16th term of the sequence to 3 significant figures.

5 Determine the general term of a geometric sequence given that its sixth term is $\frac{16}{3}$ and its tenth term is $\frac{256}{3}$.

6 Insert six numbers between 23 and 9 so that all eight numbers are in arithmetic sequence.

7 An arithmetic series has nine terms. The first term is -2 and the last term is 54. Find the sum of the series.

8 Find the sum of each of the following infinite geometric series:

   a $18 - 12 + 8 - \ldots.$ **b** $8 + 4\sqrt{2} + 4 + \ldots.$

9 Find the sum of:

   a $7 + 11 + 15 + 19 + \ldots. + 99$ **b** $35 + 33\frac{1}{2} + 32 + 30\frac{1}{2} + \ldots. + 20$

10 A ball bounces from a height of 3 metres and returns to 80% of its previous height on each bounce. Find the total distance travelled by the ball until it stops bouncing.

11 Write a formula for the general term t_n of:

   a $86, 83, 80, 77, \ldots.$ **b** $\frac{3}{4}, 1, \frac{7}{6}, \frac{9}{7}, \ldots.$ **c** $100, 90, 81, 72.9, \ldots.$

   Hint: One of these sequences is neither arithmetic nor geometric.

12 Expand and hence evaluate:

   a $\displaystyle\sum_{k=1}^{7} k^2$ **b** $\displaystyle\sum_{k=1}^{4} \frac{k+3}{k+2}$

13 The sum of the first n terms of an infinite sequence is $\dfrac{3n^2 + 5n}{2}$ for all $n \in \mathbb{Z}^+$.

   a Find the nth term.   **b** Prove that the sequence is arithmetic.

14 x, y, and z are consecutive terms of a geometric sequence.

   If $x + y + z = \frac{7}{3}$ and $x^2 + y^2 + z^2 = \frac{91}{9}$, find the values of x, y, and z.

15 £12 500 is invested in an account which pays 4.25% p.a. interest. Find the value of the investment after 5 years if the interest is compounded:

   a half-yearly   **b** monthly.

16 The sum of the first two terms of an infinite geometric series is 90. The third term is 24. Show that there are two possible series, and that both series converge.

17 An arithmetic sequence, and a geometric sequence with common ratio r, have the same first two terms. Show that the third term of the geometric sequence is $\dfrac{r^2}{2r-1}$ times the third term of the arithmetic sequence.

18 Without simplifying, write down:

    a the 4th term of $(y-3z)^9$ **b** the 5th term of $\left(2x - \dfrac{1}{\sqrt{x}}\right)^{10}$.

19 Find the value of a given:

    a the coefficient of x^5 in the expansion of $(a-2x)^7$ is -42

    b the fifth term in the expansion of $\left(\dfrac{1}{x\sqrt{2}} - a\right)^8$ is $\dfrac{1120}{x^4}$.

20 Consider the binomial expansion of $\dfrac{1}{1-x}$.

    a Write down the first 4 terms of the expansion.

    b For what values of x does the expansion converge?

    c Use the expansion to estimate $\dfrac{1}{0.95}$. Check your answer by direct calculation.

21 **a** Expand $\sqrt{1-x}$ up to the term in x^3 and state the values of x for which the expansion is valid.

    b Hence estimate $\sqrt{15}$.

Review set 1B

1 Find the first four terms of the sequence with formula $t_1 = 7$, $t_{n+1} = 2 \times t_n$, $n \geqslant 1$.

2 A sequence is defined by $t_n = 68 - 5n$.

    a Prove that the sequence is arithmetic.   **b** Find t_1 and d.

    c Find the 37th term of the sequence.

    d State the first term of the sequence which is less than -200.

3 **a** Show that the sequence $3, 12, 48, 192, \ldots.$ is geometric.

    b Find t_n and hence find t_9.

4 **a** Find the general term of the arithmetic sequence with $t_7 = 31$ and $t_{15} = -17$.

    b Hence, find the value of t_{34}.

5 Consider the sequence $24, 23\frac{1}{4}, 22\frac{1}{2}, \ldots.$

    a Which term of the sequence is -36?   **b** Find the value of t_{35}.

    c Find S_{40}, the sum of the first 40 terms of the sequence.

6 Find the sum of the first 12 terms of:

    a $3 + 9 + 15 + 21 + \ldots.$ **b** $24 + 12 + 6 + 3 + \ldots.$

7 £7000 is invested at 6% p.a. compound interest. Find the value of the investment after 3 years if interest is compounded:

    a annually   **b** quarterly   **c** monthly.

8 **a** Find k given that 4, k, and $k^2 - 1$ are consecutive terms of a geometric sequence.

    b For each value of k, find the common ratio of the sequence.

9 Find the sum of each of the following infinite geometric series:

    a $1.21 - 1.1 + 1 - ....$ **b** $\frac{14}{3} + \frac{4}{3} + \frac{8}{21} + ...$

10 Seve is training for a long distance walk. He walks for 10 km in the first week, then each week thereafter he walks 500 m further than the previous week. If he continues this pattern for a year, how far does Seve walk:

    a in the last week   **b** in total?

11 Find the first term of the sequence $24, 8, \frac{8}{3}, \frac{8}{9}, ...$ which is less than 0.001.

12 Evaluate:

    a $\displaystyle\sum_{k=1}^{8} \left(\frac{31 - 3k}{2} \right)$ **b** $\displaystyle\sum_{k=1}^{15} 50(0.8)^{k-1}$ **c** $\displaystyle\sum_{k=7}^{\infty} 5\left(\frac{2}{5}\right)^{k-1}$

13 a, b, and c are consecutive terms of an arithmetic sequence. Prove that the following are also consecutive terms of an arithmetic sequence:

    a $b + c$, $c + a$, and $a + b$ **b** $\dfrac{1}{\sqrt{b} + \sqrt{c}}$, $\dfrac{1}{\sqrt{c} + \sqrt{a}}$, and $\dfrac{1}{\sqrt{a} + \sqrt{b}}$

14 **a** Under what conditions will the series $\displaystyle\sum_{k=1}^{\infty} 50(2x - 1)^{k-1}$ converge? Explain your answer.

    b Find $\displaystyle\sum_{k=1}^{\infty} 50(2x - 1)^{k-1}$ if $x = 0.3$.

15 In 2004 there were 3000 iguanas on a Galapagos island. Since then, the population of iguanas on the island has increased by 5% each year.

    a How many iguanas were on the island in 2007?

    b In what year will the population first exceed 10 000?

16 $x + 3$ and $x - 2$ are the first two terms of a geometric series. Find the values of x for which the series converges.

17 Consider the expansion of $(3x + 4)^9$.

    a Write down the general term of the expansion.

    b Find the coefficient of:

        i x^2 **ii** x^5 **iii** x^6

18 Find the constant term in the expansion of:

    a $\left(x - \dfrac{6}{x^2}\right)^9$ **b** $\left(\sqrt{x} + \dfrac{6}{\sqrt{x}}\right)^5$

19 Find the coefficient of x^5 in the expansion of $(3x - 1)(5 + 2x)^6$.

20 Expand $\dfrac{1}{(1 + 2x)^{\frac{3}{2}}}$ up to the term in x^3 and state the values of x for which the expansion is valid.

    Hence estimate $\dfrac{1}{(1.1)^{\frac{3}{2}}}$.

21   **a** Expand $\dfrac{1}{(1 - 3x)^2}$ up to the term in x^3.

    b Hence find the coefficient of x^2 in the expansion of $\dfrac{(1 + 4x)^2}{(1 - 3x)^2}$.

2

Functions

Contents:

Opening problem

The graph of $f(x) = \sqrt{x-4}$ is shown alongside.

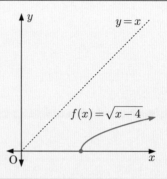

Things to think about:

a How is $f(x)$ related to the curve $g(x) = \sqrt{x}$?

b Suppose $y = f(x)$ is reflected in the line $y = x$.

  i Is the result a function?

  ii How can we find the *equation* of the resulting curve?

You should have previously seen that:

- A **relation** is any set of points which connect two variables.
    - The **domain** of a relation is the set of values which the variable on the horizontal axis can take. This variable is usually x.
    - The **range** of a relation is the set of values which the variable on the vertical axis can take. This variable is usually y.
- A **function** is a relation in which no two different ordered pairs have the same x-coordinate or first component.
    - $f(x)$ is the value of y for a given value of x, so $y = f(x)$.
    - f is the function which converts x into $f(x)$, so $f : x \mapsto f(x)$.

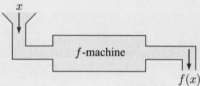

For example, if $f(x) = \sin\left(x - \frac{\pi}{3}\right)$ then:

- f converts x into $\sin\left(x - \frac{\pi}{3}\right)$
- f has the *natural domain* $x \in \mathbb{R}$
- f has the range $[-1, 1]$.

> The **natural domain** is the largest set of values of x for which $f(x)$ is defined.

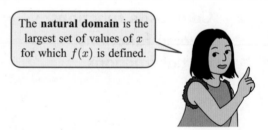

A   COMPOSITE FUNCTIONS

> Given $f : x \mapsto f(x)$ and $g : x \mapsto g(x)$, the **composite function** of g and f will convert x into $f(g(x))$.
>
> fg is used to represent the composite function of g and f. It means "f following g".
>
> $$fg(x) = f(g(x)) \qquad \text{or} \qquad fg : x \mapsto f(g(x)).$$

Consider $f : x \mapsto x^4$ and $g : x \mapsto 2x + 3$.

fg means that g converts x to $2x + 3$ and then f converts $(2x + 3)$ to $(2x + 3)^4$.

The composite function $fg(x)$ is illustrated by the two function machines below.

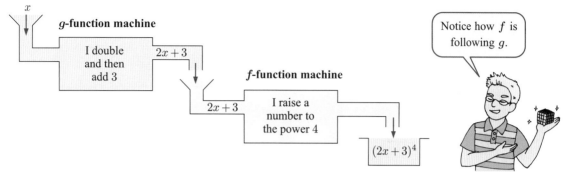

Algebraically, if $f(x) = x^4$ and $g(x) = 2x + 3$ then

$$fg(x) = f(g(x))$$
$$= f(2x + 3) \quad \{g \text{ operates on } x \text{ first}\}$$
$$= (2x + 3)^4 \quad \{f \text{ operates on } g(x) \text{ next}\}$$

and $gf(x) = g(f(x))$
$$= g(x^4) \quad \{f \text{ operates on } x \text{ first}\}$$
$$= 2(x^4) + 3 \quad \{g \text{ operates on } f(x) \text{ next}\}$$
$$= 2x^4 + 3$$

So, $f(g(x)) \neq g(f(x))$.

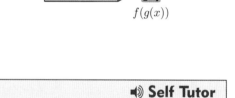

In general, $fg(x) \neq gf(x)$.

Example 1   ◀) **Self Tutor**

Given $f : x \mapsto 2x + 1$ and $g : x \mapsto 3 - 4x$, find in simplest form:

a $fg(x)$ **b** $gf(x)$

$f(x) = 2x + 1$ and $g(x) = 3 - 4x$

a $fg(x) = f(g(x))$
$$= f(3 - 4x)$$
$$= 2(3 - 4x) + 1$$
$$= 6 - 8x + 1$$
$$= 7 - 8x$$

b $gf(x) = g(f(x))$
$$= g(2x + 1)$$
$$= 3 - 4(2x + 1)$$
$$= 3 - 8x - 4$$
$$= -8x - 1$$

In the previous **Example** you should have observed how we can substitute an expression into a function.

If $f(x) = 2x + 1$ then $f(\Delta) = 2(\Delta) + 1$

and so $f(3 - 4x) = 2(3 - 4x) + 1$.

Example 2 🔊 **Self Tutor**

Given $f(x) = 6x - 5$ and $g(x) = x^2 + x$, find:

a $gf(-1)$ **b** $ff(0)$

a $gf(-1) = g(f(-1))$

     Now $f(-1) = 6(-1) - 5$
                       $= -11$

     $\therefore \ gf(-1) = g(-11)$
                        $= (-11)^2 + (-11)$
                        $= 110$

b $ff(0) = f(f(0))$

     Now $f(0) = 6(0) - 5$
                  $= -5$

     $\therefore \ ff(0) = f(-5)$
                  $= 6(-5) - 5$
                  $= -35$

EXERCISE 2A

1 Given $f : x \mapsto 2x + 3$ and $g : x \mapsto 1 - x$, find in simplest form:

     a $fg(x)$ **b** $gf(x)$ **c** $fg(-3)$ **d** $gf(0)$

2 Given $f : x \mapsto -2x$ and $g : x \mapsto 1 + x^2$, find in simplest form:

     a $fg(x)$ **b** $gf(x)$ **c** $fg(2)$ **d** $ff(-1)$

3 Given $f(x) = 3 - x^2$ and $g(x) = 2x + 4$, find in simplest form:

     a $fg(x)$ **b** $gf(x)$ **c** $gg(\frac{1}{2})$ **d** $ff(-\frac{1}{2})$

4 Given $f(x) = \sqrt{6 - x}$ and $g(x) = 5x - 7$, find:

     a $gg(x)$ **b** $fg(1)$ **c** $gf(6)$ **d** $ff(2)$

5 Suppose $f(x) = 9 - \sqrt{x}$ and $g(x) = x^2 + 4$.

     a Find $fg(x)$ and state its domain and range.   **b** Find $gf(4)$.

     c Find $ff(x)$ and state its domain and range.

6 Suppose $f(x) = 1 - 2x$ and $g(x) = 3x + 5$.

     a Find $f(g(x))$.   **b** Hence solve $fg(x) = f(x + 3)$.

7 Suppose $f : x \mapsto 2x - x^2$ and $g : x \mapsto 1 + 3x$.

     a Find in simplest form:

         i $fg(x)$ **ii** $gf(x)$

     b Find the value(s) of x such that $fg(x) = 3gf(x)$.

8 For each pair of functions, find $fg(x)$ and state its domain and range:

     a $f(x) = \dfrac{1}{x}$ and $g(x) = x - 3$ **b** $f(x) = \sqrt{x}$ and $g(x) = \sin x$

     c $f(x) = -\dfrac{1}{x}$ and $g(x) = x^2 + 3x + 2$

9 Suppose $f(x)$ and $g(x)$ are functions. $f(x)$ has domain D_f and range R_f. $g(x)$ has domain D_g and range R_g.

     a Under what circumstance will $fg(x)$ be defined?

     b Assuming $fg(x)$ is defined, find its domain.

B   INVERSE FUNCTIONS

The operations of $+$ and $-$, $\times$ and $\div$, are **inverse operations** as one "undoes" what the other does.

The function $y = 2x + 3$ can be "undone" by its *inverse* function $y = \dfrac{x-3}{2}$.

We can think of this as two machines. If the machines are inverses then the second machine *undoes* what the first machine does.

No matter what value of x enters the first machine, it is returned as the output from the second machine.

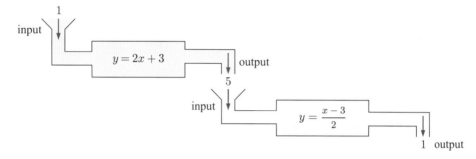

A function $y = f(x)$ *may or may not* have an inverse function. To understand which functions do have inverses, we need some more terminology.

ONE-TO-ONE AND MANY-TO-ONE FUNCTIONS

A **one-to-one** function is any function where:
- for each x there is only one value of y and
- for each y there is only one value of x.

Equivalently, a function is one-to-one if $f(a) = f(b)$ only when $a = b$.

One-to-one functions satisfy both the **vertical line test** and the **horizontal line test**.

This means that:

- no vertical line can meet the graph more than once
- no horizontal line can meet the graph more than once.

For example, $f(x) = x^3$ is one-to-one since it passes both the vertical line and horizontal line tests.

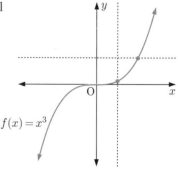

If the function $f(x)$ is **one-to-one**, it will have an inverse function which we denote $f^{-1}(x)$.

Functions that are not one-to-one are called **many-to-one**. While these functions must satisfy the vertical line test, they *do not* satisfy the horizontal line test. At least one y-value has more than one corresponding x-value.

For example, $f(x) = x^2$ fails the horizontal line test, since if $f(x) = 4$ then $x = -2$ or 2.

$f(x) = x^2$ is therefore many-to-one.

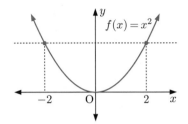

If a function $f(x)$ is **many-to-one**, it *does not* have an inverse function.

However, for a many-to-one function we can often define a new function using the same formula but with a **restricted domain** to make it a one-to-one function. This new function will have an inverse function.

For example, if we restrict $f(x) = x^2$ to the domain $x \geqslant 0$ or the domain $x \leqslant 0$, then the restricted function is one-to-one, and it has an inverse function.

PROPERTIES OF THE INVERSE FUNCTION

If $f(x)$ has an **inverse function**, this new function:

- is denoted $f^{-1}(x)$
- must satisfy the vertical line test
- has a graph which is the reflection of $y = f(x)$ in the line $y = x$
- satisfies $ff^{-1}(x) = x$ and $f^{-1}f(x) = x$.

$f^{-1}(x)$ is the **inverse** of f. In general,
$$f^{-1}(x) \neq \frac{1}{f(x)}.$$

If (x, y) lies on f, then (y, x) must lie on f^{-1}. We therefore find the formula for an inverse function by exchanging x and y. This is why the graph of $y = f^{-1}(x)$ is the reflection of $y = f(x)$ in the line $y = x$.

For example, $f : y = 5x + 2$ becomes $f^{-1} : x = 5y + 2$,

which we rearrange to obtain $f^{-1} : y = \dfrac{x - 2}{5}$.

The domain of f^{-1} is equal to the range of f.

The range of f^{-1} is equal to the domain of f.

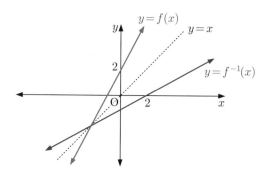

$y = f^{-1}(x)$ is the inverse of $y = f(x)$ as:

- it is also a function
- it is the reflection of $y = f(x)$ in the line $y = x$.

Example 3 ◀) **Self Tutor**

Consider $f : x \mapsto 2x + 3$.

a On the same axes, graph f and its inverse function f^{-1}.

b Find $f^{-1}(x)$ using:

  i coordinate geometry and the gradient of $y = f^{-1}(x)$ from **a**
  ii variable interchange.

c Check that $ff^{-1}(x) = f^{-1}f(x) = x$

a $f(x) = 2x + 3$ passes through $(0, 3)$ and $(2, 7)$.
∴ $f^{-1}(x)$ passes through $(3, 0)$ and $(7, 2)$.

If f includes point (a, b) then f^{-1} includes point (b, a).

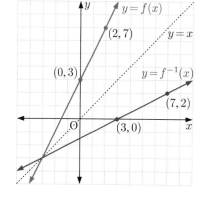

b **i** $y = f^{-1}(x)$ has gradient $\dfrac{2 - 0}{7 - 3} = \dfrac{1}{2}$

Its equation is $\dfrac{y - 0}{x - 3} = \dfrac{1}{2}$

∴ $y = \dfrac{x - 3}{2}$

∴ $f^{-1}(x) = \dfrac{x - 3}{2}$

ii f is $y = 2x + 3$,
∴ f^{-1} is $x = 2y + 3$
∴ $x - 3 = 2y$
∴ $\dfrac{x - 3}{2} = y$
∴ $f^{-1}(x) = \dfrac{x - 3}{2}$

c $ff^{-1}(x) = f(f^{-1}(x))$
  $= f\left(\dfrac{x - 3}{2}\right)$
  $= 2\left(\dfrac{x - 3}{2}\right) + 3$
  $= x$

and $f^{-1}f(x) = f^{-1}(f(x))$
  $= f^{-1}(2x + 3)$
  $= \dfrac{(2x + 3) - 3}{2}$
  $= \dfrac{2x}{2}$
  $= x$

Example 4

 Self Tutor

Consider $f : x \mapsto x^2$.

a Explain why this function does not have an inverse function.

b Does $f : x \mapsto x^2$ where $x \geqslant 0$ have an inverse function?

c Find $f^{-1}(x)$ for $f : x \mapsto x^2$, $x \geqslant 0$.

d Sketch $y = f(x)$, $y = x$, and $y = f^{-1}(x)$ for f in **b** and f^{-1} in **c**.

a From the graph, we can see that $f : x \mapsto x^2$ does not pass the horizontal line test.

∴ it is many-to-one and does not have an inverse function.

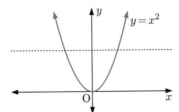

b If we restrict the domain to $x \geqslant 0$ or $x \in [0, \infty)$, the function now satisfies the horizontal line test.

∴ it is one-to-one and has an inverse function.

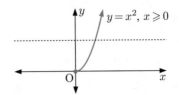

c f is defined by $y = x^2$, $x \geqslant 0$

∴ f^{-1} is defined by $x = y^2$, $y \geqslant 0$

∴ $y = \pm\sqrt{x}$, $y \geqslant 0$

∴ $y = \sqrt{x}$ {as $-\sqrt{x}$ is $\leqslant 0$}

So, $f^{-1}(x) = \sqrt{x}$

d

SELF-INVERSE FUNCTIONS

Any function which has an inverse, and whose graph is symmetrical about the line $y = x$, is a **self-inverse function**.

For example:

- The function $f(x) = x$ is the **identity function**, and is also a self-inverse function.
- The function $f(x) = \dfrac{1}{x}$, $x \neq 0$, is also a self-inverse function, as $f = f^{-1}$.

EXERCISE 2B

1 For each of the following functions f:

   i On the same set of axes, graph $y = x$, $y = f(x)$, and $y = f^{-1}(x)$.

   ii Find $f^{-1}(x)$ using coordinate geometry and the gradient of $y = f^{-1}(x)$ from **i**.

   iii Find $f^{-1}(x)$ using variable interchange.

   a $f : x \mapsto 3x + 1$

   b $f : x \mapsto \dfrac{x + 2}{4}$

2 For each of the following functions f:

  i Find $f^{-1}(x)$.

  ii Sketch $y = f(x)$, $y = f^{-1}(x)$, and $y = x$ on the same set of axes.

  iii Show that $f^{-1}f(x) = ff^{-1}(x) = x$, the identity function.

 a $f : x \mapsto 2x + 5$ **b** $f : x \mapsto \dfrac{3 - 2x}{4}$ **c** $f : x \mapsto x + 3$

3 Copy the graphs of the following functions and draw the graphs of $y = x$ and $y = f^{-1}(x)$ on the same set of axes. In each case, state the domain and range of both f and f^{-1}.

a 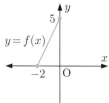   **b** 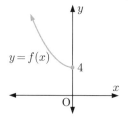   **c** 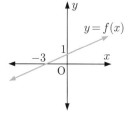   **PRINTABLE GRAPHS**

d 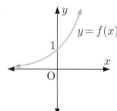   **e** 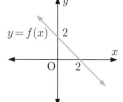   **f**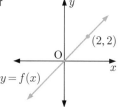

4 Given $f(x) = 2x - 5$, find $(f^{-1})^{-1}(x)$. What do you notice?

5 Find *all* linear functions which are self-inverse.

> A function is self-inverse if $f^{-1}(x) = f(x)$.

6 Which of the following functions have inverses? Where an inverse exists, write down the inverse function.

 a $\{(1, 2), (2, 4), (3, 5)\}$ **b** $\{(-1, 3), (0, 2), (1, 3)\}$

 c $\{(2, 1), (-1, 0), (0, 2), (1, 3)\}$ **d** $\{(-1, -1), (0, 0), (1, 1)\}$

7 **a** Sketch the graph of $f : x \mapsto x^2 - 4$ and reflect it in the line $y = x$.

  b Does f have an inverse function?

  c Does f with restricted domain $x \geqslant 0$ have an inverse function?

8 Sketch the graph of $f : x \mapsto x^3$ and its inverse function $f^{-1}(x)$.

9 Given $f : x \mapsto \dfrac{1}{x}$, $x \neq 0$, find f^{-1} algebraically and show that f is a self-inverse function.

10 The **horizontal line test** says: *For a function to have an inverse function, no horizontal line can cut*
its graph more than once.

    a Explain why this is a valid test for the existence of an inverse function.

    b Which of the following functions have an inverse function?

    i 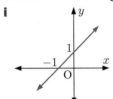   **ii** 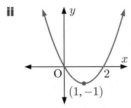   **iii**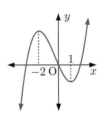

    c For the functions in **b** which do not have an inverse, specify restricted domains as wide as possible
such that the resulting function does have an inverse.

11 Consider $f : x \mapsto x^2, \ x \leqslant 0$.

    a Find $f^{-1}(x)$.

    b Sketch $y = f(x), \ y = x,$ and $y = f^{-1}(x)$ on the same set of axes.

12 **a** Explain why $f(x) = x^2 - 4x + 3$ is a function but does not have an inverse function.

    b Explain why $g(x) = x^2 - 4x + 3, \ x \geqslant 2,$ has the inverse function $g^{-1}(x) = 2 + \sqrt{1 + x}$.

    c State the domain and range of g and g^{-1}.

    d Show that $gg^{-1}(x) = g^{-1}g(x) = x,$ the identity function.

13 Consider $f : x \mapsto (x + 1)^2 + 3, \ x \geqslant -1$.

    a Find $f^{-1}(x)$.

    b Use technology to help sketch the graphs of $y = f(x), \ y = x,$ and $y = f^{-1}(x)$.

    c State the domain and range of f and f^{-1}.

14 Consider the functions $f : x \mapsto 2x + 5$ and $g : x \mapsto \dfrac{8 - x}{2}$.

    a Find $g^{-1}(-1)$.

    b Show that $f^{-1}(-3) - g^{-1}(6) = 0$.

    c Find x such that $fg^{-1}(x) = 9$.

15 Consider the functions $f : x \mapsto 5^x$ and $g : x \mapsto \sqrt{x}$.

    a Find:

        i $f(2)$ **ii** $g^{-1}(4)$.

    b Solve the equation $g^{-1}f(x) = 25$.

16 Given $f : x \mapsto 2x$ and $g : x \mapsto 4x - 3,$ show that $f^{-1}g^{-1}(x) = (gf)^{-1}(x)$.

17 Which of these functions is a self-inverse function?

    a $f(x) = 2x$ **b** $f(x) = x$ **c** $f(x) = -x$

    d $f(x) = \dfrac{2}{x}$ **e** $f(x) = -\dfrac{6}{x}$

18

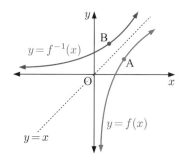

a B is the image of A under a reflection in the line $y = x$.
If A is $(x, f(x))$, find the coordinates of B.

b By substituting your result from **a** into $y = f^{-1}(x)$, show that $f^{-1}(f(x)) = x$.

c Using a similar method, show that $f(f^{-1}(x)) = x$.

19 Prove that $y = \dfrac{k}{x}$ is a self-inverse function for all $k \in \mathbb{R}$, $k \neq 0$.

20 **a** Prove that $f(x) = e^x$ and $g(x) = \ln x$ are the inverses of one another.

  b Find the inverse of:

  i $f(x) = e^{2x+1}$ **ii** $f(x) = \ln(1 - x)$

21 Show that $f : x \mapsto \dfrac{3x - 8}{x - 3}$, $x \neq 3$ is a self-inverse function by:

  a referring to its graph   **b** using algebra.

C   PARTIAL FRACTIONS

Just like fractions involving numbers, fractions involving algebra can be added by writing them with a common denominator.

For example: $\dfrac{3}{x - 2} + \dfrac{4}{x + 3}$

$= \dfrac{3}{x - 2} \times \dfrac{x + 3}{x + 3} + \dfrac{4}{x + 3} \times \dfrac{x - 2}{x - 2}$

$= \dfrac{3(x + 3) + 4(x - 2)}{(x - 2)(x + 3)}$

$= \dfrac{3x + 9 + 4x - 8}{x^2 + 3x - 2x - 6}$

$= \dfrac{7x + 1}{x^2 + x - 6}$ which is a rational function of the form $\dfrac{\text{linear}}{\text{quadratic}}$.

In integral calculus, we are more likely to want to reverse the process above. We need to **decompose** a function such as $\dfrac{7x + 1}{x^2 + x - 6}$ into the sum of **partial fractions** $\dfrac{3}{x - 2} + \dfrac{4}{x + 3}$ so it can be integrated.

To decompose a rational function of the form $\dfrac{\text{linear}}{\text{quadratic}}$ into the sum of partial fractions, we follow the following procedure:

Step 1:   Factorise the denominator.

Step 2:   Write a sum of partial fractions with the factors as denominators and unknown numerators.

Step 3:   Multiply both sides by the denominator.

Step 4:   Find the unknown numerators by substituting the zeros of the denominators.

Example 5

◀) **Self Tutor**

Write as a sum of partial fractions: $\dfrac{7x+1}{x^2+x-6}$

Step 1: $x^2+x-6=(x-2)(x+3)$

Step 2:   Let $\dfrac{7x+1}{x^2+x-6}=\dfrac{A}{x-2}+\dfrac{B}{x+3}$.

Step 3: $\therefore \ \ 7x+1=A(x+3)+B(x-2)$

Step 4:   Substituting $x=-3$, $7(-3)+1=B(-3-2)$

$\therefore \ \ -5B=-20$

$\therefore \ \ B=4$

Substituting $x=2$, $7(2)+1=A(2+3)$

$\therefore \ \ 5A=15$

$\therefore \ \ A=3$

$\therefore \ \ \dfrac{7x+1}{x^2+x-6}=\dfrac{3}{x-2}+\dfrac{4}{x+3}$

EXERCISE 2C

1 Show that:

a $\dfrac{3}{x^2+5x+4}$ can be written as the sum of partial fractions $\dfrac{-1}{x+4}+\dfrac{1}{x+1}$

b $\dfrac{9-x}{x^2+3x-10}$ can be written as the sum of partial fractions $\dfrac{-2}{x+5}+\dfrac{1}{x-2}$

c $\dfrac{2x+13}{x^2-x-12}$ can be written as the sum of partial fractions $\dfrac{-1}{x+3}+\dfrac{3}{x-4}$.

2 Write as the sum of partial fractions:

a $\dfrac{1}{x^2+3x+2}$

b $\dfrac{x}{x^2-5x+6}$

c $\dfrac{3x}{x^2-2x-8}$

d $\dfrac{3x+2}{x^2-4}$

e $\dfrac{2x-1}{x^2-2x-3}$

f $\dfrac{2-5x}{x^2+4x-12}$

Activity
Partial fractions

The technique we have used for writing rational functions of the form $\dfrac{\text{linear}}{\text{quadratic}}$ as the sum of partial fractions can be extended to more complicated rational functions.

What to do:

1 Write $\dfrac{2}{x^3-2x^2-3x}$ as the sum of partial fractions in the form $\dfrac{A}{x}+\dfrac{B}{x-3}+\dfrac{C}{x+1}$.

2 Write $\dfrac{x^2+x-2}{x^2-x-12}$ as the sum of partial fractions in the form:

a $A+\dfrac{B}{x+3}+\dfrac{C}{x-4}$

b $\dfrac{Ax+B}{x+3}+\dfrac{C}{x-4}$

c $\dfrac{A}{x+3}+\dfrac{Bx+C}{x-4}$.

3 Write $\dfrac{x^2+3}{x^3+x^2+x}$ as the sum of partial fractions in the form $\dfrac{A}{x}+\dfrac{Bx+C}{x^2+x+1}$.

4 Write $\dfrac{2x^2-4x+3}{x^3-4x^2+x+6}$ as the sum of partial fractions in the form $\dfrac{A}{x+1}+\dfrac{B}{x-2}+\dfrac{C}{x-3}$.

Review set 2A

1 If $f(x)=2x-3$ and $g(x)=x^2+2$, find in simplest form:

    a $fg(x)$ **b** $gf(x)$ **c** $ff(2)$

2 Suppose $f(x)=2x-5$ and $g(x)=3x+1$.

    a Find $fg(x)$.   **b** Solve $fg(x)=f(x+3)$.

3 If $f(x)=1-2x$ and $g(x)=\sqrt{x}$, find in simplest form:

    a $fg(x)$ **b** $gf(x)$ **c** $gg(81)$

4 Find $f^{-1}(x)$ given that $f(x)$ is: **a** $4x+2$ **b** $\dfrac{3-5x}{4}$

5 Given $f:x\mapsto 3x+6$ and $h:x\mapsto \dfrac{x}{3}$, show that $f^{-1}h^{-1}(x)=(hf)^{-1}(x)$.

6 Suppose $h(x)=(x-4)^2+3$, $x\in[4,\infty)$.

    a Find the defining equation of h^{-1}.   **b** Show that $hh^{-1}(x)=h^{-1}h(x)=x$.

7   **a** Sketch the graph of $g:x\mapsto x^2+6x+7$ for $x\in(-\infty,-3]$.

    b Explain why g has an inverse function g^{-1}.

    c Find algebraically, a formula for g^{-1}.

    d Sketch the graph of $y=g^{-1}(x)$.

    e Find the range of g.

    f Find the domain and range of g^{-1}.

8 The graph of the function $f(x)=-\frac{1}{2}x^2$, $0\leqslant x\leqslant 2$ is shown alongside.

    a Sketch the graph of $y=f^{-1}(x)$.

    b State the range of f^{-1}.

    c Solve:

        i $f(x)=-\frac{3}{2}$ **ii** $f^{-1}(x)=1$

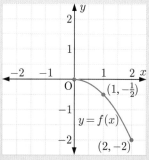

9 Show that $f:x\mapsto \dfrac{5x-1}{x-5}$, $x\neq 5$ is a self-inverse function by:

    a referring to its graph   **b** using algebra.

10 Write as the sum of partial fractions:

    a $\dfrac{x+1}{x^2+4x-5}$ **b** $\dfrac{2x-4}{2x^2+7x+3}$

Review set 2B

1 Given $f(x) = 3 - x^2$ and $g(x) = 2x - 1$, find in simplest form:

   a $fg(x)$ **b** $gf(x)$ **c** $ff(-2)$

2 For each of the following pairs of functions, find $fg(x)$ and state its domain and range:

   a $f(x) = \dfrac{1}{x^2}$ and $g(x) = x^2 - 4x + 3$ **b** $f(x) = \dfrac{1}{x}$ and $g(x) = \cos x$

3 Suppose $f(x) = 3x + 5$ and $g(x) = 2x^2 - x$.

   a Find in simplest form: **i** $fg(x)$ **ii** $gf(x)$

   b Hence solve $3fg(x) = gf(x)$.

4 Suppose $f(x) = \ln x$ and $g(x) = \sqrt{x}$.

   a Find $fg(x)$ and state its domain and range.

   b Find $gf(x)$ and state its domain and range.

5 Copy the following graphs and draw the graph of each inverse function on the same set of axes:

   a 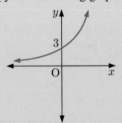   **b**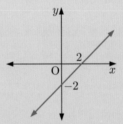

6 Find the inverse function $f^{-1}(x)$ for:

   a $f(x) = 7 - 4x$ **b** $f(x) = \dfrac{3 + 2x}{5}$

7 Consider $f : x \mapsto 2x - 7$.

   a On the same set of axes graph $y = x$, $y = f(x)$, and $y = f^{-1}(x)$.

   b Find $f^{-1}(x)$ using variable interchange.

   c Show that $ff^{-1}(x) = f^{-1}f(x) = x$, the identity function.

8 Given $f : x \mapsto 5x - 2$ and $h : x \mapsto \dfrac{3x}{4}$, show that $f^{-1}h^{-1}(x) = (hf)^{-1}(x)$.

9 Consider the functions $f(x) = 3x + 1$ and $g(x) = \dfrac{2}{x}$.

   a Find $gf(x)$.

   b Given $gf(x) = -4$, solve for x.

   c Let $h(x) = gf(x)$, $x \neq -\frac{1}{3}$.

      i Write down the equations of the asymptotes of $h(x)$.

      ii Sketch the graph of $h(x)$ for $-3 \leqslant x \leqslant 2$.

      iii State the range of $h(x)$ for the domain $-3 \leqslant x \leqslant 2$.

10 Write as the sum of partial fractions:

   a $\dfrac{2x}{-x^2 + 4x - 3}$ **b** $\dfrac{x - 7}{x^2 + 7x + 12}$

3

The unit circle and radian measure

Contents:

Opening problem

There are several theories for why one complete turn was divided into 360 degrees:

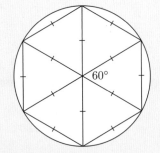

- 360 is approximately the number of days in a year.
- The Babylonians used a counting system in base 60. If they drew 6 equilateral triangles within a circle as shown, and divided each angle into 60 subdivisions, then there were 360 subdivisions in one turn. The division of an hour into 60 minutes, and a minute into 60 seconds, is from this base 60 counting system.
- 360 has 24 divisors, including every integer from 1 to 10 except 7.

Consider a circle of radius 1 unit.

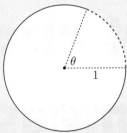

What to do:

a Find the circumference of the circle.

b Suppose the angle θ was measured as the length of the red dotted arc. What angle θ would correspond to:

   i   a right angle   ii   a straight angle   iii   a complete turn?

c Do you think this is a more *natural* way to measure angles? Are there any benefits to measuring angles this way?

A   RADIAN MEASURE

We have seen previously that one full revolution makes an angle of $360°$, and the angle on a straight line is $180°$. Hence, one **degree**, $1°$, can be defined as $\frac{1}{360}$th of one full revolution. This measure of angle is commonly used by surveyors and architects.

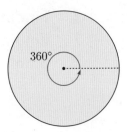

In this Chapter we consider an alternative angle measure, the **radian**, which is most commonly used by mathematicians, physicists, and engineers.

Although it was decided that a circle would be divided into $360°$, it could just as well be divided into $100°$, or $120°$, or some other number of degrees. This makes degrees a rather artificial measure of angles.

A more *natural* way to measure an angle is to compare the length of the arc formed by an angle with the radius of the circle.

Suppose the arc length formed by an angle is the same length as the radius. This angle is said to have a measure of 1 **radian** (1^c or 1 rad). The word "radian" is an abbreviation of "radial angle".

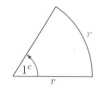

If the arc length is twice as long as the radius, the angle has a measure of 2 radians.

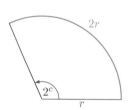

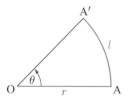

In general, if OA has length r, and A is rotated about O to A′, then the angle θ, in radians, is the ratio of the arc length l to the radius r.

DEMO

$$\theta \text{ (in radians)} = \frac{l}{r}$$

The symbol "c" is used for radian measure but is usually omitted. By contrast, the degree symbol is *always* used when the measure of an angle is given in degrees.

From the diagram to the right, it can be seen that 1^c is slightly smaller than $60°$. In fact, $1^c \approx 57.3°$.

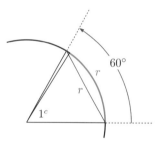

DEGREE-RADIAN CONVERSIONS

Consider a semi-circle of radius r. The arc length is πr, so there are π radians in a semi-circle.

Therefore, π radians $\equiv 180°$.

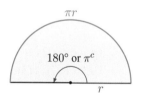

So, $1^c = \left(\frac{180}{\pi}\right)^° \approx 57.3°$ and $1° = \left(\frac{\pi}{180}\right)^c \approx 0.0175^c$.

To convert from degrees to radians, we multiply by $\frac{\pi}{180}$.

To convert from radians to degrees, we multiply by $\frac{180}{\pi}$.

We indicate degrees with a small °.
To indicate radians we use a small c or rad, or else use no symbol at all.

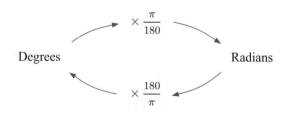

Example 1
🔊 **Self Tutor**

Convert $45°$ to radians, in terms of π.

$45° = \left(45 \times \frac{\pi}{180}\right)$ radians   *or* $180° = \pi$ radians

$\qquad = \frac{\pi}{4}$ radians $\qquad \qquad \qquad \therefore \left(\frac{180}{4}\right)° = \frac{\pi}{4}$ radians

$\qquad \qquad \qquad \qquad \qquad \qquad \therefore \quad 45° = \frac{\pi}{4}$ radians

Example 2
🔊 **Self Tutor**

Convert $126.5°$ to radians.

$\qquad 126.5°$

$= \left(126.5 \times \frac{\pi}{180}\right)$ radians

≈ 2.21 radians

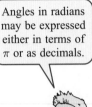

Angles in radians may be expressed either in terms of π or as decimals.

EXERCISE 3A

1 Convert to radians, in terms of π:

a 90°	**b** 60°	**c** 30°	**d** 18°	**e** 9°
f 135°	**g** 225°	**h** 270°	**i** 360°	**j** 720°
k 315°	**l** 540°	**m** 36°	**n** 80°	**o** 230°

2 Convert to radians, correct to 3 significant figures:

a 36.7°	**b** 137.2°	**c** 317.9°	**d** 219.6°	**e** 396.7°

Example 3
🔊 **Self Tutor**

Convert to degrees:

a $\frac{5\pi}{6}$ **b** 0.638 radians.

a $\frac{5\pi}{6}$ **b** 0.638 radians

$\quad = \left(\frac{5\pi}{6} \times \frac{180}{\pi}\right)°$ $\quad = \left(0.638 \times \frac{180}{\pi}\right)°$

$\quad = 150°$ $\qquad \approx 36.6°$

3 Convert to degrees:

a $\frac{\pi}{5}$	**b** $\frac{3\pi}{5}$	**c** $\frac{3\pi}{4}$	**d** $\frac{\pi}{18}$	**e** $\frac{\pi}{9}$
f $\frac{7\pi}{9}$	**g** $\frac{\pi}{10}$	**h** $\frac{3\pi}{20}$	**i** $\frac{7\pi}{6}$	**j** $\frac{\pi}{8}$

4 Convert to degrees, correct to 2 decimal places:

a 2	**b** 1.53	**c** 0.867	**d** 3.179	**e** 5.267

5 Match each angle measurement with the correct diagram:

   a 70°   **b** 2^c **c** $\frac{\pi}{2}$ **d** 200°   **e** 3^c **f** 0.5

A    **B**    **C**

D   **E**    **F**

6 Copy and complete, giving answers in terms of π:

a

Degrees	0	45	90	135	180	225	270	315	360
Radians									

b

Degrees	0	30	60	90	120	150	180	210	240	270	300	330	360
Radians													

B ARC LENGTH AND SECTOR AREA

You should be familiar with these terms relating to the parts of a circle:

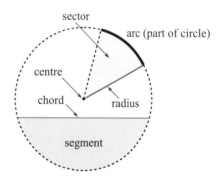

An arc, sector, or segment is described as:

- **minor** if it involves less than half the circle
- **major** if it involves more than half the circle.

For example:

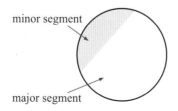

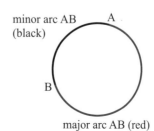

ARC LENGTH

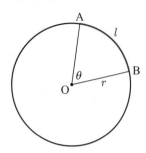

In the diagram, the **arc length** AB is l. Angle θ is measured in **radians**.

θ measures how many times longer the arc length is than the radius.

$$\therefore \quad \theta = \frac{l}{r}$$
$$\therefore \quad l = \theta r$$

> For θ in radians, arc length $l = \boldsymbol{\theta r}$.
>
> For θ in degrees, arc length $l = \frac{\theta}{360} \times \boldsymbol{2\pi r}$.

AREA OF SECTOR

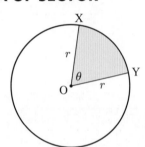

In the diagram, the area of minor sector XOY is shaded. θ is measured in **radians**.

There are 2π radians in a circle so

$$\text{area of sector} = \frac{\theta}{2\pi} \times \text{area of circle}$$
$$\therefore \quad A = \frac{\theta}{2\pi} \times \pi r^2$$
$$\therefore \quad A = \tfrac{1}{2}\theta r^2$$

> For θ in radians, area of sector $A = \tfrac{1}{2}\theta r^2$.
>
> For θ in degrees, area of sector $A = \frac{\theta}{360} \times \pi r^2$.

Example 4   ◀)) Self Tutor

A sector has radius 12 cm and angle 3 radians. Find its:

  a arc length   **b** area

$$
\begin{aligned}
\textbf{a} \quad \text{arc length} &= \theta r \\
&= 3 \times 12 \\
&= 36 \text{ cm}
\end{aligned}
\qquad
\begin{aligned}
\textbf{b} \quad \text{area} &= \tfrac{1}{2}\theta r^2 \\
&= \tfrac{1}{2} \times 3 \times 12^2 \\
&= 216 \text{ cm}^2
\end{aligned}
$$

EXERCISE 3B

1 Find the arc length of each sector:

  a

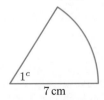

  b

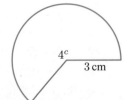

  c

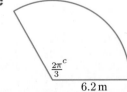

2 Find the area of each sector:

a

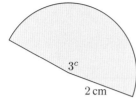

b

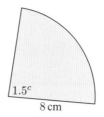

c

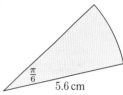

3 Find the arc length and area of a sector of a circle with:

  a radius 9 cm and angle $\frac{7\pi}{4}$

  b radius 4.93 cm and angle 4.67 radians.

Example 5   ◀) **Self Tutor**

A sector has radius 8.2 cm and arc length 12.3 cm. Find its:

  a angle   **b** area.

  a $l = \theta r$ {θ in radians}   **b** area $= \frac{1}{2}\theta r^2$

  $\therefore \ \theta = \dfrac{l}{r} = \dfrac{12.3}{8.2} = 1.5$ radians $= \frac{1}{2} \times 1.5 \times 8.2^2$

                                $= 50.43 \text{ cm}^2$

---

4 Find, in radians, the angle of a sector of:

  a radius 4.3 m and arc length 2.95 m

  b radius 10 cm and area 30 cm^2.

5 Find θ (in radians) for each of the following, and hence find the area of each figure:

a

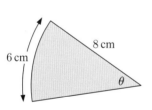

b

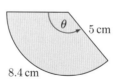

c

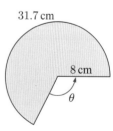

6 A sector has an angle of 107.9° and an arc length of 5.92 m. Find its:

  a radius   **b** area.

7 A sector has an angle of 1.19 radians and an area of 20.8 cm^2. Find its:

  a radius   **b** perimeter.

8

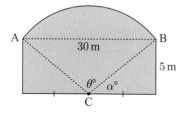

The end wall of a building has the shape illustrated, where the centre of arc AB is at C. Find:

  a α to 4 significant figures

  b θ to 4 significant figures

  c the area of the wall.

9   The cone 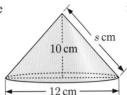 is made from this sector:

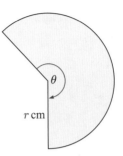

Find, correct to 3 significant figures:

   a   the slant length s cm

   c   the arc length of the sector

   b   the value of r

   d   the sector angle θ in radians.

10

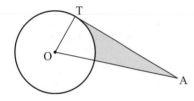

[AT] is a tangent to the given circle. $OA = 13$ cm   and the circle has radius 5 cm. Find the perimeter of the shaded region.

11   A **nautical mile** (nmi) is the distance on the Earth's surface that subtends an angle of 1 minute (or $\frac{1}{60}$ of a degree) of the Great Circle arc measured from the centre of the Earth.

A **knot** is a speed of 1 nautical mile per hour.

   a   Given that the radius of the Earth is 6370 km, show that 1 nmi is approximately 1.853 km.

   b   Calculate how long it would take a plane to fly 1490 km from London to Budapest if the plane can fly at 480 knots.

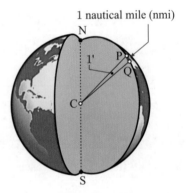

12

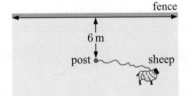

A sheep is tethered to a post which is 6 m from a long fence. The length of the rope is 9 m. Find the area which the sheep can feed on.

C | THE UNIT CIRCLE AND THE TRIGONOMETRIC RATIOS

The **unit circle** is the circle with centre $(0, 0)$ and radius 1 unit.

The equation of the unit circle is $x^2 + y^2 = 1$.

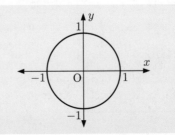

Suppose P is a point on the unit circle such that [OP] makes an angle θ measured anticlockwise from the positive x-axis.

- The **cosine of θ** or $\cos\theta$ is the x-coordinate of P.
- The **sine of θ** or $\sin\theta$ is the y-coordinate of P.
- The **tangent of θ** or $\tan\theta$ is the gradient of [OP].

$$\tan\theta = \frac{\sin\theta}{\cos\theta}$$

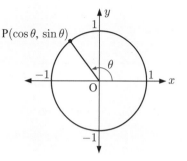

For all points on the unit circle, $-1 \leqslant x \leqslant 1$, $-1 \leqslant y \leqslant 1$, and $x^2 + y^2 = 1$.

For any angle θ:

- $-1 \leqslant \cos\theta \leqslant 1$
- $-1 \leqslant \sin\theta \leqslant 1$
- $\cos^2\theta + \sin^2\theta = 1$

PERIODICITY OF TRIGONOMETRIC RATIOS

Since there are 2π radians in a full revolution, if we add any integer multiple of 2π to θ then the position of P on the unit circle is unchanged.

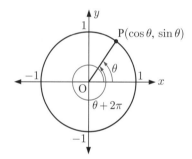

For any $k \in \mathbb{Z}$, $\cos(\theta + 2k\pi) = \cos\theta$ and $\sin(\theta + 2k\pi) = \sin\theta$.

We notice that for any point $(\cos\theta, \sin\theta)$ on the unit circle, the point directly opposite is $(-\cos\theta, -\sin\theta)$.

$\therefore$ $\cos(\theta + \pi) = -\cos\theta$

$\sin(\theta + \pi) = -\sin\theta$

and $\tan(\theta + \pi) = \dfrac{-\sin\theta}{-\cos\theta} = \tan\theta$

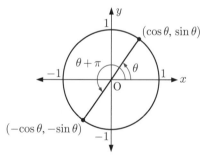

For any $k \in \mathbb{Z}$, $\tan(\theta + k\pi) = \tan\theta$.

CALCULATOR USE

When using your calculator to find trigonometric ratios for angles, you must make sure your calculator is correctly set to either **degree** or **radian** mode. Click on the icon for instructions.

GRAPHICS CALCULATOR INSTRUCTIONS

EXERCISE 3C

1 With the aid of a unit circle, complete the following table:

θ (radians)	0	$\frac{\pi}{2}$	π	$\frac{3\pi}{2}$	2π	$\frac{5\pi}{2}$
sine						
cosine						
tangent						

2 **a** Use your calculator to evaluate: **i** $\frac{1}{\sqrt{2}}$ **ii** $\frac{\sqrt{3}}{2}$

  b Copy and complete the following table. Use your calculator to evaluate the trigonometric ratios, then **a** to write them exactly.

θ (radians)	$\frac{\pi}{6}$	$\frac{\pi}{4}$	$\frac{\pi}{3}$	$\frac{3\pi}{4}$	$\frac{5\pi}{6}$	$\frac{7\pi}{6}$	$\frac{4\pi}{3}$	$\frac{7\pi}{4}$
θ (degrees)								
sine								
cosine								
tangent								

3 **a** Copy and complete:

Quadrant	Degree measure	Radian measure	$\cos\theta$	$\sin\theta$	$\tan\theta$
1	$0° < \theta < 90°$	$0 < \theta < \frac{\pi}{2}$	positive	positive	
2					
3					
4					

  b In which quadrants are the following true?
   i $\cos\theta$ is positive.   **ii** $\cos\theta$ is negative.
   iii $\cos\theta$ and $\sin\theta$ are both negative.   **iv** $\cos\theta$ is negative and $\sin\theta$ is positive.

4 Explain why:

  a $\cos\frac{7\pi}{9} = \cos\frac{25\pi}{9}$ **b** $\sin\frac{12\pi}{7} = \sin\left(-\frac{2\pi}{7}\right)$ **c** $\tan\frac{4\pi}{11} = \tan\frac{15\pi}{11}$

5 Use the diagram alongside to explain why:

  a $\sin(\pi - \theta) = \sin\theta$
  b $\cos(\pi - \theta) = -\cos\theta$.

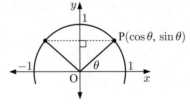

6 Use the diagram alongside to explain why:

  a $\sin(-\theta) = -\sin\theta$
  b $\cos(-\theta) = \cos\theta$.

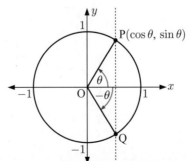

7 a Explain why:

    i P has coordinates $\left(\cos\left(\frac{\pi}{2} - \theta\right),\ \sin\left(\frac{\pi}{2} - \theta\right)\right)$

    ii $XP = \sin\theta$ **iii** $OX = \cos\theta$

b Hence, copy and complete:

    i $\cos\left(\frac{\pi}{2} - \theta\right) = \ ......$ **ii** $\sin\left(\frac{\pi}{2} - \theta\right) = \ ......$

c Check your answer to **c** by calculating:

    i $\cos\frac{\pi}{5}$ and $\sin\frac{3\pi}{10}$ **ii** $\sin\frac{2\pi}{5}$ and $\cos\frac{\pi}{10}$.

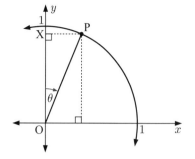

D   MULTIPLES OF $\frac{\pi}{6}$ AND $\frac{\pi}{4}$

You should have previously learnt the trigonometric ratios for angles which are multiples of $30°$ and $45°$, or equivalently $\frac{\pi}{6}$ and $\frac{\pi}{4}$.

$$\cos\frac{\pi}{6} = \frac{\sqrt{3}}{2} \qquad \sin\frac{\pi}{6} = \frac{1}{2}$$
$$\cos\frac{\pi}{4} = \frac{1}{\sqrt{2}} \qquad \sin\frac{\pi}{4} = \frac{1}{\sqrt{2}}$$
$$\cos\frac{\pi}{3} = \frac{1}{2} \qquad \sin\frac{\pi}{3} = \frac{\sqrt{3}}{2}$$

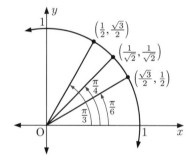

- For **multiples of** $\frac{\pi}{2}$, the coordinates of the points on the unit circle involve 0 and ± 1.

- For *other* **multiples of** $\frac{\pi}{4}$, the coordinates involve $\pm\frac{1}{\sqrt{2}}$.

- For *other* **multiples of** $\frac{\pi}{6}$, the coordinates involve $\pm\frac{1}{2}$ and $\pm\frac{\sqrt{3}}{2}$.

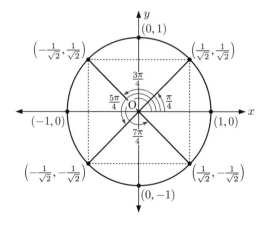

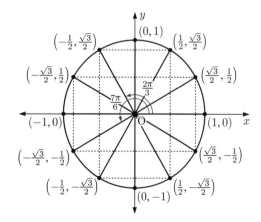

Example 6
◆) **Self Tutor**

Find the exact values of $\sin \alpha$, $\cos \alpha$, and $\tan \alpha$ for:

a $\alpha = \frac{3\pi}{4}$
b $\alpha = \frac{4\pi}{3}$

a

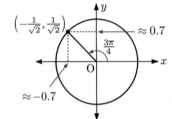

$\sin \frac{3\pi}{4} = \frac{1}{\sqrt{2}}$

$\cos \frac{3\pi}{4} = -\frac{1}{\sqrt{2}}$

$\tan \frac{3\pi}{4} = -1$

b

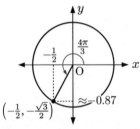

$\sin \frac{4\pi}{3} = -\frac{\sqrt{3}}{2}$

$\cos \frac{4\pi}{3} = -\frac{1}{2}$

$\tan \frac{4\pi}{3} = \dfrac{-\frac{\sqrt{3}}{2}}{-\frac{1}{2}} = \sqrt{3}$

EXERCISE 3D

1 Use a unit circle diagram to find exact values for $\sin \theta$, $\cos \theta$, and $\tan \theta$ for θ equal to:

a $\frac{\pi}{4}$　　　**b** $\frac{3\pi}{4}$　　　**c** $\frac{7\pi}{4}$　　　**d** π　　　**e** $-\frac{3\pi}{4}$

2 Use a unit circle diagram to find exact values for $\sin \beta$, $\cos \beta$, and $\tan \beta$ for β equal to:

a $\frac{\pi}{6}$　　　**b** $\frac{2\pi}{3}$　　　**c** $\frac{7\pi}{6}$　　　**d** $\frac{5\pi}{3}$　　　**e** $\frac{11\pi}{6}$

3 Find the exact values of:

a $\cos \frac{2\pi}{3}$, $\sin \frac{2\pi}{3}$, and $\tan \frac{2\pi}{3}$

b $\cos\left(-\frac{\pi}{4}\right)$, $\sin\left(-\frac{\pi}{4}\right)$, and $\tan\left(-\frac{\pi}{4}\right)$

4 **a** Find the exact values of $\cos \frac{\pi}{2}$ and $\sin \frac{\pi}{2}$.
   b What can you say about $\tan \frac{\pi}{2}$?

Example 7
◆) **Self Tutor**

Without using a calculator, show that $8 \sin \frac{\pi}{3} \cos \frac{5\pi}{6} = -6$.

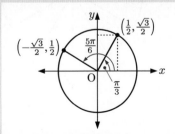

$\sin \frac{\pi}{3} = \frac{\sqrt{3}}{2}$　and　$\cos \frac{5\pi}{6} = -\frac{\sqrt{3}}{2}$

$\therefore 8 \sin \frac{\pi}{3} \cos \frac{5\pi}{6} = 8\left(\frac{\sqrt{3}}{2}\right)\left(-\frac{\sqrt{3}}{2}\right)$

$= 2(-3)$

$= -6$

5 Without using a calculator, evaluate:

a $\sin^2\left(\frac{\pi}{3}\right)$ **b** $\sin\frac{\pi}{6}\cos\frac{\pi}{3}$ **c** $4\sin\frac{\pi}{3}\cos\frac{\pi}{6}$

d $1-\cos^2\left(\frac{\pi}{6}\right)$ **e** $\sin^2\left(\frac{2\pi}{3}\right)-1$ **f** $\cos^2\left(\frac{\pi}{4}\right)-\sin\frac{7\pi}{6}$

g $\sin\frac{3\pi}{4}-\cos\frac{5\pi}{4}$ **h** $1-2\sin^2\left(\frac{7\pi}{6}\right)$ **i** $\cos^2\left(\frac{5\pi}{6}\right)-\sin^2\left(\frac{5\pi}{6}\right)$

j $\tan^2\left(\frac{\pi}{3}\right)-2\sin^2\left(\frac{\pi}{4}\right)$ **k** $2\tan\left(-\frac{5\pi}{4}\right)-\sin\frac{3\pi}{2}$ **l** $\dfrac{2\tan\frac{5\pi}{6}}{1-\tan^2\left(\frac{5\pi}{6}\right)}$

Check all answers using your calculator.

Example 8   ◀ Self Tutor

Find all angles $0 \leqslant \theta \leqslant 2\pi$ with a cosine of $\frac{1}{2}$.

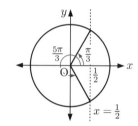

Since the cosine is $\frac{1}{2}$, we draw the vertical line $x = \frac{1}{2}$.

Because $\frac{1}{2}$ is involved, we know the required angles are multiples of $\frac{\pi}{6}$.

They are $\frac{\pi}{3}$ and $\frac{5\pi}{3}$.

---

6 Find all angles between 0 and 2π with:

a a sine of $\frac{1}{2}$ **b** a sine of $\frac{\sqrt{3}}{2}$ **c** a cosine of $\frac{1}{\sqrt{2}}$

d a cosine of $-\frac{1}{2}$ **e** a cosine of $-\frac{1}{\sqrt{2}}$ **f** a sine of $-\frac{\sqrt{3}}{2}$

7 Find all angles between 0 and 2π (inclusive) which have a tangent of:

a 1   **b** -1 **c** $\sqrt{3}$ **d** 0   **e** $\frac{1}{\sqrt{3}}$ **f** $-\sqrt{3}$

8 Find all angles between 0 and 4π with:

a a cosine of $\frac{\sqrt{3}}{2}$ **b** a sine of $-\frac{1}{2}$ **c** a sine of -1

9 Find θ if $0 \leqslant \theta \leqslant 2\pi$ and:

a $\cos\theta = \frac{1}{2}$ **b** $\sin\theta = \frac{\sqrt{3}}{2}$ **c** $\cos\theta = -1$ **d** $\sin\theta = 1$

e $\cos\theta = -\frac{1}{\sqrt{2}}$ **f** $\sin^2\theta = 1$ **g** $\cos^2\theta = 1$ **h** $\cos^2\theta = \frac{1}{2}$

i $\tan\theta = -\frac{1}{\sqrt{3}}$ **j** $\tan^2\theta = 3$

10 Find *all* values of θ for which $\tan\theta$ is:

a zero   **b** undefined.

E FINDING TRIGONOMETRIC RATIOS

The identity $\cos^2 \theta + \sin^2 \theta = 1$ is essential for finding trigonometric ratios.

Example 9
◀) **Self Tutor**

If $\sin \theta = -\frac{3}{5}$ and $\frac{3\pi}{2} < \theta < 2\pi$, find $\cos \theta$ and $\tan \theta$. Give exact values.

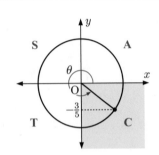

Now $\cos^2 \theta + \sin^2 \theta = 1$

$\therefore \cos^2 \theta + \frac{9}{25} = 1$

$\therefore \cos^2 \theta = \frac{16}{25}$

$\therefore \cos \theta = \pm\frac{4}{5}$

But $\frac{3\pi}{2} < \theta < 2\pi$, so θ is a quadrant 4 angle.

$\therefore \cos \theta$ is positive.

$\therefore \cos \theta = \frac{4}{5}$ and $\tan \theta = \dfrac{\sin \theta}{\cos \theta} = \dfrac{-\frac{3}{5}}{\frac{4}{5}} = -\frac{3}{4}$

EXERCISE 3E

1 Find the possible values of $\cos \theta$ for:

  a $\sin \theta = \frac{1}{3}$ **b** $\sin \theta = -\frac{2}{3}$ **c** $\sin \theta = 1$ **d** $\sin \theta = 0$

2 Find the possible values of $\sin \theta$ for:

  a $\cos \theta = \frac{1}{4}$ **b** $\cos \theta = -\frac{1}{5}$ **c** $\cos \theta = \frac{3}{5}$ **d** $\cos \theta = -1$

3 Find the exact value of:

  a $\sin \theta$ if $\cos \theta = \frac{4}{5}$ and $0 < \theta < \frac{\pi}{2}$ **b** $\cos \theta$ if $\sin \theta = -\frac{4}{5}$ and $\pi < \theta < \frac{3\pi}{2}$

  c $\tan \theta$ if $\sin \theta = -\frac{1}{4}$ and $\frac{3\pi}{2} < \theta < 2\pi$ **d** $\sin \theta$ if $\cos \theta = -\frac{3}{7}$ and $\frac{\pi}{2} < \theta < \pi$

  e $\tan \theta$ if $\cos \theta = \frac{2}{5}$ and $\pi < \theta < 2\pi$ **f** $\cos \theta$ if $\sin \theta = -\frac{6}{7}$ and $\frac{\pi}{2} < \theta < \frac{3\pi}{2}$

Example 10
◀) **Self Tutor**

If $\tan \theta = -\frac{1}{2}$ and $\frac{3\pi}{2} < \theta < 2\pi$, find $\cos \theta$ and $\sin \theta$ exactly.

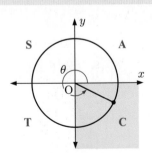

$\tan \theta = \dfrac{\sin \theta}{\cos \theta} = -\frac{1}{2}$

$\therefore \sin \theta = -\frac{1}{2} \cos \theta$

Now $\sin^2 \theta + \cos^2 \theta = 1$

$\therefore (-\frac{1}{2} \cos \theta)^2 + \cos^2 \theta = 1$

$\therefore \frac{1}{4} \cos^2 \theta + \cos^2 \theta = 1$

$\therefore \frac{5}{4} \cos^2 \theta = 1$

$\therefore \cos \theta = \pm\frac{2}{\sqrt{5}}$

But $\frac{3\pi}{2} < \theta < 2\pi$, so θ is a quadrant 4 angle. Therefore, $\cos \theta$ is positive and $\sin \theta$ is negative.

$\therefore \cos \theta = \frac{2}{\sqrt{5}}$ and $\sin \theta = -\frac{1}{\sqrt{5}}$.

4 Find exact values for $\cos\theta$ and $\sin\theta$ given that:

a $\tan\theta = 2$ and $0 < \theta < \frac{\pi}{2}$

b $\tan\theta = -\frac{1}{3}$ and $\frac{\pi}{2} < \theta < \pi$

c $\tan\theta = \frac{1}{4}$ and $\pi < \theta < 2\pi$

d $\tan\theta = -\frac{3}{\sqrt{5}}$ and θ is reflex.

F FINDING ANGLES

In **Exercise 3C** you should have discovered that:

> For any angle θ:
> $$\sin(\pi - \theta) = \sin\theta \qquad \sin(2\pi - \theta) = -\sin\theta$$
> $$\cos(\pi - \theta) = -\cos\theta \qquad \cos(2\pi - \theta) = \cos\theta$$

We need results such as these, and also the periodicity of the trigonometric ratios, to find angles which have a particular sine, cosine, or tangent.

Example 11   ◀)) **Self Tutor**

Find the two angles θ on the unit circle, with $0 \leqslant \theta \leqslant 2\pi$, such that:

a $\cos\theta = -\frac{1}{4}$ **b** $\sin\theta = \frac{2}{3}$ **c** $\tan\theta = -3.2$

a Using technology, $\cos^{-1}\left(-\frac{1}{4}\right) \approx 1.823$

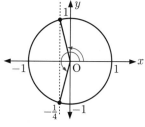

$\therefore \quad \theta \approx 1.823 \quad \text{or} \quad 2\pi - 1.823$
$\therefore \quad \theta \approx 1.82 \quad \text{or} \quad 4.46$

b Using technology, $\sin^{-1}\left(\frac{2}{3}\right) \approx 0.7297$

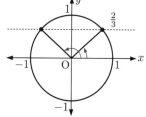

$\therefore \quad \theta \approx 0.7297 \quad \text{or} \quad \pi - 0.7297$
$\therefore \quad \theta \approx 0.730 \quad \text{or} \quad 2.41$

c Using technology, $\tan^{-1}(-3.2) \approx -1.268$

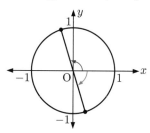

> If $\sin\theta$ or $\tan\theta$ is negative, your calculator will give θ in the domain $-\frac{\pi}{2} < \theta < 0$. For example, see the angle marked green in **c**.

$\therefore \quad \theta \approx \pi - 1.268 \quad \text{or} \quad 2\pi - 1.268$
$\therefore \quad \theta \approx 1.87 \quad\quad \text{or} \quad 5.02$

EXERCISE 3F

1 Find *two* angles θ on the unit circle, with $0 \leqslant \theta \leqslant 2\pi$, such that:

   a $\tan\theta = \frac{1}{3}$ **b** $\cos\theta = \frac{3}{7}$ **c** $\sin\theta = 0.61$

   d $\cos\theta = -\frac{1}{4}$ **e** $\tan\theta = -0.114$ **f** $\sin\theta = -\frac{1}{6}$

Your calculator should be set to *radians*.

2 Find all θ such that $-\pi \leqslant \theta \leqslant \pi$ and:

   a $\cos\theta = -\frac{1}{10}$ **b** $\sin\theta = \frac{4}{5}$ **c** $\tan\theta = \frac{3}{2}$

   d $\cos\theta = 0.8$ **e** $\tan\theta = -\frac{5}{6}$ **f** $\sin\theta = -\frac{7}{11}$

G   SMALL ANGLE APPROXIMATIONS

In this Section we look at approximations for the trigonometric ratios for small angles. These are useful when we perform calculus with trigonometric functions. They are also used directly in astronomy and other sciences.

Investigation   Small angle approximations

In the diagram opposite, we see a point P corresponding to angle θ on the unit circle, where $0 < \theta < \frac{\pi}{2}$.

Notice three shapes:

- the large shaded sector with arc AP, radius 1, and angle θ
- the small sector with radius $\cos\theta$ and angle θ
- the triangle OAP.

For any $0 < \theta < \frac{\pi}{2}$, the areas of the shapes follow the order small sector < triangle < large sector.

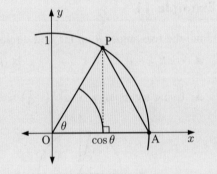

What to do:

1 Write down, in terms of θ, a formula for the area of:

   a the small sector   **b** the triangle   **c** the large sector.

2 Hence show that for all $0 < \theta < \frac{\pi}{2}$, $\cos^2\theta < \dfrac{\sin\theta}{\theta} < 1$.

3 Suppose P is made to approach A, so θ approaches 0 from the positive side. Click on the icon to observe what happens to the shapes.

DEMO

   a As θ approaches 0 from the positive side, what value does $\cos^2\theta$ approach?

   b Hence deduce the value which the ratio $\dfrac{\sin\theta}{\theta}$ approaches in this case.

4 Now consider a similar scenario for $-\frac{\pi}{2} < \theta < 0$.

   a Write expressions in terms of θ for:

      i the angle of the sectors

     ii the area of the small sector

    iii the area of the triangle

    iv the area of the large sector.

   b Hence deduce the value which the ratio $\dfrac{\sin\theta}{\theta}$

approaches as θ approaches 0 from the negative side.

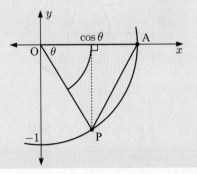

5 Using your results from **3** and **4**, explain why $\sin\theta \approx \theta$ for small θ.

6 Use the Pythagorean identity $\cos^2\theta + \sin^2\theta = 1$ to show that $\cos\theta \approx \sqrt{1-\theta^2}$ for small θ.

Hence show that $\cos\theta \approx 1 - \dfrac{\theta^2}{2}$ for small θ using the binomial expansion for rational powers.

7 Explain why θ^2 is much smaller than θ for small θ. Hence explain why $\tan\theta \approx \theta$ for small θ.

From the **Investigation**, you should have discovered that:

For small θ,

- $\sin\theta \approx \theta$
- $\cos\theta \approx 1 - \dfrac{\theta^2}{2}$
- $\tan\theta \approx \theta$

Example 12 ◄◎ Self Tutor

For a small angle θ, where θ is in radians, find an approximation for:

a $\cos\theta + 3\sin\theta$ **b** $2 + \theta\tan 2\theta - \cos 3\theta$

a $\cos\theta + 3\sin\theta$

$\approx \left(1 - \dfrac{\theta^2}{2}\right) + 3(\theta)$

$\approx 1 + 3\theta - \tfrac{1}{2}\theta^2$

b $2 + \theta\tan 2\theta - \cos 3\theta$

$\approx 2 + \theta(2\theta) - \left(1 - \dfrac{(3\theta)^2}{2}\right)$

$\approx 2 + 2\theta^2 - 1 + \tfrac{9}{2}\theta^2$

$\approx 1 + \tfrac{13}{2}\theta^2$

EXERCISE 3G

1 For a small angle θ, where θ is in radians, find an approximation for:

a $2\sin\theta + \tan^2\theta$ **b** $\theta\sin\theta - \cos\theta$ **c** $\tan 2\theta - 3\cos\theta$

d $1 + \cos 2\theta$ **e** $3 - \theta\sin 3\theta + 2\cos\theta$ **f** $2\theta - \tfrac{1}{3}\tan 4\theta + \tfrac{1}{4}\cos 2\theta$

2 When θ is small, show that $1 + \cos^2\theta - \tfrac{1}{2}\cos\theta \approx \tfrac{3}{2} - \tfrac{3}{4}\theta^2$.

3 Find an approximation for $\theta\cos\theta + \tan 3\theta + \cos 3\theta - 2$ up to the term in θ^2.

4 The "*1 in 60 rule*" is used by pilots in basic aircraft to quickly estimate errors in their flight path. The rule states that for every degree a plane travels off course over a distance of 60 nautical miles, it will be 1 nautical mile off course.

Use the small angle approximation for $\sin\theta$ to explain why this rule is reasonably accurate.

5 A ship's captain observes a lighthouse through his telescope. The telescope has graduations which show the angle of observation of the object. The captain knows the lighthouse is 31.4 m tall, and the angle of observation is $0.18°$. Estimate the ship's distance from the lighthouse.

6 In astronomy, the distance between two stars is usually written in terms of the **angular distance** between them as observed from Earth.

Suppose two stars are distance d apart, and the midpoint between the stars is distance D from the Earth.

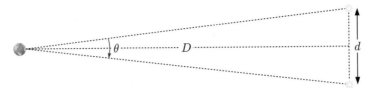

Assuming the two stars are approximately the same distance from Earth, use a small angle approximation to show that the angular distance between the stars $\theta \approx \dfrac{d}{D}$.

7 Show that a better small angle approximation for $\tan \theta$ is $\theta + \dfrac{\theta^3}{2}$.

> **Hint:** $\tan \theta = \dfrac{\sin \theta}{\cos \theta}$

Review set 3A

1 Convert to radians in terms of π:

   a 120°   **b** 225°   **c** 150°   **d** 540°

2 Convert these radian measurements to degrees:

   a $\dfrac{2\pi}{5}$ **b** $\dfrac{5\pi}{4}$ **c** $\dfrac{7\pi}{9}$ **d** $\dfrac{11\pi}{6}$

3 Illustrate the quadrants where $\sin \theta$ and $\cos \theta$ have opposite signs.

4 Find the arc length of a sector with angle 1.5 radians and radius 8 cm.

5 Use a unit circle diagram to explain why $\cos \dfrac{3\pi}{8}$ is equal to:

   a $\cos \dfrac{19\pi}{8}$ **b** $-\cos \dfrac{5\pi}{8}$ **c** $\cos\left(-\dfrac{3\pi}{8}\right)$ **d** $\sin \dfrac{\pi}{8}$

6 Find exact values for $\sin \theta$, $\cos \theta$, and $\tan \theta$ for θ equal to:

   a $\dfrac{5\pi}{6}$ **b** $\dfrac{4\pi}{3}$ **c** $\dfrac{5\pi}{4}$

7

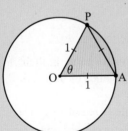

   a State the value of θ in:

       i degrees   **ii** radians.

   b State the arc length AP.

   c State the area of the minor *sector* OAP.

8 If $\sin x = -\dfrac{1}{4}$ and $\pi < x < \dfrac{3\pi}{2}$, find $\tan x$ exactly.

9 Evaluate:

   a $2 \sin \dfrac{\pi}{3} \cos \dfrac{\pi}{3}$ **b** $\tan^2 \left(\dfrac{\pi}{4}\right) - 1$ **c** $\cos^2 \left(\dfrac{\pi}{6}\right) - \sin^2 \left(\dfrac{\pi}{6}\right)$

10 Given $\tan x = \dfrac{\sqrt{3}}{2}$ and $\pi < x < 2\pi$, find:

   a $\cos x$ **b** $\sin x$.

11 Suppose $\cos \theta = \sqrt{\frac{5}{7}}$ and θ is acute. Find the exact value of $\tan \theta$.

12 Find *two* angles on the unit circle, with $0 \leqslant \theta \leqslant 2\pi$, such that:

    a $\cos \theta = -\frac{1}{3}$ **b** $\sin \theta = \frac{3}{4}$ **c** $\tan \theta = -2$

13 For a small angle θ, where θ is in radians, find an approximation for:

    a $\cos \theta + \theta \tan 2\theta$ **b** $3 - 2\sin^2 \theta + \cos 2\theta$

Review set 3B

1 Convert to radians, to 4 significant figures:

    a $71°$ **b** $124.6°$ **c** $-142°$

2 Convert to degrees, to 2 decimal places:

    a 3^c **b** 1.46 rad   **c** 0.435^c **d** -5.271

3 Determine the area of a sector with angle $\frac{5\pi}{12}$ and radius 13 cm.

4 Use a unit circle diagram to find:

    a $\cos \frac{3\pi}{2}$ and $\sin \frac{3\pi}{2}$ **b** $\cos\left(-\frac{\pi}{2}\right)$ and $\sin\left(-\frac{\pi}{2}\right)$

5 Find all angles between 0 and 2π which have:

    a a cosine of $-\frac{\sqrt{3}}{2}$ **b** a sine of $\frac{1}{\sqrt{2}}$ **c** a tangent of $-\frac{1}{\sqrt{3}}$

6 Find θ for $0 \leqslant \theta \leqslant 2\pi$ if:

    a $\cos \theta = -1$ **b** $\sin^2 \theta = \frac{3}{4}$

7 Find the perimeter and area of a sector with radius 11 cm and angle $\frac{\pi}{5}$.

8 Show that $\cos \frac{3\pi}{4} - \sin \frac{3\pi}{4} = -\sqrt{2}$.

9 If $\cos \theta = -\frac{3}{4}$, $\frac{\pi}{2} < \theta < \pi$ find the exact value of:

    a $\sin \theta$ **b** $\tan \theta$ **c** $\cos(\pi - \theta)$

10 Without using a calculator, evaluate:

    a $\tan^2\left(\frac{\pi}{3}\right) - \sin^2\left(\frac{\pi}{4}\right)$ **b** $\cos^2\left(\frac{\pi}{4}\right) + \sin \frac{\pi}{2}$ **c** $\cos \frac{5\pi}{3} - \tan \frac{5\pi}{4}$

11 Three circles with radius r are drawn as shown, each with its centre on the circumference of the other two circles. A, B, and C are the centres of the three circles.

Prove that an expression for the area of the shaded region is

$$A = \frac{r^2}{2}(\pi - \sqrt{3}).$$

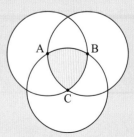

12 For a small angle θ, find an approximation for $3\sin^2 \theta + \cos \frac{\theta}{2} - 2$, giving your answer in the form $a + b\theta^2$.

13 In a physics experiment, light is diffracted as it passes through a slit. The intensity of light striking a screen behind the slit is shown by the red curve.

Suppose the slit has width d, the screen is distance D from the slit, and the first point of minimum intensity is at angle θ and distance x from the mean point.

The first point of minimum intensity satisfies $d \sin \theta = \lambda$ where λ is the wavelength of the light.

Show that $x \approx \dfrac{D\lambda}{d}$.

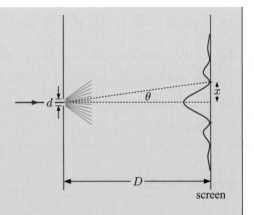

4

Trigonometric functions

Contents:

Opening problem

A Ferris wheel rotates at a constant speed. The wheel's radius is 10 m and the bottom of the wheel is 2 m above ground level. From his viewing point next to the ticket booth, Andrew is watching a green light on the perimeter of the wheel. He notices that the green light moves in a circle. It takes 100 seconds for a full revolution.

Click on the icon to visit a simulation of the Ferris wheel. You will be able to view the light from:

DEMO

- in front of the wheel
- a side-on position
- above the wheel.

You can then observe graphs of the green light's position as the wheel rotates at a constant rate.

Things to think about:

a Andrew estimates how high the light is above ground level at two second intervals. What will a scatter diagram of this data look like?

b Andrew then estimates the horizontal position of the light at two second intervals. What will the scatter diagram of this data look like?

c What similarities and differences are there between your two scatter diagrams?

In previous courses you should have had an introduction to **periodic functions** and **waves**.

PERIODIC FUNCTIONS

A **periodic function** is one which repeats itself over and over in a horizontal direction, in intervals of the same length.

The **period** of a periodic function is the length of one repetition or cycle.

$f(x)$ is a periodic function with period p if $f(x + p) = f(x)$ for all x, and p is the smallest positive value for this to be true.

WAVES

The most important periodic functions in physics and engineering are waves.

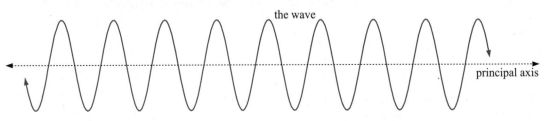

the wave

principal axis

The wave oscillates about a horizontal line called the **principal axis** or **mean line** which has equation $\quad y = \dfrac{\textbf{max} + \textbf{min}}{\textbf{2}}$.

A **maximum point** occurs at the top of a crest, and a **minimum point** at the bottom of a trough.

The **amplitude** is the distance between a maximum (or minimum) point and the principal axis.

$$\textbf{amplitude} = \frac{\textbf{max} - \textbf{min}}{\textbf{2}}$$

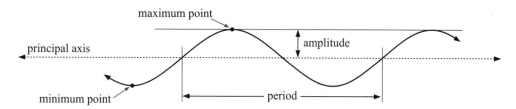

TRIGONOMETRIC FUNCTIONS

A **trigonometric function** is a function which involves one of the trigonometric ratios.

The sine and cosine functions are particularly important, since their graphs are waves.

For example, consider the Ferris wheel in the **Opening Problem** which has radius 10 m and which revolves at constant speed. We let P represent the green light on the wheel.

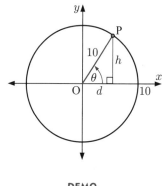

- The **height** of P relative to the centre of the wheel can be determined using right angled triangle trigonometry:

$$\sin \theta = \frac{h}{10}$$
$$\therefore \quad h = 10 \sin \theta$$

- The **horizontal displacement** d of the light on the wheel can also be determined:

$$\cos \theta = \frac{d}{10}$$
$$\therefore \quad d = 10 \cos \theta$$

DEMO

From the demonstration you will see that both graphs are waves.

A   THE SINE AND COSINE FUNCTIONS

Consider the point $P(\cos \theta, \sin \theta)$ on the unit circle.

As θ increases, the point P moves around the unit circle, and the values of $\cos \theta$ and $\sin \theta$ change.

We can draw the graphs of $y = \sin \theta$ and $y = \cos \theta$ by plotting the values of $\sin \theta$ and $\cos \theta$ against θ.

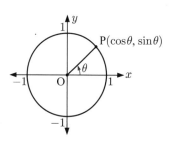

THE GRAPH OF $y = \sin\theta$

The diagram alongside gives the y-coordinates for all points on the unit circle at intervals of $30°$.

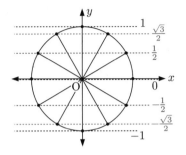

A table for $\sin\theta$ can be constructed from these values:

θ	0	$\frac{\pi}{6}$	$\frac{\pi}{3}$	$\frac{\pi}{2}$	$\frac{2\pi}{3}$	$\frac{5\pi}{6}$	π	$\frac{7\pi}{6}$	$\frac{4\pi}{3}$	$\frac{3\pi}{2}$	$\frac{5\pi}{3}$	$\frac{11\pi}{6}$	2π
$\sin\theta$	0	$\frac{1}{2}$	$\frac{\sqrt{3}}{2}$	1	$\frac{\sqrt{3}}{2}$	$\frac{1}{2}$	0	$-\frac{1}{2}$	$-\frac{\sqrt{3}}{2}$	-1	$-\frac{\sqrt{3}}{2}$	$-\frac{1}{2}$	0

Plotting $\sin\theta$ against θ gives:

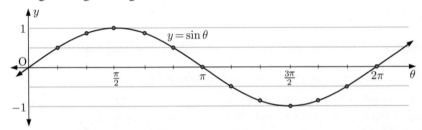

The graph of $y = \sin\theta$ shows the y-coordinate of P as P moves around the unit circle.

Once we reach 2π, P has completed a full revolution of the unit circle, and so this pattern repeats itself.

DEMO

THE GRAPH OF $y = \cos\theta$

By considering the x-coordinates of the points on the unit circle at intervals of $\frac{\pi}{6}$, we can create a table of values for $\cos\theta$:

θ	0	$\frac{\pi}{6}$	$\frac{\pi}{3}$	$\frac{\pi}{2}$	$\frac{2\pi}{3}$	$\frac{5\pi}{6}$	π	$\frac{7\pi}{6}$	$\frac{4\pi}{3}$	$\frac{3\pi}{2}$	$\frac{5\pi}{3}$	$\frac{11\pi}{6}$	2π
$\cos\theta$	1	$\frac{\sqrt{3}}{2}$	$\frac{1}{2}$	0	$-\frac{1}{2}$	$-\frac{\sqrt{3}}{2}$	-1	$-\frac{\sqrt{3}}{2}$	$-\frac{1}{2}$	0	$\frac{1}{2}$	$\frac{\sqrt{3}}{2}$	1

Plotting $\cos\theta$ against θ gives:

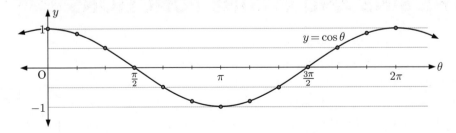

USING TRANSFORMATIONS TO GRAPH TRIGONOMETRIC FUNCTIONS

Now that we are familiar with the graphs of $y = \sin\theta$ and $y = \cos\theta$, we can use transformations to graph more complicated trigonometric functions.

Instead of using θ, we will now use x to represent the angle variable. This is just for convenience, so we are dealing with the familiar function form $y = f(x)$.

For the graphs of $y = \sin x$ and $y = \cos x$:

- The **period** is 2π.
- The **amplitude** is 1.
- The **principal axis** is the line $y = 0$.

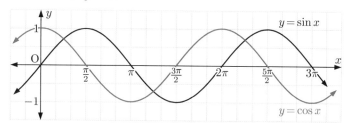

We immediately notice that $y = \sin x$ is a horizontal translation of $y = \cos x$ by $\frac{\pi}{2}$ units to the right. We conclude that:

$$\text{For all } x: \quad \sin x = \cos\left(x - \tfrac{\pi}{2}\right)$$
$$\cos x = \sin\left(x + \tfrac{\pi}{2}\right)$$

Consider the **general sine function**

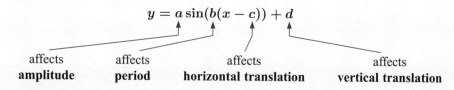

$$y = a\sin(b(x - c)) + d$$

affects	affects	affects	affects
amplitude	**period**	**horizontal translation**	**vertical translation**

- the amplitude is $|a|$
- the period is $\dfrac{2\pi}{b}$ for $b > 0$
- the principal axis is $y = d$
- $y = a\sin(b(x - c)) + d$ is obtained from $y = \sin x$ by a vertical dilation with scale factor a and a horizontal dilation with scale factor $\dfrac{1}{b}$, followed by a horizontal translation of c units and a vertical translation of d units.

The properties of the **general cosine function** $y = a\cos(b(x - c)) + d$ are the same as those of the general sine function.

DEMO

Example 1

◀)) **Self Tutor**

Sketch the graphs of the following on $0 \leqslant x \leqslant 2\pi$:

a $y = \sin\left(x - \frac{\pi}{3}\right)$ **b** $y = \cos 3x$ **c** $y = \cos\left(x + \frac{\pi}{6}\right) + 1$ **d** $y = -\sin x$

a We translate $y = \sin x$ horizontally $\frac{\pi}{3}$ units to the right.

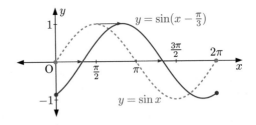

b We dilate $y = \cos x$ horizontally with scale factor $\frac{1}{3}$.

$\therefore$ $y = \cos 3x$ has period $\frac{2\pi}{3}$.

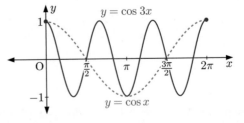

c We translate $y = \cos x$ horizontally $\frac{\pi}{6}$ units to the left, and 1 unit upwards.

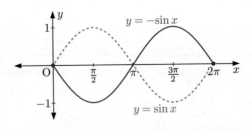

d We reflect $y = \sin x$ in the x-axis.

EXERCISE 4A

1 State the transformation which maps $y = \sin x$ onto:

a $y = \sin x - 1$ **b** $y = \sin\left(x - \frac{\pi}{4}\right)$ **c** $y = 2\sin x$ **d** $y = \sin 4x$

e $y = \frac{1}{2}\sin x$ **f** $y = \sin\frac{x}{4}$ **g** $y = -\sin x$ **h** $y = \sin\left(x - \frac{\pi}{3}\right) + 2$

2 State the period of:

a $y = \sin 5x$ **b** $y = \sin(0.6x)$ **c** $y = \sin \pi x$

d $y = \cos 3x$ **e** $y = \cos\frac{x}{3}$ **f** $y = \cos\frac{\pi x}{50}$

3 Find b given that the function $y = \sin bx$, $b > 0$ has period:

    a 5π **b** $\frac{2\pi}{3}$ **c** 12π **d** 4 **e** 100

4 State the maximum and minimum value of:

    a $y = 4\cos 2x$ **b** $y = 3\cos x + 5$ **c** $y = -2\cos(x - 3) - 4$

5 The general cosine function is $y = a\cos(b(x - c)) + d$.
    State the geometrical significance of a, b, c, and d.

6 Sketch the graphs of the following for $0 \leqslant x \leqslant 4\pi$:

    a $y = \sin x - 2$ **b** $y = \sin x + 3$ **c** $y = \sin x - 0.5$

    d $y = \sin(x - 2)$ **e** $y = \sin(x + 2)$ **f** $y = \sin\left(x - \frac{\pi}{4}\right)$

    g $y = \sin\left(x - \frac{\pi}{6}\right) + 1$ **h** $y = \sin(x - 1) - 2$ **i** $y = \sin\left(x + \frac{\pi}{4}\right) + 2$

    j $y = 3\sin x$ **k** $y = \frac{1}{2}\sin x$ **l** $y = \frac{3}{2}\sin x$

    m $y = \sin 3x$ **n** $y = \sin\frac{x}{2}$ **o** $y = \sin 4x$

7 Sketch the graphs of the following for $-2\pi \leqslant x \leqslant 2\pi$:

    a $y = \cos x + 2$ **b** $y = \cos\left(x - \frac{\pi}{4}\right)$ **c** $y = \cos\left(x + \frac{\pi}{6}\right)$

    d $y = \frac{3}{2}\cos x$ **e** $y = -\cos x$ **f** $y = \cos\left(x - \frac{\pi}{6}\right) + 1$

    g $y = \cos\left(x + \frac{\pi}{4}\right) - 1$ **h** $y = \cos 2x$ **i** $y = \cos\frac{x}{2}$

8 **a** Sketch the curve $y = 4\sin x$ for $0 \leqslant x \leqslant 2\pi$.

    b Find the value of y when: **i** $x = \frac{5\pi}{6}$ **ii** $x = \frac{7\pi}{4}$

        Mark these points on your graph in **a**.

9 For what values of d does the graph of $y = 3\cos x + d$ lie:

    a entirely above the x-axis   **b** entirely below the x-axis

    c partially above and partially below the x-axis?

Example 2   ◀) Self Tutor

Sketch the graph of $y = 3\cos 2x$ for $0 \leqslant x \leqslant 2\pi$.

$a = 3$, so the amplitude is $|3| = 3$.

$b = 2$, so the period is $\dfrac{2\pi}{b} = \dfrac{2\pi}{2} = \pi$.

We dilate $y = \cos x$ vertically with scale factor 3 to give $y = 3\cos x$, then dilate $y = 3\cos x$ horizontally with scale factor $\frac{1}{2}$ to give $y = 3\cos 2x$.

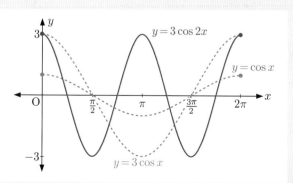

10 State the transformations which map:

    a $y = \sin x$ onto $y = 2\sin 3x$

    b $y = \cos x$ onto $y = -2\cos x$

    c $y = \sin x$ onto $y = 3\sin x - 5$

    d $y = \cos x$ onto $y = \cos\left(2\left(x + \frac{\pi}{6}\right)\right)$

11 Sketch the graph of the following for $-2\pi \leqslant x \leqslant 2\pi$:

    a $y = -3\sin x$

    b $y = \cos 2x + 1$

    c $y = \frac{1}{2}\sin\left(x + \frac{\pi}{6}\right) - \frac{1}{3}$

    d $y = \frac{1}{3}\cos\left(x + \frac{\pi}{4}\right) + 1$

    e $y = 3\sin\left(x - \frac{\pi}{3}\right) - 1$

    f $y = -\cos\left(\frac{1}{2}\left(x - \frac{\pi}{4}\right)\right)$

12 Consider the general sine function $y = a\sin(b(x - c)) + d$.   State which of the variables a, b, c, and d can be changed to always produce a change in:

    a the x-intercepts of the function

    b the y-intercept of the function

    c the range of the function.

Example 3   ◀)) Self Tutor

Find the equation of this sine function.

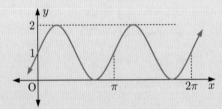

The amplitude is 1, so $a = 1$.

The period is π, so $\dfrac{2\pi}{b} = \pi$ and $\therefore$ $b = 2$.

There is no horizontal translation, so $c = 0$.

The principal axis is $y = 1$, so $d = 1$.

The equation of the function is $y = \sin 2x + 1$.

---

13 Find the equation of each sine function:

    a

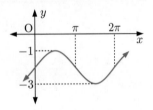

    b

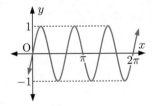

    c

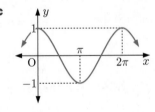

    d

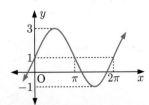

    e

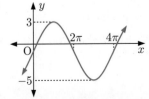

    f

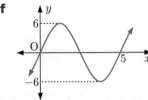

14 Find the cosine function shown in the graph:

a 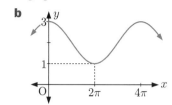   **b** 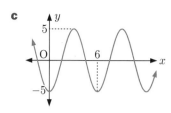   **c**

Investigation 1   Modelling with trigonometric functions

Periodic phenomena occur all around us. If we are given a set of periodic data, we can use a trigonometric function to *model* this behaviour. The model can then be used to predict future behaviour.

Task 1: The undamped spring

An object is suspended from a spring. If the object is pulled below its resting position and then released, it will oscillate up and down.

The data below shows the height of the object relative to its rest position, at different times.

Time (t seconds)	0	0.1	0.2	0.3	0.4	0.5	0.6	0.7	0.8	0.9
Height (H cm)	−15	−13	−7.5	0	7.5	13	15	13	7.5	0

Time (t seconds)	1	1.1	1.2	1.3	1.4	1.5	1.6	1.7	1.8	1.9	2.0
Height (H cm)	−7.5	−13	−15	−13	−7.5	0	7.5	13	15	13	7.5

We will attempt to model the data with a trigonometric function of the form $H = A \sin B(t - C)$.

What to do:

1 **a** Draw a scatter diagram of the data.

    b What features of the data suggest that a trigonometric model might be appropriate?

2 **a** State the *amplitude* of the oscillation.

    b Hence determine the value of A.

3 **a** State the *period* of the oscillation.

    b Hence determine the value of B.

4 Explain why $C = \frac{\pi}{2}$.

5 Write down the trigonometric function which models the height of the object over time.

6 Use your model to predict the height of the object after 4.25 seconds.

7 What do you think is unrealistic about this model? What would happen differently   **VIDEO**
in reality? Watch the video to find out.

Task 2: The pendulum

This task is best performed in small groups.

You will need: string, sticky tape, ruler, a stopwatch, and a tennis ball.

What to do:

1 Cut a piece of string of length 30 cm. Attach one end
of the string to the tennis ball, and the other end to your
desk.

2 Hold the ball to one side, then release it, causing the ball
to swing back and forth like a pendulum.

3 Using your stopwatch and ruler, measure the maximum
and minimum horizontal displacement reached by the ball,
and the times at which they occurred. You may need to
repeat the experiment several times, but make sure the
ball is released from the same position each time.

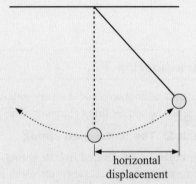

horizontal
displacement

4 Use your data to find a trigonometric function which models the horizontal displacement of the ball
over time.

5 What part of the function affects the *period* of the pendulum?

6 Repeat the experiment with strings of different length. Explore the relationship between the length
of the string and the period of the pendulum.

B   THE TANGENT FUNCTION

If P($\cos\theta$, $\sin\theta$) is a point on the unit circle,
then $\tan\theta$ is the **gradient** of the line (OP).

The **tangent function** is defined by

$$\tan\theta = \frac{\sin\theta}{\cos\theta}.$$

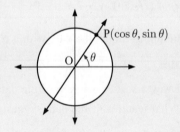

Click on the icon to explore how the tangent function is produced from the unit circle.

**TANGENT
FUNCTION**

THE GRAPH OF $y = \tan x$

$\tan x$ is zero whenever $\sin x = 0$

$\therefore$ the **zeros** of $y = \tan x$ are $k\pi$, $k \in \mathbb{Z}$.

$\tan x$ is undefined whenever $\cos x = 0$

$\therefore$ the **vertical asymptotes** of $y = \tan x$ are $x = \frac{\pi}{2} + k\pi$, $k \in \mathbb{Z}$.

$\tan x$ has period $= \pi$ and range $y \in \mathbb{R}$.

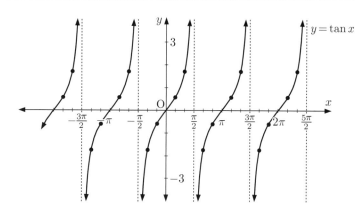

DEMO

THE GENERAL TANGENT FUNCTION

The **general tangent function** is $y = a\tan(b(x - c)) + d,\quad a \neq 0,\ b > 0$.

- The **principal axis** is $y = d$.
- The **period** of this function is $\dfrac{\pi}{b}$.
- The **amplitude** of this function is undefined.
- There are infinitely many vertical asymptotes.

Click on the icon to explore the properties of this function.

DYNAMIC TANGENT FUNCTION

Example 4
◀) **Self Tutor**

Sketch the graph of the following for $-\pi \leqslant x \leqslant \pi$:   **a** $\tan\left(x - \frac{\pi}{3}\right)$ **b** $-\tan\frac{x}{2}$

a   We translate $y = \tan x$ horizontally $\frac{\pi}{3}$ units to the right.

b   We dilate $y = \tan x$ horizontally with scale factor 2 to give $y = \tan\frac{x}{2}$, then reflect this in the x-axis to give $y = -\tan\frac{x}{2}$.

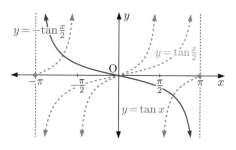

EXERCISE 4B

1 State the transformations which map $y = \tan x$ onto:

    a $y = \tan\left(x - \frac{\pi}{4}\right)$ **b** $y = \tan 2x - 1$ **c** $y = -\frac{1}{2}\tan x$ **d** $y = \tan(x - \pi)$

2 State the period of:

    a $y = \tan 3x$ **b** $y = \tan \pi x$ **c** $y = \tan\left(\frac{2x}{3} - \frac{\pi}{3}\right)$ **d** $y = \tan nx, \ n \neq 0$

3 For each function, write down the:

    i zeros   **ii** vertical asymptotes.

    a $y = \tan 2x$ **b** $y = \tan\left(x + \frac{\pi}{3}\right)$ **c** $y = \frac{1}{2}\tan\left(\frac{x - \frac{\pi}{6}}{2}\right)$

4 Sketch the graph of the following for $-2\pi \leqslant x \leqslant 2\pi$:

    a $y = \tan\left(x - \frac{\pi}{4}\right)$ **b** $y = \frac{1}{2}\tan\frac{x}{4}$ **c** $y = 3\tan\left(x - \frac{\pi}{9}\right)$

5 Find p and q given the following graph is of the function $y = \tan pt + q$.

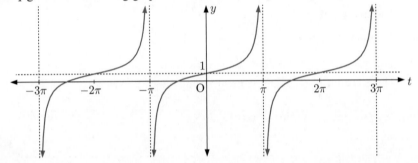

6 Find the possible values of a and b given the following graph is of the function $y = \tan a(x - b)$.

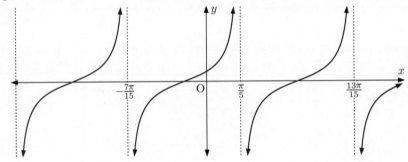

7 **a** Describe the sequence of transformations used to transform $y = \tan x$ into $y = 2\tan\left(x + \frac{\pi}{4}\right) - 1$.

    b Sketch $y = 2\tan\left(x + \frac{\pi}{4}\right) - 1$ for $-2\pi \leqslant x \leqslant 2\pi$.

8 Consider the functions $f(x) = \tan x$ and $g(x) = 2x - \frac{\pi}{2}$.

    a Find:

        i $fg(x)$ **ii** $gf(x)$

    b Find the value of:

        i $fg\left(\frac{\pi}{3}\right)$ **ii** $gf(\pi)$

    c Write down the period and vertical asymptotes of:

        i $fg(x)$ **ii** $gf(x)$

    d Sketch the graphs of $fg(x)$ and $gf(x)$ for $-2\pi \leqslant x \leqslant 2\pi$.

Investigation 2   Errors in small angle approximations

We now have formulae which can be used to approximate the trigonometric ratios for small angles.

- $\sin\theta \approx \theta$
- $\cos\theta \approx 1 - \frac{\theta^2}{2}$
- $\tan\theta \approx \theta$

However, we also need an understanding of *how small* an angle must be before it is reasonable to use these approximations.

What to do:

1 Use the graphing package to plot $y = \sin x$, $y = \tan x$, and $y = x$ on the same set of axes.

GRAPHING PACKAGE

Use the solver on the graphing package to calculate the values of θ for which the approximations $\sin\theta \approx \theta$ and $\tan\theta \approx \theta$ are accurate to within 0.1%, 0.5%, and 1%.

Record your results in the following table, in both radians and degrees:

Approximation	0.1% accuracy	0.5% accuracy	1% accuracy
$\sin\theta \approx \theta$			
$\cos\theta \approx 1 - \frac{\theta^2}{2}$			
$\tan\theta \approx \theta$			

2 Use the graphing package to plot $y = \cos x$ and $y = 1 - \dfrac{x^2}{2}$ on the same set of axes. Hence complete the table in **1**.

3 Discuss the values of θ for which a small angle approximation is reasonable.

Activity 1

Click on the icon to run a card game for trigonometric functions.

CARD GAME

C   RECIPROCAL TRIGONOMETRIC FUNCTIONS

We define the reciprocal trigonometric functions cosec x, secant x, and cotangent x as:

$$\csc x = \frac{1}{\sin x}, \qquad \sec x = \frac{1}{\cos x}, \qquad \text{and} \qquad \cot x = \frac{1}{\tan x} = \frac{\cos x}{\sin x}$$

For example, $\csc \frac{\pi}{6} = \dfrac{1}{\sin \frac{\pi}{6}} = \dfrac{1}{\frac{1}{2}} = 2.$

EXERCISE 4C

1 Without using a calculator, find:

    a $\csc \frac{\pi}{3}$
    b $\cot \frac{2\pi}{3}$
    c $\sec \frac{5\pi}{6}$
    d $\cot \pi$

2 Without using a calculator, find $\csc x$, $\sec x$, and $\cot x$ given:

    a $\sin x = \frac{3}{5}$, $0 \leqslant x \leqslant \frac{\pi}{2}$
    b $\cos x = \frac{2}{3}$, $\frac{3\pi}{2} < x < 2\pi$

3 Find the other *five* trigonometric ratios if:

    a $\cos x = \frac{3}{4}$ and $\frac{3\pi}{2} < x < 2\pi$
    b $\sin x = -\frac{2}{3}$ and $\pi < x < \frac{3\pi}{2}$

    c $\sec x = 2\frac{1}{2}$ and $0 < x < \frac{\pi}{2}$
    d $\csc x = 2$ and $\frac{\pi}{2} < x < \pi$

    e $\tan \beta = \frac{1}{2}$ and $\pi < \beta < \frac{3\pi}{2}$
    f $\cot \theta = \frac{4}{3}$ and $\pi < \theta < \frac{3\pi}{2}$

Example 5   ◀) Self Tutor

Using the graph of $y = \sin x$, sketch the graph of $y = \dfrac{1}{\sin x} = \csc x$ for $x \in [-2\pi, 2\pi]$.
Check your answer using technology.

- The zeros of $y = \sin x$ become vertical asymptotes of $y = \dfrac{1}{\sin x}$.

- The local maxima of $y = \sin x$ become local minima of $y = \dfrac{1}{\sin x}$.

- The local minima of $y = \sin x$ become local maxima of $y = \dfrac{1}{\sin x}$.

- When $\sin x = 1$, $\dfrac{1}{\sin x} = 1$ and when $\sin x = -1$, $\dfrac{1}{\sin x} = -1$.

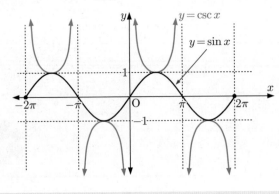
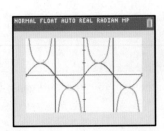

---

4 Using the graph of $y = \cos x$, sketch the graph of $y = \dfrac{1}{\cos x} = \sec x$ for $x \in [-2\pi, 2\pi]$.
Check your answer using technology.

5 Using the graph of $y = \tan x$, sketch the graph of $y = \dfrac{1}{\tan x} = \cot x$ for $x \in [-2\pi, 2\pi]$.

6 Use technology to sketch $y = \sec x$ and $y = \csc\left(x + \frac{\pi}{2}\right)$ on the same set of axes for $x \in [-2\pi, 2\pi]$.
Explain your answer.

Example 6

◀) **Self Tutor**

Find the small angle approximation for $\sec^2 x$.

$$\sec^2 x = (\cos x)^{-2}$$

$$\approx \left(1 - \frac{x^2}{2}\right)^{-2} \qquad\qquad \{\text{small angle approximation}\}$$

$$\approx 1 + (-2)\left(-\frac{x^2}{2}\right) + \frac{(-2)(-3)}{2!}\left(-\frac{x^2}{2}\right)^2 + ... \quad \{\text{binomial expansion}\}$$

$$\approx 1 + x^2 + \tfrac{3}{4}x^4$$

7 Show that for small x:

a $\sec x \tan x \approx x + x^3$

b $\dfrac{\sec x \tan x - \csc^2 x}{\cot^2 x} \approx -1 - x^2 + x^3$

Historical note   **Astronomy and trigonometry**

The Greek astronomer **Hipparchus** (140 BC) is credited with being the founder of trigonometry. To aid his astronomical calculations, he produced a table of numbers in which the lengths of chords of a circle were related to the length of the radius.

Ptolemy, another great Greek astronomer of the time, extended this table in his major published work *Almagest*, which was used by astronomers for the next 1000 years. In fact, much of Hipparchus' work is known through the writings of Ptolemy. These writings found their way to Hindu and Arab scholars.

Aryabhata, a Hindu mathematician in the 5th and 6th Century AD, constructed a table of the lengths of half-chords of a circle with radius one unit. This was the first table of **sine** values.

In the late 16th century, **Georg Joachim de Porris**, also known as **Rheticus**, produced comprehensive and remarkably accurate tables of all six trigonometric ratios. These involved a tremendous number of tedious calculations, all without the aid of calculators or computers.

Rheticus was the only student of **Nicolaus Copernicus**, and helped his tutor publish his work *De revolutionibus orbium coelestium* (On the Revolutions of the Heavenly Spheres).

Nicolaus Copernicus

D   INVERSE TRIGONOMETRIC FUNCTIONS

In many problems we need to know what angle results in a particular trigonometric ratio. We have previously seen this for right angled triangle problems and when using the cosine and sine rules.

For example:

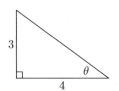

$$\tan\theta = \frac{\text{OPP}}{\text{ADJ}} = \frac{3}{4}$$

$$\therefore\ \theta = \tan^{-1}\tfrac{3}{4}$$

$$\therefore\ \theta \approx 36.9°$$

Rather than writing the inverse trigonometric functions as $\sin^{-1}$, $\cos^{-1}$, and $\tan^{-1}$, which can be confused with reciprocals, mathematicians more formally refer to these functions as the **inverse trigonometric functions arcsine, arccosine,** and **arctangent**.

Investigation 3   Inverse trigonometric functions

The function $y = \sin x$ fails the horizontal line test on its natural domain. For example, when $y = \frac{1}{2}$, there is more than one corresponding value of x, since $\sin x = \frac{1}{2}$ has infinitely many solutions, including $x = \frac{\pi}{6}$ and $x = \frac{5\pi}{6}$.

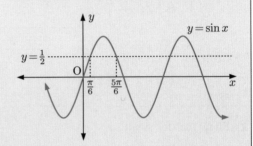

Therefore, if we want to find inverse functions for $\sin x$, $\cos x$, and $\tan x$, we will need to apply suitable domain restrictions.

What to do:

1 Draw the graphs of $y = \sin x$, $y = \cos x$, and $y = \tan x$ on the domain $-3\pi \leqslant x \leqslant 3\pi$.

2 Copy and complete the following table, indicating whether each function is one-to-one on the given restricted domain.

Restricted domain	$\sin x$	$\cos x$	$\tan x$
$0 \leqslant x \leqslant 2\pi$			
$-\pi \leqslant x \leqslant \pi$			
$-\frac{\pi}{2} \leqslant x \leqslant \frac{\pi}{2}$			
$0 \leqslant x \leqslant \pi$			
$-\frac{\pi}{2} \leqslant x \leqslant \frac{3\pi}{2}$			
$\pi \leqslant x \leqslant 2\pi$			

3 Discuss which domain restriction you think would be most suitable for the inverse function of:

   a $\sin x$ **b** $\cos x$ **c** $\tan x$.

4 The function $f(x) = \sin x$, $-\frac{\pi}{2} \leqslant x \leqslant \frac{\pi}{2}$ has inverse $f^{-1}(x) = \arcsin x$.

   Sketch the graph of $y = f(x)$, and hence sketch the graph of $f^{-1}(x) = \arcsin x$ on the same set of axes.

5 The function $f(x) = \cos x$, $0 \leqslant x \leqslant \pi$ has inverse $f^{-1}(x) = \arccos x$.

   Sketch the graph of $y = f(x)$, and hence sketch the graph of $f^{-1}(x) = \arccos x$ on the same set of axes.

6 The function $f(x) = \tan x$, $-\frac{\pi}{2} \leqslant x \leqslant \frac{\pi}{2}$ has inverse $f^{-1}(x) = \arctan x$.

   Sketch the graph of $y = f(x)$, and hence sketch the graph of $f^{-1}(x) = \arctan x$ on the same set of axes.

The **inverse trigonometric functions** are defined as:

Function	Definition	Range
$y = \arcsin x$	$x = \sin y, \quad -1 \leqslant x \leqslant 1$	$-\frac{\pi}{2} \leqslant y \leqslant \frac{\pi}{2}$
$y = \arccos x$	$x = \cos y, \quad -1 \leqslant x \leqslant 1$	$0 \leqslant y \leqslant \pi$
$y = \arctan x$	$x = \tan y, \quad x \in \mathbb{R}$	$-\frac{\pi}{2} < y < \frac{\pi}{2}$

The graphs of these functions are illustrated below, along with the corresponding graphs of $\sin x$, $\cos x$, and $\tan x$ on their restricted domains.

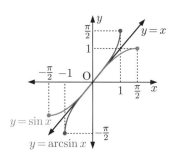
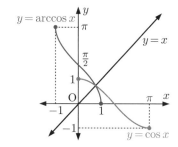
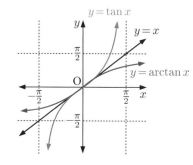

EXERCISE 4D

1 Find, giving your answer in radians:

   a $\arccos 1$ **b** $\arcsin(-1)$ **c** $\arctan 1$ **d** $\arctan(-1)$

   e $\arcsin \frac{1}{2}$ **f** $\arccos\left(-\frac{\sqrt{3}}{2}\right)$ **g** $\arctan \sqrt{3}$ **h** $\arccos\left(-\frac{1}{\sqrt{2}}\right)$

   i $\arctan\left(-\frac{1}{\sqrt{3}}\right)$ **j** $\sin^{-1}(-0.767)$ **k** $\cos^{-1} 0.327$ **l** $\tan^{-1}(-50)$

2 Find the invariant point for the inverse transformation from:

   a $y = \sin x$ to $y = \arcsin x$

   b $y = \tan x$ to $y = \arctan x$

   c $y = \cos x$ to $y = \arccos x$.

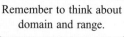

An **invariant point** is a point on the function that does not move when a transformation is applied.

3 **a** State the equations of the asymptotes of $y = \arctan x$.

   b Do the functions $y = \arcsin x$ and $y = \arccos x$ have vertical asymptotes? Explain your answer.

4 Simplify:

   a $\arcsin\left(\sin \frac{\pi}{3}\right)$ **b** $\arccos\left(\cos\left(-\frac{\pi}{6}\right)\right)$

   c $\tan(\arctan(0.3))$ **d** $\cos\left(\arccos\left(-\frac{1}{2}\right)\right)$

Remember to think about domain and range.

   e $\arctan(\tan \pi)$ **f** $\arcsin\left(\sin \frac{4\pi}{3}\right)$

Investigation 4   arctan x

Carl Friedrich Gauss used his **Gaussian hypergeometric series** to analyse the continued fraction:

$$\cfrac{x}{1+\cfrac{(1x)^2}{3+\cfrac{(2x)^2}{5+\cfrac{(3x)^2}{7+\cfrac{(4x)^2}{9+....}}}}}$$

What to do:

1 Evaluate the fraction with $x = 1$ for as many levels as necessary for the answer to be accurate to 5 decimal places. You may wish to use a spreadsheet.

2 Compare your result with $\arctan 1$.

3 Compare the continued fraction and $\arctan x$ for another value of x of your choosing.

E TRIGONOMETRIC EQUATIONS

In this Section we consider equations involving trigonometric functions.

Trigonometric equations generally have infinitely many solutions unless a restricted domain such as $0 \leqslant x \leqslant 3\pi$ is given.

In the **Opening Problem**, the height of the green light after t seconds is given by

$H(t) = 10\sin\left(\frac{\pi}{50}(t - 25)\right) + 12$ metres. So, the light will be 16 metres above the ground when

$10\sin\left(\frac{\pi}{50}(t - 25)\right) + 12 = 16$ metres. This trigonometric equation has infinitely many solutions provided the wheel keeps rotating. For this reason we would normally specify a time interval for the solution. For example, if we are interested in the first three minutes of its rotation, we specify the domain $0 \leqslant t \leqslant 180$.

SOLVING TRIGONOMETRIC EQUATIONS USING ALGEBRA

Exact solutions obtained using algebra are called **analytic** solutions. We can find analytic solutions to *some* trigonometric equations, but only if they correspond to angles for which the trigonometric ratios can be expressed exactly.

We use the *periodicity* of the trigonometric functions to give us all solutions in the required domain. Remember that $\sin x$ and $\cos x$ both have period 2π, and $\tan x$ has period π.

> When solving trigonometric equations, you must find all of the solutions in the required domain.

For an equation such as $\sin 2x = \frac{1}{2}$ on the domain $0 \leqslant x \leqslant 2\pi$, we need to understand that if $0 \leqslant x \leqslant 2\pi$ then $0 \leqslant 2x \leqslant 4\pi$. So, when we consider points on the unit circle with sine $\frac{1}{2}$, we need to consider angles from 0 to 4π.

Example 7

◆) **Self Tutor**

Solve for x on the domain $0 \leqslant x \leqslant 2\pi$:

a $\cos x = -\frac{\sqrt{3}}{2}$

b $2\sin x - 1 = 0$

c $\tan x + \sqrt{3} = 0$

a $\cos x = -\frac{\sqrt{3}}{2}$

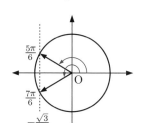

$\therefore \quad x = \frac{5\pi}{6}$ or $\frac{7\pi}{6}$

b $2\sin x - 1 = 0$

$\therefore \quad \sin x = \frac{1}{2}$

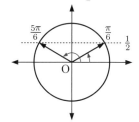

$\therefore \quad x = \frac{\pi}{6}$ or $\frac{5\pi}{6}$

c $\tan x + \sqrt{3} = 0$

$\therefore \quad \tan x = -\sqrt{3}$

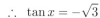
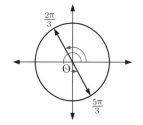

$\therefore \quad x = \frac{2\pi}{3}$ or $\frac{5\pi}{3}$

EXERCISE 4E

1 Solve for x on the domain $0 \leqslant x \leqslant 2\pi$:

   a $\cos x = \frac{1}{2}$ **b** $\sin x = -\frac{1}{\sqrt{2}}$ **c** $\tan x = \frac{1}{\sqrt{3}}$

   d $\sin x = -1$ **e** $\cos x = 0$ **f** $\tan x = 0$

2 Solve for x on the domain $0 \leqslant x \leqslant 2\pi$:

   a $2\sin x = \sqrt{3}$ **b** $3\cos x + 3 = 0$ **c** $2\tan x - 2 = 0$

3 Solve for x on the domain $0 \leqslant x \leqslant 4\pi$:

   a $2\cos x + 1 = 0$ **b** $\sqrt{2}\sin x = 1$ **c** $\tan x = 1$

4 Solve for x on the domain $-2\pi \leqslant x \leqslant 2\pi$:

   a $2\sin x + \sqrt{3} = 0$ **b** $\sqrt{2}\cos x + 1 = 0$ **c** $\tan x = -1$

Example 8

◆) **Self Tutor**

Solve $\cos^2 x = \frac{1}{2}$ on $0 \leqslant x \leqslant 2\pi$.

$\cos^2 x = \frac{1}{2}$

$\therefore \quad \cos x = \pm\frac{1}{\sqrt{2}}$

$\therefore \quad x = \frac{\pi}{4}, \frac{3\pi}{4}, \frac{5\pi}{4},$ or $\frac{7\pi}{4}$

5 Solve for x on $0 \leqslant x \leqslant 2\pi$:

   a $\cos^2 x = \frac{3}{4}$ **b** $\sin^2 x = 1$ **c** $\tan^2 x = 3$

Example 9
◀)) **Self Tutor**

Solve exactly for $0 \leqslant x \leqslant 3\pi$:

a $\sin x = -\frac{1}{2}$
b $\sin 2x = -\frac{1}{2}$
c $\sin\left(x - \frac{\pi}{6}\right) = -\frac{1}{2}$

The three equations all have the form $\sin \theta = -\frac{1}{2}$.

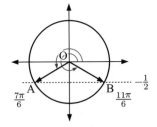

a $0 \leqslant x \leqslant 3\pi$
$\therefore \quad x = \frac{7\pi}{6}$ or $\frac{11\pi}{6}$

b In this case θ is $2x$.
If $0 \leqslant x \leqslant 3\pi$ then $0 \leqslant 2x \leqslant 6\pi$.
$\therefore \quad 2x = \frac{7\pi}{6}, \frac{11\pi}{6}, \frac{19\pi}{6}, \frac{23\pi}{6}, \frac{31\pi}{6}$, or $\frac{35\pi}{6}$
$\therefore \quad x = \frac{7\pi}{12}, \frac{11\pi}{12}, \frac{19\pi}{12}, \frac{23\pi}{12}, \frac{31\pi}{12}$, or $\frac{35\pi}{12}$

c In this case θ is $x - \frac{\pi}{6}$.
If $0 \leqslant x \leqslant 3\pi$ then $-\frac{\pi}{6} \leqslant x - \frac{\pi}{6} \leqslant \frac{17\pi}{6}$.
$\therefore \quad x - \frac{\pi}{6} = -\frac{\pi}{6}, \frac{7\pi}{6}$, or $\frac{11\pi}{6}$
$\therefore \quad x = 0, \frac{4\pi}{3}$, or 2π

> Start at $-\frac{\pi}{6}$ and work around to $\frac{17\pi}{6}$, recording the angle every time you reach points A and B.

6 If $0 \leqslant x \leqslant 2\pi$, state the domain of:

a $2x$
b $\frac{x}{4}$
c $x + \frac{\pi}{2}$
d $x - \frac{\pi}{6}$
e $2\left(x - \frac{\pi}{4}\right)$
f $-x$

7 If $-\pi \leqslant x \leqslant \pi$, state the domain of:

a $3x$
b $\frac{x}{4}$
c $x - \frac{\pi}{2}$
d $2x + \frac{\pi}{2}$
e $-2x$
f $\pi - x$

8 Solve exactly for $0 \leqslant x \leqslant 3\pi$:

a $\cos x = \frac{1}{2}$
b $\cos 2x = \frac{1}{2}$
c $\cos\left(x + \frac{\pi}{3}\right) = \frac{1}{2}$

9 Solve for x on $0 \leqslant x \leqslant 2\pi$:

a $\sin 2x = -\frac{1}{2}$
b $\cos 3x = \frac{\sqrt{3}}{2}$
c $\tan 2x - \sqrt{3} = 0$
d $\sin \frac{x}{2} = \frac{1}{\sqrt{2}}$
e $2\cos \frac{x}{2} + 1 = 0$
f $3\tan \frac{x}{3} - 3 = 0$

10 Solve for x on $0 \leqslant x \leqslant 2\pi$:

a $\cos^2 3x = \frac{1}{4}$
b $\sin^2 2x = 1$
c $\tan^2 \left(\frac{x}{2}\right) = \frac{1}{3}$

11 Find the exact solutions for $0 \leqslant x \leqslant 2\pi$:

a $\sin x = -\cos x$
b $\sin 3x = \cos 3x$
c $\sin 2x = \sqrt{3}\cos 2x$

Example 10   ◀ᴷ) **Self Tutor**

Solve $\sqrt{2}\cos\left(x - \frac{3\pi}{4}\right) + 1 = 0$ for $0 \leqslant x \leqslant 6\pi$.

$\sqrt{2}\cos\left(x - \frac{3\pi}{4}\right) + 1 = 0$

$\therefore \cos\left(x - \frac{3\pi}{4}\right) = -\frac{1}{\sqrt{2}}.$

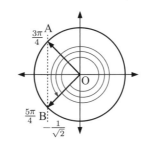

Start at $-\frac{3\pi}{4}$ and work around to $\frac{21\pi}{4}$, recording the angle every time you reach points A and B.

Since $0 \leqslant x \leqslant 6\pi$,

$-\frac{3\pi}{4} \leqslant x - \frac{3\pi}{4} \leqslant \frac{21\pi}{4}$

So, $x - \frac{3\pi}{4} = -\frac{3\pi}{4}, \frac{3\pi}{4}, \frac{5\pi}{4}, \frac{11\pi}{4}, \frac{13\pi}{4}, \frac{19\pi}{4},$ or $\frac{21\pi}{4}$

$\therefore x = 0, \frac{3\pi}{2}, 2\pi, \frac{7\pi}{2}, 4\pi, \frac{11\pi}{2},$ or 6π

12 Solve exactly:

   a $\cos\left(x - \frac{2\pi}{3}\right) = \frac{1}{2}, \ -2\pi \leqslant x \leqslant 2\pi$ **b** $\sqrt{2}\sin\left(x - \frac{\pi}{4}\right) + 1 = 0, \ 0 \leqslant x \leqslant 3\pi$

   c $\sin\left(4\left(x - \frac{\pi}{4}\right)\right) = 0, \ 0 \leqslant x \leqslant \pi$ **d** $2\sin\left(2\left(x - \frac{\pi}{3}\right)\right) = -\sqrt{3}, \ 0 \leqslant x \leqslant 2\pi$

13 Find the exact solutions of $\tan x = \sqrt{3}$ for $0 \leqslant x \leqslant 2\pi$. Hence solve the following equations for $0 \leqslant x \leqslant 2\pi$:

   a $\tan\left(x - \frac{\pi}{6}\right) = \sqrt{3}$ **b** $\tan 4x = \sqrt{3}$ **c** $\tan^2 x = 3$

14 Solve exactly for $x \in [0, 2\pi]$:

   a $\sec x = 2$ **b** $\csc x = -\sqrt{2}$ **c** $\sqrt{3}\sec 2x = -2$

   d $\csc\left(x + \frac{\pi}{6}\right) + \sqrt{2} = 0$ **e** $\cot x + 1 = 0$ **f** $\cot\left(2x - \frac{\pi}{4}\right) - \sqrt{3} = 0$

Example 11   ◀ᴷ) **Self Tutor**

Solve for x on $0 \leqslant x \leqslant 2\pi$, giving your answers as exact values:

   a $2\sin^2 x + \sin x = 0$ **b** $2\cos^2 x + \cos x - 1 = 0$

  a $2\sin^2 x + \sin x = 0$

     $\therefore \ \sin x(2\sin x + 1) = 0$

         $\therefore \ \sin x = 0$ or $-\frac{1}{2}$

  b $2\cos^2 x + \cos x - 1 = 0$

     $\therefore \ (2\cos x - 1)(\cos x + 1) = 0$

         $\therefore \ \cos x = \frac{1}{2}$ or -1

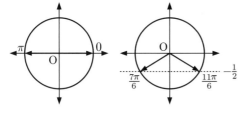

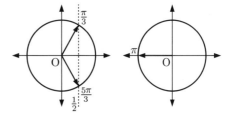

$\therefore \ x = 0, \pi, \frac{7\pi}{6}, \frac{11\pi}{6},$ or 2π. $\therefore \ x = \frac{\pi}{3}, \pi,$ or $\frac{5\pi}{3}$.

15 Solve for $0 \leqslant x \leqslant 2\pi$ giving your answers as exact values:

  a $2\sin^2 x - \sin x = 0$ **b** $2\cos^2 x = \cos x$

  c $2\cos^2 x - \cos x - 1 = 0$ **d** $2\sin^2 x + 3\sin x + 1 = 0$

  e $2\sin x + \csc x = 3$ **f** $3\tan x = \cot x$

  g $\tan^4 x - 2\tan^2 x - 3 = 0$

Example 12 　　　　　　　　　　　　　　　　◄» Self Tutor

Find, where possible, the exact solutions of:

  a $\arctan x = \frac{\pi}{3}$ **b** $\arccos(x - 1) = \frac{2\pi}{3}$ **c** $\arcsin x = \frac{2\pi}{3}$

  a The range of $y = \arctan x$ is $-\frac{\pi}{2} < y < \frac{\pi}{2}$. $\frac{\pi}{3}$ is within the range.

      $\therefore \quad x = \tan \frac{\pi}{3} = \sqrt{3}$

  b The range of $y = \arccos(x - 1)$ is $0 \leqslant y \leqslant \pi$. $\frac{2\pi}{3}$ is within the range.

        $\therefore \quad x - 1 = \cos \frac{2\pi}{3}$

        $\therefore \quad x - 1 = -\frac{1}{2}$

         $\therefore \quad x = \frac{1}{2}$

  c The range of $y = \arcsin x$ is $-\frac{\pi}{2} \leqslant y \leqslant \frac{\pi}{2}$, and $\frac{2\pi}{3}$ is outside this range.

    $\therefore$ no solution exists, even though we can find $\sin \frac{2\pi}{3}$.

---

16 Find, where possible, the exact solutions of:

  a $\arctan x = \frac{\pi}{4}$ **b** $\arcsin x = -\frac{\pi}{3}$

  c $\arccos x = \frac{3\pi}{4}$ **d** $\arcsin(x + 1) = \frac{\pi}{6}$

  e $\arccos x = -\frac{\pi}{4}$ **f** $\arctan(x - \sqrt{3}) = -\frac{\pi}{3}$

Activity 2 　　　　　Solving trigonometric equations using technology

Trigonometric equations may be solved **numerically** using either a **graphing package** or a **graphics calculator**. In most cases the answers will not be exact, but rather given as decimal numbers.

GRAPHING PACKAGE

When using a graphics calculator make sure that the **mode** is set to **radians**.

For example, consider the equation $2\sin x - \cos x = 4 - x$ for $0 \leqslant x \leqslant 2\pi$.

We graph the functions $Y_1 = 2\sin X - \cos X$ and $Y_2 = 4 - X$ on the same set of axes.

GRAPHICS CALCULATOR INSTRUCTIONS

We need to use **window** settings just larger than the domain.

In this case, $X\min = -\frac{\pi}{6}$ $X\max = \frac{13\pi}{6}$ $X\text{scale} = \frac{\pi}{6}$

Casio fx-9860G PLUS	Casio fx-CG20	TI-84 Plus CE
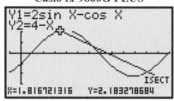	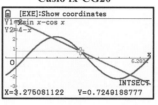	

The solutions are $x \approx 1.82$, 3.28, and 5.81 .

What to do:

1 Solve for x on the domain $0 < x < 12$:

    a $\sin x = 0.431$ **b** $\cos x = -0.814$ **c** $3 \tan x - 2 = 0$

2 Solve for x on the domain $-5 \leqslant x \leqslant 5$:

    a $5 \cos x - 4 = 0$ **b** $2 \tan x + 13 = 0$ **c** $8 \sin x + 3 = 0$

3 Solve each of the following for $0 \leqslant x \leqslant 2\pi$:

    a $\sin(x + 2) = 0.0652$ **b** $\sin^2 x + \sin x - 1 = 0$

4 Solve for x:

    $\cos(x - 1) + \sin(x + 1) = 6x + 5x^2 - x^3$ for $-2 \leqslant x \leqslant 6$.

Make sure you find *all* the solutions on the given domain.

F | TRIGONOMETRIC MODELS

When patterns of variation can be identified and quantified using a formula or equation, predictions may be made about behaviour in the future. Examples of this include tidal movement which can be predicted many months ahead, and the date of a future full moon.

In this Section we use trigonometric functions to model periodic biological and physical phenomena.

Example 13   ◀)) **Self Tutor**

The height of the tide above mean sea level on January 24th at Cape Town is modelled by $h(t) = 3 \sin \frac{\pi t}{6}$ metres, where t is the number of hours after midnight.

  a Graph $y = h(t)$ for $0 \leqslant t \leqslant 24$.

  b When is high tide and what is the maximum height?

  c What is the height of the tide at 2 pm?

  d A ship can cross the harbour provided the tide is at least 2 m above mean sea level. When is crossing possible on January 24th?

a $h(0) = 0$

$h(t) = 3\sin\frac{\pi t}{6}$ has period $= \frac{2\pi}{\frac{\pi}{6}} = 2\pi \times \frac{6}{\pi} = 12$ hours

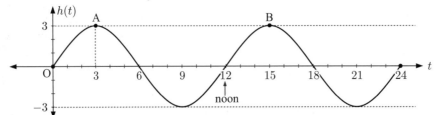

b   High tide is at 3 am and 3 pm. The maximum height is 3 m above the mean as seen at points A and B.

c   At 2 pm, $t = 14$ and $h(14) = 3\sin\frac{14\pi}{6} \approx 2.60$ m.

So, the tide is 2.6 m above the mean.

d

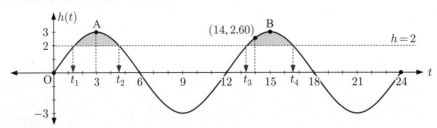

We need to solve $h(t) = 2$, so $3\sin\frac{\pi t}{6} = 2$.

Using a graphics calculator with $Y_1 = 3\sin\frac{\pi X}{6}$ and $Y_2 = 2$

we obtain $t_1 \approx 1.39$, $t_2 \approx 4.61$, $t_3 \approx 13.39$, $t_4 \approx 16.61$

Now 1.39 hours $= 1$ hour 23 minutes, and so on.

So, the ship can cross between 1:23 am and 4:37 am or 1:23 pm and 4:37 pm.

EXERCISE 4F

1   The population of grasshoppers after t weeks where $0 \leqslant t \leqslant 12$ is estimated by
$P(t) = 7500 + 3000\sin\frac{\pi t}{8}$.

   a   Find: **i** the initial estimate   **ii** the estimate after 5 weeks.

   b   What is the greatest population size over this interval and when does it occur?

   c   When is the population: **i** 9000   **ii** 6000?

   d   During what time interval(s) does the population size exceed 10 000?

2   The model for the height of a passenger on a Ferris wheel is $H(t) = 20 - 19\cos\frac{2\pi t}{3}$, where H is the height in metres above the ground, and t is in minutes.

   a   Where is the passenger at time $t = 0$?

   b   At what time is the passenger at the maximum height in the first revolution of the wheel?

   c   How long does the wheel take to complete one revolution?

   d   Sketch the graph of the function $H(t)$ over one revolution.

   e   The passenger can see his friend when he is at least 13 m above the ground. During what times in the first revolution can the passenger see his friend?

3 The population of water buffalo is given by
$P(t) = 400 + 250 \sin \frac{\pi t}{2}$ where t is the number of years since the first estimate was made.

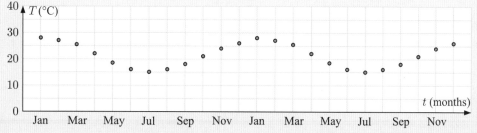

a What was the initial estimate?

b What was the population size after:
  i 6 months
  ii two years?

c Find $P(1)$. What is the significance of this value?

d Find the smallest population size and when it first occurred.

e Find the first time when the herd exceeded 500.

4 Over a 28 day period, the cost per litre of petrol was modelled by
$C(t) = 9.2 \sin\left(\frac{\pi}{7}(t-4)\right) + 107.8$ pence L^{-1}.

a True or false?
  i "The cost per litre oscillates about 107.8 pence with maximum price £1.17 per litre."
  ii "Every 14 days, the cycle repeats itself."

b What was the cost of petrol on day 7, to the nearest tenth of a pence per litre?

c On which days was the petrol priced at £1.10 per litre?

d What was the minimum cost per litre and when did it occur?

Example 14 ◄⟩ **Self Tutor**

The data below shows the mean monthly maximum temperature for Buenos Aires, Argentina over 12 months:

Month t	Jan	Feb	Mar	Apr	May	Jun	Jul	Aug	Sep	Oct	Nov	Dec
Temperature T (°C)	28	27	$25\frac{1}{2}$	22	$18\frac{1}{2}$	16	15	16	18	$21\frac{1}{2}$	24	26

The graph over a two year period is shown below:

Model this data using the general sine function $T = a \sin(b(t-c)) + d$.

The period is 12 months, so $\frac{2\pi}{b} = 12$ and $\therefore$ $b = \frac{\pi}{6}$.

The amplitude $= \frac{\text{max} - \text{min}}{2} \approx \frac{28 - 15}{2} \approx 6.5$, so $a \approx 6.5$.

The principal axis is midway between the maximum and minimum, so $d \approx \frac{28 + 15}{2} \approx 21.5$.

So, the model is $T \approx 6.5 \sin\left(\frac{\pi}{6}(t-c)\right) + 21.5$ for some constant c.

The first point at which the sine function starts a new period is $(10, 21.5)$.

$\therefore$ $c = 10$

The model is therefore $T \approx 6.5 \sin\left(\frac{\pi}{6}(t - 10)\right) + 21.5$.

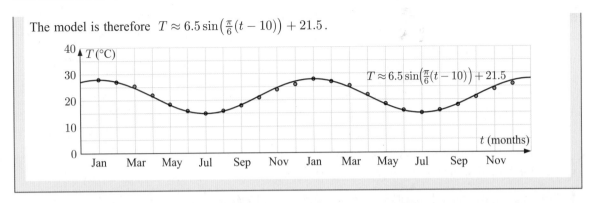

5   Below is a table which shows the mean monthly maximum temperatures for a city in Greece.

Month	Jan	Feb	Mar	Apr	May	Jun	July	Aug	Sept	Oct	Nov	Dec
Temperature (°C)	15	14	15	18	21	25	27	26	24	20	18	16

    a   Find a sine model for this data in the form $T \approx a \sin(b(t - c)) + d$ without using technology. Let Jan $\equiv 1$, Feb $\equiv 2$, and so on.

    b   Use technology to check your answer to **a**. How well does your model fit?

GRAPHICS CALCULATOR INSTRUCTIONS

6   The data in the table shows the mean monthly temperatures for Christchurch, New Zealand.

Month	Jan	Feb	Mar	Apr	May	Jun	July	Aug	Sept	Oct	Nov	Dec
Temperature (°C)	15	16	$14\frac{1}{2}$	12	10	$7\frac{1}{2}$	7	$7\frac{1}{2}$	$8\frac{1}{2}$	$10\frac{1}{2}$	$12\frac{1}{2}$	14

    a   Find a cosine model for this data in the form $T \approx a \cos(b(t - c)) + d$. Assume Jan $\equiv 1$, Feb $\equiv 2$, and so on.

    b   Use technology to check your answer to **a**.

7   Some of the largest tides in the world are observed in Canada's Bay of Fundy. The difference between high and low tides is 14 metres, and the average time difference between high tides is about 12.4 hours. Let H be the height of the tide in metres and t be the time in hours after the first low tide.

    a   Find a sine model for H in terms of t.

    b   Sketch the graph of the model over one period.

8   At the Mawson base in Antarctica, the mean monthly temperatures for the last 30 years are:

Month	Jan	Feb	Mar	Apr	May	Jun	July	Aug	Sept	Oct	Nov	Dec
Temperature (°C)	0	0	-4	-9	-14	-17	-18	-19	-17	-13	-6	-2

    a   Find a sine model for this data without using technology. Use Jan $\equiv 1$, Feb $\equiv 2$, and so on.

    b   How well does the model fit the data?

9 A paint spot X lies on the outer rim of the wheel of a paddle-steamer.

The wheel has radius 3 m.

It rotates anticlockwise at a constant rate, and X is seen entering the water every 4 seconds.

H is the distance of X above the bottom of the boat.

At time $t = 0$, X is at its highest point.

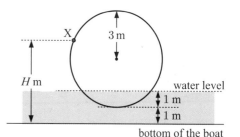

a Find a cosine model for H in the form
$H(t) = a\cos(b(t+c)) + d$.

b At what time t does X first enter the water?

Review set 4A

1 State the minimum and maximum values of:

   a $1 + \sin x$ **b** $-2\cos 3x$

2 State the period of:

   a $y = 4\sin\frac{x}{5}$ **b** $y = -2\cos 4x$ **c** $y = 4\cos\frac{x}{2} + 4$ **d** $y = \frac{1}{2}\tan 3x$

3 Copy and complete:

Function	Period	Amplitude	Range
$y = -3\sin\frac{x}{4} + 1$			
$y = 3\cos\pi x$			

4 **a** Draw the graph of $y = \cos 3x$ for $0 \leqslant x \leqslant 2\pi$.

   b Find the value of y when $x = \frac{3\pi}{4}$. Mark this point on your graph.

5 Sketch the graphs of the following for $-2\pi \leqslant x \leqslant 2\pi$:

   a $y = 4\sin x$ **b** $y = \sin\left(x - \frac{\pi}{3}\right) + 2$ **c** $y = \sin\frac{3x}{2}$

   d $y = \cos\left(x + \frac{\pi}{4}\right)$ **e** $y = \frac{3}{4}\cos x$ **f** $y = \cos 3x$

6 State the transformations which map:

   a $y = \sin x$ onto $y = 3\sin 2x$

   b $y = \cos x$ onto $y = \cos\left(x - \frac{\pi}{3}\right) - 1$

   c $y = \tan x$ onto $y = -\tan 2x$

   d $y = \sin x$ onto $y = 2\sin\left(\frac{x}{2} - \frac{\pi}{4}\right) + \frac{1}{2}$

7 Find the cosine function represented in each of the following graphs:

   a 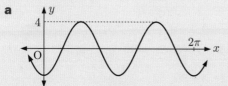   **b**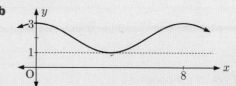

8 Sketch for $0 \leqslant x \leqslant 4\pi$:

   a $y = \tan\frac{x}{4}$ **b** $y = \frac{1}{4}\tan\frac{x}{2}$

9 **a** Describe the sequence of transformations used to transform $y = \tan x$ into $y = \tan 3x + 2$.

   b State the period of $y = \tan 3x + 2$.

   c Sketch $y = \tan 3x + 2$ for $-\pi \leqslant x \leqslant \pi$.

10 Find the remaining five trigonometric ratios from sin, cos, tan, csc, sec, and cot, if:

   a $\cos x = \frac{1}{3}$ and $0 < x < \pi$ **b** $\tan x = \frac{4}{5}$ and $\pi < x < 2\pi$.

11 Solve for $0 \leqslant x \leqslant 2\pi$:

   a $2\sin x = -1$ **b** $\sqrt{2}\cos x - 1 = 0$ **c** $2\cos 2x + 1 = 0$ **d** $\sec x = \sqrt{2}$

12 Solve algebraically for $0 \leqslant x \leqslant 2\pi$, giving answers in terms of π:

   a $\tan^2 2x = 1$ **b** $\sin^2 x - \sin x - 2 = 0$ **c** $4\sin^2 x = 1$

13 Find the exact solutions of:

   a $\sqrt{2}\cos\left(x + \frac{\pi}{4}\right) - 1 = 0, \quad x \in [0, 4\pi]$ **b** $\tan 2x - \sqrt{3} = 0, \quad x \in [0, 2\pi]$

14 An ecologist studying a species of water beetle estimates the population of a colony over an eight week period. If t is the number of weeks after the initial estimate is made, then the population in thousands can be modelled by $P(t) = 5 + 2\sin\frac{\pi t}{3}$ where $0 \leqslant t \leqslant 8$.

   a What was the initial population?

   b What were the smallest and largest populations?

   c During what time interval(s) did the population exceed 6000?

15 A robot on Mars records the temperature every Mars day. A summary series, showing every one hundredth Mars day, is shown in the table below.

Number of Mars days	0	100	200	300	400	500	600	700	800	900	1000	1100	1200	1300
Temp. (°C)	−43	−15	−5	−21	−59	−79	−68	−50	−27	−8	−15	−70	−78	−68

   a Find the maximum and minimum temperatures recorded by the robot.

   b Find a sine model for the temperature T in terms of the number of Mars days n.

   c Use this information to estimate the length of a Mars year.

Review set 4B

1 State the transformation which maps:

   a $y = \cos x$ onto $y = \cos\left(x - \frac{\pi}{3}\right) + 1$ **b** $y = \sin x$ onto $y = \sin 3x$

2 State the period of:

   a $y = 4\sin\frac{x}{3}$ **b** $y = \tan 4x$

3 Find b given that the function $y = \sin bx$, $b > 0$ has period:

   a 6π **b** $\frac{\pi}{12}$ **c** 9

4 State the minimum and maximum values of:

   a $y = 5\sin x - 3$ **b** $y = \frac{1}{3}\cos x + 1$

5 Sketch the graphs of the following for $0 \leqslant x \leqslant 2\pi$:

   a $y = 2\cos 3x$ **b** $y = 2\sin\left(x - \frac{\pi}{3}\right) + 3$ **c** $y = -\cos\left(x + \frac{\pi}{4}\right)$

   d $y = 2\sin x - \frac{1}{2}$ **e** $y = \frac{3}{2}\tan\left(x - \frac{\pi}{6}\right)$ **f** $y = 2\tan\frac{x}{2}$

6 **a** Find the sine function shown in this graph.

   b Write down the equivalent cosine function for this graph.

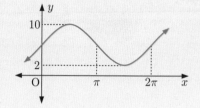

7 Find a and b given the graph of $y = \tan ax + b$ shown.

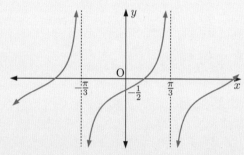

8 **a** Sketch the graphs of $y = \sec x$ and $y = \csc x$ on the same set of axes for $-2\pi \leqslant x \leqslant 2\pi$.

   b State a transformation which maps $y = \sec x$ onto $y = \csc x$ for all $x \in \mathbb{R}$.

9 Find, giving your answer in radians:

   a $\arccos\frac{1}{\sqrt{2}}$ **b** $\arctan\frac{1}{\sqrt{3}}$ **c** $\arcsin\left(-\frac{1}{2}\right)$

10 Solve for $0 \leqslant x \leqslant 2\pi$:

   a $2\sin 3x = -\sqrt{3}$ **b** $\sqrt{3}\tan\frac{x}{2} = -1$ **c** $\cos 2x = \sqrt{3}\sin 2x$

11 Find exact solutions for x given $-\pi \leqslant x \leqslant \pi$:

   a $\tan\left(x + \frac{\pi}{6}\right) = -\sqrt{3}$ **b** $\tan 2x = -\sqrt{3}$ **c** $\tan^2 x - 3 = 0$ **d** $\cot x = \sqrt{3}$

12 Find the x-intercepts of:

   a $y = 2\sin 3x + \sqrt{3}$ for $0 \leqslant x \leqslant 2\pi$ **b** $y = \sqrt{2}\sin\left(x + \frac{\pi}{4}\right)$ for $0 \leqslant x \leqslant 3\pi$

13 Solve exactly:

   a $\arcsin x = \frac{\pi}{3}$ **b** $\arctan(x - 2) = \frac{\pi}{6}$

14 In an industrial city, the amount of pollution in the air becomes greater during the working week when factories are operating, and lessens over the weekend. The number of milligrams of pollutants in a cubic metre of air is given by $P(t) = 40 + 12\sin\left(\frac{2\pi}{7}\left(t - \frac{37}{12}\right)\right)$ where t is the number of days after midnight on Saturday night.

   a What is the minimum level of pollution?

   b At what time during the week does this minimum level occur?

15 The table below gives the mean monthly maximum temperature for Manchester Airport:

Month	Jan	Feb	Mar	Apr	May	Jun	Jul	Aug	Sep	Oct	Nov	Dec
Temperature (°C)	7.7	8.3	10.1	12.4	15.5	18.8	21.5	21.8	19.5	15.4	11.5	8.8

    a A sine function of the form $T \approx a\sin(b(t-c)) + d$ is used to model the data.
       Find good estimates of the constants a, b, c, and d without using technology.
       Use Jan $\equiv 1$, Feb $\equiv 2$, and so on.

    b Check your answer to **a** using technology. How well does your model fit?

5

Trigonometric identities

Contents:

Opening problem

Consider the triangle alongside.

Things to think about:

a Can you find $\tan\theta$ using a combination of Pythagoras' theorem, the cosine rule, and your knowledge of the unit circle?

b Write down the ratios $\tan\phi$ and $\tan(\theta+\phi)$. Is there an easier way to calculate $\tan\theta$ exactly, which involves these ratios?

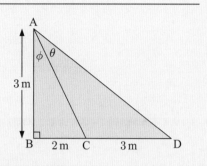

In mathematics, an **identity** is a result which is true for all values of a variable.

For example, we have seen that for any value of x:

- $\cos^2 x + \sin^2 x = 1$
- $\sin(x + \frac{\pi}{2}) = \cos x$
- $\sin(-x) = -\sin x$
- $\cos(x - \frac{\pi}{2}) = \sin x$
- $\cos(\pi - x) = -\cos x$

These equations are all examples of identities.

There are a vast number of trigonometric identities. However, we only need to remember a few because we can obtain the rest by rearrangement or substitution. This requires trigonometric *algebra*.

A   ALGEBRA WITH TRIGONOMETRIC FUNCTIONS

For any given angle θ, $\sin\theta$ and $\cos\theta$ are real numbers. $\tan\theta$ is also real whenever it is defined. The algebra of trigonometry is therefore identical to the algebra of real numbers.

An expression like $2\sin\theta + 3\sin\theta$ compares with $2x + 3x$ when we wish to do simplification, and so $2\sin\theta + 3\sin\theta = 5\sin\theta$.

Example 1
◆) **Self Tutor**

Simplify: **a** $3\cos\theta + 4\cos\theta$ **b** $\tan\alpha - 3\tan\alpha$

a $3\cos\theta + 4\cos\theta = 7\cos\theta$
{compare with $3x + 4x = 7x$}

b $\tan\alpha - 3\tan\alpha = -2\tan\alpha$
{compare with $x - 3x = -2x$}

ANGLE RELATIONSHIPS

The **negative angle formulae** are established by reflection in the x-axis:
$$\sin(-\theta) = -\sin\theta \qquad \cos(-\theta) = \cos\theta$$

The **supplementary angle formulae** are established by reflection in the y-axis:
$$\sin(\pi - \theta) = \sin\theta \qquad \cos(\pi - \theta) = -\cos\theta$$

The **complementary angle formulae** are established by reflection in the line $y = x$:
$$\sin(\frac{\pi}{2} - \theta) = \cos\theta \qquad \cos(\frac{\pi}{2} - \theta) = \sin\theta$$

The tangent definition $\tan \theta = \dfrac{\sin \theta}{\cos \theta}$ enables us to calculate the tangent ratio in each case.

Example 2
◀)) **Self Tutor**

Simplify: **a** $\sin(-\theta) + 2\sin \theta$ **b** $\dfrac{\cos(\frac{\pi}{2} - \theta)}{\cos(\pi - \theta)}$

a $\sin(-\theta) + 2\sin \theta$
$= -\sin \theta + 2\sin \theta$
$= \sin \theta$

b $\dfrac{\cos(\frac{\pi}{2} - \theta)}{\cos(\pi - \theta)} = \dfrac{\sin \theta}{-\cos \theta}$
$= -\tan \theta$

EXERCISE 5A.1

1 Simplify:

   a $\sin \theta + \sin \theta$ **b** $2\cos \theta + \cos \theta$ **c** $3\sin \theta - \sin \theta$

   d $3\sin \theta - 2\sin \theta$ **e** $\tan \theta - 3\tan \theta$ **f** $2\cos^2 \theta - 5\cos^2 \theta$

2 Simplify:

   a $3\tan x - \dfrac{\sin x}{\cos x}$ **b** $\dfrac{\sin^2 x}{\cos^2 x}$ **c** $\tan x \cos x$

   d $\dfrac{\sin x}{\tan x}$ **e** $3\sin x + 2\cos x \tan x$ **f** $\dfrac{2\tan x}{\sin x}$

3 Simplify:

   a $\tan x \cot x$ **b** $\sin x \csc x$ **c** $\csc x \cot x$

   d $\sin x \cot x$ **e** $\dfrac{\cot x}{\csc x}$

4 Simplify:

   a $3\cos \theta - \cos(-\theta)$ **b** $\sin(-\theta) + \cos\left(\frac{\pi}{2} - \theta\right)$ **c** $\tan(\pi - \theta)$

   d $\tan\left(\frac{\pi}{2} - \theta\right)$ **e** $\sin\left(\frac{\pi}{2} - \theta\right) - \cos(\pi - \theta)$ **f** $\dfrac{\sin(-\theta)}{\cos(\pi - \theta)}$

   g $\dfrac{\cos(\frac{\pi}{2} - \theta)}{\cos(-\theta)}$ **h** $\dfrac{\sin(\pi - \theta) - \sin(-\theta)}{\cos(-\theta)}$

THE PYTHAGOREAN IDENTITY

The **Pythagorean identity** is established by applying Pythagoras' theorem on the unit circle:

$$\sin^2 \theta + \cos^2 \theta = 1$$

We commonly use rearrangements of these formulae such as:

$$\sin^2 \theta = 1 - \cos^2 \theta \qquad\qquad \cos^2 \theta = 1 - \sin^2 \theta$$

Using these definitions we can also derive the identities:

$$\tan^2 \theta + 1 = \sec^2 \theta \quad \text{and} \quad 1 + \cot^2 \theta = \csc^2 \theta$$

Proof (for the first case):

Using $\sin^2 \theta + \cos^2 \theta = 1$,

$$\frac{\sin^2 \theta}{\cos^2 \theta} + \frac{\cos^2 \theta}{\cos^2 \theta} = \frac{1}{\cos^2 \theta} \qquad \{\text{dividing each term by } \cos^2 \theta\}$$

$$\therefore \quad \tan^2 \theta + 1 = \sec^2 \theta$$

Example 3
◆) **Self Tutor**

Simplify: **a** $2 - 2\sin^2 \theta$ **b** $\cos^2 \theta \sin \theta + \sin^3 \theta$

a $2 - 2\sin^2 \theta$
$= 2(1 - \sin^2 \theta)$
$= 2\cos^2 \theta$
$\{\text{as } \cos^2 \theta + \sin^2 \theta = 1\}$

b $\cos^2 \theta \sin \theta + \sin^3 \theta$
$= \sin \theta (\cos^2 \theta + \sin^2 \theta)$
$= \sin \theta \times 1$
$= \sin \theta$

Example 4
◆) **Self Tutor**

Expand and simplify: $(\cos \theta - \sin \theta)^2$

$(\cos \theta - \sin \theta)^2$
$= \cos^2 \theta - 2\cos \theta \sin \theta + \sin^2 \theta \qquad \{\text{using } (a-b)^2 = a^2 - 2ab + b^2\}$
$= \cos^2 \theta + \sin^2 \theta - 2\cos \theta \sin \theta$
$= 1 - 2\cos \theta \sin \theta$

EXERCISE 5A.2

1 Simplify:

a $3\sin^2 \theta + 3\cos^2 \theta$

b $-2\sin^2 \theta - 2\cos^2 \theta$

c $-\cos^2 \theta - \sin^2 \theta$

d $3 - 3\sin^2 \theta$

e $4 - 4\cos^2 \theta$

f $\cos^3 \theta + \cos \theta \sin^2 \theta$

g $\cos^2 \theta - 1$

h $\sin^2 \theta - 1$

i $2\cos^2 \theta - 2$

j $\dfrac{1 - \sin^2 \theta}{\cos^2 \theta}$

k $\dfrac{1 - \cos^2 \theta}{\sin \theta}$

l $\dfrac{\cos^2 \theta - 1}{-\sin \theta}$

2 Prove that $1 + \cot^2 \theta = \csc^2 \theta$.

3 Expand and simplify, if possible:

a $(1 + \sin \theta)^2$

b $(\sin \alpha - 2)^2$

c $(\tan \alpha - 1)^2$

d $(\sin \alpha + \cos \alpha)^2$

e $(\sin \beta - \cos \beta)^2$

f $-(2 - \cos \alpha)^2$

4 Simplify:

a $1 - \sec^2 \beta$

b $\dfrac{\tan^2 \theta (\cot^2 \theta + 1)}{\tan^2 \theta + 1}$

c $\cos^2 \alpha (\sec^2 \alpha - 1)$

d $(\sin x + \tan x)(\sin x - \tan x)$

e $(2\sin \theta + 3\cos \theta)^2 + (3\sin \theta - 2\cos \theta)^2$

f $(1 + \csc \theta)(\sin \theta - \sin^2 \theta)$

g $\sec A - \sin A \tan A - \cos A$

FACTORISING TRIGONOMETRIC EXPRESSIONS

Example 5
🔊 **Self Tutor**

Factorise:

a $\cos^2 \alpha - \sin^2 \alpha$

b $\tan^2 \theta - 3\tan\theta + 2$

a $\cos^2 \alpha - \sin^2 \alpha$

$= (\cos\alpha + \sin\alpha)(\cos\alpha - \sin\alpha)$ $\{a^2 - b^2 = (a+b)(a-b)\}$

b $\tan^2 \theta - 3\tan\theta + 2$

$= (\tan\theta - 2)(\tan\theta - 1)$ $\{x^2 - 3x + 2 = (x-2)(x-1)\}$

Example 6
🔊 **Self Tutor**

Simplify:

a $\dfrac{2 - 2\cos^2 \theta}{1 + \cos\theta}$

b $\dfrac{\cos\theta - \sin\theta}{\cos^2 \theta - \sin^2 \theta}$

a $\dfrac{2 - 2\cos^2 \theta}{1 + \cos\theta}$

$= \dfrac{2(1 - \cos^2 \theta)}{1 + \cos\theta}$

$= \dfrac{2\cancel{(1 + \cos\theta)}(1 - \cos\theta)}{\cancel{(1 + \cos\theta)}}$

$= 2(1 - \cos\theta)$

b $\dfrac{\cos\theta - \sin\theta}{\cos^2 \theta - \sin^2 \theta}$

$= \dfrac{\cancel{(\cos\theta - \sin\theta)}}{(\cos\theta + \sin\theta)\cancel{(\cos\theta - \sin\theta)}}$

$= \dfrac{1}{\cos\theta + \sin\theta}$

EXERCISE 5A.3

1 Factorise:

a $1 - \sin^2 \theta$

b $\sin^2 \alpha - \cos^2 \alpha$

c $\tan^2 \alpha - 1$

d $2\sin^2 \beta - \sin\beta$

e $2\cos\phi + 3\cos^2 \phi$

f $3\sin^2 \theta - 6\sin\theta$

g $\tan^2 \theta + 5\tan\theta + 6$

h $2\cos^2 \theta + 7\cos\theta + 3$

i $6\cos^2 \alpha - \cos\alpha - 1$

j $3\tan^2 \alpha - 2\tan\alpha$

k $\sec^2 \beta - \csc^2 \beta$

l $2\cot^2 x - 3\cot x + 1$

m $2\sin^2 x + 7\sin x \cos x + 3\cos^2 x$

2 Simplify:

a $\dfrac{1 - \sin^2 \alpha}{1 - \sin\alpha}$

b $\dfrac{\tan^2 \beta - 1}{\tan\beta + 1}$

c $\dfrac{\cos^2 \phi - \sin^2 \phi}{\cos\phi + \sin\phi}$

d $\dfrac{\cos^2 \phi - \sin^2 \phi}{\cos\phi - \sin\phi}$

e $\dfrac{\sin\alpha + \cos\alpha}{\sin^2 \alpha - \cos^2 \alpha}$

f $\dfrac{3 - 3\sin^2 \theta}{6\cos\theta}$

g $1 - \dfrac{\cos^2 \theta}{1 + \sin\theta}$

h $\dfrac{1 + \cot\theta}{\csc\theta} - \dfrac{\sec\theta}{\tan\theta + \cot\theta}$

i $\dfrac{\tan^2 \theta}{\sec\theta - 1}$

3 Show that:

a $(\cos\theta + \sin\theta)^2 + (\cos\theta - \sin\theta)^2 = 2$

b $(2\sin\theta + 3\cos\theta)^2 + (3\sin\theta - 2\cos\theta)^2 = 13$

c $(1 - \cos\theta)\left(1 + \dfrac{1}{\cos\theta}\right) = \tan\theta\sin\theta$

d $\left(1 + \dfrac{1}{\sin\theta}\right)(\sin\theta - \sin^2\theta) = \cos^2\theta$

e $\sec A - \cos A = \tan A \sin A$

f $\dfrac{\cos\theta}{1 - \sin\theta} = \sec\theta + \tan\theta$

g $\dfrac{\cos\alpha}{1 - \tan\alpha} + \dfrac{\sin\alpha}{1 - \cot\alpha} = \sin\alpha + \cos\alpha$

h $\dfrac{\sin\theta}{1 + \cos\theta} + \dfrac{1 + \cos\theta}{\sin\theta} = 2\csc\theta$

i $\dfrac{\sin\theta}{1 - \cos\theta} - \dfrac{\sin\theta}{1 + \cos\theta} = 2\cot\theta$

j $\dfrac{1}{1 - \sin\theta} + \dfrac{1}{1 + \sin\theta} = 2\sec^2\theta$

Use a graphing package to check these simplifications by graphing each function on the same set of axes.

GRAPHING PACKAGE

4 Solve for $x \in [0, 2\pi]$:

a $2\cos^2 x = \sin x + 1$

b $\sin^2 x = 2 - \cos x$

c $2\cos^2 x = 3\sin x$

d $2\tan^2 x + 3\sec^2 x = 7$

B | DOUBLE ANGLE IDENTITIES

Investigation 1   Double angle identities

What to do:

1 Copy and complete the table below using your calculator. Include extra lines for angles of your choice.

θ	$\sin 2\theta$	$2\sin\theta$	$2\sin\theta\cos\theta$	$\cos 2\theta$	$2\cos\theta$	$\cos^2\theta - \sin^2\theta$
0.631						
57.81°						
−3.697						
⋮						

2 Write down any discoveries from your table of values.

3 In the diagram alongside, the semi-circle has radius 1 unit, and $\widehat{PAB} = \theta$.

$\widehat{APO} = \theta$ {$\triangle AOP$ is isosceles}

$\widehat{PON} = 2\theta$ {exterior angle of a triangle}

a Find in terms of θ, the lengths of:

   i [OM]   **ii** [AM]

  iii [ON]   **iv** [PN]

b Use $\triangle ANP$ and the lengths in **a** to show that:

   i $\cos\theta = \dfrac{\sin 2\theta}{2\sin\theta}$

  ii $\cos\theta = \dfrac{1 + \cos 2\theta}{2\cos\theta}$

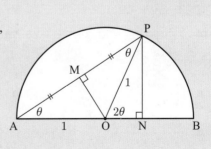

c Hence deduce that:

    i $\sin 2\theta = 2\sin\theta\cos\theta$ **ii** $\cos 2\theta = 2\cos^2\theta - 1$

d For what values of θ have we proven the identities in **c**?

From the **Investigation** you should have deduced two of the **double angle identities**:
$\sin 2\theta = 2\sin\theta\cos\theta$ and $\cos 2\theta = 2\cos^2\theta - 1$.

These formulae are in fact true for all angles θ.

Using $\cos^2\theta = 1 - \sin^2\theta$, we find $\cos 2\theta = \cos^2\theta - \sin^2\theta$
$$\text{and} \quad \cos 2\theta = 1 - 2\sin^2\theta.$$

Using the definition of tangent, we find $\tan 2\theta = \dfrac{\sin 2\theta}{\cos 2\theta} = \dfrac{2\sin\theta\cos\theta}{\cos^2\theta - \sin^2\theta}$

$$= \dfrac{\dfrac{2\sin\theta\cos\theta}{\cos^2\theta}}{\dfrac{\cos^2\theta}{\cos^2\theta} - \dfrac{\sin^2\theta}{\cos^2\theta}}$$

$$= \dfrac{2\tan\theta}{1 - \tan^2\theta}$$

So, the **double angle identities** are:

$$\boldsymbol{\sin 2\theta = 2\sin\theta\cos\theta}$$
$$\boldsymbol{\cos 2\theta = \cos^2\theta - \sin^2\theta}$$
$$\boldsymbol{= 1 - 2\sin^2\theta}$$
$$\boldsymbol{= 2\cos^2\theta - 1}$$
$$\boldsymbol{\tan 2\theta = \dfrac{2\tan\theta}{1 - \tan^2\theta}}$$

GRAPHING PACKAGE

Example 7
🔊 **Self Tutor**

Given that $\sin\alpha = \frac{3}{5}$ and $\cos\alpha = -\frac{4}{5}$ find:

  a $\sin 2\alpha$ **b** $\cos 2\alpha$

  a $\sin 2\alpha$ **b** $\cos 2\alpha$

        $= 2\sin\alpha\cos\alpha$ $= \cos^2\alpha - \sin^2\alpha$

        $= 2\left(\frac{3}{5}\right)\left(-\frac{4}{5}\right)$ $= \left(-\frac{4}{5}\right)^2 - \left(\frac{3}{5}\right)^2$

        $= -\frac{24}{25}$ $= \frac{7}{25}$

EXERCISE 5B

1 For $\theta = 30°$, verify that:

  a $\sin 2\theta = 2\sin\theta\cos\theta$ **b** $\cos 2\theta = \cos^2\theta - \sin^2\theta$ **c** $\tan 2\theta = \dfrac{2\tan\theta}{1 - \tan^2\theta}$

2 If $\sin\theta = \frac{4}{5}$ and $\cos\theta = \frac{3}{5}$ find the exact values of:

  a $\sin 2\theta$ **b** $\cos 2\theta$ **c** $\tan 2\theta$

3 **a** If $\cos A = \frac{1}{3}$, find $\cos 2A$.   **b** If $\sin\phi = -\frac{2}{3}$, find $\cos 2\phi$.

Example 8
◀ᴵⁱ **Self Tutor**

If $\sin \alpha = \frac{5}{13}$ where $\frac{\pi}{2} < \alpha < \pi$, find the exact value of: **a** $\sin 2\alpha$ **b** $\tan 2\alpha$.

α is in quadrant 2, so $\cos \alpha$ is negative.

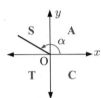

Now $\cos^2 \alpha + \sin^2 \alpha = 1$

$$\therefore \cos^2 \alpha + \frac{25}{169} = 1$$

$$\therefore \cos^2 \alpha = \frac{144}{169}$$

$$\therefore \cos \alpha = \pm \frac{12}{13}$$

$$\therefore \cos \alpha = -\frac{12}{13}$$

a $\sin 2\alpha = 2 \sin \alpha \cos \alpha$

$$\therefore \sin 2\alpha = 2\left(\frac{5}{13}\right)\left(-\frac{12}{13}\right)$$

$$= -\frac{120}{169}$$

b $\tan \alpha = \dfrac{\sin \alpha}{\cos \alpha} = -\dfrac{5}{12}$

$$\therefore \tan 2\alpha = \frac{2 \tan \alpha}{1 - \tan^2 \alpha}$$

$$= \frac{2\left(-\frac{5}{12}\right)}{1 - \left(-\frac{5}{12}\right)^2}$$

$$= \frac{-\frac{5}{6}}{1 - \frac{25}{144}}$$

$$= -\frac{120}{119}$$

4 If $\sin \alpha = -\frac{2}{3}$ where $\pi < \alpha < \frac{3\pi}{2}$, find the exact value of:

    a $\cos \alpha$ **b** $\sin 2\alpha$

5 If $\cos \beta = \frac{2}{5}$ where $270° < \beta < 360°$, find the exact value of:

    a $\sin \beta$ **b** $\sin 2\beta$

Example 9
◀ᴵⁱ **Self Tutor**

If α is acute and $\cos 2\alpha = \frac{3}{4}$ find the exact values of: **a** $\cos \alpha$ **b** $\sin \alpha$.

a $\cos 2\alpha = 2 \cos^2 \alpha - 1$

$$\therefore \frac{3}{4} = 2 \cos^2 \alpha - 1$$

$$\therefore \cos^2 \alpha = \frac{7}{8}$$

$$\therefore \cos \alpha = \pm \frac{\sqrt{7}}{2\sqrt{2}}$$

$$\therefore \cos \alpha = \frac{\sqrt{7}}{2\sqrt{2}}$$

    {as α is acute, $\cos \alpha > 0$}

b $\sin \alpha = \sqrt{1 - \cos^2 \alpha}$

    {as α is acute, $\sin \alpha > 0$}

$$\therefore \sin \alpha = \sqrt{1 - \frac{7}{8}}$$

$$\therefore \sin \alpha = \sqrt{\frac{1}{8}}$$

$$\therefore \sin \alpha = \frac{1}{2\sqrt{2}}$$

6 If α is acute and $\cos 2\alpha = -\frac{7}{9}$, find without a calculator: **a** $\cos \alpha$ **b** $\sin \alpha$.

7 Find the exact value of $\tan A$ if:

    a $\tan 2A = \frac{21}{20}$ and A is obtuse   **b** $\tan 2A = -\frac{12}{5}$ and A is acute.

8 Find the exact value of $\tan \frac{\pi}{8}$.

Example 10 ◀)) **Self Tutor**

Use an appropriate double angle identity to simplify:

a $3\sin\theta\cos\theta$ **b** $4\cos^2 2B - 2$

a $\quad 3\sin\theta\cos\theta$

$\quad = \frac{3}{2}(2\sin\theta\cos\theta)$

$\quad = \frac{3}{2}\sin 2\theta$

b $\quad 4\cos^2 2B - 2$

$\quad = 2(2\cos^2 2B - 1)$

$\quad = 2\cos 2(2B)$

$\quad = 2\cos 4B$

9 Find the exact value of $\left[\cos\frac{\pi}{12} + \sin\frac{\pi}{12}\right]^2$.

10 Use an appropriate double angle identity to simplify:

   a $2\sin\alpha\cos\alpha$ **b** $4\cos\alpha\sin\alpha$ **c** $\sin\alpha\cos\alpha$

   d $2\cos^2\beta - 1$ **e** $1 - 2\cos^2\phi$ **f** $1 - 2\sin^2 N$

   g $2\sin^2 M - 1$ **h** $\cos^2\alpha - \sin^2\alpha$ **i** $\sin^2\alpha - \cos^2\alpha$

   j $2\sin 2A\cos 2A$ **k** $2\cos 3\alpha\sin 3\alpha$ **l** $2\cos^2 4\theta - 1$

   m $1 - 2\cos^2 3\beta$ **n** $1 - 2\sin^2 5\alpha$ **o** $2\sin^2 3D - 1$

   p $\cos^2 2A - \sin^2 2A$ **q** $\cos^2\left(\frac{\alpha}{2}\right) - \sin^2\left(\frac{\alpha}{2}\right)$ **r** $2\sin^2 3P - 2\cos^2 3P$

11 Show that:

   a $(\sin\theta + \cos\theta)^2 = 1 + \sin 2\theta$ **b** $\cos^4\theta - \sin^4\theta = \cos 2\theta$

GRAPHING PACKAGE

12 Solve exactly for x where $0 \leqslant x \leqslant 2\pi$:

   a $\sin 2x + \sin x = 0$ **b** $\sin 2x - 2\cos x = 0$ **c** $\sin 2x + 3\sin x = 0$

13   **a** Use a double angle identity to show that:

     i $\sin^2\theta = \frac{1}{2} - \frac{1}{2}\cos 2\theta$ **ii** $\cos^2\theta = \frac{1}{2} + \frac{1}{2}\cos 2\theta$

   b Hence show that:

     i $\sin^2\left(\frac{\theta}{2}\right) = \frac{1}{2} - \frac{1}{2}\cos\theta$ **ii** $\cos^2\left(\frac{\theta}{2}\right) = \frac{1}{2} + \frac{1}{2}\cos\theta$

14 Find the exact value of $\cos A$ in the diagram:

   a

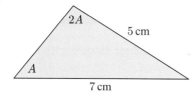

   b

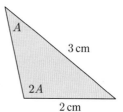

15 Prove that:

   a $\dfrac{\sin 2\theta}{1 - \cos 2\theta} = \cot\theta$ **b** $\dfrac{\sin\theta + \sin 2\theta}{1 + \cos\theta + \cos 2\theta} = \tan\theta$ **c** $\dfrac{\sin 2\theta}{1 + \cos 2\theta} = \tan\theta$

   d $\csc 2\theta = \tan\theta + \cot 2\theta$ **e** $\dfrac{\sin 2\theta}{\sin\theta} - \dfrac{\cos 2\theta}{\cos\theta} = \sec\theta$

16 Solve for $0 \leqslant x \leqslant 2\pi$, giving exact answers:

   a $\cos 2x - \cos x = 0$ **b** $\cos 2x + 3 \cos x = 1$ **c** $\cos 2x + \sin x = 0$

   d $\sin 4x = \sin 2x$ **e** $2 \cos 2x + 9 \sin x = 7$ **f** $\sin x + \cos x = \sqrt{2}$

   g $\sin 2x + \cos x - 2 \sin x - 1 = 0$

17 Use the double angle identity $\cos 2x = 1 - 2 \sin^2 x$ with $x = \frac{\theta}{2}$ to obtain the small-angle
approximation for $\cos \theta$.

Investigation 2   **Parametric equations**

Usually we write functions in the form $y = f(x)$.

For example: $y = 3x + 7$, $y = x^2 - 6x + 8$, $y = \sin x$

However, when we describe relations, it is often useful to express **both** x and y in terms of another
variable t, called the **parameter**. In this case we say we have **parametric equations**.

What to do:

1 **a** Use technology to plot $\{(x, y) : x = \cos t, \ y = \sin t, \ 0 \leqslant t \leqslant 2\pi\}$.
          Use the same scale on both axes.
          Note: Your calculator will need to be set to radians.

   b Describe the resulting graph. Is it the graph of a function?

   c Evaluate $x^2 + y^2$. Hence determine the equation of this graph in
          terms of x and y only.

GRAPHICS
CALCULATOR
INSTRUCTIONS

2 Use technology to plot:

   a $\{(x, y) : x = 2 \cos t, \ y = \sin 2t, \ 0 \leqslant t \leqslant 2\pi\}$

   b $\{(x, y) : x = 2 \cos t, \ y = 2 \sin 3t, \ 0 \leqslant t \leqslant 2\pi\}$

   c $\{(x, y) : x = 2 \cos t, \ y = \cos t - \sin t, \ 0 \leqslant t \leqslant 2\pi\}$

   d $\{(x, y) : x = \cos^2 t + \sin 2t, \ y = \cos t, \ 0 \leqslant t \leqslant 2\pi\}$

   e $\{(x, y) : x = \cos^3 t, \ y = \sin t, \ 0 \leqslant t \leqslant 2\pi\}$

PARAMETRIC
GRAPHING
PACKAGE

C   ANGLE SUM AND DIFFERENCE IDENTITIES

Investigation 3   **Angle sum and difference identities**

What to do:

1 Copy and complete for angles A and B in radians or degrees. Include some angles of your own
choosing.

A	B	$\cos A$	$\cos B$	$\cos(A - B)$	$\cos A - \cos B$	$\cos A \cos B + \sin A \sin B$
47°	24°					
138°	49°					
3^c	2^c					
⋮	⋮					

2 Write down any discoveries from your table of values.

3 Copy and complete for four pairs of angles A and B of your own choosing:

A	B	$\sin A$	$\sin B$	$\sin(A+B)$	$\sin A + \sin B$	$\sin A \cos B + \cos A \sin B$
$\vdots$	$\vdots$					

4 Write down any discoveries from your table of values.

If A and B are **any** two angles then:

$$\cos(A \pm B) = \cos A \cos B \mp \sin A \sin B$$
$$\sin(A \pm B) = \sin A \cos B \pm \cos A \sin B$$
$$\tan(A \pm B) = \frac{\tan A \pm \tan B}{1 \mp \tan A \tan B}$$

These are known as the angle sum and difference identities. There are many ways of establishing them, but many are unsatisfactory as the arguments limit the angles A and B to being acute.

Proof:

Consider $P(\cos A, \sin A)$ and $Q(\cos B, \sin B)$ as any two points on the unit circle, as shown.

Angle POQ is $A - B$.

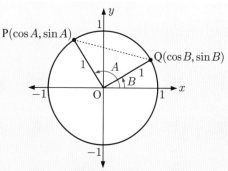

Using the distance formula:

$$PQ = \sqrt{(\cos A - \cos B)^2 + (\sin A - \sin B)^2}$$
$$\therefore \ (PQ)^2 = \cos^2 A - 2\cos A \cos B + \cos^2 B + \sin^2 A - 2\sin A \sin B + \sin^2 B$$
$$= (\cos^2 A + \sin^2 A) + (\cos^2 B + \sin^2 B) - 2(\cos A \cos B + \sin A \sin B)$$
$$= 2 - 2(\cos A \cos B + \sin A \sin B) \quad \text{.... (1)}$$

But, by the *cosine rule* in $\triangle POQ$,

$$(PQ)^2 = 1^2 + 1^2 - 2(1)(1)\cos(A - B)$$
$$= 2 - 2\cos(A - B) \quad \text{.... (2)}$$
$$\therefore \ \cos(A - B) = \cos A \cos B + \sin A \sin B \quad \text{\{comparing (1) and (2)\}}$$

From this formula the other formulae can be established:

- $\cos(A + B) = \cos(A - (-B))$
  $$= \cos A \cos(-B) + \sin A \sin(-B)$$
  $$= \cos A \cos B + \sin A(-\sin B)$$
  $$\qquad \text{\{}\cos(-\theta) = \cos \theta \ \text{ and } \ \sin(-\theta) = -\sin \theta\text{\}}$$
  $$= \cos A \cos B - \sin A \sin B$$

- $\sin(A - B)$
  $$= \cos\left(\tfrac{\pi}{2} - (A - B)\right)$$
  $$= \cos\left(\left(\tfrac{\pi}{2} - A\right) + B\right)$$
  $$= \cos\left(\tfrac{\pi}{2} - A\right)\cos B - \sin\left(\tfrac{\pi}{2} - A\right)\sin B$$
  $$= \sin A \cos B - \cos A \sin B$$

- $\sin(A + B)$
  $$= \sin(A - (-B))$$
  $$= \sin A \cos(-B) - \cos A \sin(-B)$$
  $$= \sin A \cos B - \cos A(-\sin B)$$
  $$= \sin A \cos B + \cos A \sin B$$

- $\tan(A + B)$

$= \dfrac{\sin(A + B)}{\cos(A + B)}$

$= \dfrac{\sin A \cos B + \cos A \sin B}{\cos A \cos B - \sin A \sin B}$

$= \dfrac{\frac{\sin A \cos B}{\cos A \cos B} + \frac{\cos A \sin B}{\cos A \cos B}}{\frac{\cos A \cos B}{\cos A \cos B} - \frac{\sin A \sin B}{\cos A \cos B}}$

$= \dfrac{\tan A + \tan B}{1 - \tan A \tan B}$

- $\tan(A - B)$

$= \tan(A + (-B))$

$= \dfrac{\tan A + \tan(-B)}{1 - \tan A \tan(-B)}$

$= \dfrac{\tan A - \tan B}{1 + \tan A \tan B}$

$\{\tan(-B) = -\tan B\}$

Example 11   ◀) Self Tutor

Expand and simplify $\sin(270° + \alpha)$.

$\sin(270° + \alpha)$

$= \sin 270° \cos \alpha + \cos 270° \sin \alpha \qquad \{\text{angle sum identity}\}$

$= -1 \times \cos \alpha + 0 \times \sin \alpha$

$= -\cos \alpha$

EXERCISE 5C

1 Use your calculator to verify all six of the angle sum and difference identities for $A = 57°$ and $B = 21°$.

2 Expand and simplify:

   a $\sin(90° + \theta)$ **b** $\cos(90° + \theta)$ **c** $\sin(180° - \theta)$

   d $\cos(\pi + \alpha)$ **e** $\sin(2\pi - A)$ **f** $\cos\left(\frac{3\pi}{2} - \theta\right)$

   g $\tan\left(\frac{\pi}{4} + \theta\right)$ **h** $\tan\left(\theta - \frac{3\pi}{4}\right)$ **i** $\tan(\pi + \theta)$

3 Expand, then simplify and write your answer in the form $A \sin \theta + B \cos \theta$:

   a $\sin\left(\theta + \frac{\pi}{3}\right)$ **b** $\cos\left(\frac{2\pi}{3} - \theta\right)$ **c** $\cos\left(\theta + \frac{\pi}{4}\right)$ **d** $\sin\left(\frac{\pi}{6} - \theta\right)$

Example 12   ◀) Self Tutor

Simplify $\cos 3\theta \cos \theta - \sin 3\theta \sin \theta$.

$\cos 3\theta \cos \theta - \sin 3\theta \sin \theta$

$= \cos(3\theta + \theta) \qquad \{\text{angle sum identity in reverse}\}$

$= \cos 4\theta$

4 Simplify using the angle sum and difference identities in reverse:

   a $\cos 2\theta \cos \theta + \sin 2\theta \sin \theta$ **b** $\sin 2A \cos A + \cos 2A \sin A$

   c $\cos A \sin B - \sin A \cos B$ **d** $\sin \alpha \sin \beta + \cos \alpha \cos \beta$

   e $\sin \phi \sin \theta - \cos \phi \cos \theta$ **f** $2 \sin \alpha \cos \beta - 2 \cos \alpha \sin \beta$

   g $\dfrac{\tan 2\theta - \tan \theta}{1 + \tan 2\theta \tan \theta}$ **h** $\dfrac{\tan 2A + \tan A}{1 - \tan 2A \tan A}$

5 Use the angle sum identities to prove the **double angle identities**:

   a $\sin 2\theta = 2 \sin \theta \cos \theta$ **b** $\cos 2\theta = \cos^2 \theta - \sin^2 \theta$ **c** $\tan 2\theta = \dfrac{2 \tan \theta}{1 - \tan^2 \theta}$

Example 13   ◀) **Self Tutor**

Without using your calculator, show that $\sin 75° = \dfrac{\sqrt{6} + \sqrt{2}}{4}$.

$\sin 75° = \sin(45° + 30°)$

$\qquad = \sin 45° \cos 30° + \cos 45° \sin 30°$

$\qquad = \left(\tfrac{1}{\sqrt{2}}\right)\left(\tfrac{\sqrt{3}}{2}\right) + \left(\tfrac{1}{\sqrt{2}}\right)\left(\tfrac{1}{2}\right)$

$\qquad = \left(\dfrac{\sqrt{3} + 1}{2\sqrt{2}}\right) \dfrac{\sqrt{2}}{\sqrt{2}}$

$\qquad = \dfrac{\sqrt{6} + \sqrt{2}}{4}$

---

6 Without using your calculator, show that:

   a $\cos 75° = \dfrac{\sqrt{6} - \sqrt{2}}{4}$ **b** $\sin 105° = \dfrac{\sqrt{6} + \sqrt{2}}{4}$ **c** $\cos \dfrac{13\pi}{12} = \dfrac{-\sqrt{6} - \sqrt{2}}{4}$

7 Find the exact value of:

   a $\tan \dfrac{5\pi}{12}$ **b** $\tan 105°$

8 If $\tan A = \tfrac{2}{3}$ and $\tan B = -\tfrac{1}{5}$, find the exact value of $\tan(A + B)$.

9 If $\tan A = \tfrac{3}{4}$, find $\tan\left(A + \tfrac{\pi}{4}\right)$.

10 If $\sin A = -\tfrac{1}{3}$, $\pi \leqslant A \leqslant \tfrac{3\pi}{2}$ and $\cos B = \tfrac{1}{\sqrt{5}}$, $0 \leqslant B \leqslant \tfrac{\pi}{2}$, find $\tan(A + B)$.

11 Simplify, giving your answer exactly: $\dfrac{\tan 80° - \tan 20°}{1 + \tan 80° \tan 20°}$

12 If $\tan(A + B) = \tfrac{3}{5}$ and $\tan B = \tfrac{2}{3}$, find the exact value of $\tan A$.

13 Find the exact value of $\tan \alpha$:

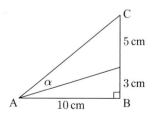

14 Find exactly the tangent of the acute angle between two lines with gradients $\tfrac{1}{2}$ and $\tfrac{2}{3}$.

15 Simplify:

    a $\cos(\alpha + \beta)\cos(\alpha - \beta) - \sin(\alpha + \beta)\sin(\alpha - \beta)$

    b $\sin(\theta - 2\phi)\cos(\theta + \phi) - \cos(\theta - 2\phi)\sin(\theta + \phi)$

    c $\cos\alpha\cos(\beta - \alpha) - \sin\alpha\sin(\beta - \alpha)$

    d $\tan\left(A + \frac{\pi}{4}\right)\tan\left(A - \frac{\pi}{4}\right)$

    e $\dfrac{\tan(A + B) + \tan(A - B)}{1 - \tan(A + B)\tan(A - B)}$

16 If $\sin A = \frac{2}{3}$, $\frac{\pi}{2} \leqslant A \leqslant \pi$ and $\cos B = -\frac{4}{5}$, $\pi \leqslant B \leqslant \frac{3\pi}{2}$, find:

    a $\tan(A + B)$ **b** $\tan 2A$.

17 Find $\tan A$ if $\tan(A - B)\tan(A + B) = 1$.

18 Express $\tan(A + B + C)$ in terms of $\tan A$, $\tan B$, and $\tan C$.
Hence show that if A, B, and C are the angles of a triangle, then
$$\tan A + \tan B + \tan C = \tan A\tan B\tan C.$$

19 Show that:

    a $\sqrt{2}\cos\left(\theta + \frac{\pi}{4}\right) = \cos\theta - \sin\theta$ **b** $2\cos\left(\theta - \frac{\pi}{3}\right) = \cos\theta + \sqrt{3}\sin\theta$

    c $\cos(\alpha + \beta) - \cos(\alpha - \beta) = -2\sin\alpha\sin\beta$ **d** $\cos(\alpha + \beta)\cos(\alpha - \beta) = \cos^2\alpha - \sin^2\beta$.

20 Prove that, in the given figure, $\alpha + \beta = \frac{\pi}{4}$.

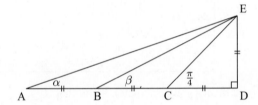

21   **a** Prove that $\cos 3\theta = 4\cos^3\theta - 3\cos\theta$ by replacing 3θ by $(2\theta + \theta)$.

    b Hence solve the equation $8\cos^3\theta - 6\cos\theta + 1 = 0$ for $\theta \in [-\pi, \pi]$.

22   **a** Write $\sin 3\theta$ in the form $a\sin^3\theta + b\sin\theta$ where $a, b \in \mathbb{Z}$.

    b Hence solve the equation $\sin 3\theta = \sin\theta$ for $\theta \in [0, 3\pi]$.

Example 14   ◀)) Self Tutor

Suppose $\sin x - \sqrt{3}\cos x = k\cos(x + b)$ for $k > 0$ and $0 < b < 2\pi$. Find k and b.

$\sin x - \sqrt{3}\cos x = k\cos(x + b)$
$$\qquad\qquad\qquad = k[\cos x\cos b - \sin x\sin b]$$
$$\qquad\qquad\qquad = k\cos x\cos b - k\sin x\sin b$$

Equating coefficients of $\cos x$ and $\sin x$,

$$k\cos b = -\sqrt{3} \quad \text{.... (1)} \qquad \text{and} \qquad -k\sin b = 1 \quad \text{.... (2)}$$
$$\therefore \; k^2\cos^2 b = 3 \qquad\qquad\qquad \text{and} \qquad k^2\sin^2 b = 1 \quad \{\text{squaring both sides}\}$$
$$\therefore \; k^2(\cos^2 b + \sin^2 b) = 4 \quad \{\text{adding the two equations}\}$$
$$\therefore \; k^2 = 4$$
$$\therefore \; k = 2 \quad \{\text{since } k > 0\}$$

Substituting $k = 2$ into (1) gives $\cos b = -\frac{\sqrt{3}}{2}$

and into (2) gives $\sin b = -\frac{1}{2}$

$\therefore \quad b = \frac{7\pi}{6}$

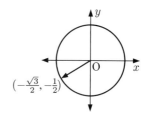

$\left(-\frac{\sqrt{3}}{2}, -\frac{1}{2}\right)$

23 Suppose $\sqrt{3}\sin x + \cos x = k\sin(x + a)$ for $k > 0$ and $0 < a < 2\pi$.   Find k and a.

24 **a** Write $2\cos x + 2\sin x$ in the form $k\cos(x + a)$ for $k > 0$, $0 < a < 2\pi$.

 b Hence solve the equation $2\cos x + 2\sin x = \sqrt{2}$ for $0 \leqslant x \leqslant 2\pi$.

25 **a** Find a sequence of transformations which map the graph of $y = \sin x$ onto the graph of $y = \cos x + 3\sin x$.

 b Find the greatest and least values of $(\cos x + 3\sin x)^2 + 2$.

26 Use the basic definition of periodicity to show algebraically that the period of $f(x) = \sin(nx)$ is $\dfrac{2\pi}{n}$, for all $n > 0$.

27 **a** Write $2\cos x - 5\sin x$ in the form $k\cos(x + b)$ for $k > 0$, $0 < b < 2\pi$.

 b Hence solve the equation $2\cos x - 5\sin x = -2$ for $0 \leqslant x \leqslant \pi$.

 c Given that $t = \tan\frac{x}{2}$, prove that $\sin x = \dfrac{2t}{1 + t^2}$ and $\cos x = \dfrac{1 - t^2}{1 + t^2}$.

 d Hence solve $2\cos x - 5\sin x = -2$ for $0 \leqslant x \leqslant \pi$.

28 **a** Show that $\sin(A + B) + \sin(A - B) = 2\sin A \cos B$.

 b Hence show that $\sin A \cos B = \frac{1}{2}\sin(A + B) + \frac{1}{2}\sin(A - B)$.

 c Hence write the following as sums:

   i $\sin 3\theta \cos \theta$ **ii** $\sin 6\alpha \cos \alpha$ **iii** $2\sin 5\beta \cos \beta$

   iv $4\cos \theta \sin 4\theta$ **v** $6\cos 4\alpha \sin 3\alpha$ **vi** $\frac{1}{3}\cos 5A \sin 3A$

29 **a** Show that $\cos(A + B) + \cos(A - B) = 2\cos A \cos B$.

 b Hence show that $\cos A \cos B = \frac{1}{2}\cos(A + B) + \frac{1}{2}\cos(A - B)$.

 c Hence write the following as a *sum* of cosines:

   i $\cos 4\theta \cos \theta$ **ii** $\cos 7\alpha \cos \alpha$ **iii** $2\cos 3\beta \cos \beta$

   iv $6\cos x \cos 7x$ **v** $3\cos P \cos 4P$ **vi** $\frac{1}{4}\cos 4x \cos 2x$

30 **a** Show that $\cos(A - B) - \cos(A + B) = 2\sin A \sin B$.

 b Hence show that $\sin A \sin B = \frac{1}{2}\cos(A - B) - \frac{1}{2}\cos(A + B)$.

 c Hence write the following as a *difference* of cosines:

   i $\sin 3\theta \sin \theta$ **ii** $\sin 6\alpha \sin \alpha$ **iii** $2\sin 5\beta \sin \beta$

   iv $4\sin \theta \sin 4\theta$ **v** $10\sin 2A \sin 8A$ **vi** $\frac{1}{5}\sin 3M \sin 7M$

31 The **products to sums formulae** are:

$$\sin A \cos B = \tfrac{1}{2}\sin(A+B) + \tfrac{1}{2}\sin(A-B) \quad \text{.... (1)}$$

$$\cos A \cos B = \tfrac{1}{2}\cos(A+B) + \tfrac{1}{2}\cos(A-B) \quad \text{.... (2)}$$

$$\sin A \sin B = \tfrac{1}{2}\cos(A-B) - \tfrac{1}{2}\cos(A+B) \quad \text{.... (3)}$$

a What formulae result if we replace B by A in each of these formulae?

b Suppose $A + B = S$ and $A - B = D$.

  i Show that $A = \dfrac{S+D}{2}$ and $B = \dfrac{S-D}{2}$.

  ii Using the substitutions $A + B = S$ and $A - B = D$, show that equation (1) becomes
$$\sin S + \sin D = 2\sin\left(\tfrac{S+D}{2}\right)\cos\left(\tfrac{S-D}{2}\right) \quad \text{.... (4)}$$

  iii By replacing D by $(-D)$ in (4), show that $\sin S - \sin D = 2\cos\left(\tfrac{S+D}{2}\right)\sin\left(\tfrac{S-D}{2}\right)$.

c What formula results when the substitution $A = \dfrac{S+D}{2}$ and $B = \dfrac{S-D}{2}$ is made into (2)?

d What formula results when the substitution $A = \dfrac{S+D}{2}$ and $B = \dfrac{S-D}{2}$ is made into (3)?

32 The **factor formulae** are:

$$\sin S + \sin D = 2\sin\left(\tfrac{S+D}{2}\right)\cos\left(\tfrac{S-D}{2}\right) \qquad \cos S + \cos D = 2\cos\left(\tfrac{S+D}{2}\right)\cos\left(\tfrac{S-D}{2}\right)$$

$$\sin S - \sin D = 2\cos\left(\tfrac{S+D}{2}\right)\sin\left(\tfrac{S-D}{2}\right) \qquad \cos S - \cos D = -2\sin\left(\tfrac{S+D}{2}\right)\sin\left(\tfrac{S-D}{2}\right)$$

Use these formulae to convert the following to products:

a $\sin 5x + \sin x$ **b** $\cos 8A + \cos 2A$ **c** $\cos 3\alpha - \cos \alpha$

d $\sin 5\theta - \sin 3\theta$ **e** $\cos 7\alpha - \cos \alpha$ **f** $\sin 3\alpha + \sin 7\alpha$

g $\cos 2B - \cos 4B$ **h** $\sin(x+h) - \sin x$ **i** $\cos(x+h) - \cos x$

Example 15   ◀) **Self Tutor**

For small angle x, show that: $\cos\left(x + \tfrac{\pi}{4}\right) \approx \tfrac{1}{\sqrt{2}} - \tfrac{1}{\sqrt{2}}x - \tfrac{1}{2\sqrt{2}}x^2$

$$\cos\left(x + \tfrac{\pi}{4}\right) = \cos x \cos\tfrac{\pi}{4} - \sin x \sin\tfrac{\pi}{4}$$

$$= \tfrac{1}{\sqrt{2}}\cos x - \tfrac{1}{\sqrt{2}}\sin x$$

$$\approx \tfrac{1}{\sqrt{2}}\left(1 - \tfrac{x^2}{2}\right) - \tfrac{1}{\sqrt{2}}x \qquad \text{\{small angle approximations\}}$$

$$\approx \tfrac{1}{\sqrt{2}} - \tfrac{1}{\sqrt{2}}x - \tfrac{1}{2\sqrt{2}}x^2$$

---

33 For small angle x, show that:

a $\sin\left(x - \tfrac{\pi}{6}\right) \approx -\tfrac{1}{2} + \tfrac{\sqrt{3}}{2}x + \tfrac{1}{4}x^2$ **b** $\cos^2\left(x + \tfrac{\pi}{3}\right) \approx \tfrac{1}{4} - \tfrac{\sqrt{3}}{2}x + \tfrac{1}{2}x^2$

c $\tan^2\left(\tfrac{\pi}{3} + x\right) \approx 3 + 8\sqrt{3}x + 40x^2$

Historical note

In the late sixteenth century, there were no calculators. Instead, values of trigonometric functions for different angles were calculated by hand and recorded in tables. The values could then readily be used for trigonometric applications, and also in other surprising ways. For example, the angle sum and difference identities could be used to quickly multiply two numbers together.

Using the angle sum and difference identities, we can show that

$$\cos A \times \cos B = \frac{\cos(A+B) + \cos(A-B)}{2} \quad \dots (*)$$

Now, suppose you need to find 0.1362×0.4573. By consulting trigonometric tables, it could be found that $\cos 82.172° \approx 0.1362$ and $\cos 62.787° \approx 0.4573$.

So, 0.1362×0.4573

$\approx \cos 82.172° \times \cos 62.787°$

$\approx \dfrac{\cos(82.172° + 62.787°) + \cos(82.172° - 62.787°)}{2}$ {using $(*)$}

$\approx \dfrac{\cos(144.959°) + \cos(19.385°)}{2}$

$\approx \dfrac{-0.8187 + 0.9433}{2}$ {using trigonometric tables}

≈ 0.0623

This method may seem complicated, but in the late sixteenth century it was much faster than performing long multiplication!

Investigation 4

A **trigonometric series** is the sum of a sequence of trigonometric expressions which follow a rule.

In this Investigation we will explore patterns formed by a trigonometric series.

What to do:

1 Consider the function $f(x) = \sin x + \dfrac{\sin 3x}{3}$.

   a Show that $f(x) = \frac{2}{3} \sin x (2\cos^2 x + 1)$.

   Hint: Write $\sin 3x$ as $\sin(2x + x)$.

   b Hence find the x-intercepts of $y = f(x)$ on $-4\pi \leqslant x \leqslant 4\pi$.

   c Use the graphing package to sketch $y = f(x)$ on $-4\pi \leqslant x \leqslant 4\pi$. Discuss the similarities and differences between this graph and the graph of $y = \sin x$.

GRAPHING PACKAGE

2   **a** Write the function $f(x) = \sin x + \dfrac{\sin 3x}{3} + \dfrac{\sin 5x}{5}$ in terms of $\sin x$ and $\cos x$.

   b Hence find the x-intercepts of this function on $-4\pi \leqslant x \leqslant 4\pi$.

   c Use the graphing package to sketch the graph of the function on $-4\pi \leqslant x \leqslant 4\pi$. Compare your graph with the graphs in **1**.

3 Use the graphing package to graph on $-4\pi \leqslant x \leqslant 4\pi$:

   a $f(x) = \sin x + \dfrac{\sin 3x}{3} + \dfrac{\sin 5x}{5} + \dfrac{\sin 7x}{7}$

   b $f(x) = \sin x + \dfrac{\sin 3x}{3} + \dfrac{\sin 5x}{5} + \dfrac{\sin 7x}{7} + \dfrac{\sin 9x}{9} + \dfrac{\sin 11x}{11}$

4 Predict the graph of $\;f(x) = \sin x + \dfrac{\sin 3x}{3} + \dfrac{\sin 5x}{5} + \dfrac{\sin 7x}{7} + ... + \dfrac{\sin 1001x}{1001}$

Review set 5A

1 Simplify:

   a $3\cos(-\theta) - 2\cos\theta$ **b** $\cos\left(\dfrac{3\pi}{2} - \theta\right)$ **c** $\sin\left(\theta + \dfrac{\pi}{2}\right)$

   d $\dfrac{1 - \cos^2\theta}{1 + \cos\theta}$ **e** $\dfrac{\sin\alpha - \cos\alpha}{\sin^2\alpha - \cos^2\alpha}$ **f** $\dfrac{4\sin^2\alpha - 4}{8\cos\alpha}$

2 If $\sin\alpha = -\frac{3}{4}$, $\pi \leqslant \alpha \leqslant \frac{3\pi}{2}$, find the exact value of:

   a $\cos\alpha$ **b** $\sin 2\alpha$ **c** $\cos 2\alpha$

   d $\tan 2\alpha$ **e** $\cos\frac{\alpha}{2}$ **f** $\sin\frac{\alpha}{2}$

3 Show that:

   a $\sqrt{2}\sin\left(\theta - \frac{\pi}{4}\right) = \sin\theta - \cos\theta$ **b** $\sin\theta\cos(\phi - \theta) + \cos\theta\sin(\phi - \theta) = \sin\phi$

   c $\dfrac{\sin 2\alpha - \sin\alpha}{\cos 2\alpha - \cos\alpha + 1} = \tan\alpha$

4 Prove that $\;\csc 2x + \cot 2x = \cot x$.

5 Find the exact value of: **a** $\cos 165°$ **b** $\tan\frac{\pi}{12}$

6 From ground level, a shooter is aiming at targets on a vertical brick wall. At the current angle of elevation of his rifle, he will hit a target 20 m above ground level. If he doubles the angle of elevation of the rifle, he will hit a target 45 m above ground level. How far is the shooter from the wall?

7 Solve for $0 \leqslant x \leqslant 2\pi$: $\sec^2 x = \tan x + 1$

8

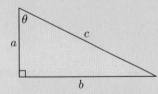

For the diagram alongside, prove that:

   a $\sin 2\theta = \dfrac{2ab}{c^2}$ **b** $\cos 2\theta = \dfrac{a^2 - b^2}{c^2}$

9 Consider triangle ABC shown.

   a Show that $\cos\alpha = \frac{5}{6}$.

   b Show that x is a solution of $3x^2 - 25x + 48 = 0$.

   c Find x by solving the equation in **b**.

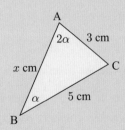

10 If α and β are the other angles of a right angled triangle, show that $\sin 2\alpha = \sin 2\beta$.

11 Write $3\sin x - 5\cos x$ in the form $k\cos(x+a)$ where $k > 0$ and $0 < a < 2\pi$.

12 The graph of $y = 2\sin x + \sqrt{3}\cos x$ is shown alongside.

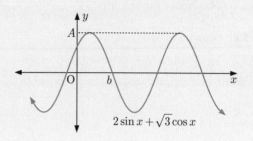

a Write $2\sin x + \sqrt{3}\cos x$ in the form $k\sin(x+a)$ where $k > 0$ and $0 < a < 2\pi$.

b Hence find:

  i the exact value of A

  ii b correct to 2 decimal places.

Review set 5B

1 Simplify:

a $\csc x \tan x$
  b $\dfrac{\tan x}{\sec x}$
  c $\sec x - \tan x \sin x$

2 Simplify:

a $\cos^3 \theta + \sin^2 \theta \cos \theta$
  b $\dfrac{\cos^2 \theta - 1}{\sin \theta}$
  c $5 - 5\sin^2 \theta$

d $\dfrac{\sin^2 \theta - 1}{\cos \theta}$
  e $\dfrac{\tan \theta + \cot \theta}{\sec \theta}$
  f $\cos^2 \theta (\tan \theta + 1)^2 - 1$

3 If $\sin A = \frac{5}{13}$ and $\cos A = \frac{12}{13}$, find:

a $\sin 2A$
  b $\cos 2A$
  c $\tan 2A$

4 Show that:

a $\dfrac{\cos \theta}{1 + \sin \theta} + \dfrac{1 + \sin \theta}{\cos \theta} = 2\sec \theta$
  b $\left(1 + \dfrac{1}{\cos \theta}\right)(\cos \theta - \cos^2 \theta) = \sin^2 \theta$

5 Show that $\sin \frac{\pi}{8} = \frac{1}{2}\sqrt{2 - \sqrt{2}}$ using a suitable double angle identity.

6 Solve for x: $\sqrt{3}\cos x + \sin 2x = 0$ for $-\pi \leqslant x \leqslant \pi$.

7 Given $\sin \theta = \frac{3}{4}$ and $\frac{\pi}{2} < \theta < \pi$, find $\sin\left(\theta + \frac{\pi}{6}\right)$.

8 For the figure in the **Opening Problem**, find $\tan \theta$ using the ratios $\tan \phi$ and $\tan(\theta + \phi)$.

9 Write $3\sin x + 4\cos x$ in the form $k\sin(x+a)$, where $k > 0$ and $0 < a < 2\pi$.

10 Find exactly the length of [BC].

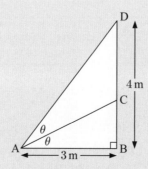

11 **a** Show that $\dfrac{1}{1+\sqrt{2}\sin x} + \dfrac{1}{1-\sqrt{2}\sin x} = 2\sec 2x$.

    b Hence explain why $\dfrac{1}{1+\sqrt{2}\sin x} + \dfrac{1}{1-\sqrt{2}\sin x} = 1$ has no solutions.

12 For small angle x, show that:

    a $\sin\left(x+\frac{\pi}{3}\right) \approx \frac{\sqrt{3}}{2} + \frac{1}{2}x - \frac{\sqrt{3}}{4}x^2$
        b $\tan\left(x-\frac{\pi}{6}\right) \approx -\frac{1}{\sqrt{3}} + \frac{4}{3}x - \frac{4\sqrt{3}}{9}x^2$

6

Reasoning and proof

Contents:

Opening problem

A Sudoku is a puzzle played on a 9×9 grid. The goal is to fill the grid with digits so that each column, each row, and each of the nine 3×3 subgrids contains each of the digits from 1 to 9.

The person posing the puzzle will provide a starting grid with some numbers. A puzzle is *well-posed* if it has a unique solution.

Things to think about:

Suppose the numbers shown are *not* part of the starting grid. Can these numbers be part of a solution to a well-posed Sudoku? Explain your answer.

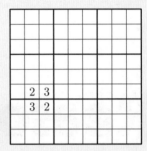

You should have seen previously how mathematical arguments are built out of mathematical statements linked by **logical connectives**:

Connective	Symbol	Formal name
and	$\wedge$	conjunction
or	$\vee$	disjunction
not	$\neg$	negation
if .... then	$\Rightarrow$	implication
if and only if	$\Leftrightarrow$	equivalence

For example:

- The implication $A \Rightarrow B$ means "if A then B", or more specifically "if A is true then B is true".
  A and B are themselves statements or variables. The statement A is known as the *hypothesis*. The statement B is known as the *conclusion*.
  A chain of implications is used to form a **proof by deduction**.

- The equivalence $A \Leftrightarrow B$ means "A if and only if B". This means that both $A \Rightarrow B$ and that $B \Rightarrow A$.
  To prove an equivalence statement, we need to make sure the proof works in both directions. Sometimes we need to prove $A \Rightarrow B$ and $B \Rightarrow A$ separately.
  A chain of equivalence statements can be used to form a **proof by equivalence**.

Both deduction and equivalence are said to be **direct proofs** because we start with our assumptions or **hypotheses** and move directly along the chain to the **conclusion**.

In this Chapter we consider **proof by contradiction**. This comes in a variety of forms which are all related. They are said to be **indirect proofs** because they do *not* start with our assumptions or hypotheses, but rather the *opposite* of the *conclusion*.

A   PROOF BY CONTRAPOSITIVE

The **contrapositive** of $A \Rightarrow B$ is the statement $\neg B \Rightarrow \neg A$.

For example, the contrapositive of "if x is rational then $2^x \neq 3$" is "if $2^x = 3$ then x is not rational".

The implication $A \Rightarrow B$ and its contrapositive are said to be *logically equivalent* because they tell us exactly the same information.

We can explain this using a **truth table** which considers every possibility for whether A and B are true.

As an example, suppose A is "Sam is in the library" and B is "Sam is reading".

We will mark $A \Rightarrow B$ and $\neg B \Rightarrow \neg A$ as true if they are consistent with the truth values of A and B, and false if they are not.

A	B	$A \Rightarrow B$	$\neg B \Rightarrow \neg A$
T	T	Sam is in the library and he is reading. This is consistent with $A \Rightarrow B$. T	Since Sam is reading, $\neg B$ is false. The hypothesis is false, so the implication cannot be inconsistent. T
T	F	Sam is in the library but he is *not* reading. This is inconsistent with $A \Rightarrow B$. F	Sam is not reading but he *is* in the library. This is inconsistent with $\neg B \Rightarrow \neg A$. F
F	T	Sam is not in the library. The hypothesis is false, so the implication cannot be inconsistent. T	Since Sam is reading, $\neg B$ is false. The hypothesis is false, so the implication cannot be inconsistent. T
F	F	Sam is not in the library. The hypothesis is false, so the implication cannot be inconsistent. T	Sam is not reading and he is not in the library. This is consistent with $\neg B \Rightarrow \neg A$. T

An implication only says something if the hypothesis is true. If the hypothesis is false, it cannot be inconsistent.

The implication $A \Rightarrow B$ and its contrapositive $\neg B \Rightarrow \neg A$ are logically equivalent because they always have the same truth value.

You can learn more about truth tables and mathematical logic by clicking on the icon.

MATHEMATICAL LOGIC

The important conclusion for us here is that because the implication and its contrapositive are logically equivalent, proving one is equivalent to proving the other.

If the implication $A \Rightarrow B$ is proven by proving its contrapositive $\neg B \Rightarrow \neg A$, we call this a **proof by contrapositive**.

Proof by contrapositive is worth considering because proving a contrapositive may be significantly easier than proving the original implication directly.

Example 1
◀) **Self Tutor**

Let n be an integer. Prove that if $n^2 - 4n + 5$ is even, then n is odd.

Proof by contrapositive:

The contrapositive is: If n is even then $n^2 - 4n + 5$ is odd.

If n is even then there exists an integer k for which $n = 2k$.

Hence
$$n^2 - 4n + 5 = (2k)^2 - 4(2k) + 5$$
$$= 4k^2 - 8k + 5$$
$$= 2(2k^2 - 4k + 2) + 1 \quad \text{which is odd.}$$

This proves the contrapositive and hence the original statement.

A proof by contrapositive still involves a proof, and so you can use all the normal proof techniques. In the following Example we use proof by exhaustion to prove the contrapositive statement.

Example 2
◀) **Self Tutor**

Let n be a positive integer. Prove that if $n \equiv 2 \bmod 3$ then n is not a perfect square.

$n \equiv 2 \bmod 3$ means that n has remainder 2 when divided by 3.

Proof by contrapositive:

The contrapositive is: If n is a perfect square then $n \equiv 0$ or $1 \bmod 3$.
{if the remainder when divided by 3 is not 2, then it must be 0 or 1}

Suppose n is a perfect square, so $n = k^2$ for some integer k.
There are three cases to consider:

If $k \equiv 0 \bmod 3$ then $k = 3q$ for some integer q.

$\quad \therefore \quad n = k^2 = (3q)^2 = 9q^2 = 3(3q^2)$

$\quad \therefore \quad n \equiv 0 \bmod 3$

If $k \equiv 1 \bmod 3$ then $k = 3q + 1$ for some integer q.

$\quad \therefore \quad n = k^2 = (3q + 1)^2$

$\qquad\qquad = 9q^2 + 6q + 1$

$\qquad\qquad = 3(3q^2 + 2q) + 1$

$\therefore \quad n \equiv 1 \bmod 3$

If $k \equiv 2 \bmod 3$ then $k = 3q + 2$ for some integer q.

$$\therefore \quad n = k^2 = (3q + 2)^2$$
$$= 9q^2 + 12q + 4$$
$$= 3(3q^2 + 4q + 1) + 1$$

$\therefore \quad n \equiv 1 \bmod 3$

$\therefore$ if n is a perfect square then $n \equiv 0$ or $1 \bmod 3$.

This proves the original statement: If $n \equiv 2 \bmod 3$ then n is not a perfect square.

EXERCISE 6A

1 Write the contrapositive statement of each of the following:

   a If it is raining then you get wet.
   b If penguins can fly then it is Thursday.
   c If you drink alcohol then you must be aged 18 or over.
   d If k is divisible by 5 then k is not prime.
   e If $a^2 + b^2 = c^2$ then a or b is even.

2 Let m and n be integers. Prove, by constructing the contrapositive, that:

   a if n^2 is even then n is even
   b if mn is even than at least one of n or m is even
   c if $m + n$ is even then m and n have the same *parity* (are either both even or both odd).

3 Let a and b be real numbers. Prove that if the product ab is irrational then either a or b must be irrational.

4 Suppose n is a positive integer. Prove that if $n \equiv 2$ or $3 \bmod 4$ then n is not a perfect square.

B PROOF BY CONTRADICTION: *REDUCTIO AD ABSURDUM*

When people talk about proof by contradiction, they are normally referring to *reductio ad absurdum* which is Latin for "reduction to absurdity". In this form of argument we show that attempting to disprove a statement inevitably leads to a ridiculous, absurd, or impractical conclusion. For example, if we end up with $1 = 0$, or that an odd number is also an even number, we regard this as absurd. If the result is absurd or impossible, and all the reasoning is correct, then the only conclusion is that the hypothesis is false.

Historical note

Godfrey Harold Hardy (1877 - 1947) was an English mathematician famous for his contributions to number theory and mathematical analysis.

A prominent lecturer of Trinity College, Cambridge, Hardy collaborated with **John Littlewood** and the Indian mathematician **Srinivasa Ramanujan**.

He wrote of proof by contradiction:

It is a far finer gambit than any chess gambit: a chess player may offer the sacrifice of a pawn or even a piece, but a mathematician offers the game.

Godfrey Harold Hardy

To use *reductio ad absurdum*, we:
- start with the hypothesis
- also assume the *negation* (opposite) of the conclusion
- correctly prove that this leads to a contradiction
- hence deduce that the assumption was false, so the conclusion must be true.

Reductio ad absurdum is only valid if *correct reasoning* leads to a contradiction!

For a hypothesis A and conclusion B, we show that

$$(A \wedge \neg B) \Rightarrow (C \wedge \neg C)$$

where C is some statement which is contradicted by $\neg C$.

Since $C \wedge \neg C$ is always false, $A \wedge \neg B$ is false.

So, if A is true, then B must also be true. This is the implication $A \Rightarrow B$.

Example 3   ◀ᵰ **Self Tutor**

Prove that $\log_2 3$ is irrational.

Proof by contradiction:

Suppose $\log_2 3$ is rational

$\therefore$ $\log_2 3 = \dfrac{p}{q}$ where $p, q \in \mathbb{Z}$, $q \neq 0$

$\therefore$ $3 = 2^{\frac{p}{q}}$

$\therefore$ $3^q = 2^p$

The left hand side is always odd and the right hand side is always even. This is a contradiction.

$\therefore$ our original supposition is false, and $\log_2 3$ is irrational.

Example 4   ◀ᵰ **Self Tutor**

The number $\sqrt{2}$ is the length of a diagonal of a square with side length 1. Prove that $\sqrt{2}$ is irrational.

Proof by contradiction:

Suppose $\sqrt{2}$ is rational, so $\sqrt{2} = \dfrac{p}{q}$ for some (positive) integers p and q, $q \neq 0$.

We assume this fraction has been written in lowest terms, so p and q have no common factors.

Squaring both sides, $2 = \dfrac{p^2}{q^2}$

$\therefore$ $p^2 = 2q^2$ .... (1) $\therefore$ p^2 is even, and so p must be even.

Thus $p = 2k$ for some $k \in \mathbb{Z}^+$.

Substituting into (1), $4k^2 = 2q^2$

$\therefore$ $q^2 = 2k^2$ $\therefore$ q^2 is even, and so q must be even.

Here we have a contradiction, as p and q have no common factors.

Thus our original supposition is false, and $\sqrt{2}$ is irrational.

Example 5

🔊 **Self Tutor**

Prove that there are infinitely many prime numbers.

Proof by contradiction:

Suppose there are only a finite number n of prime numbers, which we label p_1, p_2, ..., p_n.

Every positive integer greater than 1 is either a member of this list, or is divisible by a member of this list.

Let $N = p_1 \times p_2 \times .... \times p_n + 1$.

Notice that:

- $N > p_k$ for all $k = 1$, ..., n
  $\therefore$ N is not a member of the list.
- If we divide N by any p_k, $k = 1$, ..., n, then the remainder is 1.
  $\therefore$ N is not divisible by any p_k.

This is a contradiction, so our supposition is false, and there are infinitely many prime numbers.

Notice that this proof does *not* prove that $N = p_1 \times p_2 \times .... \times p_n + 1$ is actually a prime number. It only proves that N is not in the list, and not divisible by members of the list. In fact, N is not always prime. For example, consider $N = 2 \times 3 \times 5 \times 7 \times 11 \times 13 + 1 = 30\,031 = 59 \times 509$.

EXERCISE 6B

1 Prove that $\log_2 5$ is irrational.

2 Prove that $\log_3 5$ is irrational.

3 Prove that $\sqrt{3}$ is irrational.

4 Attempt to prove that $\sqrt{4}$ is irrational using the method in **Example 4**. Explain why the argument fails.

5 Prove that $\sqrt[3]{2}$ is irrational.

6 Suppose a, b, $c \in \mathbb{Z}$. Prove that if $a^2 + b^2 = c^2$ then a or b is even.

7 Answer the **Opening Problem** on page **134**.

8 In the game Noughts and Crosses, two players O (noughts) and X (crosses) take turns marking their symbol in the spaces in a 3×3 grid. The first player who places three of their marks in a horizontal, vertical, or diagonal row wins the game.

Can the second player devise a winning strategy S which guarantees they will win?

> The area of mathematics which examines the mathematics of games is called **game theory**.

9 **Dirichlet's theorem on arithmetic progressions** states that for any two positive integers a and d which have no common factors, there are infinitely many prime numbers of the form $a + nd$ where $n \in \mathbb{Z}^+$. Prove Dirichlet's theorem on arithmetic progressions for the case where $a = 3$ and $d = 4$.

C   FERMAT'S METHOD OF INFINITE DESCENT

The French lawyer and councillor **Pierre de Fermat** (1601 - 1665) is most famous for his contributions as an amateur mathematician. He developed a technique called *infinite descent* which he used primarily to prove negative assertions. For example, he used it to prove that:

- the equation $x^2 + y^4 = z^4$ has no integer solutions (other than the trivial $x = y = z = 0$)
- $\sqrt{2}$ is irrational.

To understand the method of infinite descent, consider a statement that "P_n is true for all $n \in \mathbb{Z}^+$". We:

- Start with the assumption that P_n is *false* for at least one value of n. There must therefore exist a *smallest* value of n for which P_n is false.
- Seek to prove that if a value of n exists for which P_n is false, there must also be a *smaller* value n for which it is false. If we can prove this, then there cannot be a *smallest* value of n. This would present a contradiction, implying that P_n can never be false at all.

Remember $n \in \mathbb{Z}^+$.

Example 6
◀⑴ **Self Tutor**

Prove using infinite descent that $\sqrt{2}$ is irrational.

Proof by infinite descent:

Assume $\sqrt{2}$ is rational, so $\sqrt{2} = \dfrac{p_1}{q_1}$ for some $p_1, q_1 \in \mathbb{Z}^+$.

We know that $1 < \sqrt{2} < 2$, so $p_1 > q_1$ and $2q_1 > p_1$ .... $(*)$

Now
$$\sqrt{2} = \sqrt{2} \times \frac{\sqrt{2} - 1}{\sqrt{2} - 1}$$

$$= \frac{2 - \sqrt{2}}{\sqrt{2} - 1}$$

$$= \frac{2 - \frac{p_1}{q_1}}{\frac{p_1}{q_1} - 1}$$

$$= \frac{2q_1 - p_1}{p_1 - q_1}$$

Letting $p_2 = 2q_1 - p_1$ and $q_2 = p_1 - q_1$,

$\sqrt{2} = \dfrac{p_2}{q_2}$ where $p_2, q_2 \in \mathbb{Z}^+$, $p_2 > q_2$ ⟨since $\sqrt{2} > 1$⟩

and $p_2 < p_1$ and $q_2 < q_1$ ⟨using $(*)$⟩

This process can be repeated an infinite number of times, so there is an infinite descent through the positive integers p, q satisfying $\sqrt{2} = \dfrac{p}{q}$.

$\therefore$ no *smallest* p, q exist satisfying $\sqrt{2} = \dfrac{p}{q}$

$\therefore$ $\sqrt{2}$ cannot be rational.

EXERCISE 6C

1 Use the method of infinite descent to prove that $\sqrt{3}$ is irrational.

2 Starting from a vertex of an acute triangle, the perpendicular is drawn to meet the opposite side at A_1. From A_1, a perpendicular is drawn to meet another side at A_2. This process is repeated so the sides are each visited in turn.

Prove that points A_1, A_2, A_3, ... are all distinct.

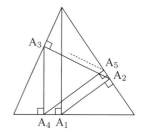

Activity

As a class, find all positive primes p for which there exist positive integers x, y, and n such that $p^n = x^3 + y^3$.

You need to *prove* you have found them all!

Discussion

A proof is a useful way to present a solution. However, sometimes it is not clear from the proof how someone came up with the idea. The mathematician **Carl Gauss** (1777 - 1855) supposedly said "the architect does not leave scaffolding behind".

For example, consider this proof that an irrational number raised to an irrational power can be rational.

Proof:

Let $I = \sqrt{2}^{\sqrt{2}}$. I is either rational or irrational.

If I is rational then since $\sqrt{2}$ is irrational, I is itself an example of an irrational number to an irrational power being rational.

If I is irrational, consider $I^{\sqrt{2}} = \left(\sqrt{2}^{\sqrt{2}}\right)^{\sqrt{2}} = \sqrt{2}^{\sqrt{2} \times \sqrt{2}} = \sqrt{2}^2 = 2$.

In this case $I^{\sqrt{2}}$ is an example of an irrational number to an irrational power being rational.

By exhaustion, there must be an irrational number which raised to an irrational power, is rational.

In this proof, we do not actually establish whether I is rational. We do not need to.

List the different forms of proof you have learned. Discuss what features of a problem point you towards using particular forms of proof.

Review set 6A

1 Let n be an integer. Prove, by constructing the contrapositive, that if n^3 is even then n is even.

2 Prove that $\sqrt[3]{4}$ is irrational.

3 **a** Let n be an integer. Prove by contrapositive that if n^2 is divisible by 6 then n is divisible by 6.

    b Prove by contradiction that $\sqrt{6}$ is irrational.

    c Prove by contradiction that $\sqrt{2} + \sqrt{3}$ is irrational.

4 The following problem is called the Wason THOG task, named after **Peter Wason**. In front of you are four designs: a black diamond, a white diamond, a black square, and a white square.

Meanwhile, I have written down one of the colours (black or white) and one of the shapes (diamond or square). Read the following rule carefully:

> If, and only if, any of the designs includes either the colour I have written down, or the shape I have written down, but not both, then it is called a THOG.

I will tell you that the black diamond is a THOG. Classify each of the following designs into one of the categories: definitely a THOG, insufficient information to decide, or definitely not a THOG. Justify your answers.

    a ◆   **b** ◇   **c** ■   **d** ☐

Review set 6B

1 Let a, b, and n be integers. Prove that if ab is not divisible by n then a is not divisible by n and b is not divisible by n.

2 Prove that there are infinitely many prime numbers.

3 Prove that $\log_3 2$ is irrational.

4 Prove by contradiction that there exist no integers n and m for which:

    a $14n + 7m = 1$ **b** $4n^2 - m^2 = 1$

5 Use the method of infinite descent to prove that $\sqrt{5}$ is irrational.

7

Differential calculus

Contents:

Opening problem

Consider the functions $f(x) = 6x^2$ and $g(x) = 2x$.

Things to think about:

a What are the *derivative* functions $f'(x)$ and $g'(x)$?

b Find:

  i $f(x) + g(x)$ **ii** $f(x) - g(x)$ **iii** $f(x) \times g(x)$ **iv** $\dfrac{f(x)}{g(x)}$

c Find the derivatives of the functions you found in **b**.

d Which of the following statements do you think might be true?

  i The derivative of $f(x) + g(x)$ is $f'(x) + g'(x)$.

  ii The derivative of $f(x) - g(x)$ is $f'(x) - g'(x)$.

  iii The derivative of $f(x) \times g(x)$ is $f'(x) \times g'(x)$.

  iv The derivative of $\dfrac{f(x)}{g(x)}$ is $\dfrac{f'(x)}{g'(x)}$.

Differential calculus is a branch of mathematics that deals with **rates of change**.

You should have seen previously that the **instantaneous rate of change** of a function at a particular point is the gradient of the tangent to the function at that point.

The gradient of the tangent to $f(x)$ at $x = a$ is denoted $f'(a)$.

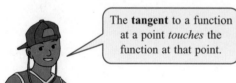

The **tangent** to a function at a point *touches* the function at that point.

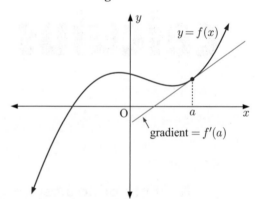

More generally, for a function $f(x)$ we can define a **derivative function** or **gradient function** $f'(x)$ which allows us to calculate the gradient of the tangent at any point on the function.

The process of finding the derivative function is called **differentiation**.

We have previously differentiated simple functions involving powers of x. In this Chapter we will explore techniques for differentiating more complicated functions.

A   FIRST PRINCIPLES

To differentiate a function, we can use the theory of **limits**.

LIMITS

If $f(x)$ is as close as we like to some real number A for all x sufficiently close to (but not equal to) a, then we say that $f(x)$ has a **limit** of A as x approaches a, and we write

$$\lim_{x \to a} f(x) = A.$$

In this case, $f(x)$ is said to **converge** to A as x approaches a.

For example, $\lim\limits_{x \to 3} x^2 = 9$, since x^2 approaches 9 as x gets closer and closer to 3.

When considering $\lim\limits_{x \to a} f(x)$, it is not important whether $f(x)$ is defined at $x = a$ or not. What *is* important is the behaviour of the function as x gets very close to a.

FINDING THE DERIVATIVE FUNCTION BY FIRST PRINCIPLES

Consider the function $y = f(x)$ where A is the point $(x, f(x))$, and B is the point $(x + h, \; f(x + h))$.

The gradient of the chord [AB] is $\dfrac{f(x + h) - f(x)}{h}$.

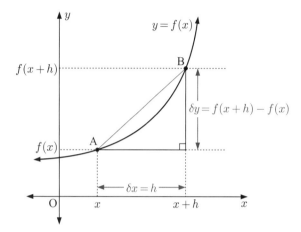

Another way to express this gradient is using the **Leibniz notation** $\dfrac{\delta y}{\delta x}$

where $\delta y = f(x + h) - f(x)$ is the vertical step
and $\delta x = h$ is the horizontal step.

δx reads as "delta x", and refers to the *change* in x.

To calculate the gradient of the tangent at A, we let the point B get closer and closer to A. This means that the horizontal step $\delta x = h$ becomes infinitely small, and the gradient of the tangent at the general point $A(x, \; f(x))$ is $\lim\limits_{h \to 0} \dfrac{f(x + h) - f(x)}{h}$.

The **derivative function** or simply **derivative** of $y = f(x)$ is defined as

$$f'(x) = \lim_{h \to 0} \frac{f(x + h) - f(x)}{h}$$

In Leibniz notation, we write the derivative function as $\dfrac{dy}{dx}$.
It is read "dee y by dee x".

The derivative function can also be written as y'.

Example 1

◀) Self Tutor

a Use first principles to find the derivative function $f'(x)$ for $f(x) = x^3$.

b Find $f'(1)$ and interpret your answer.

a $f'(x) = \lim\limits_{h \to 0} \dfrac{f(x + h) - f(x)}{h}$

$ = \lim\limits_{h \to 0} \dfrac{(x + h)^3 - x^3}{h}$

$ = \lim\limits_{h \to 0} \dfrac{x^3 + 3x^2h + 3xh^2 + h^3 - x^3}{h}$ {binomial expansion}

$ = \lim\limits_{h \to 0} \dfrac{h(3x^2 + 3xh + h^2)}{h}$

$ = \lim\limits_{h \to 0} (3x^2 + 3xh + h^2)$ {since $h \neq 0$}

$ = 3x^2$ {as $h \to 0, \; 3x^2 + 3xh + h^2 \to 3x^2$}

b $f'(1) = 3(1)^2$
$\quad = 3$

The gradient of the tangent to $f(x) = x^3$ at the point
where $x = 1$, is 3.

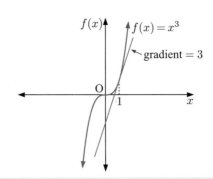

Example 2   ◀)) **Self Tutor**

Given $f(x) = \dfrac{1}{x}$, use first principles to find $f'(x)$.

$$f'(x) = \lim_{h \to 0} \frac{f(x+h) - f(x)}{h}$$

$$= \lim_{h \to 0} \left(\frac{\frac{1}{x+h} - \frac{1}{x}}{h} \right) \times \left(\frac{x(x+h)}{x(x+h)} \right)$$

$$= \lim_{h \to 0} \frac{x - (x+h)}{hx(x+h)}$$

$$= \lim_{h \to 0} \frac{-1\cancel{h}}{\cancel{h}x(x+h)} \qquad \{\text{since } h \neq 0\}$$

$$= -\frac{1}{x^2} \qquad \{\text{as } h \to 0, \ x+h \to x\}$$

EXERCISE 7A

1 a Use first principles to find the derivative function $f'(x)$ for $f(x) = x^2$.

   b Find $f'(3)$ and interpret your answer.

2 Find $f'(x)$ from first principles, given that $f(x)$ is:

   a x **b** 1 **c** $-x^2$ **d** $2x^3$

3 Find $f'(x)$ from first principles, given that $f(x)$ is:

   a $\dfrac{3}{x}$ **b** $\dfrac{1}{x^2}$ **c** $-\dfrac{2}{x^3}$

4 The graph of $f(x) = \dfrac{2}{x^2}$ is shown alongside.

   a Use first principles to find $f'(x)$.

   b Hence find the gradient of the tangent to $f(x)$ at:

      i $x = -1$ **ii** $x = 2$.

   c Copy the graph, and include the information
      from **b**.

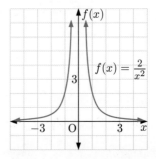

PRINTABLE
GRAPH

Example 3   ◀) **Self Tutor**

Given $f(x) = \sqrt{x}$, use first principles to find $f'(x)$.

$$f'(x) = \lim_{h \to 0} \frac{f(x+h) - f(x)}{h}$$

$$= \lim_{h \to 0} \frac{\sqrt{x+h} - \sqrt{x}}{h}$$

$$= \lim_{h \to 0} \left(\frac{\sqrt{x+h} - \sqrt{x}}{h} \right) \times \left(\frac{\sqrt{x+h} + \sqrt{x}}{\sqrt{x+h} + \sqrt{x}} \right)$$

$$= \lim_{h \to 0} \frac{x + h - x}{h(\sqrt{x+h} + \sqrt{x})}$$

$$= \lim_{h \to 0} \frac{1\!\!\!/h}{1\!\!\!/h(\sqrt{x+h} + \sqrt{x})} \qquad \{\text{since } h \neq 0\}$$

$$= \frac{1}{\sqrt{x} + \sqrt{x}} \qquad \{\text{as } h \to 0, \ x + h \to x\}$$

$$= \frac{1}{2\sqrt{x}}$$

5 Find $f'(x)$ from first principles, given that $f(x)$ is:

   a $2\sqrt{x}$ **b** $x\sqrt{x}$ **c** $x^2\sqrt{x}$

6 The graph of $y = \dfrac{1}{\sqrt{x}}$ is shown alongside.

   a Use first principles to find $\dfrac{dy}{dx}$.

   b Hence find the gradient of the illustrated tangent.

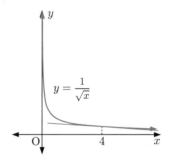

7   **a** Use the previous results to copy and complete each table.

$f(x)$	$f'(x)$
$1 = x^0$	
$x = x^1$	
x^2	
x^3	

$f(x)$	$f'(x)$
$\dfrac{1}{x} = x^{-1}$	
$\dfrac{1}{x^2} = x^{-2}$	
$\sqrt{x} = x^{\frac{1}{2}}$	
$x\sqrt{x} = x^{\frac{3}{2}}$	

   b Copy and complete:

       If $f(x) = x^n$, then $f'(x) = $ ......

8 Find $\dfrac{dy}{dx}$ from first principles, given:

   a $y = x^2 - 3x + 1$ **b** $y = 2x - \dfrac{1}{x}$ **c** $y = 3x + 2\sqrt{x}$

B   SIMPLE RULES OF DIFFERENTIATION

We have previously explored several **rules of differentiation** which can be applied so we do not need to use first principles.

$f(x)$	$f'(x)$	*Name of rule*
c (a constant)	0	**differentiating a constant**
x^n	nx^{n-1}	**differentiating x^n**
$c\,u(x)$	$c\,u'(x)$	**constant times a function**
$u(x) + v(x)$	$u'(x) + v'(x)$	**addition rule**

Example 4   ◀)) **Self Tutor**

Find $f'(x)$ for $f(x)$ equal to:

a $2x^3 + 7x^2 - 5x + 4$ **b** $\dfrac{5}{x} - \dfrac{6}{x^2}$

a $f(x) = 2x^3 + 7x^2 - 5x + 4$
$\therefore\ f'(x) = 2(3x^2) + 7(2x) - 5(1)$
$\qquad\quad = 6x^2 + 14x - 5$

b $f(x) = \dfrac{5}{x} - \dfrac{6}{x^2}$
$\qquad\quad = 5x^{-1} - 6x^{-2}$
$\therefore\ f'(x) = 5(-1x^{-2}) - 6(-2x^{-3})$
$\qquad\quad = -5x^{-2} + 12x^{-3}$
$\qquad\quad = -\dfrac{5}{x^2} + \dfrac{12}{x^3}$

With practice you should not need the intermediate steps.

EXERCISE 7B

1 Find $f'(x)$ given that $f(x)$ is:

a x^5

b x^9

c $3x^6$

d $-2x^7$

e $\frac{1}{2}x^4$

f $x^3 + 4x$

g $x^2 - 6x + 2$

h $4x^2 + 3x - 5$

i $8x^2 - 1$

j $9 - 3x^2$

k $x^3 - 2x + 5$

l $\frac{1}{3}x^3 - 5x^2 + 4x$

m $6 - 5x - 2x^2$

n $x^6 + \frac{1}{3}x^4 - 7x$

o $x^2(x - 4)$

p $(x + 3)(x - 1)$

2 Find $f'(t)$ for:

a $f(t) = t^6$

b $f(t) = 3t^3$

c $f(t) = t^2 - 4t + 1$

d $f(t) = \frac{1}{2}t^4 - \frac{1}{3}t$

e $f(t) = 1 - t - \dfrac{t^5}{2}$

f $f(t) = (t + 2)(t - 6)$

3 Find $\dfrac{dy}{dx}$ for:

a $y = \dfrac{1}{x^3}$ **b** $y = \dfrac{1}{x^4}$ **c** $y = \dfrac{1}{x^7}$

d $y = \dfrac{4}{x}$ **e** $y = -\dfrac{2}{x^5}$ **f** $y = \dfrac{5}{x^2}$

g $y = 3x + \dfrac{2}{x}$ **h** $y = \dfrac{1}{x} - \dfrac{4}{x^2}$ **i** $y = \dfrac{3}{x^2} - \dfrac{7}{x^4}$

j $y = \dfrac{1}{2x}$ **k** $y = -\dfrac{2}{3x^2}$ **l** $y = 7x^2 - \dfrac{1}{4x^5}$

m $y = \dfrac{x^2 + 6}{x}$ **n** $y = \dfrac{x - 5}{x^3}$ **o** $y = \dfrac{x^2 - 4x + 3}{x^2}$

Remember that
$\dfrac{1}{x^n} = x^{-n}$.

4 Find the gradient of the tangent to:

a $y = 3x^2 + \dfrac{1}{x}$ at $x = 2$

b $y = \tfrac{1}{2}x - \dfrac{2}{x^2}$ at $x = 1$

c $y = \dfrac{3 - x}{x^2}$ at the point $(2, \tfrac{1}{4})$

d $y = (x^2 - 2)\left(x + \dfrac{1}{x}\right)$ at the point $(-1, 2)$.

Example 5 🔊 Self Tutor

Find $f'(x)$ for $f(x)$ equal to:

a $6\sqrt{x} + \dfrac{3}{x}$

b $\dfrac{9}{\sqrt[3]{x}}$

a $\quad f(x) = 6\sqrt{x} + \dfrac{3}{x}$

$\qquad = 6x^{\frac{1}{2}} + 3x^{-1}$

$\therefore \ f'(x) = 6(\tfrac{1}{2}x^{-\frac{1}{2}}) + 3(-1x^{-2})$

$\qquad = 3x^{-\frac{1}{2}} - 3x^{-2}$

$\qquad = \dfrac{3}{\sqrt{x}} - \dfrac{3}{x^2}$

b $\quad f(x) = \dfrac{9}{\sqrt[3]{x}}$

$\qquad = 9x^{-\frac{1}{3}}$

$\therefore \ f'(x) = 9(-\tfrac{1}{3}x^{-\frac{4}{3}})$

$\qquad = -3x^{-\frac{4}{3}}$

$\qquad = -\dfrac{3}{\sqrt[3]{x^4}}$

---

5 Find $f'(x)$ for $f(x)$ equal to:

a $\sqrt{x}$ **b** $4\sqrt{x}$ **c** $\sqrt[5]{x}$ **d** $-x\sqrt{x}$

e $\dfrac{2}{\sqrt{x}}$ **f** $2x + 8\sqrt{x}$ **g** $6x^2\sqrt{x}$ **h** $\dfrac{7}{x} - \dfrac{4}{\sqrt{x}}$

i $\dfrac{3x - 8}{\sqrt{x}}$ **j** $\dfrac{6}{\sqrt[3]{x}}$ **k** $-\dfrac{3}{x^2\sqrt{x}}$ **l** $\dfrac{5}{x\sqrt{x}} + x^2$

6 Find:

a $\dfrac{dy}{dx}$ for $y = 8\sqrt{x}$

b $\dfrac{dy}{dt}$ for $y = 7 - \dfrac{6}{\sqrt{t}}$

c $\dfrac{dT}{dx}$ for $T = \sqrt[3]{x} - \dfrac{2}{x^2}$

d $\dfrac{dP}{du}$ for $P = \dfrac{5}{u} - 10u\sqrt{u}$

7 **a** Find $\dfrac{dy}{dx}$ for $y = 2x - 4\sqrt{x}$.

    b Hence find the gradient of the tangent to $y = 2x - 4\sqrt{x}$ at the point $(9, 6)$.

8 The graph of $f(x) = x^2 - 7x + \dfrac{16}{\sqrt{x}}$ is shown alongside.

    a Find $f'(x)$.

    b Hence show that the tangent to the graph at $x = 4$ is horizontal.

    c Copy the graph, and include the information from **b**.

PRINTABLE
GRAPH

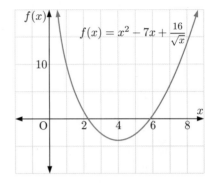

Activity   The increments formula

For a function y expressed in terms of x, the **increments formula** can be used to estimate the change in y for a given change in x.

The increments formula states that the change in y, δy, can be estimated by

$$\delta y \approx \frac{dy}{dx} \times \delta x, \quad \text{where } \delta x \text{ is the change in } x.$$

For example, suppose $y = x^5$, and therefore $\dfrac{dy}{dx} = 5x^4$.

To estimate the value of 2.01^5, we let $x = 2$ and $\delta x = 0.01$.

Now $\delta y \approx \dfrac{dy}{dx} \times \delta x$

     $\approx 5x^4 \times \delta x$

     $\approx 5 \times 2^4 \times 0.01$

     ≈ 0.8

Since $2^5 = 32$, we estimate that $2.01^5 \approx 32 + 0.8 \approx 32.8$.

What to do:

1 Use the increments formula to estimate the value of:

    a 5.01^2 **b** 2.01^6 **c** 2.98^3 **d** 1.95^4

    Use your calculator to check your estimates.

2 **a** Use the diagram to explain what is actually calculated when we use the increments formula to estimate δy.

    b Explain why the formula only provides accurate estimates for small values of δx.

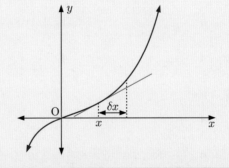

C   THE CHAIN RULE

In **Chapter 2** we saw that the **composite** of two functions f and g is $gf(x)$ or $g(f(x))$.

We can often write complicated functions as the composite of two or more simpler functions.

For example $y = (x^2 + 3x)^4$ could be rewritten as $y = u^4$ where $u = x^2 + 3x$, or as $y = g(f(x))$ where $g(x) = x^4$ and $f(x) = x^2 + 3x$.

Example 6   ◀)) Self Tutor

Find:

a $g(f(x))$ if $g(x) = \sqrt{x}$ and $f(x) = 2 - 3x$

b $g(x)$ and $f(x)$ such that $g(f(x)) = \dfrac{1}{x - x^2}$.

a $g(f(x)) = g(2 - 3x)$
$\qquad\qquad = \sqrt{2 - 3x}$

b If we let $f(x) = x - x^2$ then

$$g(f(x)) = \frac{1}{f(x)}$$

$$\therefore \quad g(x) = \frac{1}{x} \quad \text{and} \quad f(x) = x - x^2$$

There are several possible answers for **b**.

EXERCISE 7C.1

1 Find $g(f(x))$ if:

    a $g(x) = x^2$ and $f(x) = 2x + 7$ **b** $g(x) = 2x + 7$ and $f(x) = x^2$

    c $g(x) = \sqrt{x}$ and $f(x) = 3 - 4x$ **d** $g(x) = 3 - 4x$ and $f(x) = \sqrt{x}$

    e $g(x) = \dfrac{2}{x}$ and $f(x) = x^2 + 3$ **f** $g(x) = x^2 + 3$ and $f(x) = \dfrac{2}{x}$

2 Find $g(x)$ and $f(x)$ such that $g(f(x))$ is:

    a $(3x + 10)^3$ **b** $(7 - 2x)^5$ **c** $\dfrac{1}{2x + 4}$

    d $\sqrt{x^2 - 3x}$ **e** $\dfrac{1}{(5x - 1)^4}$ **f** $\dfrac{10}{(3x - x^2)^3}$

DERIVATIVES OF COMPOSITE FUNCTIONS

The reason we are interested in writing complicated functions as composite functions is to make finding derivatives easier.

Investigation 1   Differentiating composite functions

In this Investigation we want to learn how to differentiate composite functions.

Based on the rule "if $y = x^n$ then $\dfrac{dy}{dx} = nx^{n-1}$", we might suspect that if $y = (2x+1)^2$ then $\dfrac{dy}{dx} = 2(2x+1)^1$. But is this so?

What to do:

1 Expand $y = (2x+1)^2$, and hence find $\dfrac{dy}{dx}$. How does this compare with $2(2x+1)^1$?

2 Expand $y = (3x+1)^2$, and hence find $\dfrac{dy}{dx}$. How does this compare with $2(3x+1)^1$?

3 Expand $y = (ax+1)^2$ where a is a constant, and hence find $\dfrac{dy}{dx}$. How does this compare with $2(ax+1)^1$?

4 Suppose $y = u^2$.

   a Find $\dfrac{dy}{du}$.

   b Now suppose $u = ax + 1$, so $y = (ax+1)^2$.

      i Find $\dfrac{du}{dx}$.   **ii** Write $\dfrac{dy}{du}$ from **a** in terms of x.

      iii Hence find $\dfrac{dy}{du} \times \dfrac{du}{dx}$.   **iv** Compare your answer to the result in **3**.

   c If $y = u^2$ where u is a function of x, what do you suspect $\dfrac{dy}{dx}$ will be equal to?

5 Expand $y = (x^2 + 3x)^2$ and hence find $\dfrac{dy}{dx}$.

   Does your answer agree with the rule you suggested in **4 c**?

6 Consider $y = (2x+1)^3$.

   a Expand the brackets and hence find $\dfrac{dy}{dx}$.

   b If we let $u = 2x+1$, then $y = u^3$.

      i Find $\dfrac{du}{dx}$.   **ii** Find $\dfrac{dy}{du}$, and write it in terms of x.

      iii Hence find $\dfrac{dy}{du} \times \dfrac{du}{dx}$.   **iv** Compare your answer to the result in **a**.

7 Copy and complete: "If y is a function of u, and u is a function of x, then $\dfrac{dy}{dx} = \ldots\ldots$"

THE CHAIN RULE

$$\text{If } y = g(u) \text{ where } u = f(x) \text{ then } \frac{dy}{dx} = \frac{dy}{du}\frac{du}{dx}.$$

Proof: Consider $y = g(u)$ where $u = f(x)$.

For a small change of δx in x, there is a small change of $f(x + \delta x) - f(x) = \delta u$
in u and a small change of δy in y.

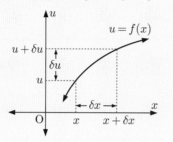   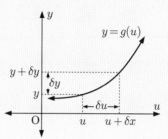

Now $\dfrac{\delta y}{\delta x} = \dfrac{\delta y}{\delta u} \times \dfrac{\delta u}{\delta x}$ {fraction multiplication}

As $\delta x \to 0$, $\delta u \to 0$ also.

$\therefore \quad \lim\limits_{\delta x \to 0} \dfrac{\delta y}{\delta x} = \lim\limits_{\delta u \to 0} \dfrac{\delta y}{\delta u} \times \lim\limits_{\delta x \to 0} \dfrac{\delta u}{\delta x}$ {limit rule}

$\therefore \quad \dfrac{dy}{dx} = \dfrac{dy}{du} \dfrac{du}{dx}$

This rule is extremely important and allows us to differentiate complicated functions much faster.

For example, for any function $f(x)$:   If $y = [f(x)]^n$ then $\dfrac{dy}{dx} = n[f(x)]^{n-1} \times f'(x)$.

Example 7   ◀ **Self Tutor**

Find $\dfrac{dy}{dx}$ if: **a** $y = (x^2 - 2x)^4$ **b** $y = \dfrac{4}{\sqrt{1 - 2x}}$

a $y = (x^2 - 2x)^4$

$\therefore \ y = u^4$ where $u = x^2 - 2x$

Now $\dfrac{dy}{dx} = \dfrac{dy}{du} \dfrac{du}{dx}$ {chain rule}

$\qquad = 4u^3(2x - 2)$

$\qquad = 4(x^2 - 2x)^3(2x - 2)$

> The brackets around $2x - 2$ are essential.

b $y = \dfrac{4}{\sqrt{1 - 2x}}$

$\therefore \ y = 4u^{-\frac{1}{2}}$ where $u = 1 - 2x$

Now $\dfrac{dy}{dx} = \dfrac{dy}{du} \dfrac{du}{dx}$ {chain rule}

$\qquad = 4 \times \left(-\tfrac{1}{2}u^{-\frac{3}{2}}\right) \times (-2)$

$\qquad = 4u^{-\frac{3}{2}}$

$\qquad = 4(1 - 2x)^{-\frac{3}{2}}$

EXERCISE 7C.2

1 Write in the form au^n, clearly stating what u is:

  a $\dfrac{1}{(2x-1)^2}$

  b $\sqrt{x^2 - 3x}$

  c $\dfrac{2}{\sqrt{2-x^2}}$

  d $\sqrt[3]{x^3 - x^2}$

  e $\dfrac{4}{(3-x)^3}$

  f $\dfrac{10}{x^2 - 3}$

2 Differentiate $y = (2x+3)^2$ by:

  a using the chain rule with $u = 2x+3$

  b expanding $y = (2x+3)^2$ then differentiating term-by-term.

3 Find the derivative function $\dfrac{dy}{dx}$ for:

  a $y = (4x-5)^2$

  b $y = \dfrac{1}{5-2x}$

  c $y = \sqrt{3x - x^2}$

  d $y = (1-3x)^4$

  e $y = 6(5-x)^3$

  f $y = \sqrt[3]{2x^3 - x^2}$

  g $y = \dfrac{6}{(5x-4)^2}$

  h $y = (x^2 - 5x + 8)^5$

  i $y = 2\left(x^2 - \dfrac{2}{x}\right)^3$

4 Find the gradient of the tangent to:

  a $y = \sqrt{1 - x^2}$ at $x = \frac{1}{2}$

  b $y = (3x+2)^6$ at $x = -1$

  c $y = \dfrac{1}{(2x-1)^4}$ at $x = 1$

  d $y = 6 \times \sqrt[3]{1-2x}$ at $x = 0$

  e $y = \dfrac{4}{x + 2\sqrt{x}}$ at $x = 4$

  f $y = \left(x + \dfrac{1}{x}\right)^3$ at $x = 1$.

5 The gradient function of $f(x) = (2x - b)^a$ is $f'(x) = 24x^2 - 24x + 6$.
Find the constants a and b.

6 Suppose $y = \dfrac{a}{\sqrt{1 + bx}}$ where a and b are constants. When $x = 3$, $y = 1$ and $\dfrac{dy}{dx} = -\dfrac{1}{8}$.
Find a and b.

7 Suppose $f(x) = 3\left(ax - \dfrac{b}{x}\right)^3$. Given that $f(\frac{3}{2}) = 3$ and $f'(\frac{3}{2}) = 30$, find a and b.

8 If $y = x^3$ then $x = y^{\frac{1}{3}}$.

  a Find $\dfrac{dy}{dx}$ and $\dfrac{dx}{dy}$, and hence show that $\dfrac{dy}{dx} \times \dfrac{dx}{dy} = 1$.

  b Explain why $\dfrac{dy}{dx} \times \dfrac{dx}{dy} = 1$ whenever these derivatives exist for any general function $y = f(x)$.

9 A curve has equation $x = y^3 + y$.

  a Find $\dfrac{dx}{dy}$ in terms of y.

  b Find the gradient of the tangent to the curve
at the point where $y = 2$.

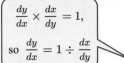

$\dfrac{dy}{dx} \times \dfrac{dx}{dy} = 1,$

so $\dfrac{dy}{dx} = 1 \div \dfrac{dx}{dy}$

D   THE PRODUCT RULE

We have seen the addition rule which allows us to differentiate term-by-term:

$$\text{If } f(x) = u(x) + v(x) \text{ then } f'(x) = u'(x) + v'(x).$$

If we now consider the case $f(x) = u(x)\,v(x)$, we might wonder if $f'(x) = u'(x)\,v'(x)$.

In other words, does the derivative of a product of two functions equal the product of their derivatives?

Investigation 2   The product rule

Suppose $u(x)$ and $v(x)$ are two functions of x, and that $f(x) = u(x)\,v(x)$ is the product of these functions.

The purpose of this Investigation is to find a rule for determining $f'(x)$.

What to do:

1 Suppose $u(x) = x$ and $v(x) = x$, so $f(x) = x^2$.

    a Find $f'(x)$ by direct differentiation.   **b** Find $u'(x)$ and $v'(x)$.

    c Does $f'(x) = u'(x)\,v'(x)$?

2 Suppose $u(x) = x$ and $v(x) = \sqrt{x}$, so $f(x) = x\sqrt{x} = x^{\frac{3}{2}}$.

    a Find $f'(x)$ by direct differentiation.   **b** Find $u'(x)$ and $v'(x)$.

    c Does $f'(x) = u'(x)\,v'(x)$?

3 Copy and complete the following table, finding $f'(x)$ by direct differentiation.

$f(x)$	$f'(x)$	$u(x)$	$v(x)$	$u'(x)$	$v'(x)$	$u'(x)\,v(x) + u(x)\,v'(x)$
x^2		x	x			
$x^{\frac{3}{2}}$		x	$\sqrt{x}$			
$x(x+1)$		x	$x+1$			
$(x-1)(2-x^2)$		$x-1$	$2-x^2$			

4 Copy and complete: "If $f(x) = u(x)\,v(x)$ then $f'(x) = \ldots\ldots$"

THE PRODUCT RULE

If $f(x) = u(x)\,v(x)$ then $f'(x) = u'(x)\,v(x) + u(x)\,v'(x)$.

Alternatively, if $y = uv$ where u and v are functions of x, then $\dfrac{dy}{dx} = \dfrac{du}{dx}\,v + u\,\dfrac{dv}{dx}$.

Proof: Let $y = u(x)\,v(x)$. Suppose there is a small change of δx in x which causes corresponding changes of δu in u, δv in v, and δy in y.

Since $y = uv$, $\quad y + \delta y = (u + \delta u)(v + \delta v)$

$\therefore \quad y + \delta y = uv + (\delta u)v + u(\delta v) + \delta u \delta v$

$\therefore \quad \delta y = (\delta u)v + u(\delta v) + \delta u \delta v$

$\therefore \quad \dfrac{\delta y}{\delta x} = \left(\dfrac{\delta u}{\delta x}\right)v + u\left(\dfrac{\delta v}{\delta x}\right) + \left(\dfrac{\delta u}{\delta x}\right)\delta v \qquad$ {dividing each term by δx}

$\therefore \quad \lim\limits_{\delta x \to 0} \dfrac{\delta y}{\delta x} = \left(\lim\limits_{\delta x \to 0} \dfrac{\delta u}{\delta x}\right)v + u\left(\lim\limits_{\delta x \to 0} \dfrac{\delta v}{\delta x}\right) + 0 \qquad$ {as $\delta x \to 0$, $\delta v \to 0$ also}

$\therefore \quad \dfrac{dy}{dx} = \dfrac{du}{dx}v + u\dfrac{dv}{dx}$

Example 8   ◀)) Self Tutor

Find $\dfrac{dy}{dx}$ if:

a $y = \sqrt{x}(2x + 1)^3$

b $y = x^2(x^2 - 2x)^4$

---

a $y = \sqrt{x}(2x + 1)^3$ is the product of $u = x^{\frac{1}{2}}$ and $v = (2x + 1)^3$

$\therefore \quad u' = \tfrac{1}{2}x^{-\frac{1}{2}}$ and $v' = 3(2x + 1)^2 \times 2 \qquad$ {chain rule}

$\qquad\qquad\qquad\qquad\qquad = 6(2x + 1)^2$

Now $\dfrac{dy}{dx} = u'v + uv' \qquad$ {product rule}

Used together, the product rule and chain rule are extremely powerful.

$\qquad = \tfrac{1}{2}x^{-\frac{1}{2}}(2x + 1)^3 + x^{\frac{1}{2}} \times 6(2x + 1)^2$

$\qquad = \tfrac{1}{2}x^{-\frac{1}{2}}(2x + 1)^3 + 6x^{\frac{1}{2}}(2x + 1)^2$

b $y = x^2(x^2 - 2x)^4$ is the product of $u = x^2$ and $v = (x^2 - 2x)^4$

$\therefore \quad u' = 2x$ and $v' = 4(x^2 - 2x)^3(2x - 2) \qquad$ {chain rule}

Now $\dfrac{dy}{dx} = u'v + uv' \qquad$ {product rule}

$\qquad = 2x(x^2 - 2x)^4 + x^2 \times 4(x^2 - 2x)^3(2x - 2)$

$\qquad = 2x(x^2 - 2x)^4 + 4x^2(x^2 - 2x)^3(2x - 2)$

EXERCISE 7D

1 Use the product rule to differentiate:

a $f(x) = x(x - 1)$

b $f(x) = 2x(x + 1)$

c $f(x) = x^2\sqrt{x + 1}$

d $f(x) = (x + 3)(x - 1)$

e $f(x) = x\sqrt{x^2 - 1}$

f $f(x) = x(x + 1)^2$

2 Find $\dfrac{dy}{dx}$ using the product rule:

a $y = x^2(2x - 1)$

b $y = 4x(2x + 1)^3$

c $y = x^2\sqrt{3 - x}$

d $y = \sqrt{x}(x - 3)^2$

e $y = 5x^2(3x^2 - 1)^2$

f $y = \sqrt{x}(x - x^2)^3$

3 Find the gradient of the tangent to:

a $y = x^4(1-2x)^2$ at $x = -1$

b $y = \sqrt{x}(x^2 - x + 1)^2$ at $x = 4$

c $y = x\sqrt{1-2x}$ at $x = -4$

d $y = x^3\sqrt{5-x^2}$ at $x = 1$.

4 Consider $y = \sqrt{x}(3-x)^2$.

a Show that $\dfrac{dy}{dx} = \dfrac{(3-x)(3-5x)}{2\sqrt{x}}$.

b Find the x-coordinates of all points on $y = \sqrt{x}(3-x)^2$ where the tangent is horizontal.

c For what values of x is $\dfrac{dy}{dx}$ undefined?

5 Suppose $y = -2x^2(x+4)$. For what values of x does $\dfrac{dy}{dx} = 10$?

6 Find the value of x for which the tangent to $f(x) = ax\sqrt{1-x}$ has gradient:

a 0

b a.

E THE QUOTIENT RULE

Expressions like $\dfrac{x^2+1}{2x-5}$, $\dfrac{\sqrt{x}}{1-3x}$, and $\dfrac{x^3}{(x-x^2)^4}$ are called **quotients** because they represent the division of one expression by another.

Quotient functions have the form $Q(x) = \dfrac{u(x)}{v(x)}$.

Notice that $u(x) = Q(x)\,v(x)$

$\therefore\ u'(x) = Q'(x)\,v(x) + Q(x)\,v'(x)$ {product rule}

$\therefore\ u'(x) - Q(x)\,v'(x) = Q'(x)\,v(x)$

$\therefore\ Q'(x)\,v(x) = u'(x) - \dfrac{u(x)}{v(x)}\,v'(x)$

$\therefore\ Q'(x)\,v(x) = \dfrac{u'(x)\,v(x) - u(x)\,v'(x)}{v(x)}$

$\therefore\ Q'(x) = \dfrac{u'(x)\,v(x) - u(x)\,v'(x)}{[v(x)]^2}$ when this exists.

THE QUOTIENT RULE

If $Q(x) = \dfrac{u(x)}{v(x)}$ then $Q'(x) = \dfrac{u'(x)\,v(x) - u(x)\,v'(x)}{[v(x)]^2}$.

Alternatively, if $y = \dfrac{u}{v}$ where u and v are functions of x, then $\dfrac{dy}{dx} = \dfrac{u'v - uv'}{v^2}$.

Example 9

◀)) **Self Tutor**

Use the quotient rule to find $\dfrac{dy}{dx}$ if:

a $y = \dfrac{1 + 3x}{x^2 + 1}$

b $y = \dfrac{\sqrt{x}}{(1 - 2x)^2}$

a $y = \dfrac{1 + 3x}{x^2 + 1}$ is a quotient with $\quad u = 1 + 3x \quad$ and $\quad v = x^2 + 1$

$$\therefore \quad u' = 3 \qquad \text{and} \quad v' = 2x$$

Now $\dfrac{dy}{dx} = \dfrac{u'v - uv'}{v^2}$ {quotient rule}

$$= \frac{3(x^2 + 1) - (1 + 3x)2x}{(x^2 + 1)^2}$$

$$= \frac{3x^2 + 3 - 2x - 6x^2}{(x^2 + 1)^2}$$

$$= \frac{3 - 2x - 3x^2}{(x^2 + 1)^2}$$

b $y = \dfrac{\sqrt{x}}{(1 - 2x)^2}$ is a quotient with

$$u = x^{\frac{1}{2}} \qquad \text{and} \quad v = (1 - 2x)^2$$

$$\therefore \quad u' = \tfrac{1}{2} x^{-\frac{1}{2}} \quad \text{and} \quad v' = 2(1 - 2x)^1 \times (-2) \qquad \text{\{chain rule\}}$$

$$= -4(1 - 2x)$$

Now $\dfrac{dy}{dx} = \dfrac{u'v - uv'}{v^2}$ {quotient rule}

$$= \frac{\tfrac{1}{2} x^{-\frac{1}{2}} (1 - 2x)^2 - x^{\frac{1}{2}} \times (-4(1 - 2x))}{(1 - 2x)^4}$$

$$= \frac{\tfrac{1}{2} x^{-\frac{1}{2}} (1 - 2x)^2 + 4x^{\frac{1}{2}} (1 - 2x)}{(1 - 2x)^4}$$

$$= \frac{(1 - 2x) \left[\frac{1 - 2x}{2\sqrt{x}} + 4\sqrt{x} \left(\frac{2\sqrt{x}}{2\sqrt{x}} \right) \right]}{(1 - 2x)^{4\,3}} \qquad \text{\{look for common factors\}}$$

$$= \frac{1 - 2x + 8x}{2\sqrt{x}(1 - 2x)^3}$$

$$= \frac{6x + 1}{2\sqrt{x}(1 - 2x)^3}$$

Simplification of $\dfrac{dy}{dx}$ is unnecessary
if you simply want the gradient of
a tangent at a given point. In such cases,
substitute the value for x without
simplifying the derivative function first.

EXERCISE 7E

1 Use the quotient rule to find $\dfrac{dy}{dx}$ if:

a $y = \dfrac{1+3x}{2-x}$

b $y = \dfrac{x^2}{2x+1}$

c $y = \dfrac{x}{x^2-3}$

d $y = \dfrac{\sqrt{x}}{1-2x}$

e $y = \dfrac{x^2-3}{3x-x^2}$

f $y = \dfrac{x}{\sqrt{1-3x}}$

2 Find the gradient of the tangent to:

a $y = \dfrac{x}{1-2x}$ at $x=1$

b $y = \dfrac{x^3}{x^2+1}$ at $x=-1$

c $y = \dfrac{\sqrt{x}}{2x+1}$ at $x=4$

d $y = \dfrac{x^2}{\sqrt{x^2+5}}$ at $x=-2$.

3 Suppose $f(x) = \dfrac{x}{\sqrt{x-1}}$. Find $f'(x)$ using the quotient rule.

Check your answer by writing $f(x) = \dfrac{x-1}{\sqrt{x-1}} + \dfrac{1}{\sqrt{x-1}}$ and then differentiating.

4 Consider the graph of $y = \dfrac{2x+3}{x+1}$ alongside.

a Find $\dfrac{dy}{dx}$.

b Hence show that the illustrated tangents are parallel.

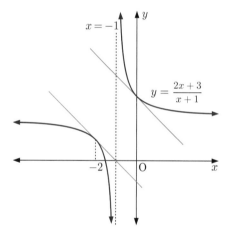

5 **a** If $y = \dfrac{2\sqrt{x}}{1-x}$, show that $\dfrac{dy}{dx} = \dfrac{x+1}{\sqrt{x}(1-x)^2}$.

b For what values of x is $\dfrac{dy}{dx}$: **i** zero   **ii** undefined?

6 **a** If $y = \dfrac{x^2+6}{2x+1}$, show that $\dfrac{dy}{dx} = \dfrac{2x^2+2x-12}{(2x+1)^2}$.

b For what values of x is $\dfrac{dy}{dx}$: **i** zero   **ii** undefined?

7 **a** If $y = \dfrac{x^2-3x+1}{x+2}$, show that $\dfrac{dy}{dx} = \dfrac{x^2+4x-7}{(x+2)^2}$.

b For what values of x is $\dfrac{dy}{dx}$: **i** zero   **ii** undefined?

8 A curve has equation $x = \dfrac{y^2+1}{y-5}$.

a Find $\dfrac{dx}{dy}$ in terms of y.

b Find the gradient of the tangent to the curve at the point where $y=-1$.

9 Prove the quotient rule by applying the product rule to $Q(x) = u(x)[v(x)]^{-1}$.

F   DERIVATIVES OF EXPONENTIAL FUNCTIONS

We have seen previously that the simplest **exponential functions** have the form $f(x) = b^x$ where b is any positive constant, $b \neq 1$.

The graphs of all members of the exponential family $f(x) = b^x$:

- pass through the point $(0, 1)$
- are asymptotic to the x-axis at one end
- lie above the x-axis for all x.

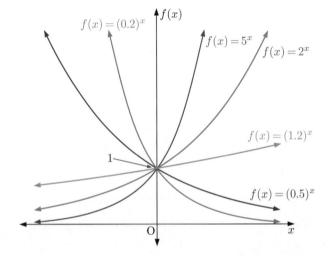

Investigation 3   The derivative of b^x

The purpose of this Investigation is to observe the nature of the derivatives of $f(x) = b^x$ for various values of b.

What to do:

1 Use the software provided to help fill in the table for $y = 2^x$:

x	y	$\dfrac{dy}{dx}$	$\dfrac{dy}{dx} \div y$
0			
0.5			
1			
1.5			
2			

CALCULUS DEMO

2 Repeat **1** for the following functions:

    a $y = 3^x$ **b** $y = 5^x$ **c** $y = (0.5)^x$

3 Use your observations from **1** and **2** to write a statement about the derivative of the general exponential $y = b^x$ for $b > 0$, $b \neq 1$.

From the **Investigation** you should have found that:

$$\text{If } f(x) = b^x \text{ then } f'(x) = f'(0) \times b^x.$$

Proof: If $f(x) = b^x$,

then $f'(x) = \lim\limits_{h \to 0} \dfrac{b^{x+h} - b^x}{h}$ {definition of the derivative}

$= \lim\limits_{h \to 0} \dfrac{b^x(b^h - 1)}{h}$

$= b^x \times \left(\lim\limits_{h \to 0} \dfrac{b^h - 1}{h} \right)$ {as b^x is independent of h}

But $f'(0) = \lim\limits_{h \to 0} \dfrac{f(0 + h) - f(0)}{h}$

$= \lim\limits_{h \to 0} \dfrac{b^h - 1}{h}$

$\therefore \ f'(x) = b^x \times f'(0)$

Given this result, if we can find a value of b such that $f'(0) = 1$, then we will have found *a function which is its own derivative*!

Investigation 4   **Solving for b if $f(x) = b^x$ and $f'(x) = b^x$**

Click on the icon to graph $f(x) = b^x$ and its derivative function $y = f'(x)$.   **DEMO**

Experiment with different values of b until the graphs of $f(x) = b^x$ and $y = f'(x)$ appear the same.

Estimate the corresponding value of b to 3 decimal places.

You should have discovered that $f(x) = f'(x) = b^x$ when $b \approx 2.718$.

To find this value of b more accurately we return to the algebraic approach:

We have already shown that if $f(x) = b^x$ then $f'(x) = b^x \left(\lim\limits_{h \to 0} \dfrac{b^h - 1}{h} \right)$.

So, to find the value of b such that $f'(x) = b^x$, we require

$$\lim\limits_{h \to 0} \dfrac{b^h - 1}{h} = 1$$

Now $\dfrac{b^h - 1}{h} = 1 \ \Leftrightarrow \ b = (1 + h)^{\frac{1}{h}}$,   so taking the limit on both sides gives

$$b = \lim\limits_{h \to 0} (1 + h)^{\frac{1}{h}}$$

Letting $h = \dfrac{1}{n}$, we notice that $h \to 0$ as $n \to \infty$.

$\therefore \ b = \lim\limits_{n \to \infty} \left(1 + \dfrac{1}{n} \right)^n$ if this limit exists.

We have in fact already seen this limit previously in an **Investigation** on continuously compounding interest.

We found that as $n \to \infty$,

$$\left(1 + \frac{1}{n}\right)^n \to 2.718\,281\,828\,459\,045\,235\,....$$

and this irrational number is the natural exponential e.

Check this for yourself by evaluating

$$\left(1 + \frac{1}{n}\right)^n$$

for very large values of n.

We therefore conclude: If $f(x) = e^x$ then $f'(x) = e^x$.

THE DERIVATIVE OF $e^{f(x)}$

The functions e^{-x}, e^{2x+3}, and e^{-x^2} all have the form $e^{f(x)}$.

Suppose $y = e^{f(x)} = e^u$ where $u = f(x)$.

Now $\dfrac{dy}{dx} = \dfrac{dy}{du}\dfrac{du}{dx}$ {chain rule}

$\qquad = e^u \dfrac{du}{dx}$

$\qquad = e^{f(x)} \times f'(x)$

Function	Derivative
e^x	e^x
$e^{f(x)}$	$e^{f(x)} \times f'(x)$

Example 10   ◀⦚ **Self Tutor**

Find the gradient function for y equal to:

a $2e^x + e^{-3x}$ **b** $x^2 e^{-x}$ **c** $\dfrac{e^{2x}}{x}$

a If $y = 2e^x + e^{-3x}$ then $\dfrac{dy}{dx} = 2e^x + e^{-3x}(-3)$ {addition rule}

$\qquad\qquad\qquad\qquad\qquad = 2e^x - 3e^{-3x}$

b If $y = x^2 e^{-x}$ then $\dfrac{dy}{dx} = 2xe^{-x} + x^2 e^{-x}(-1)$ {product rule}

$\qquad\qquad\qquad\qquad\quad = 2xe^{-x} - x^2 e^{-x}$

c If $y = \dfrac{e^{2x}}{x}$ then $\dfrac{dy}{dx} = \dfrac{e^{2x}(2)x - e^{2x}(1)}{x^2}$ {quotient rule}

$\qquad\qquad\qquad\qquad\quad = \dfrac{e^{2x}(2x - 1)}{x^2}$

EXERCISE 7F

1 Find the gradient function for $f(x)$ equal to:

a e^{4x} **b** $e^x + 3$ **c** e^{-2x} **d** $e^{\frac{x}{2}}$

e $2e^{-\frac{x}{2}}$ **f** $1 - 2e^{-x}$ **g** $4e^{\frac{x}{2}} - 3e^{-x}$ **h** $\dfrac{e^x + e^{-x}}{2}$

i e^{-x^2} **j** $e^{\frac{1}{x}}$ **k** $10(1 + e^{2x})$ **l** $20(1 - e^{-2x})$

m e^{2x+1} **n** $e^{\frac{x}{4}}$ **o** e^{1-2x^2} **p** $e^{-0.02x}$

2 Find the derivative of:

   a xe^x **b** x^3e^{-x} **c** $\dfrac{e^x}{x}$ **d** $\dfrac{x}{e^x}$

   e x^2e^{3x} **f** $\dfrac{e^x}{\sqrt{x}}$ **g** $20xe^{-0.5x}$ **h** $\dfrac{e^x+2}{e^{-x}+1}$

Example 11   ◆) Self Tutor

Find the gradient function for y equal to:

  a $(e^x-1)^3$ **b** $\dfrac{1}{\sqrt{2e^{-x}+1}}$

  a $y=(e^x-1)^3$

      $=u^3$ where $u=e^x-1$

    $\dfrac{dy}{dx}=\dfrac{dy}{du}\dfrac{du}{dx}$ {chain rule}

      $=3u^2\dfrac{du}{dx}$

      $=3(e^x-1)^2\times e^x$

      $=3e^x(e^x-1)^2$

  b $y=(2e^{-x}+1)^{-\frac12}$

      $=u^{-\frac12}$ where $u=2e^{-x}+1$

    $\dfrac{dy}{dx}=\dfrac{dy}{du}\dfrac{du}{dx}$ {chain rule}

      $=-\tfrac12 u^{-\frac32}\dfrac{du}{dx}$

      $=-\tfrac12(2e^{-x}+1)^{-\frac32}\times 2e^{-x}(-1)$

      $=e^{-x}(2e^{-x}+1)^{-\frac32}$

---

3 Find the gradient function for y equal to:

   a $(2+e^x)^4$ **b** $\sqrt{e^x-1}$ **c** $(e^x+e^{-x})^{\frac32}$ **d** $\dfrac{1}{\sqrt{e^{2x}+2}}$

4 Find the gradient of the tangent to:

   a $y=(e^x+2)^4$ at $x=0$ **b** $y=\dfrac{1}{2-e^{-x}}$ at $x=0$

   c $y=\sqrt{e^{2x}+10}$ at $x=\ln 3$ **d** $y=\dfrac{2-x}{e^{3x}}$ at $x=1$.

5 The graph of $f(x)=\sqrt{6-e^x}$ is shown alongside.

   a State the domain of the function.

   b The point P has y-coordinate 2. Find exactly:

     i the coordinates of P

     ii the gradient of the tangent at P.

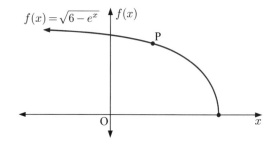

6 Given $f(x)=e^{kx}+x$ and $f'(0)=-8$, find k.

7 **a** By substituting $e^{\ln 2}$ for 2 in $y=2^x$, find $\dfrac{dy}{dx}$.

   b Show that if $y=b^x$ where $b>0$, $b\neq 1$, then $\dfrac{dy}{dx}=b^x\times\ln b$.

   c Find $\dfrac{dy}{dx}$ for: **i** $y=5^x$ **ii** $y=8\times 10^x$

8 The tangent to $f(x)=x^2e^{-x}$ at point P is horizontal. Find the possible coordinates of P.

G   DERIVATIVES OF LOGARITHMIC FUNCTIONS

Investigation 5   The derivative of $\ln x$

What to do:

1 Click on the icon to see the graph of $y = \ln x$. Observe the gradient function being drawn as the point moves from left to right along the graph.

CALCULUS
DEMO

2 Predict a formula for the gradient function of $y = \ln x$.

3 Find the gradient of the tangent to $y = \ln x$ for $x = 0.25$, 0.5, 1, 2, 3, 4, and 5. Do your results confirm your prediction in **2**?

From the **Investigation** you should have observed: If $y = \ln x$ then $\dfrac{dy}{dx} = \dfrac{1}{x}$, $x > 0$.

The simplest way to prove this result is using implicit differentiation, which we will study shortly. However, we can also prove the result by comparing the logarithmic function with its inverse, the exponential function.

Proof:

Since $y = e^x$ is the inverse of $y = \ln x$, the graphs are symmetric in the line $y = x$.

$\therefore$ the point $(a, \ln a)$ on $y = \ln x$ corresponds to the point $(\ln a, a)$ on the curve $y = e^x$.

For the curve $y = e^x$, $\dfrac{dy}{dx} = e^x$

$\therefore$ the tangent at $(\ln a, a)$ has gradient $= e^{\ln a} = a$

$\therefore$ its equation has the form $y = ax + c$ for some c.

The tangent to $y = \ln x$ at $(a, \ln a)$ is the reflection of this tangent in the line $y = x$.

$\therefore$ its equation is $x = ay + c$

$$\therefore \quad y = \frac{1}{a}x - \frac{c}{a}$$

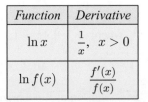

Since the gradient to the tangent to $y = \ln x$ at any point $(a, \ln a)$ has gradient $\dfrac{1}{a}$, the gradient function

$\dfrac{dy}{dx} = \dfrac{1}{x}$, $x > 0$.

THE DERIVATIVE OF $\ln f(x)$

Suppose $y = \ln f(x)$

$\qquad \therefore \; y = \ln u$ where $u = f(x)$.

Now $\dfrac{dy}{dx} = \dfrac{dy}{du}\dfrac{du}{dx}$ {chain rule}

$\therefore \quad \dfrac{dy}{dx} = \dfrac{1}{u}\dfrac{du}{dx}$

$\qquad\quad = \dfrac{f'(x)}{f(x)}$

Function	Derivative
$\ln x$	$\dfrac{1}{x}$, $x > 0$
$\ln f(x)$	$\dfrac{f'(x)}{f(x)}$

Example 12

◀)) **Self Tutor**

Find the gradient function of:

a $y = \ln(kx)$, k a constant   **b** $y = \ln(1 - 3x)$ **c** $y = x^3 \ln x$

a $y = \ln(kx)$ **b** $y = \ln(1 - 3x)$ **c** $y = x^3 \ln x$

$\therefore \ \dfrac{dy}{dx} = \dfrac{k}{kx}$ $\therefore \ \dfrac{dy}{dx} = \dfrac{-3}{1 - 3x}$ $\therefore \ \dfrac{dy}{dx} = 3x^2 \ln x + x^3 \left(\dfrac{1}{x} \right)$

$\qquad = \dfrac{1}{x}$ $\qquad = \dfrac{3}{3x - 1}$ $\qquad\qquad\qquad$ {product rule}

$\qquad\qquad\qquad\qquad = 3x^2 \ln x + x^2$

$\qquad\qquad\qquad\qquad = x^2(3 \ln x + 1)$

> $\ln(kx) = \ln k + \ln x$
> $\qquad = \ln x + \text{constant}$
> so $\ln(kx)$ and $\ln x$
> both have derivative $\dfrac{1}{x}$.

The laws of logarithms can help us to differentiate some logarithmic functions more easily.

For $a > 0$, $b > 0$, $n \in \mathbb{R}$: $\ln(ab) = \ln a + \ln b$

$$\ln\left(\frac{a}{b}\right) = \ln a - \ln b$$

$$\ln(a^n) = n \ln a$$

Example 13

◀)) **Self Tutor**

Differentiate with respect to x:

a $y = \ln(xe^{-x})$ **b** $y = \ln\left[\dfrac{x^2}{(x + 2)(x - 3)} \right]$

a $y = \ln(xe^{-x})$

$\qquad = \ln x + \ln e^{-x}$ $\{ \ln(ab) = \ln a + \ln b \}$

$\qquad = \ln x - x$ $\{ \ln e^a = a \}$

$\therefore \ \dfrac{dy}{dx} = \dfrac{1}{x} - 1$

> A derivative function
> will only be valid on
> *at most* the domain of
> the original function.

b $y = \ln\left[\dfrac{x^2}{(x + 2)(x - 3)} \right]$

$\qquad = \ln x^2 - \ln[(x + 2)(x - 3)]$ $\left\{ \ln\left(\dfrac{a}{b}\right) = \ln a - \ln b \right\}$

$\qquad = 2 \ln x - [\ln(x + 2) + \ln(x - 3)]$

$\qquad = 2 \ln x - \ln(x + 2) - \ln(x - 3)$

$\therefore \ \dfrac{dy}{dx} = \dfrac{2}{x} - \dfrac{1}{x + 2} - \dfrac{1}{x - 3}$

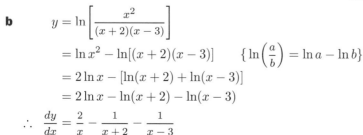

EXERCISE 7G

1 Find the gradient function of:

  a $y = \ln(7x)$

  b $y = \ln(2x + 1)$

  c $y = \ln(x - x^2)$

  d $y = 3 - 2\ln x$

  e $y = x^2 \ln x$

  f $y = \dfrac{\ln x}{2x}$

  g $y = e^x \ln x$

  h $y = (\ln x)^2$

  i $y = \sqrt{\ln x}$

  j $y = e^{-x} \ln x$

  k $y = \sqrt{x} \ln(2x)$

  l $y = \dfrac{2\sqrt{x}}{\ln x}$

  m $y = 3 - 4\ln(1 - x)$

  n $y = x\ln(x^2 + 1)$

  o $y = \dfrac{\ln x}{x^2}$

2 Find $\dfrac{dy}{dx}$ for:

  a $y = x\ln 5$

  b $y = \ln(x^3)$

  c $y = \ln(x^4 + x)$

  d $y = \ln(10 - 5x)$

  e $y = [\ln(2x + 1)]^3$

  f $y = \dfrac{\ln(4x)}{x}$

  g $y = \ln\left(\dfrac{1}{x}\right)$

  h $y = \ln(\ln x)$

  i $y = \dfrac{1}{\ln x}$

3 Use the laws of logarithms to help differentiate with respect to x:

  a $y = \ln\sqrt{1 - 2x}$

  b $y = \ln\left(\dfrac{1}{2x + 3}\right)$

  c $y = \ln\left(e^x \sqrt{x}\right)$

  d $y = \ln\left(x\sqrt{2 - x}\right)$

  e $y = \ln\left(\dfrac{x + 3}{x - 1}\right)$

  f $y = \ln\left(\dfrac{x^2}{3 - x}\right)$

4 Differentiate with respect to x:

  a $f(x) = \ln\left((3x - 4)^3\right)$

  b $f(x) = \ln\left(x(x^2 + 1)\right)$

  c $f(x) = \ln\left(\dfrac{x^2 + 2x}{x - 5}\right)$

  d $f(x) = \ln\left(\dfrac{x^3}{(x + 4)(x - 1)}\right)$

5 Find the gradient of the tangent to:

  a $y = x\ln x$ at the point where $x = e$

  b $y = \ln\left(\dfrac{x + 2}{x^2}\right)$ at the point where $x = 1$.

6 Suppose $x = (y + 2)\ln y$.

  a Find $\dfrac{dx}{dy}$ in terms of y.

  b Find the gradient of the tangent to the curve at the point where the curve crosses the y-axis.

7 Suppose $f(x) = a\ln(bx^2)$ where $f(e) = 3$ and $f'(1) = 6$. Find the constants a and b.

8 Suppose $f(x) = ax\ln(bx)$ where $f(1) = 12$ and $f'(1) = 16$. Find the constants a and b.

H   DERIVATIVES OF TRIGONOMETRIC FUNCTIONS

In **Chapter 4** we saw that sine and cosine curves naturally arise from circular motion.

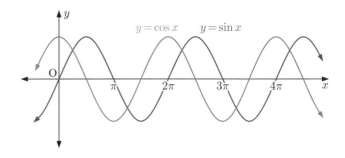

Investigation 6   **Derivatives of** $\sin x$ **and** $\cos x$

What to do:

1 Click on the icon to observe the graph of $y = \sin x$. A tangent with x-step of length 1 unit moves across the curve, and its y-step is translated onto the gradient graph. Predict the derivative of the function $y = \sin x$.

DERIVATIVES DEMO

2 Repeat the process in **1** for the graph of $y = \cos x$. Hence predict the derivative of the function $y = \cos x$.

3

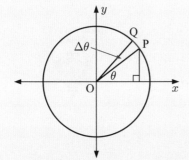

Suppose P and Q are points on the unit circle corresponding to angles θ and $\theta + \Delta\theta$ respectively from the positive x-axis.

a Explain why $PR = \sin(\theta + \Delta\theta) - \sin\theta$.

b If P and Q are close together then $\Delta\theta$ is very small. Explain why as Q approaches P:

   i the arc PQ resembles line segment [PQ]

   ii the length of the line segment $PQ \approx \Delta\theta$

   iii $Q\widehat{P}O$ approaches a right angle

   iv $Q\widehat{P}R \approx \theta$.

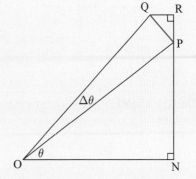

c Use right angled triangle trigonometry in $\triangle QRP$ to show that $\cos\theta \approx \dfrac{\sin(\theta + \Delta\theta) - \sin\theta}{\Delta\theta}$.

Hence explain why in the limit as $\Delta\theta \to 0$, $\cos\theta = \dfrac{d}{d\theta}(\sin\theta)$.

From the **Investigation** you should have deduced that:

For x in radians: If $f(x) = \sin x$ then $f'(x) = \cos x$.

If $f(x) = \cos x$ then $f'(x) = -\sin x$.

Proof for the case $f(x) = \sin x$

Using first principles,

$$f'(x) = \lim_{h \to 0} \frac{f(x+h) - f(x)}{h}$$

$$= \lim_{h \to 0} \frac{\sin(x+h) - \sin x}{h}$$

$$= \lim_{h \to 0} \frac{2\cos\left(\frac{x+h+x}{2}\right)\sin\left(\frac{x+h-x}{2}\right)}{h} \qquad \left\{ \sin S - \sin D = 2\cos\left(\frac{S+D}{2}\right)\sin\left(\frac{S-D}{2}\right) \text{ from \textbf{Exercise 5C}} \right\}$$

$$= \lim_{h \to 0} \frac{2\cos\left(x + \frac{h}{2}\right)\sin\left(\frac{h}{2}\right)}{h}$$

We use the small angle approximation for $\sin x$.

$$= \lim_{h \to 0} \frac{2\cos\left(x + \frac{h}{2}\right)}{2} \times \frac{\sin\left(\frac{h}{2}\right)}{\frac{h}{2}}$$

$$= \cos x \times 1 \qquad \left\{ \text{as } h \to 0, \ \frac{h}{2} \to 0, \ \frac{\sin\left(\frac{h}{2}\right)}{\frac{h}{2}} \to 1 \right\}$$

$$= \cos x$$

THE DERIVATIVE OF $\tan x$

Consider $y = \tan x = \dfrac{\sin x}{\cos x}$

We let $u = \sin x$ and $v = \cos x$

$\therefore \ \dfrac{du}{dx} = \cos x$ and $\dfrac{dv}{dx} = -\sin x$

$\therefore \ \dfrac{dy}{dx} = \dfrac{u'v - uv'}{v^2}$ {quotient rule}

$$= \frac{\cos x \cos x - \sin x(-\sin x)}{[\cos x]^2}$$

$$= \frac{\cos^2 x + \sin^2 x}{\cos^2 x}$$

$$= \frac{1}{\cos^2 x} \qquad \{\text{since } \sin^2 x + \cos^2 x = 1\}$$

$$= \sec^2 x$$

DERIVATIVES DEMO

Function	Derivative
$\sin x$	$\cos x$
$\cos x$	$-\sin x$
$\tan x$	$\sec^2 x$

THE DERIVATIVES OF $\sin[f(x)]$, $\cos[f(x)]$, AND $\tan[f(x)]$

Suppose $y = \sin[f(x)]$

If we let $u = f(x)$, then $y = \sin u$.

But $\dfrac{dy}{dx} = \dfrac{dy}{du}\dfrac{du}{dx}$ {chain rule}

$\therefore \ \dfrac{dy}{dx} = \cos u \times f'(x)$

$$= \cos[f(x)] \times f'(x)$$

We can perform the same procedure for $\cos[f(x)]$ and $\tan[f(x)]$, giving the following results:

Function	Derivative
$\sin[f(x)]$	$\cos[f(x)]\,f'(x)$
$\cos[f(x)]$	$-\sin[f(x)]\,f'(x)$
$\tan[f(x)]$	$\sec^2[f(x)]\,f'(x)$

Example 14

◀) **Self Tutor**

Differentiate with respect to x:

a $x \sin x$ **b** $4\tan^2 3x$

a
$$y = x \sin x$$
$$\therefore \ \frac{dy}{dx} = (1)\sin x + (x)\cos x$$
$$\text{\{product rule\}}$$
$$= \sin x + x \cos x$$

b
$$y = 4\tan^2 3x$$
$$= 4u^2 \quad \text{where} \quad u = \tan 3x$$
$$\frac{dy}{dx} = \frac{dy}{du}\frac{du}{dx} \quad \text{\{chain rule\}}$$
$$\therefore \ \frac{dy}{dx} = 8u \times \frac{du}{dx}$$
$$= 8\tan 3x \times \frac{3}{\cos^2 3x}$$
$$= \frac{24\sin 3x}{\cos^3 3x}$$

EXERCISE 7H

1 Use first principles to prove that if $f(x) = \cos x$ then $f'(x) = -\sin x$.

   Hint: Use the identity $\cos S - \cos D = -2\sin\left(\frac{S+D}{2}\right)\sin\left(\frac{S-D}{2}\right)$ from **Exercise 5C**.

2 Find $\dfrac{dy}{dx}$ for:

  a $y = \sin 2x$ **b** $y = \sin x + \cos x$ **c** $y = \cos 3x - \sin x$

  d $y = \sin(x+1)$ **e** $y = \cos(3 - 2x)$ **f** $y = \tan 5x$

  g $y = \sin\frac{x}{2} - 3\cos x$ **h** $y = 3\tan \pi x$ **i** $y = 4\sin x - \cos 2x$

3 Differentiate with respect to x:

  a $x^2 + \cos x$ **b** $\tan x - 3\sin x$ **c** $e^x \cos x$

  d $e^{-x}\sin x$ **e** $\ln(\sin x)$ **f** $e^{2x}\tan x$

  g $\sin 3x + 4\cos 2x$ **h** $\cos\frac{x}{2}$ **i** $3\tan 2x$

  j $x \cos x$ **k** $\dfrac{\sin x}{x}$ **l** $x \tan x$

4 Use the definitions of the reciprocal trigonometric functions to find the derivative of:

  a $\sec x$ **b** $\operatorname{cosec} x$ **c** $\cot x$

5 Differentiate with respect to x:

  a $\sin(x^2)$ **b** $\cos(\sqrt{x})$ **c** $\sqrt{\cos x}$ **d** $\sin^2 x$

  e $\cos^3 x$ **f** $\cos x \sin 2x$ **g** $\cos(\cos x)$ **h** $\cos^3 4x$

  i $\operatorname{cosec} 3x$ **j** $\sec 2x$ **k** $\dfrac{2}{\sin^2 2x}$ **l** $\dfrac{8}{\tan^3 \frac{x}{2}}$

6 Find the gradient of the tangent to:

   a $f(x) = \sin^3 x$ at the point where $x = \frac{2\pi}{3}$

   b $f(x) = \cos x \sin x$ at the point where $x = \frac{\pi}{4}$.

7 Consider the function $f(x) = 2\cos^2 x + 2\sin^2 x + 1$.

   a Find $f'(x)$.

   b Explain your answer to **a**.

8 The graph of $y = \cos x + 2\sin 2x$ is shown alongside.

   a Which tangent appears to have the steeper gradient?

   b Find $\dfrac{dy}{dx}$, and hence check your answer to **a**.

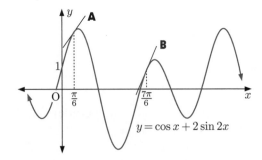

9 Use first principles to prove that if $f(x) = \sec x$, then $f'(x) = \sec x \tan x$.

I   SECOND DERIVATIVES

Given a function $f(x)$, the derivative $f'(x)$ is known as the **first derivative**.

The **second derivative** of $f(x)$ is the derivative of $f'(x)$, or **the derivative of the first derivative**.

We use $f''(x)$, y'', or $\dfrac{d^2y}{dx^2}$ to represent the second derivative.

$f''(x)$ reads "f *double dashed* x".

$\dfrac{d^2y}{dx^2} = \dfrac{d}{dx}\left(\dfrac{dy}{dx}\right)$ reads "*dee two y by dee x squared*".

> $f''(x)$ is the rate of change of $f'(x)$ with respect to x.

Example 15   ◀) **Self Tutor**

Find $f''(x)$ given that $f(x) = x\cos x + \dfrac{1}{x}$.

$\quad f(x) = x\cos x + x^{-1}$

$\therefore\ f'(x) = (1)\cos x + (x)(-\sin x) - x^{-2}$ {product rule}

$\qquad\quad = \cos x - x\sin x - x^{-2}$

$\therefore\ f''(x) = -\sin x - (1)\sin x - (x)\cos x + 2x^{-3}$ {product rule}

$\qquad\quad = -2\sin x - x\cos x + \dfrac{2}{x^3}$

EXERCISE 7I

1 Find $f''(x)$ given that:

   a $f(x) = 4x^2 + 3x - 5$ **b** $f(x) = -x^3 + 2x^2 + 7$ **c** $f(x) = x\sqrt{x} + \sqrt{x}$

   d $f(x) = (2 - 3x)^3$ **e** $f(x) = (x + 5)^2(3 - 2x)$ **f** $f(x) = \dfrac{x + 2}{2x - 1}$

2 Find $\dfrac{d^2y}{dx^2}$ given that:

   a $y = \dfrac{5 - x^2}{2x}$ **b** $y = (3 - x^2)^3$ **c** $y = x^2 - x + \dfrac{1}{1 - x}$

   d $y = 3e^x - 2x$ **e** $y = \dfrac{1 - e^{-x}}{x}$ **f** $y = \dfrac{3 - x}{xe^x}$

3 Given $f(x) = 3e^x - 2x$, find:

   a $f(1)$ **b** $f'(1)$ **c** $f''(1)$

4 Find the value(s) of x such that $f''(x) = 0$, given:

   a $f(x) = x^3 - 3x^2 - 72x + 18$ **b** $f(x) = \dfrac{4x - x^4}{1 + x^2}$

5 Find $\dfrac{d^2y}{dx^2}$ given that:

   a $y = x \sin x$ **b** $y = \dfrac{\cos^2 x - x}{x^2}$ **c** $y = e^{-x} \sin x$

6 Suppose $y = Ae^{kx}$ where A and k are constants. Show that:

   a $\dfrac{dy}{dx} = ky$ **b** $\dfrac{d^2y}{dx^2} = k^2y$

7 Suppose $f(x) = 2\sin^3 x - 3\sin x$.

   a Show that $f'(x) = -3\cos x \cos 2x$.   **b** Find $f''(x)$.

8 Find $\dfrac{d^2y}{dx^2}$ given:

   a $y = -\ln x$ **b** $y = x \ln x$ **c** $y = (\ln x)^2$

9 If $y = 2e^{3x} + 5e^{4x}$, show that $\dfrac{d^2y}{dx^2} - 7\dfrac{dy}{dx} + 12y = 0$.

10 If $y = \sin(2x + 3)$, show that $\dfrac{d^2y}{dx^2} + 4y = 0$.

11 If $y = 2\sin x + 3\cos x$, show that $y'' + y = 0$.

J   IMPLICIT DIFFERENTIATION

For relations such as $y^3 + 3xy^2 - xy + 11 = 0$, it is often difficult or impossible to write y as a function of x. Such relationships between x and y are called **implicit relations**.

Although we cannot write $y = f(x)$ and find a derivative *function* $f'(x)$, we can still find $\dfrac{dy}{dx}$, the rate of change in y with respect to x.

For example, the circle with centre $(0, 0)$ and radius 2 has equation $x^2 + y^2 = 4$.

One way of finding $\dfrac{dy}{dx}$ for the circle is to split the relation into two parts.

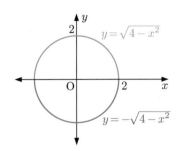

$$x^2 + y^2 = 4,$$
$$\therefore \quad y^2 = 4 - x^2$$
$$\therefore \quad y = \pm\sqrt{4 - x^2}$$

For the (yellow) top half:

$$y = \sqrt{4 - x^2} = \left(4 - x^2\right)^{\frac{1}{2}}$$

$$\therefore \quad \frac{dy}{dx} = \tfrac{1}{2}\left(4 - x^2\right)^{-\frac{1}{2}} \times (-2x)$$

$$= \frac{-x}{\sqrt{4 - x^2}}$$

$$= \frac{-x}{y}$$

For the (green) bottom half:

$$y = -\sqrt{4 - x^2} = -\left(4 - x^2\right)^{\frac{1}{2}}$$

$$\therefore \quad \frac{dy}{dx} = -\tfrac{1}{2}\left(4 - x^2\right)^{-\frac{1}{2}} \times (-2x)$$

$$= \frac{x}{\sqrt{4 - x^2}}$$

$$= \frac{x}{-y}$$

So, in both cases $\dfrac{dy}{dx} = -\dfrac{x}{y}$.

However, this method is time-consuming and relies on us being able to rearrange the relation into pieces where y is a function of x.

A more useful method for solving this type of problem is **implicit differentiation**.

If we are given an implicit relation between y and x and we want to find $\dfrac{dy}{dx}$, we can differentiate both sides of the equation with respect to x, applying the chain, product, quotient, and any other rules as appropriate. This will generate terms containing $\dfrac{dy}{dx}$, and we then proceed to make $\dfrac{dy}{dx}$ the subject of the equation.

A useful property using the chain rule is:
$$\frac{d}{dx}(y^n) = ny^{n-1}\frac{dy}{dx}.$$

For example, if $x^2 + y^2 = 4$, then $\dfrac{d}{dx}(x^2 + y^2) = \dfrac{d}{dx}(4)$

$$\therefore \quad 2x + 2y\frac{dy}{dx} = 0$$

$$\therefore \quad \frac{dy}{dx} = -\frac{x}{y}$$

Example 16   ◀) **Self Tutor**

If y is a function of x, find:

a $\dfrac{d}{dx}(y^3)$ **b** $\dfrac{d}{dx}\left(\dfrac{1}{y}\right)$ **c** $\dfrac{d}{dx}(xy^2)$

a $\dfrac{d}{dx}(y^3)$

$= 3y^2 \dfrac{dy}{dx}$

b $\dfrac{d}{dx}\left(\dfrac{1}{y}\right)$

$= \dfrac{d}{dx}(y^{-1})$

$= -y^{-2}\dfrac{dy}{dx}$

$= -\dfrac{1}{y^2}\dfrac{dy}{dx}$

c $\dfrac{d}{dx}(xy^2)$

$= 1 \times y^2 + x \times 2y\dfrac{dy}{dx}$ {product rule}

$= y^2 + 2xy\dfrac{dy}{dx}$

Example 17

◀)) **Self Tutor**

Find $\dfrac{dy}{dx}$ if:

a $x^2 + y^3 = 8$

b $x + x^2 y + y^3 = 100$

a
$$x^2 + y^3 = 8$$
$$\therefore \ \dfrac{d}{dx}(x^2) + \dfrac{d}{dx}(y^3) = \dfrac{d}{dx}(8)$$
$$\therefore \ 2x + 3y^2\dfrac{dy}{dx} = 0$$
$$\therefore \ \dfrac{dy}{dx} = \dfrac{-2x}{3y^2}$$

b
$$x + x^2 y + y^3 = 100$$
$$\therefore \ \dfrac{d}{dx}(x) + \dfrac{d}{dx}(x^2 y) + \dfrac{d}{dx}(y^3) = \dfrac{d}{dx}(100)$$
$$\therefore \ 1 + \underbrace{\left[2xy + x^2\dfrac{dy}{dx}\right]}_{\text{\{product rule\}}} + 3y^2\dfrac{dy}{dx} = 0$$
$$\therefore \ (x^2 + 3y^2)\dfrac{dy}{dx} = -1 - 2xy$$
$$\therefore \ \dfrac{dy}{dx} = \dfrac{-1 - 2xy}{x^2 + 3y^2}$$

When dealing with implicit relationships, $\dfrac{dy}{dx}$ is usually found in terms of x *and* y. Therefore, to find the gradient of the tangent to the relation at a particular point, we need to know both the x- and y-coordinates of the point.

Example 18

◀)) **Self Tutor**

Find the gradient of the tangent to $x^2 + y^3 = 5$ at the point where $x = 2$.

We first find $\dfrac{dy}{dx}$: $2x + 3y^2\dfrac{dy}{dx} = 0$

{implicit differentiation}

$$\therefore \ \dfrac{dy}{dx} = \dfrac{-2x}{3y^2}$$

When $x = 2$, $4 + y^3 = 5$

$$\therefore \ y = 1$$

$\therefore$ at the point $(2, 1)$, $\dfrac{dy}{dx} = \dfrac{-2(2)}{3(1)^2} = -\dfrac{4}{3}$

$\therefore$ the gradient of the tangent at $x = 2$ is $-\dfrac{4}{3}$.

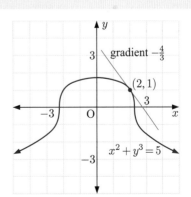

EXERCISE 7J

1 If y is a function of x, find:

a $\dfrac{d}{dx}(2y)$ **b** $\dfrac{d}{dx}(-3y)$ **c** $\dfrac{d}{dx}(2y^3)$ **d** $\dfrac{d}{dx}\left(-\dfrac{4}{y}\right)$

e $\dfrac{d}{dx}(y^4)$ **f** $\dfrac{d}{dx}(\sqrt{y})$ **g** $\dfrac{d}{dx}\left(\dfrac{1}{y^2}\right)$ **h** $\dfrac{d}{dx}(xy)$

i $\dfrac{d}{dx}(x^2y)$ **j** $\dfrac{d}{dx}(xy^3)$

2 Find $\dfrac{dy}{dx}$ if:

a $x^2 + y^2 = 25$ **b** $x^2 + 3y^2 = 9$ **c** $y^2 - x^2 = 8$

d $x^2 - y^3 = 10$ **e** $x^2 + xy = 4$ **f** $x^3 - 2xy = 5$

g $xy + \dfrac{2}{x} = 12$ **h** $x + y^2 + 2xy = 7$ **i** $x^3 - 2y^2 + xy^2 = x$

j $\dfrac{x}{y} + y = 0$ **k** $x + \cos y = 1$ **l** $\sin y + xe^y = 2y$

3 Use implicit differentiation to prove that for $y = \ln x$, $\dfrac{dy}{dx} = \dfrac{1}{x}$.

Hint: If $y = \ln x$, then $x = e^y$.

4 Consider the relation $xy^2 - 3x - y = 0$.

a Find $\dfrac{dy}{dx}$.

b Hence find the gradient of the tangent to the relation at:

  i $(0, 0)$ **ii** $(2, 2)$

5 Find the gradient of the tangent to:

a $x + y^2 = 5$ at $(1, 2)$ **b** $x^3 - y^2 = -1$ at $(2, -3)$

c $x + y^3 = 4y$ at $y = 1$ **d** $x + y = 8xy$ at $x = \frac{1}{2}$

e $\dfrac{x}{y^2} - x = 2$ at $y = 2$ **f** $x^3 - xy^3 = y - 1$ at $x = 1$.

6 The graph of $\dfrac{2x^2}{y} - \dfrac{y}{x} = 1$ is shown alongside.

a Find $\dfrac{dy}{dx}$.

b The points P and Q both have x-coordinate 1.

  i Find the coordinates of P and Q.

  ii Find the gradient of the tangent to the graph at each point.

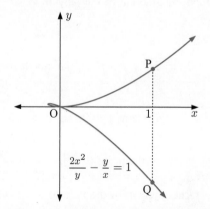

Review set 7A

1 Given $f(x) = \frac{2}{x}$, find $f'(x)$ from first principles.

2 Find $f'(x)$ given that $f(x)$ is:

    a $x^6 - 5x$ **b** $4x^3 + \frac{1}{x^2}$ **c** $2x\sqrt{x}$

3 Find $\frac{dy}{dx}$ for:

    a $y = 3x^2 - x^4$ **b** $y = \frac{x^3 - x}{x^2}$ **c** $y = x^2\sqrt{x-2}$

4   **a** Find $f'(x)$ given $f(x) = \dfrac{x}{\sqrt{x^2 + 1}}$.

    b At what point on the curve $f(x) = \dfrac{x}{\sqrt{x^2 + 1}}$ does the tangent have gradient 1?

5 Find $\frac{dy}{dx}$ if:

    a $y = e^{x^3 + 2}$ **b** $y = \ln\left(\dfrac{x+3}{x^2}\right)$ **c** $y = x^3 e^{2x}$

6 Given $y = 3e^x - e^{-x}$, show that $\frac{d^2y}{dx^2} = y$.

7 Differentiate with respect to x:

    a $5x - 3x^{-1}$ **b** $(3x^2 + \sqrt{x})^4$ **c** $(x^2 + 1)(1 - x^2)^3$

8 Find all points on the curve $y = 2x^3 + 3x^2 - 10x + 3$ where the gradient of the tangent is 2.

9 Find the gradient of the tangent to:

    a $y = (2 - 3x)^5$ at $x = 1$ **b** $y = 5\ln(e^{3x} + 4)$ at $x = 0$.

10 Differentiate with respect to x:

    a $\sin 5x \ln x$ **b** $\sin x \cos 2x$ **c** $e^{-2x} \tan x$

11 Find the gradient of the tangent to $y = \sin^2 x$ at the point where $x = \frac{\pi}{3}$.

12 Consider the function $f(x) = \dfrac{x^2 - 4x - 1}{e^x}$.

    a Find $f'(x)$.

    b Find the gradient of the tangent to $y = f(x)$ at $x = 1$.

    c For what values of x is the tangent to $y = f(x)$ horizontal?

13 Find the derivative with respect to x of:

    a $f(x) = (x^2 + 3)^4$ **b** $g(x) = \dfrac{\sqrt{x+5}}{x^2}$ **c** $h(x) = \dfrac{e^{4x}}{1 - 2x}$

14 For $f(x) = 2\sin x + \cos 2x$, find:

    a $f\left(\frac{\pi}{2}\right)$ **b** $f'\left(\frac{\pi}{2}\right)$ **c** $f''\left(\frac{\pi}{2}\right)$

15 The graph of $y = \sin x \cos x$ is shown alongside.

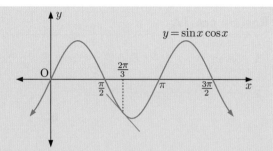

   a Find $\dfrac{dy}{dx}$.

   b Show that $y = \frac{1}{2}\sin 2x$ has the same derivative.

   c Find the gradient of the illustrated tangent.

16 Find $\dfrac{d^2y}{dx^2}$ for:

   a $y = \frac{1}{8}x^4 + \frac{1}{6}x^3 - \frac{1}{4}x^2$ **b** $y = xe^{-x}$

17 If $y = \tan x$, find $\dfrac{dy}{dx}$ from first principles.

18 Find $\dfrac{dy}{dx}$ if:

   a $y^2 + 3x^3 = 7$ **b** $2y - xy^2 = x$ **c** $\dfrac{3x}{y} = x - y$

19 The graph of $x^2 - y^2 = 15$ is shown alongside.

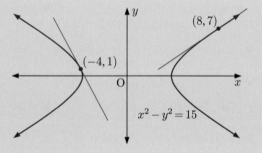

   a Find $\dfrac{dy}{dx}$.

   b Find the gradient of each illustrated tangent.

Review set 7B

1 Find $f'(x)$ from first principles, given that $f(x)$ is:

   a $3x$ **b** $4x\sqrt{x}$

2 Find $\dfrac{dy}{dx}$ for:

   a $y = 2x^3 - 6x^2 + 7x - 4$ **b** $y = \dfrac{3}{x} - \dfrac{5}{x^3}$ **c** $y = \dfrac{15}{\sqrt[3]{x}}$

3 Differentiate with respect to x:

   a $y = x^3\sqrt{1 - x^2}$ **b** $y = \dfrac{x^2 - 3x}{\sqrt{x + 1}}$

4 **a** Find $\dfrac{dy}{dx}$ for $y = xe^x$.

   b Find all points on the curve $y = xe^x$ where the gradient of the tangent is $2e$.

5 Differentiate with respect to x:

   a $f(x) = \ln(e^x + 3)$ **b** $f(x) = \ln\left[\dfrac{(x + 2)^3}{x}\right]$

6 The graph of $f(x) = \dfrac{x^2 + 2}{x}$ is shown alongside.

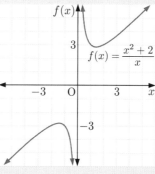

PRINTABLE GRAPH

a Find $f'(x)$.

b Hence find the gradient of the tangent at:

　　i $x = 1$　　　**ii** $x = -2$

c Copy the graph, and include the information from **b**.

7 Suppose $y = \left(x - \dfrac{1}{x}\right)^4$. Find $\dfrac{dy}{dx}$ at the point where $x = 1$.

8 Find $\dfrac{dy}{dx}$ if:

a $y = \ln(x^3 - 3x)$　　　**b** $y = \dfrac{e^x}{x^2}$　　　**c** $y = e^{2x}\sin x$

9 Suppose $f(x) = 2x^4 - 4x^3 - 9x^2 + 4x + 7$.

a Find $f''(x)$.　　　**b** Find x such that $f''(x) = 0$.

10 Differentiate with respect to x:

a $10x - \sin 10x$　　　**b** $\ln\left(\dfrac{1}{\cos x}\right)$　　　**c** $\sin 5x \ln(2x)$

11 Find the gradient of the tangent to:

a $y = \dfrac{x^3}{x + 1}$ at $x = 2$　　　**b** $y = \tan 2x$ at $x = \dfrac{\pi}{6}$.

12 Suppose $f(x) = a\ln(bx)$ where $f(e) = 12$ and $f'(2) = 2$. Find the constants a and b.

13 The graph of $y = \dfrac{\cos x}{\sin x + 2}$ is shown alongside.

a Find the gradient of the illustrated tangent.

b Show that it is impossible to draw a tangent to the graph of gradient $-\dfrac{1}{2}$.

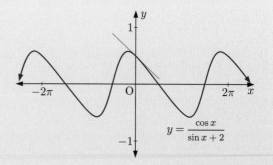

14 a If $y = \dfrac{e^x}{\sqrt{x}}$, show that $\dfrac{dy}{dx} = \dfrac{e^x(2x - 1)}{2x\sqrt{x}}$.

b For what values of x is $\dfrac{dy}{dx}$:　　**i** zero　　**ii** undefined?

15 Find $\dfrac{d^2y}{dx^2}$ for:

a $y = \dfrac{3x^2 - 2}{1 - 2x}$　　　**b** $y = x^3 - x + \dfrac{1}{\sqrt{x}}$

16 Suppose $y = 3\sin 2x + 2\cos 2x$. Show that $4y + \dfrac{d^2y}{dx^2} = 0$.

17 Find $\dfrac{dy}{dx}$ if:

   a $x^4 + 5y^3 = 17$ **b** $e^x + e^{2y} = x^2$ **c** $x \cos x = \sin y$

18 The ellipse with equation $\dfrac{x^2}{48} + \dfrac{y^2}{24} = 1$ is graphed alongside.

   a Find $\dfrac{dy}{dx}$.

   b The point P has x-coordinate 4, and the point Q has x-coordinate 3.

      i Find the y-coordinates of P and Q.

      ii Find the gradients of the tangents to the ellipse at P and Q.

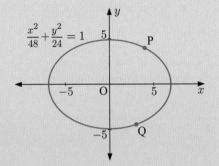

8

Applications of differential calculus

Contents:

Opening problem

A 5 m ladder rests against a vertical wall at point B. Its feet are at point A on horizontal ground.

Suppose the ladder slips and slides down the wall. As this happens, the value of x increases.

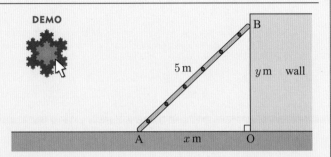

Things to think about:

a   Why *must* the value of y change when the value of x changes?

b   As x increases will y increase or decrease?

c   What is the relationship between x and y?

d   How can we establish a relationship between the *rate of change* in x and the *rate of change* in y?

In the previous Chapter we saw how to differentiate many types of functions.

In this Chapter we will use derivatives to find:

Minima and maxima are the plurals of minimum and maximum.

- tangents to curves
- turning points, which are local minima and maxima
- inflection points where the curve changes shape.

We will then look at applying these techniques to real world problems including:

- rates of change
- optimisation (finding maxima and minima).

A   TANGENTS AND NORMALS

TANGENTS

> The **tangent** to a curve at point A is the best approximating straight line to the curve at A.

The tangent to $y = f(x)$ at $x = a$ has gradient $f'(a)$.

So, the equation of the tangent at $(a, f(a))$ is
$\dfrac{y - f(a)}{x - a} = f'(a)$,   which rearranges to:

$$y = f'(a)(x - a) + f(a)$$

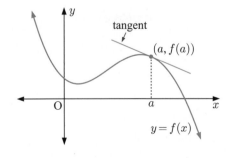

Example 1   ◀) Self Tutor

Find the equation of the tangent to $f(x) = \sqrt{x^2 + 5}$ at the point $(2, 3)$.

$$f(x) = \sqrt{x^2 + 5}$$
$$= (x^2 + 5)^{\frac{1}{2}}$$
$$\therefore \ f'(x) = \tfrac{1}{2}(x^2 + 5)^{-\frac{1}{2}}(2x) \qquad \{\text{chain rule}\}$$
$$= \frac{x}{\sqrt{x^2 + 5}}$$
$$\therefore \ f'(2) = \frac{2}{\sqrt{2^2 + 5}} = \frac{2}{3}$$

So, the tangent has equation $\ y = \tfrac{2}{3}(x - 2) + 3$
$$\therefore \ y = \tfrac{2}{3}x + \tfrac{5}{3}$$

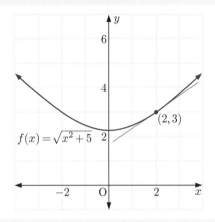

NORMALS

A **normal** to a curve is a line which is perpendicular to the tangent at the point of contact.

The gradients of perpendicular lines are negative reciprocals of each other, so:

The gradient of the normal to the curve at $x = a$ is $\ -\dfrac{1}{f'(a)}$.

The equation of the normal to the curve at $x = a$ is
$$y = -\frac{1}{f'(a)}(x - a) + f(a).$$

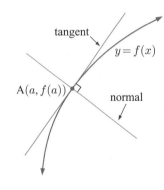

Example 2   ◀) Self Tutor

Find the equation of the normal to $y = e^{x-2}$ at the point where $x = 3$.

When $x = 3$, $y = e^{3-2} = e$.   So, the point of contact is $(3, e)$.

Now as $y = e^{x-2}$, $\dfrac{dy}{dx} = e^{x-2}$

$\therefore \ $ when $x = 3$, $\dfrac{dy}{dx} = e^{3-2} = e$

$\therefore \ $ the normal at $(3, e)$ has gradient $-\dfrac{1}{e}$.

$\therefore \ $ the equation of the normal is
$$y = -\frac{1}{e}(x - 3) + e$$
$$\therefore \ x + ey = 3 + e^2$$

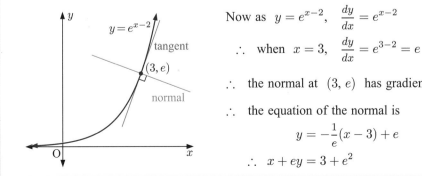

EXERCISE 8A

1 The graph of $f(x) = x^2 - 4x$ is shown alongside.

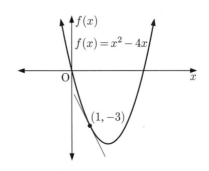

   a Find $f'(x)$.

   b Hence find the equation of the illustrated tangent.

2 Find the equation of the tangent to:

   a $y = \dfrac{2}{x^2}$ at $(-1, 2)$

   b $y = x\sqrt{x}$ at $(4, 8)$

   c $y = x^3 - 2x + 5$ at $(2, 9)$

   d $y = x - \dfrac{3}{\sqrt{x}}$ at $(9, 8)$

   e $y = \sqrt{2x + 1}$ at $x = 4$

   f $y = \dfrac{1}{2 - x}$ at $x = -1$

   g $y = \dfrac{1}{(x^2 + 1)^2}$ at $(1, \frac{1}{4})$

   h $y = \dfrac{1}{\sqrt{3 - 2x}}$ at $x = -3$.

3 Find the equation of the normal to:

   a $f(x) = \frac{1}{2}x^3 - \frac{1}{4}x^2$ at $(2, 3)$

   b $y = e^{-x}$ at $x = 0$

   c $y = \dfrac{2}{\sqrt{3 - x}}$ at $x = 2$

   d $f(x) = x\sqrt{x + 1}$ at $x = 3$

   e $f(x) = \dfrac{x}{1 - 3x}$ at $(-1, -\frac{1}{4})$

   f $f(x) = \dfrac{x^2}{1 - x}$ at $(2, -4)$

   g $y = \sqrt{x}(1 - x)^2$ at $x = 4$

   h $y = \dfrac{x^2 - 1}{2x + 3}$ at $x = -1$.

4 Suppose $f(x) = x^2 - \dfrac{8}{x}$.

   a Find the equation of the tangent to $y = f(x)$ at $x = -2$.

   b Find the equation of the normal to $y = f(x)$ at $x = 3$.

5 The tangent to $f(x) = (2x - 1)^4$ at $x = k$ has gradient 8.

   a Find k.

   b Find the equation of this tangent.

   c Hence find the x-intercept of this tangent.

6 Consider the curve $y = a\sqrt{1 - bx}$ where a and b are constants. The tangent to this curve at the point where $x = -1$ is $3x + y = 5$. Find the values of a and b.

7 The graph of $x^3 + y^2 = 100$ is shown alongside.

   a Find $\dfrac{dy}{dx}$.

   b Find the equation of the illustrated tangent.

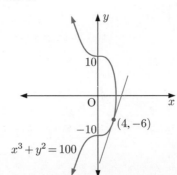

Example 3   ◀) **Self Tutor**

Show that the equation of the tangent to $y = \ln x$ at the point where $y = -1$ is $y = ex - 2$.

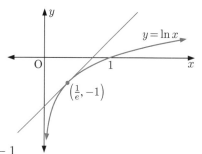

When $y = -1$, $\ln x = -1$
$$\therefore \quad x = e^{-1} = \tfrac{1}{e}$$

$\therefore$ the point of contact is $\left(\tfrac{1}{e}, -1\right)$.

Now $f(x) = \ln x$ has derivative $f'(x) = \dfrac{1}{x}$

$\therefore$ the tangent at $\left(\tfrac{1}{e}, -1\right)$ has gradient $\dfrac{1}{\frac{1}{e}} = e$

$\therefore$ the tangent has equation $y = e\left(x - \tfrac{1}{e}\right) - 1$

which is $y = ex - 2$

8 Find the equation of the tangent to:

  a $f(x) = e^{-x}$ at the point where $x = 2$ **b** $y = \ln(2 - x)$ at the point where $x = -1$

  c $y = (x + 2)e^x$ at the point where $x = 1$ **d** $y = \ln \sqrt{x}$ at the point where $y = -1$

  e $y = e^{3x-5}$ at the point where $y = e$.

9   **a** Find the y-intercept of the tangent to $f(x) = x \ln x$ at the point where:

    i $x = 1$ **ii** $x = 2$ **iii** $x = 3$.

  b Make a conjecture about the y-intercept of the tangent to $f(x) = x \ln x$ at the point where $x = a$, $a > 0$.

  c Prove your conjecture algebraically.

10 Find the axes intercepts of the tangent to $y = x^2 e^x$ at $x = 1$.

11 Suppose $x = (y + 1)e^y$.

  a Find $\dfrac{dx}{dy}$ in terms of y.

  b Find the equation of the tangent to the curve at the point where $y = 2$.

Example 4   ◀) **Self Tutor**

Find the equation of the tangent to $y = \tan x$ at the point where $x = \tfrac{\pi}{4}$.

When $x = \tfrac{\pi}{4}$, $y = \tan \tfrac{\pi}{4} = 1$

$\therefore$ the point of contact is $\left(\tfrac{\pi}{4}, 1\right)$.

Now $f(x) = \tan x$ has derivative $f'(x) = \sec^2 x$

$\therefore$ the tangent at $\left(\tfrac{\pi}{4}, 1\right)$ has gradient $\dfrac{1}{\cos^2(\frac{\pi}{4})} = \dfrac{1}{\frac{1}{2}} = 2$

$\therefore$ the tangent has equation $y = 2\left(x - \tfrac{\pi}{4}\right) + 1$

which is $y = 2x + \left(1 - \tfrac{\pi}{2}\right)$

12 Find the equation of the tangent to:

    a $y = \sin x$ at the origin

    b $y = \tan x$ at the origin

    c $y = \cos x$ at the point where $x = \frac{\pi}{6}$

    d $y = \dfrac{1}{\sin 2x}$ at the point where $x = \frac{\pi}{4}$

    e $y = \cos 2x + 3\sin x$ at the point where $x = \frac{\pi}{2}$.

13 Show that the curve with equation $y = \dfrac{\cos x}{1 + \sin x}$ does not have any horizontal tangents.

14 The graph of $y = e^{\cos x}$ is shown alongside.
Find the area of the shaded triangle.

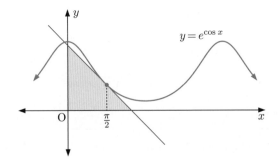

15 Find the equation of the tangent to:

    a $y = \sec x$ at $x = \frac{\pi}{4}$

    b $y = \cot \frac{x}{2}$ at $x = \frac{\pi}{3}$

16 Find the equation of the normal to:

    a $y = \csc x$ at $x = \frac{\pi}{6}$

    b $y = \sqrt{\sec \frac{x}{3}}$ at $x = \pi$

Example 5
    ◀ᴺ) **Self Tutor**

Find where the tangent to $y = x^3 - 4x - 2$ at $(2, -2)$ meets the curve again.

Let $f(x) = x^3 - 4x - 2$

$\therefore \ f'(x) = 3x^2 - 4$ and $\therefore \ f'(2) = 12 - 4 = 8$

$\therefore$ the equation of the tangent at $(2, -2)$ is $\quad y = 8(x - 2) - 2$
                                   or $\quad y = 8x - 18$.

The curve meets the tangent again when $\quad x^3 - 4x - 2 = 8x - 18$

                                    $\therefore \ x^3 - 12x + 16 = 0$

                                 $\therefore \ (x - 2)^2(x + 4) = 0$

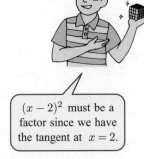

$(x - 2)^2$ must be a factor since we have the tangent at $x = 2$.

When $x = -4$, $y = (-4)^3 - 4(-4) - 2 = -50$

$\therefore$ the tangent meets the curve again at $(-4, -50)$.

---

17 Find where the tangent to the curve $y = -x^3 + x^2$ at the point where $x = 1$, meets the curve again.

18 Find where the tangent to the curve $y = x^3 + 3x^2 - 4$ at $(-3, -4)$ meets the curve again.

19 Find where the tangent to the curve $y = \dfrac{1}{x} - \dfrac{1}{x^2}$ at the point where $x = 1$, meets the curve again.

20 Find the points where the normal to $y = x^3 - 2x^2 + 1$ at $x = 1$, meets the curve again.

21 Let $P(x) = x^3 - 3x^2 - x + 3$.

    a Show that $x = 1$ is a zero of $P(x)$, and find the three real zeros.

    b Sketch the graph of $y = P(x)$.

    c Find the equation of the tangent to $y = P(x)$ at the point where $x = 2$.

    d Find where the tangent in **c** crosses the curve again.

    e Suppose a cubic has zeros a, b, and c with $a < b < c$. Prove that the tangent to the cubic at
$x = \dfrac{a + b}{2}$ meets the cubic again at $x = c$.

Example 6
🔊 **Self Tutor**

Find the equations of the tangents to $y = -x^2 + x + 2$ which pass through $(1, 3)$.

Let $(a, -a^2 + a + 2)$ be a general point on the curve.

Now $\dfrac{dy}{dx} = -2x + 1$

∴ the gradient of the tangent when $x = a$ is $-2a + 1$

∴ the equation of the tangent at $(a, -a^2 + a + 2)$ is $y = (-2a + 1)(x - a) + (-a^2 + a + 2)$

which is $y = (1 - 2a)x + a^2 + 2$

The tangents which pass through $(1, 3)$ must satisfy $(1 - 2a)(1) + a^2 + 2 = 3$

$$\therefore \quad a^2 - 2a = 0$$
$$\therefore \quad a(a - 2) = 0$$
$$\therefore \quad a = 0 \text{ or } 2$$

∴ two tangents pass through the external point $(1, 3)$.

If $a = 0$, the tangent has equation $y = x + 2$ with point of contact $(0, 2)$.

If $a = 2$, the tangent has equation $y = -3x + 6$ with point of contact $(2, 0)$.

---

22 **a** Find the equation of the tangent to $y = x^2 + 4x$ at the point where $x = a$.

    b Hence find the equations of the tangents to $y = x^2 + 4x$ which pass through the external point
$(1, -4)$. State the coordinates of the points of contact.

23 Find the equations of the tangents to $y = x^2 - 3x + 1$ which pass through $(1, -10)$.

24 **a** Find the equation of the tangent to $y = e^x$ at the point where $x = a$.

    b Hence find the equation of the tangent to $y = e^x$ which passes through the origin.

25 Find the equation of the normal to $f(x) = \cos x$ which passes through the origin.

26 The graphs of $y = \sqrt{x + a}$ and $y = \sqrt{2x - x^2}$ have the same gradient at their point of intersection.
Find a and the point of intersection.

27 Suppose P is $(-2, 3)$ and Q is $(6, -3)$. The line (PQ) is a tangent to $y = \dfrac{b}{(x + 1)^2}$. Find b.

28 Find, correct to 2 decimal places, the angle between the tangents to $y = 3e^{-x}$ and $y = 2 + e^x$ at
their point of intersection.

29 A quadratic of the form $y = ax^2$, $a > 0$, touches the logarithmic function $y = \ln x$ as shown.

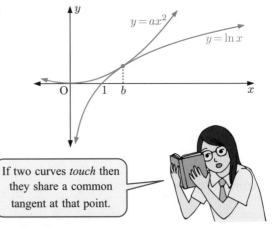

   a If the x-coordinate of the point of contact is b, explain why $ab^2 = \ln b$ and $2ab = \dfrac{1}{b}$.

   b Deduce that the point of contact is $\left(\sqrt{e}, \frac{1}{2}\right)$.

   c Find the value of a.

   d Find the equation of the common tangent.

> If two curves *touch* then they share a common tangent at that point.

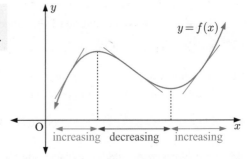

30 **a** Find the tangents to the unit circle $x^2 + y^2 = 1$ at the points where $x = \frac{1}{2}$.

   b Find the point on the x-axis where these tangents intersect.

31 Let $p(x) = ax^2$, $a \neq 0$.

   a Find the equations of the tangents to the curve at $x = s$ and $x = t$.

   b Prove that the two tangent lines intersect at $x = \dfrac{s+t}{2}$.

   c Prove that if the tangent lines are perpendicular then they intersect at $y = -\dfrac{1}{4a}$.

B   INCREASING AND DECREASING

When we draw a graph of a function, we may notice that the function is **increasing** or **decreasing** over particular intervals.

> Suppose S is an interval in the domain of $f(x)$, so $f(x)$ is defined for all x in S.
> - $f(x)$ is **increasing** on $S \iff f(a) \leqslant f(b)$ for all $a, b \in S$ such that $a < b$.
> - $f(x)$ is **decreasing** on $S \iff f(a) \geqslant f(b)$ for all $a, b \in S$ such that $a < b$.

We can determine intervals where a function $y = f(x)$ is increasing or decreasing by considering a sign diagram of the derivative function $f'(x)$.

For most functions that we deal with in this course:

- $f(x)$ is **increasing** on $S \iff f'(x) \geqslant 0$ for all $x \in S$
- $f(x)$ is **decreasing** on $S \iff f'(x) \leqslant 0$ for all $x \in S$.

Important: People often get confused about points where $f'(x) = 0$, wondering how a function can be both increasing and decreasing at the same point. We need to remember that we are saying the function is increasing or decreasing on each *interval*, rather than saying increasing or decreasing is a property of a point.

Under these definitions, constant functions such as $f(x) = 5$ are both increasing *and* decreasing on $x \in \mathbb{R}$. This is a consequence of using the non-strict inequalities $\geqslant$ and $\leqslant$. However, we can also describe intervals where a function is **strictly increasing** or **strictly decreasing**:

- $f(x)$ is **strictly increasing** on $S \iff f'(x) > 0$ for all $x \in S$
- $f(x)$ is **strictly decreasing** on $S \iff f'(x) < 0$ for all $x \in S$.

Example 7   ◀) Self Tutor

Consider $f(x) = \dfrac{2x - 3}{x^2 + 2x - 3}$.

a Show that $f'(x) = \dfrac{-2x(x - 3)}{(x - 1)^2(x + 3)^2}$, and draw its sign diagram.

b Hence find the intervals where $y = f(x)$ is increasing or decreasing.

a $f(x) = \dfrac{2x - 3}{x^2 + 2x - 3}$

$\therefore\ f'(x) = \dfrac{2(x^2 + 2x - 3) - (2x - 3)(2x + 2)}{(x^2 + 2x - 3)^2}$ {quotient rule}

$= \dfrac{2x^2 + 4x - 6 - (4x^2 - 2x - 6)}{((x - 1)(x + 3))^2}$

$= \dfrac{-2x^2 + 6x}{(x - 1)^2(x + 3)^2}$

$= \dfrac{-2x(x - 3)}{(x - 1)^2(x + 3)^2}$ which has sign diagram:

b $f(x)$ is increasing for $0 \leqslant x < 1$
and for $1 < x \leqslant 3$.

$f(x)$ is decreasing for $x < -3$
and for $-3 < x \leqslant 0$
and for $x \geqslant 3$.

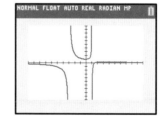

EXERCISE 8B

1

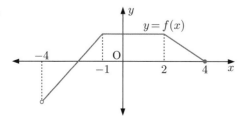

For the graph $y = f(x)$ shown, state the largest interval for which $f(x)$ is:

a increasing

b decreasing

c strictly increasing

d strictly decreasing.

2 The graph of $f(x) = x^3 - 6x^2 + 10$ is shown alongside.

   a Find $f'(x)$, and draw its sign diagram.

   b Find the intervals where $f(x)$ is increasing or decreasing.

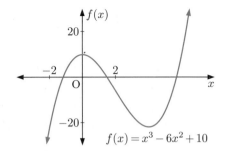

3 Find the intervals where $f(x)$ is increasing or decreasing:

   a $f(x) = x^2 + 6x$

   b $f(x) = -x^3 + 3x$

   c $f(x) = \sqrt{x}$

   d $f(x) = 3x^2 + x - 2$

   e $f(x) = \dfrac{5}{\sqrt{x}}$

   f $f(x) = x^3 - 3x^2$

4 To find the intervals where $f(x) = \ln x$ is increasing or decreasing, Kenneth states that $f'(x) = \dfrac{1}{x}$,

which has sign diagram

$$\xleftarrow{\quad\underset{\textstyle 0}{-}\quad\Big|\quad\underset{\textstyle x}{+}\quad f'(x)\quad}$$

He therefore concludes that $f(x) = \ln x$ is increasing

for $x > 0$, and decreasing for $x < 0$. Explain the mistake Kenneth has made.

5 Find the intervals where $f(x)$ is increasing or decreasing:

   a $f(x) = e^x$

   b $f(x) = \ln(x + 2)$

   c $f(x) = 3 + e^{-x}$

   d $f(x) = xe^x$

   e $f(x) = x - 2\sqrt{x}$

   f $f(x) = x^3 \ln x$

6 Consider $f(x) = \dfrac{4x}{x^2 + 1}$.

   a Show that $f'(x) = \dfrac{-4(x + 1)(x - 1)}{(x^2 + 1)^2}$ and draw its sign diagram.

   b Hence find intervals where $y = f(x)$ is increasing or decreasing.

7 Consider $f(x) = \dfrac{4x}{(x - 1)^2}$.

   a Show that $f'(x) = \dfrac{-4(x + 1)}{(x - 1)^3}$ and draw its sign diagram.

   b Hence find intervals where $y = f(x)$ is increasing or decreasing.

8 Consider $f(x) = \dfrac{-x^2 + 4x - 7}{x - 1}$.

   a Show that $f'(x) = \dfrac{-(x + 1)(x - 3)}{(x - 1)^2}$ and draw its sign diagram.

   b Hence find intervals where $y = f(x)$ is increasing or decreasing.

9 Find intervals where $f(x)$ is increasing or decreasing:

   a $f(x) = \dfrac{x^3}{x^2 - 1}$

   b $f(x) = e^{-x^2}$

   c $f(x) = (3x^2 + 1)^4$

   d $f(x) = x^2 + \dfrac{4}{x - 1}$

   e $f(x) = \ln(x^2 + 4)$

   f $f(x) = \dfrac{e^{-x}}{x}$

   g $f(x) = \sqrt{x} \ln x$

   h $f(x) = \sin x$

   i $f(x) = x + \cos x$

10 Suppose $f(x) = \dfrac{x + k}{x^2 + k}$ is never increasing. What range of values could the constant k have?

C ▌ STATIONARY POINTS

A **stationary point** of a function is a point where the tangent is horizontal, and so $f'(x) = 0$.

The sign diagram of $f'(x)$ tells us whether the point is a **local maximum**, **local minimum**, or **stationary inflection**.

At a **local maximum**, the gradient of the tangent changes from positive to negative, so the sign diagram of the derivative around

a local maximum is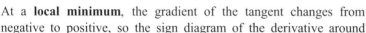

At a **local minimum**, the gradient of the tangent changes from negative to positive, so the sign diagram of the derivative around

a local minimum is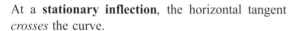

Local maxima and minima are also called **turning points**.

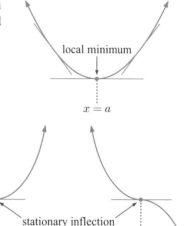

At a **stationary inflection**, the horizontal tangent *crosses* the curve.

The gradient of the tangent has the same sign either side of the stationary inflection, so the sign diagram of the derivative around a stationary inflection is

either ![+ | + f'(x)] or ![− | − f'(x)]

Example 8   ◀)) **Self Tutor**

Find the exact position and nature of the stationary point of $y = (x - 2)e^{-x}$.

$\dfrac{dy}{dx} = (1)e^{-x} + (x - 2)e^{-x}(-1)$ {product rule}

$\quad = e^{-x}(1 - (x - 2))$

$\quad = \dfrac{3 - x}{e^x}$ where e^x is positive for all x.

So, $\dfrac{dy}{dx} = 0$ when $x = 3$.

The sign diagram of $\dfrac{dy}{dx}$ is: ![+ | − at 3]

$\therefore$ at $x = 3$ we have a local maximum.

When $x = 3$, $y = (1)e^{-3} = \dfrac{1}{e^3}$ $\therefore$ the local maximum is at $\left(3, \dfrac{1}{e^3}\right)$.

EXERCISE 8C

1 The graph of $f(x) = -x^3 + 3x^2 + 9x - 7$ is shown alongside.

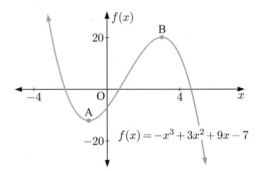

   a Classify stationary points A and B.

   b Find $f'(x)$, and draw its sign diagram.

   c Find the coordinates of A and B.

2 Consider the function $f(x) = x^4 - 6x^2 - 8x + 10$.

   a Show that $f'(x) = 4(x+1)^2(x-2)$.

   b Draw the sign diagram of $f'(x)$.

   c Find intervals where the function is increasing and decreasing.

   d Find and classify any stationary points.

   e Sketch the graph of $y = f(x)$, showing all important features.

3 For each function, find and classify any stationary points, and hence sketch its graph.

   a $f(x) = x^2 + 6x + 7$ **b** $f(x) = x^3 - 3x^2 - 24x + 30$

   c $f(x) = (2x - 1)^3 + 4$ **d** $f(x) = \sqrt{x^2 - 4x + 5}$

   e $f(x) = \dfrac{x^2}{x - 1}$ **f** $f(x) = \dfrac{x}{x^2 + 1}$

4 Find the position and nature of the stationary point(s) of:

   a $y = xe^{-x}$ **b** $y = x^2 e^x$ **c** $y = \dfrac{e^x}{x}$ **d** $y = e^{-x}(x + 2)$

5 $y = \dfrac{e^{ax}}{bx}$ has a stationary point at $\left(\dfrac{1}{3}, \dfrac{e}{2}\right)$.

   a Find a and b.   **b** State the nature of the stationary point.

6 Consider $f(x) = x \ln x$.

   a For what values of x is $f(x)$ defined?   **b** Show that the minimum value of $f(x)$ is $-\dfrac{1}{e}$.

7 For each of the following, determine the position and nature of the stationary points on the interval $0 \leqslant x \leqslant 2\pi$, then show them on a graph of the function.

   a $f(x) = \sin x$ **b** $f(x) = \cos 2x$ **c** $f(x) = \sin^2 x$

   d $f(x) = e^{\sin x}$ **e** $f(x) = \cos x - \sin x$ **f** $f(x) = \sin 2x + 2\cos x$

8 Show that $y = 4e^{-x} \sin x$ has a local maximum when $x = \frac{\pi}{4}$.

9 $f(t) = ate^{bt^2}$ has a maximum value of 1 when $t = 2$. Find constants a and b.

10 Prove that $\dfrac{\ln x}{x} \leqslant \dfrac{1}{e}$ for all $x > 0$.

11 Consider the function $f(x) = x - \ln x$.

   a Show that $y = f(x)$ has a local minimum and that this is the only turning point.

   b Hence prove that $\ln x \leqslant x - 1$ for all $x > 0$.

12 Consider the function $f(x) = \dfrac{1}{\cos x}$.

    a For what values of x is $f(x)$ undefined on the interval $0 \leqslant x \leqslant 2\pi$?

    b Find the position and nature of any stationary points on the interval $0 \leqslant x \leqslant 2\pi$.

    c Prove that $f(x + 2\pi) = f(x)$ for all $x \in \mathbb{R}$ where $f(x)$ is defined. Explain the geometrical significance of this result.

    d Sketch the graph of $y = \dfrac{1}{\cos x}$ for $-\dfrac{\pi}{2} \leqslant x \leqslant \dfrac{5\pi}{2}$. Clearly mark its stationary points.

13

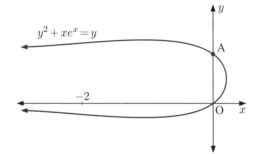

Consider the relation $y^2 + xe^x = y$.

    a Find $\dfrac{dy}{dx}$.

    b Find the equation of the normal to the relation at A.

    c Find the exact coordinates of the stationary points of the curve.

D ┃ SHAPE

We have seen that the first derivative $f'(x)$ gives the gradient of the curve $y = f(x)$ for any value of x.

The second derivative $f''(x)$ tells us the rate of change of the gradient $f'(x)$. It therefore gives us information about the **shape** or **curvature** of the curve $y = f(x)$.

- A curve is **concave downwards** on an interval $S \Leftrightarrow f''(x) \leqslant 0$ for all $x \in S$.

- A curve is **concave upwards** on an interval $S \Leftrightarrow f''(x) \geqslant 0$ for all $x \in S$.

Example 9
◀◎ **Self Tutor**

Find intervals where the curve is concave up or concave down:

    a $y = 2x^3 - 3x^2 + 4x - 6$
        b $f(x) = \dfrac{x^2 - 5x + 4}{x - 3}$

    a
$$y = 2x^3 - 3x^2 + 4x - 6$$
$$\therefore \quad \frac{dy}{dx} = 6x^2 - 6x + 4$$
$$\therefore \quad \frac{d^2y}{dx^2} = 12x - 6$$
$$= 6(2x - 1)$$

The curve is concave up for $x \geqslant \frac{1}{2}$ and concave down for $x \leqslant \frac{1}{2}$.

b $f(x) = \dfrac{x^2 - 5x + 4}{x - 3}$

$\therefore \; f'(x) = \dfrac{(2x - 5)(x - 3) - (x^2 - 5x + 4)(1)}{(x - 3)^2}$

$\qquad = \dfrac{2x^2 - 6x - 5x + 15 - x^2 + 5x - 4}{(x - 3)^2}$

$\qquad = \dfrac{x^2 - 6x + 11}{(x - 3)^2}$

$\therefore \; f''(x) = \dfrac{(2x - 6)(x - 3)^{\cancel{2}^{1}} - (x^2 - 6x + 11) \times 2\cancel{(x - 3)}}{(x - 3)^{\cancel{4}\,3}}$

$\qquad = \dfrac{2x^2 - 6x - 6x + 18 - 2x^2 + 12x - 22}{(x - 3)^3}$

$\qquad = -\dfrac{4}{(x - 3)^3}$

$$\overset{\displaystyle +}{\xleftarrow{\hspace{3cm}}} \overset{}{\underset{3}{\vdots}} \quad \overset{\displaystyle -}{\hspace{1.5cm}} \overset{f''(x)}{\underset{x}{\longrightarrow}}$$

The curve is concave up for $x < 3$ and concave down for $x > 3$.

EXERCISE 8D

1   Find intervals where the curve is concave up or concave down:

    a $f(x) = 3x^2 - 4$ **b** $f(x) = -x^3 + 3x - 2$

    c $y = -\dfrac{4}{x^2}$ **d** $y = e^{-x}$

    e $f(x) = \dfrac{1}{x} + \ln x$ **f** $f(x) = \dfrac{3 - x}{x + 2}$

    g $y = \dfrac{x^2 + x - 3}{x + 1}$ **h** $f(x) = \dfrac{\ln x}{x - 2}$

2   Find intervals where the curve $f(x) = \dfrac{x}{x^2 + 1}$ is:

    a increasing   **b** decreasing   **c** concave up   **d** concave down.

E | INFLECTION POINTS

A **point of inflection** is a point on a curve at which there is a change of **curvature**.
At a point of inflection, $f''(x) = 0$.

DEMO

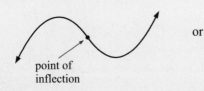

point of
inflection

or

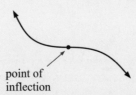

point of
inflection

If the tangent at a point of inflection is horizontal, then this is a **stationary inflection point**.

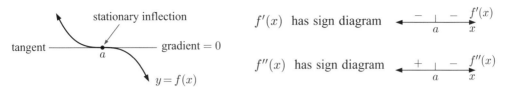

$f'(x)$ has sign diagram

$f''(x)$ has sign diagram

If the tangent at a point of inflection is *not* horizontal, then this is a **non-stationary inflection point**.

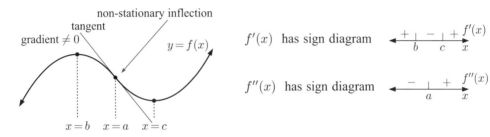

$f'(x)$ has sign diagram

$f''(x)$ has sign diagram

The tangent at the point of inflection, also called the **inflecting tangent**, crosses the curve at that point.

> There is a **point of inflection** at $x = a$ if $f''(a) = 0$ **and** the sign of $f''(x)$ changes at $x = a$.
>
> The point of inflection is a:
> - **stationary inflection** if $f'(a) = 0$
> - **non-stationary inflection** if $f'(a) \neq 0$.

Notice that if $f(x) = x^4$ then $f'(x) = 4x^3$
and $f''(x) = 12x^2$.

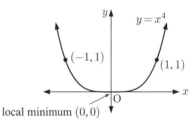

$f''(x)$ has sign diagram

Although $f''(0) = 0$ we do not have a point of inflection at $(0, 0)$ because the sign of $f''(x)$ does not change at $x = 0$.

SUMMARY

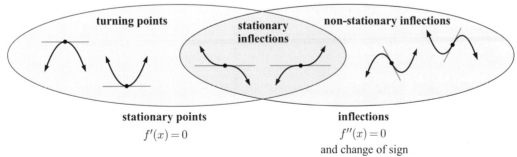

Click on the demo icon to examine some standard functions for turning points, points of inflection, and intervals where the function is increasing, decreasing, and concave up or down.

DEMO

Example 10   ◆) Self Tutor

Find and classify all points of inflection of $f(x) = x^4 - 4x^3 + 5$.

$f(x) = x^4 - 4x^3 + 5$

$\therefore \ f'(x) = 4x^3 - 12x^2 = 4x^2(x - 3)$

$\therefore \ f''(x) = 12x^2 - 24x$

$\qquad\qquad = 12x(x - 2)$

$\therefore \ f''(x) = 0$ when $x = 0$ or 2

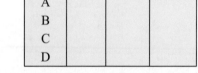

Since the signs of $f''(x)$ change about $x = 0$ and $x = 2$, these two points are points of inflection.

$f(0) = 5$ and $f'(0) = 0$.

$\therefore \ (0, 5)$ is a stationary inflection.

$f(2) = 16 - 32 + 5 = -11$ and $f'(2) = 32 - 48 \neq 0$.

$\therefore \ (2, -11)$ is a non-stationary inflection.

EXERCISE 8E

1 In the diagram alongside, each labelled point corresponds to a zero of $f(x)$, $f'(x)$, or $f''(x)$.

a Complete the table by indicating whether each value is zero, positive, or negative:

Point	$f(x)$	$f'(x)$	$f''(x)$
A			
B			
C			
D			

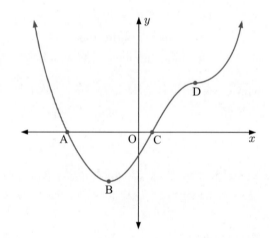

b Describe the turning point of $y = f(x)$.

c Describe the inflection points of $y = f(x)$.

2 Find and classify all points of inflection of:

a $f(x) = x^2 + 3$

b $f(x) = 2 - x^3$

c $f(x) = x^3 - 6x^2 + 9x + 1$

d $f(x) = -3x^4 - 8x^3 + 2$

e $f(x) = 3 - \dfrac{1}{\sqrt{x}}$

f $f(x) = x^3 + 6x^2 + 12x + 5$

g $f(x) = x^2 + 8\sqrt{x}$

h $f(x) = x^4 - 6x^2 + 10$

i $f(x) = 2x^2 - \dfrac{6}{x^2}$

j $f(x) = \dfrac{1}{\ln x}$

Example 11   ◀ᴏ) **Self Tutor**

Consider $f(x) = 3x^4 - 16x^3 + 24x^2 - 9$.

a Find and classify all points where $f'(x) = 0$.

b Find and classify all points of inflection.

c Find intervals where the function is increasing or decreasing.

d Find intervals where the function is concave up or down.

e Sketch the function showing the features you have found.

a $f(x) = 3x^4 - 16x^3 + 24x^2 - 9$

∴ $f'(x) = 12x^3 - 48x^2 + 48x$ ∴ $f'(x)$ has sign diagram:

$\qquad = 12x(x^2 - 4x + 4)$

$\qquad = 12x(x - 2)^2$

Now $f(0) = -9$ and $f(2) = 7$

∴ $(0, -9)$ is a local minimum and $(2, 7)$ is a stationary inflection.

b $f''(x) = 36x^2 - 96x + 48$ ∴ $f''(x)$ has sign diagram:

$\qquad = 12(3x^2 - 8x + 4)$

$\qquad = 12(x - 2)(3x - 2)$

Now $f(\frac{2}{3}) \approx -2.48$

∴ $(2, 7)$ is a stationary inflection and $(\frac{2}{3}, -2.48)$ is a non-stationary inflection.

c $f(x)$ is decreasing for $x \leqslant 0$

$f(x)$ is increasing for $x \geqslant 0$.

d $f(x)$ is concave up for $x \leqslant \frac{2}{3}$ and $x \geqslant 2$

$f(x)$ is concave down for $\frac{2}{3} \leqslant x \leqslant 2$.

e

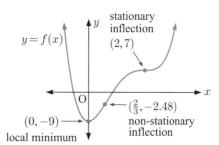

---

3 For each of the following functions:

    i Find and classify all turning points.

    ii Find and classify all points of inflection.

    iii Find intervals where the function is increasing or decreasing.

    iv Find intervals where the function is concave up or down.

    v Sketch the function showing the features you have found.

a $f(x) = x^2 - 5x + 4$ **b** $f(x) = x^3 + 4x^2$

c $f(x) = \sqrt{x}$ **d** $f(x) = x^3 - 3x^2 - 24x + 1$

e $f(x) = 3x^4 + 4x^3 - 2$ **f** $f(x) = (x - 1)^4$

g $f(x) = x^4 - 4x^2 + 3$ **h** $f(x) = 3 - \dfrac{4}{\sqrt{x}}$

Example 12

◀) **Self Tutor**

Consider the function $y = 2 - e^{-x}$.

a Find the x-intercept.
b Find the y-intercept.
c Show algebraically that the function is increasing for all x.
d Show algebraically that the function is concave down for all x.
e Explain why $y = 2$ is a horizontal asymptote.
f Sketch $y = 2 - e^{-x}$, showing the features you have found.

a When $y = 0$, $e^{-x} = 2$
$$\therefore \quad -x = \ln 2$$
$$\therefore \quad x = -\ln 2$$
$\therefore$ the x-intercept is $-\ln 2 \approx -0.693$

b When $x = 0$, $y = 2 - e^0 = 1$
$\therefore$ the y-intercept is 1.

c $\dfrac{dy}{dx} = 0 - e^{-x}(-1) = e^{-x} = \dfrac{1}{e^x}$

Now $e^x > 0$ for all x,

so $\dfrac{dy}{dx} > 0$ for all x.

$\therefore$ the function is increasing for all x.

d $\dfrac{d^2y}{dx^2} = e^{-x}(-1)$
$$= -\dfrac{1}{e^x} \quad \text{which is} \quad < 0 \text{ for all } x.$$
$\therefore$ the function is concave down for all x.

e As $x \to \infty$, $e^{-x} \to 0$
$\therefore \quad y \to 2$
Hence, the horizontal asymptote is $y = 2$.

f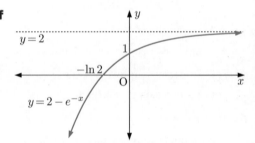

4 Consider the function $f(x) = e^{2x} - 3$.
a Find the x and y-intercepts.
b Show algebraically that the function is increasing for all x.
c Find $f''(x)$, and hence explain why $f(x)$ is concave up for all x.
d Explain why $y = -3$ is a horizontal asymptote.
e Sketch $y = e^{2x} - 3$, showing the features you have found.

5 Suppose $f(x) = e^x - 3$ and $g(x) = 3 - 5e^{-x}$.
a Find the x and y-intercepts of both functions.
b Discuss $f(x)$ and $g(x)$ as $x \to \infty$ and as $x \to -\infty$.
c Draw the sign diagrams of $f'(x)$, $f''(x)$, $g'(x)$, and $g''(x)$ and give a geometrical interpretation of each.
d Find algebraically the point(s) of intersection of the functions.
e Sketch the graphs of both functions on the same set of axes. Show all important features on your graph.

6 Consider the function $y = e^x - 3e^{-x}$.

    a Determine the x and y-intercepts.

    b Prove that the function is increasing for all x.

    c Show that $\dfrac{d^2y}{dx^2} = y$. What can be deduced about the concavity of the function above and below
    the x-axis?

    d Use technology to help graph $y = e^x - 3e^{-x}$. Show the features you have found.

7 Consider $f(x) = \ln(2x - 1) - 3$.

    a Find the x-intercept.

    b Can $f(0)$ be found? What is the significance of this result?

    c Find the domain of f.

    d Find the gradient of the tangent to the curve at $x = 1$.

    e Find $f''(x)$, and hence explain why $f(x)$ is concave down for all x in the domain of f.

    f Graph the function, showing the features you have found.

8 Consider $f(x) = \ln(3x^2)$.

    a For what values of x is $f(x)$ defined?

    b Draw the sign diagrams of $f'(x)$ and $f''(x)$, and give a geometrical interpretation of each.

    c Sketch the graph of $f(x) = \ln(3x^2)$.

9 Consider the function $f(x) = \dfrac{e^x}{x}$.

    a Does the graph of $y = f(x)$ have any x or y-intercepts?

    b Discuss $f(x)$ as $x \to \infty$ and as $x \to -\infty$.

    c Find and classify any stationary points of $y = f(x)$.

    d Find the intervals where $f(x)$ is:

        i concave up   **ii** concave down.

    e Sketch the graph of $y = f(x)$, showing all important features.

    f Find the equation of the tangent to $f(x) = \dfrac{e^x}{x}$ at the point where $x = -1$.

10 A function commonly used in statistics is the *standard normal curve* $f(x) = \dfrac{1}{\sqrt{2\pi}} e^{-\frac{1}{2}x^2}$.

    a Find the turning points of the function, and find the intervals where the function is increasing and decreasing.

    b Find all points of inflection.

    c Discuss $f(x)$ as $x \to \infty$ and as $x \to -\infty$.

    d Sketch the graph of $y = f(x)$, showing all important features.

11 Consider the function $f(x) = \cos x$.

    a Show that $f''(x) = -f(x)$. What does this tell us about the location of the inflection points?

    b Find and classify the inflection points of $f(x)$ on $0 \leqslant x \leqslant 2\pi$.

    c Find the intervals on $0 \leqslant x \leqslant 2\pi$ where $f(x)$ is:

        i increasing   **ii** decreasing   **iii** concave up   **iv** concave down.

    d Sketch the graph of $y = f(x)$ on $0 \leqslant x \leqslant 2\pi$, showing all important features.

12 Consider the *surge function* $f(t) = Ate^{-bt}$, $t \geqslant 0$, where A and b are positive constants.

    a Prove that the function has:

        i a local maximum at $t = \dfrac{1}{b}$ **ii** a point of inflection at $t = \dfrac{2}{b}$.

    b Sketch the function, showing the features you have found.

13 Consider the *logistic function* $f(t) = \dfrac{C}{1 + Ae^{-bt}}$, $t \geqslant 0$, where A, b, and C are positive constants.

    a Find the y-intercept.

    b Prove that:

        i $y = C$ is its horizontal asymptote

        ii if $A > 1$, there is a point of inflection with y-coordinate $\dfrac{C}{2}$.

    c Sketch the function, showing the features you have found.

F   RATES OF CHANGE

There are countless examples in the real world where quantities vary with time, or with respect to some other variable.

For example:

- temperature varies continuously
- the prices of stocks and shares vary with each day's trading.
- the height of a tree varies as it grows

We have already seen that if $y = f(x)$ then $f'(x)$ or $\dfrac{dy}{dx}$ gives the gradient of the tangent to $y = f(x)$ for any value of x.

> $\dfrac{dy}{dx}$ gives the **rate of change in y with respect to x.**

We can therefore use the derivative of a function to tell us the **rate** at which something is happening.

For example:

- $\dfrac{dH}{dt}$ or $H'(t)$ could be the instantaneous rate of ascent of a person in a Ferris wheel.

  It might have units metres per second or $m\,s^{-1}$.

- $\dfrac{dC}{dt}$ or $C'(t)$ could be a person's instantaneous rate of change in lung capacity.

  It might have units litres per second or $L\,s^{-1}$.

Example 13   ◀)) **Self Tutor**

According to a psychologist, the ability of a child to understand spatial concepts is given by $A = \frac{1}{3}\sqrt{t}$ where t is the age in years, $5 \leqslant t \leqslant 18$.

    a Find the rate of improvement in ability to understand spatial concepts when a child is:

        i 9 years old   **ii** 16 years old.

    b Show that $\dfrac{dA}{dt} > 0$ for $5 \leqslant t \leqslant 18$.   Comment on the significance of this result.

    c Show that $\dfrac{d^2 A}{dt^2} < 0$ for $5 \leqslant t \leqslant 18$.   Comment on the significance of this result.

a $A = \frac{1}{3}\sqrt{t} = \frac{1}{3}t^{\frac{1}{2}}$

$\therefore \quad \dfrac{dA}{dt} = \frac{1}{6}t^{-\frac{1}{2}} = \dfrac{1}{6\sqrt{t}}$

    i When $t = 9$, $\dfrac{dA}{dt} = \frac{1}{18}$ **ii** When $t = 16$, $\dfrac{dA}{dt} = \frac{1}{24}$

      $\therefore$ the rate of improvement is $\frac{1}{18}$ units $\therefore$ the rate of improvement is $\frac{1}{24}$ units

        per year for a 9 year old child.   per year for a 16 year old child.

b   Since $\sqrt{t}$ is never negative, $\dfrac{1}{6\sqrt{t}}$ is never negative

$\therefore \quad \dfrac{dA}{dt} > 0$ for all $5 \leqslant t \leqslant 18$.

This means that the ability to understand spatial concepts increases with age.

c $\dfrac{dA}{dt} = \frac{1}{6}t^{-\frac{1}{2}}$

$\therefore \quad \dfrac{d^2 A}{dt^2} = -\frac{1}{12}t^{-\frac{3}{2}} = -\dfrac{1}{12t\sqrt{t}}$

$\therefore \quad \dfrac{d^2 A}{dt^2} < 0$ for all $5 \leqslant t \leqslant 18$.

This means that while the ability to understand spatial concepts increases with age, the rate of increase slows down with age.

You are encouraged to use technology to graph each function you need to consider. This is often useful in interpreting results.

GRAPHING PACKAGE

EXERCISE 8F

1   The quantity of a chemical in human skin which is responsible for its 'elasticity' is given by $Q(t) = 100 - 10\sqrt{t}$ where t is the age of a person in years.

    a Find $Q(t)$ when:

        i $t = 0$ **ii** $t = 25$ **iii** $t = 100$ years.

    b At what rate is the quantity of the chemical changing when the person is aged:

        i 25 years   **ii** 50 years?

    c Show that the quantity of the chemical is decreasing for all $t > 0$.

2   The height of *pinus sylvestris* is given by $H = 35 - \dfrac{172.5}{t + 5}$ metres, where t is the number of years after the tree was planted from an established seedling.

    a How high was the tree when it was planted?

    b Find the height of the tree after:

        i 4 years   **ii** 8 years   **iii** 12 years.

    c Find the rate at which the tree was growing after 0, 5, and 10 years.

    d Show that $\dfrac{dH}{dt} > 0$ for all $t \geqslant 0$. What is the significance of this result?

3

The total cost of running a train from Paris to Marseille is given by $C(v) = \frac{1}{5}v^2 + \frac{200\,000}{v}$ euros where v is the average speed of the train in km h^{-1}.

a Find the total cost of the journey if the average speed is:

    i 50 km h^{-1} **ii** 100 km h^{-1}.

b Find the rate of change in the cost of running the train for the average speed:

    i 30 km h^{-1} **ii** 90 km h^{-1}.

c At what speed will the cost be a minimum?

4 A tank contains $50\,000$ litres of water. The tap is left fully on and all the water drains from the tank in 80 minutes. The volume of water remaining in the tank after t minutes is given by

$$V = 50\,000\left(1 - \frac{t}{80}\right)^2 \text{ litres where } 0 \leqslant t \leqslant 80.$$

a Find $\dfrac{dV}{dt}$, and draw the graph of $\dfrac{dV}{dt}$ against t.

b At what time was the outflow fastest?

c Show that $\dfrac{d^2V}{dt^2}$ is always constant and positive.

Interpret this result.

5 Alongside is a land and sea profile where the x-axis is sea level. The function $y = \frac{1}{10}x(x-2)(x-3)$ km gives the height of the land or sea bed relative to sea level at distance x km from the shore line.

a Find where the lake is located relative to the shore line of the sea.

b Find $\dfrac{dy}{dx}$ and interpret its value when $x = \frac{1}{2}$ km and when $x = 1\frac{1}{2}$ km.

c Find the deepest point of the lake, and the depth at this point.

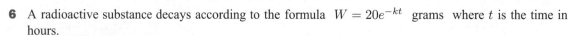

6 A radioactive substance decays according to the formula $W = 20e^{-kt}$ grams where t is the time in hours.

a Find k given that the weight is 10 grams after 50 hours.

b Find the weight of radioactive substance present:

    i initially   **ii** after 24 hours   **iii** after 1 week.

c How long will it take for the weight to reach 1 gram?

d Find the rate of radioactive decay after:

    i 100 hours   **ii** 1000 hours.

e Show that $\dfrac{dW}{dt} = bW$ for some constant b.

7 The temperature of a liquid after being placed in a refrigerator is given by $T = 5 + 95e^{-kt}$ °C where k is a positive constant and t is the time in minutes.

    a Find k if the temperature of the liquid is 20°C after 15 minutes.

    b What was the temperature of the liquid when it was first placed in the refrigerator?

    c Show that $\dfrac{dT}{dt} = c(T - 5)$ for some constant c.

    d At what rate is the temperature changing:

        i initially   **ii** after 10 minutes   **iii** after 20 minutes?

8 The height of a shrub t years after it was planted is given by $H(t) = 20\ln(3t + 2) + 30$ cm, $t \geqslant 0$.

    a How high was the shrub when it was planted?

    b How long will it take for the shrub to reach a height of 1 m?

    c At what rate is the shrub's height changing:

        i 3 years after being planted

        ii 10 years after being planted?

9 In the conversion of sugar solution to alcohol, the amount of alcohol produced t hours after the reaction commenced is given by $A = s(1 - e^{-kt})$ litres, where s is the original sugar concentration (%), $t \geqslant 0$.

    a Find A when $t = 0$.

    b Suppose $s = 10$, and $A = 5$ after 3 hours.

        i Find k.

        ii Find the speed of the reaction after 5 hours.

10 On the Indonesian coast, the depth of water t hours after midnight is given by $D = 9.3 + 6.8\cos(0.507t)$ metres.

    a Find the depth of water at 8:00 am.   **b** Is the tide rising or falling at this time?

11 The voltage in a circuit is given by $V(t) = 340\sin(100\pi t)$ volts where t is the time in seconds.

    a Find the voltage in the circuit:

        i initially   **ii** after 0.125 seconds.

    b At what rate is the voltage changing:

        i when $t = 0.01$ **ii** when $V(t)$ is a maximum?

12 The number of bees in a hive after t months is modelled by $B(t) = \dfrac{3000}{1 + 0.5e^{-1.73t}}$.

    a Find the initial bee population.

    b Find the percentage increase in the population after 1 month.

    c Is there a limit to the population size? If so, what is it?

    d Find $B'(t)$, and use it to explain why the population is increasing over time.

    e Find the rate at which the population is increasing after 6 months.

    f Sketch the graph of $B(t)$.

G | OPTIMISATION

Optimisation is the process of finding the **maximum** or **minimum** value of a function. The solution is often referred to as the **optimal solution**.

In general, we find the maximum or minimum value of a function $f(x)$ by solving $f'(x) = 0$. However, we must also examine the values of the function at the end points of the interval under consideration.

For example:

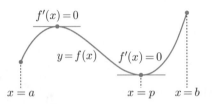

The maximum value of y occurs at the end point where $x = b$.

The minimum value of y occurs at the local minimum corresponding to $x = p$.

OPTIMISATION PROBLEM SOLVING METHOD

Step 1: Draw a large, clear **diagram** of the situation.

Step 2: Construct a **formula** with the variable to be optimised as the subject. It should be written in terms of one convenient variable, for example x. You should write down what domain restrictions there are on x.

Step 3: Find the **first derivative** and find the value(s) of x which make the first derivative **zero**.

Step 4: For each stationary point, use a sign diagram to determine if you have a local maximum or local minimum.

Step 5: Identify the optimal solution, also considering end points where appropriate.

Step 6: Write your answer in a sentence, making sure you specifically answer the question.

Example 14   ◀) **Self Tutor**

Infinitely many rectangles can be inscribed in a semi-circle of diameter 20 cm.

Find the shape of the largest rectangle which can be inscribed.

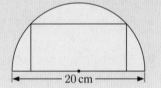

Step 1: Let $OB = x$ cm, $0 < x < 10$

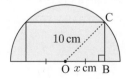

In $\triangle OBC$, $BC^2 + x^2 = 10^2$ {Pythagoras}

$\therefore BC = \sqrt{100 - x^2}$ {as $BC > 0$}

Step 2: The rectangle has area $A = \text{length} \times \text{width}$

$\therefore A = 2x\sqrt{100 - x^2}$

Step 3: $A = 2x(100 - x^2)^{\frac{1}{2}}$

$$\therefore \quad \frac{dA}{dx} = 2(100 - x^2)^{\frac{1}{2}} + 2x \times \tfrac{1}{2}(100 - x^2)^{-\frac{1}{2}} \times (-2x) \qquad \text{\{product rule\}}$$

$$= \frac{2\sqrt{100 - x^2}}{1} - \frac{2x^2}{\sqrt{100 - x^2}}$$

$$= \frac{2(100 - x^2) - 2x^2}{\sqrt{100 - x^2}}$$

$$= \frac{200 - 4x^2}{\sqrt{100 - x^2}}$$

$$= \frac{4(50 - x^2)}{\sqrt{100 - x^2}}$$

So, $\dfrac{dA}{dx} = 0$ when $x^2 = 50$

$$\therefore \quad x = \sqrt{50} \qquad \text{\{as } x > 0\text{\}}$$

Step 4: $\dfrac{dA}{dx}$ has sign diagram:

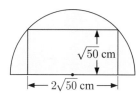

Step 5: The area is maximised when $x = \sqrt{50}$ and $BC = \sqrt{100 - 50}$
$$= \sqrt{50} \text{ cm}$$

Step 6: The largest rectangle which can be inscribed is $2\sqrt{50}$ cm long and $\sqrt{50}$ cm wide.

EXERCISE 8G

1 60 metres of fencing is used to build a rectangular enclosure along an existing fence. Suppose the sides adjacent to the existing fence are x m long.

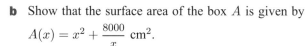

   a Show that the area A of the enclosure is given by $A(x) = x(60 - 2x)$ m^2.

   b Find the dimensions which maximise the area of the enclosure.

2 A 2 litre box has a square base and an open top. Its dimensions are labelled in the diagram.

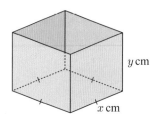

   a Explain why $y = \dfrac{2000}{x^2}$.

   b Show that the surface area of the box A is given by
   $$A(x) = x^2 + \frac{8000}{x} \text{ cm}^2.$$

   c Find the dimensions of the box which minimise its surface area.

   d Find the minimum surface area of the box.

3 Infinitely many rectangles can be inscribed in a circle of diameter
10 cm. In the diagram alongside, suppose ON $= x$ cm.

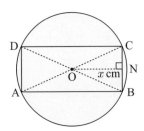

  a Find the area of ABCD in terms of x only.

  b Find the dimensions of ABCD which maximises its area.

4 A manufacturer of electric kettles performs a cost control study. They discover that to produce x kettles

per day, the cost per kettle is given by $C(x) = 4\ln x + \left(\dfrac{30 - x}{10}\right)^2$ pounds with a minimum production

capacity of 10 kettles per day.

How many kettles should be manufactured to keep the cost per kettle to a minimum?

5 Infinitely many rectangles which sit on the x-axis can be
inscribed under the curve $y = e^{-x^2}$.

Determine the coordinates of C such that rectangle ABCD
has maximum area.

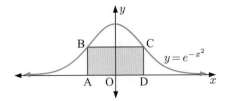

6

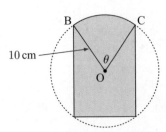

A circular piece of tinplate with radius 10 cm has 3 segments
removed as illustrated. The angle θ is measured in radians.

  a Show that the remaining area is given by
    $A = 50(\theta + 3\sin\theta)$ cm^2.

  b Find θ such that the area A is a maximum, and find the
    area A in this case.

7 When a new anaesthetic is administered, the effect is modelled by $E(t) = 750te^{-1.5t}$ units, where
$t \geqslant 0$ is the time in hours after the injection.

  a Find $E'(t)$.

  b At what time is the anaesthetic most effective?

8 A symmetrical gutter is made from a sheet of metal 30 cm wide
by bending it twice as shown.

  a Deduce that the cross-sectional area of the gutter is given by
    $A = 100\cos\theta(1 + \sin\theta)$ cm^2.

  b Show that $\dfrac{dA}{d\theta} = 0$ when $\sin\theta = \frac{1}{2}$ or -1.

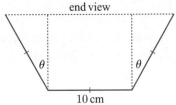

  c For what value of θ does the gutter have maximum carrying capacity? Find the cross-sectional
    area for this value of θ.

9

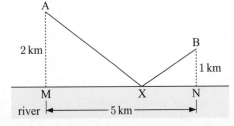

A pumphouse is to be placed at some point X along a
river.

Two pipelines will then connect the pumphouse to
homesteads A and B.

How far should point X be from M so that the total
length of pipeline is minimised?

10 Hieu can row a boat at 3 km h^{-1}, and can walk at 6 km h^{-1}. He is currently at point P on the shore of a lake 2 km in radius. He will row to point Q, then walk around the shore to point R which is opposite P.

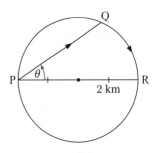

   a Show that PQ = $4 \cos \theta$ km.

   b Show that the time taken for Hieu's journey is given by
$$T = \tfrac{4}{3} \cos \theta + \frac{2\theta}{3} \text{ hours where } 0 \leqslant \theta \leqslant \tfrac{\pi}{2}.$$

   c Find θ such that $\dfrac{dT}{d\theta} = 0$ on $0 \leqslant \theta \leqslant \tfrac{\pi}{2}$.   **d** Draw a sign diagram for $\dfrac{dT}{d\theta}$.

   e What route should Hieu take to travel from P to R in:

      i the longest time   **ii** the shortest time?

11 B is a boat 5 km out at sea from A. [AC] is a straight sandy beach, 6 km long. Peter can row the boat at 8 km h^{-1} and run along the beach at 17 km h^{-1}. Suppose Peter rows directly from B to point X on [AC] such that AX = x km.

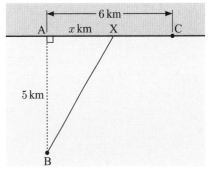

   a Explain why $0 \leqslant x \leqslant 6$.

   b Show that the *total time* Peter takes to row to X and then run along the beach to C, is given by
$$T = \frac{\sqrt{x^2 + 25}}{8} + \frac{6 - x}{17} \text{ hours, } 0 \leqslant x \leqslant 6.$$

   c Find x such that $\dfrac{dT}{dx} = 0$. Explain the significance of this value.

12 A company constructs rectangular seating arrangements for pop concerts on sports grounds. The oval shown has equation $\dfrac{x^2}{a^2} + \dfrac{y^2}{b^2} = 1$, where a and b are the lengths of the semi-major and semi-minor axes.

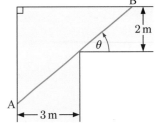

   a Show that $y = \dfrac{b}{a} \sqrt{a^2 - x^2}$ for A as shown.

   b Show that the seating area is given by
$$A(x) = \frac{4bx}{a} \sqrt{a^2 - x^2}.$$

   c Prove that the seating area is maximised when $x = \dfrac{a}{\sqrt{2}}$.

   d Given that the area of the ellipse is πab, what percentage of the ground is occupied by the seats in the optimal case?

13 Two corridors meet at right angles and are 2 m and 3 m wide respectively. [AB] is a thin metal tube which must be kept horizontal and cannot be bent as it moves around the corner from one corridor to the other.

   a Show that the length AB is given by $L = \dfrac{3}{\cos \theta} + \dfrac{2}{\sin \theta}$.

   b Show that $\dfrac{dL}{d\theta} = 0$ when $\theta = \tan^{-1}\left(\sqrt[3]{\tfrac{2}{3}}\right) \approx 41.1°$.

   c Find L when $\theta = \tan^{-1}\left(\sqrt[3]{\tfrac{2}{3}}\right)$, and comment on the significance of this value.

H | RELATED RATES

In the **Opening Problem** on page **180**, we observed a ladder slide down a wall.

Because the ladder has a fixed length, when the distance x m from the wall to the foot of the ladder increases, the height y m that the ladder reaches up the wall must decrease.

The variables $\dfrac{dx}{dt}$ and $\dfrac{dy}{dt}$ are **related rates**.

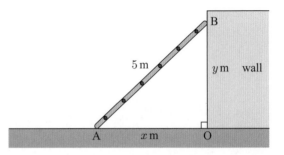

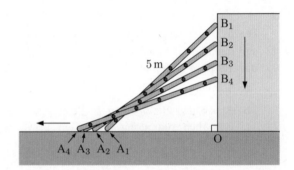

The method for solving related rates problems is:

Step 1: Draw a large, clear **diagram** of the situation.

Step 2: Write down the information, label the diagram, and make sure you distinguish between the **variables** and the **constants**.

Step 3: Write an **equation** connecting the variables. You may need to use:
- Pythagoras' theorem
- similar triangles where corresponding sides are in proportion
- right angled triangle trigonometry
- sine and cosine rules.

Step 4: **Differentiate** the equation with respect to time t.

Step 5: Substitute the values for the **particular case** corresponding to some instant in time, and solve to find the required unknown.

Warning:

> You **must not** substitute values for the particular case too early. Otherwise you will incorrectly treat variables as constants. The differentiated equation in fully generalised form must be established first.

Example 15

◀》 **Self Tutor**

A 5 m long ladder rests against a vertical wall with its feet on horizontal ground. The feet on the ground slip, and at the instant when they are 3 m from the wall, they are moving at 10 m s^{-1}.

At what speed is the other end of the ladder moving at this instant?

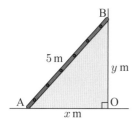

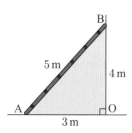

Let $OA = x$ m and $OB = y$ m

$\therefore \; x^2 + y^2 = 5^2$ {Pythagoras}

Differentiating with respect to t gives

$$2x \frac{dx}{dt} + 2y \frac{dy}{dt} = 0$$

$$\therefore \; x \frac{dx}{dt} + y \frac{dy}{dt} = 0$$

Particular case:

At the instant when $x = 3$, we find $y = 4$ and $\frac{dx}{dt} = 10$ m s^{-1}.

$$\therefore \; 3(10) + 4 \frac{dy}{dt} = 0$$

$$\therefore \; \frac{dy}{dt} = -\frac{30}{4} = -7.5 \text{ m s}^{-1}$$

The other end of the ladder is moving down the wall at 7.5 m s^{-1} at this instant.

EXERCISE 8H

1 a and b are variables related by the equation $ab^3 = 40$.

    a Differentiate this equation with respect to t.

    b At the instant when $a = 5$, b is increasing at 1 unit per second. What is happening to a at this instant?

> Differentiate **before** you substitute values for the particular case.

2 A square has side length x cm and area A cm^2. The side length of the square is increasing at 2 cm per second.

    a State the relationship between A and x.

    b Differentiate the relationship in **a** with respect to t.

    c At what rate is the area increasing when the side length is 6 cm?

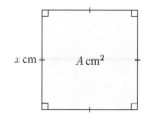

3 The length of a rectangle is decreasing at 1 cm per minute. However, the area of the rectangle remains constant at 100 cm^2.

    Let x cm be the length of the rectangle, and y cm be the width of the rectangle.

    a State the relationship between x and y.

    b Differentiate the relationship in **a** with respect to t.

    c At what rate is the width of the rectangle increasing at the instant when:

        i the length is 20 cm   **ii** the rectangle is a square?

4 A stone is thrown into a lake, and a circular ripple moves out at a constant speed of 1 ms^{-1}. Find the rate at which the circle's area is increasing at the instant when:

    a $t = 2$ seconds   **b** $t = 4$ seconds.

5 Air is pumped into a spherical weather balloon at a constant rate of $6\pi \text{ m}^3$ per minute. Find the rate of change in the balloon's radius at the instant when the radius is 2 m.

6

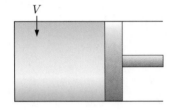

For a given mass of gas in a piston, $pV^{1.5} = 400$ where p is the pressure in N m^{-2}, and V is the volume in m^3.

Suppose the pressure increases at a constant rate of 3 N m^{-2} per minute. Find the rate at which the volume is changing at the instant when the pressure is 50 N m^{-2}.

7 Wheat runs from a hole in a silo at a constant rate and forms a conical heap whose base radius is treble its height. After 1 minute, the height of the heap is 20 cm. Find the rate at which the height is rising at this instant.

8 Two jet aeroplanes fly on parallel courses which are 12 km apart. Their air speeds are 200 ms^{-1} and 250 ms^{-1}. How fast is the distance between them changing at the instant when the slower jet is 5 km ahead of the faster one?

9 A trough of length 6 m has a uniform cross-section which is a trapezium with side lengths shown. Water leaks from the bottom of the trough at a constant rate of 0.1 m^3 per minute.

Find the rate at which the water level is falling at the instant when the water is 20 cm deep.

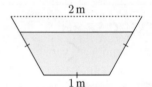

10 A ground-level floodlight located 40 m from the foot of a building shines in the direction of the building.

A 2 m tall person walks directly from the floodlight towards the building at 1 ms^{-1}. How fast is the person's shadow on the building shortening at the instant when the person is:

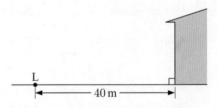

    a   20 m   **b**   10 m   from the building?

Example 16   ◆) **Self Tutor**

Triangle ABC is right angled at A, and $AB = 20$ cm. $\widehat{ABC}$ increases at a constant rate of $1°$ per minute. At what rate is BC changing at the instant when $\widehat{ABC}$ measures $30°$?

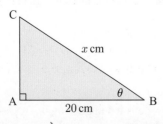

Let $\widehat{ABC} = \theta$ and $BC = x$ cm

Now $\cos\theta = \dfrac{20}{x} = 20x^{-1}$

$\therefore \ -\sin\theta\,\dfrac{d\theta}{dt} = -20x^{-2}\,\dfrac{dx}{dt}$

Particular case:

When $\theta = 30°$, $\cos 30° = \dfrac{20}{x}$

$\therefore \dfrac{\sqrt{3}}{2} = \dfrac{20}{x}$

$\therefore x = \dfrac{40}{\sqrt{3}}$

$\dfrac{d\theta}{dt}$ must be measured in **radians** per time unit.

Also, $\dfrac{d\theta}{dt} = 1°$ per minute

$= \dfrac{\pi}{180}$ radians per minute

Thus $-\sin 30° \times \dfrac{\pi}{180} = -20 \times \dfrac{3}{1600} \times \dfrac{dx}{dt}$

$\therefore -\dfrac{1}{2} \times \dfrac{\pi}{180} = -\dfrac{3}{80} \dfrac{dx}{dt}$

$\therefore \dfrac{dx}{dt} = \dfrac{\pi}{360} \times \dfrac{80}{3}$ cm per minute

≈ 0.233 cm per minute

$\therefore$ BC is increasing at approximately 0.233 cm per minute.

11 Consider a right angled triangle PQR in which QP = 10 cm and $P\widehat{Q}R = \theta$. $P\widehat{Q}R$ is increasing at $2°$ per minute.

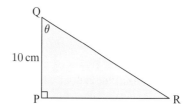

a As $P\widehat{Q}R$ increases, would you expect QR to increase or decrease?

b At what rate is QR changing when $P\widehat{Q}R$ measures $60°$?

12 A right angled triangle ABC has a fixed hypotenuse [AC] of length 10 cm. Side [AB] increases in length at 0.1 cm per second. At what rate is $C\widehat{A}B$ decreasing at the instant when the triangle is isosceles?

13 An aeroplane passes directly overhead, then flies horizontally away from an observer with altitude of 5000 m and air speed 200 m s^{-1}. At what rate is its angle of elevation to the observer changing at the instant when the angle of elevation is:

a $60°$ **b** $30°$?

14 Rectangle PQRS has [PQ] of fixed length 20 cm, and [QR] increases in length at a constant rate of 2 cm s^{-1}. At what rate is the acute angle between the diagonals of the rectangle changing at the instant when [QR] is 15 cm long?

15 Triangle PQR is right angled at Q, and [PQ] is 6 cm long. [QR] increases in length at 2 cm per minute. Find the rate of change in $Q\widehat{P}R$ at the instant when [QR] is 8 cm long.

16 Two cyclists A and B leave X simultaneously at $120°$ to one another, with constant speeds of 12 m s^{-1} and 16 m s^{-1} respectively. Find the rate at which the distance between them is changing after 2 minutes.

17 [AB] is a fixed diameter of a circle of radius 5 cm. Point P moves around the circle at a constant rate of 1 revolution in 10 seconds. Find the rate at which the distance AP is changing at the instant when:

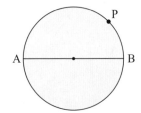

    a AP = 5 cm and increasing

    b P is at B.

18

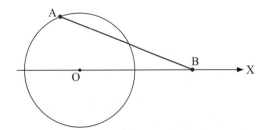

Shaft [AB] is 30 cm long and is attached to a flywheel at A. B is confined to motion along [OX]. The radius of the wheel is 15 cm, and the wheel rotates clockwise at 100 revolutions per second. Find the rate of change in $\widehat{ABO}$ when $\widehat{AOX}$ is:

    a   120°   **b**   180°

19 A farmer has a water trough of length 8 m which has a semi-circular cross-section of diameter 1 m. Water is pumped into the trough at a constant rate of 0.1 m³ per minute.

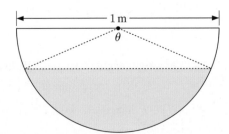

    a Show that the volume of water in the trough is given by $V = \theta - \sin\theta$, where θ is the angle illustrated (in radians).

    b Find the rate at which the water level is rising at the instant when the water is 25 cm deep.

        Hint: First find $\dfrac{d\theta}{dt}$ and then find $\dfrac{dh}{dt}$ at the given instant.

Review set 8A

1 Find the equation of the tangent to:

    a $y = x^3 - 5x + 2$ at $(2, 0)$ **b** $y = \dfrac{1 - 2x}{x^2}$ at $(1, -1)$

    c $f(x) = e^{3x-1}$ at the point where $x = 0$ **d** $f(x) = \ln(x^2)$ at the point where $x = e$.

2 Find the equation of the normal to:

    a $y = \sqrt{3x + 4}$ at $(4, 4)$ **b** $y = 3e^{2x}$ at the point where $x = 1$.

3 Show that the equation of the tangent to $y = x\tan x$ at $x = \frac{\pi}{4}$ is $(2 + \pi)x - 2y = \dfrac{\pi^2}{4}$.

4 The graph of $x^2 - y^3 = (x - 2)y + 1$ is shown alongside.

    a Find $\dfrac{dy}{dx}$.

    b Find the equation of the illustrated tangent.

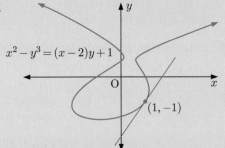

$x^2 - y^3 = (x - 2)y + 1$

$(1, -1)$

5 Consider the function $f(x) = \dfrac{3x - 2}{x + 3}$.

    a For what values of x is $f(x)$ defined?   **b** Find the axes intercepts.

    c Find $f'(x)$ and draw its sign diagram.   **d** Does the function have any stationary points?

6 Find where the tangent to $y = 2x^3 + 4x - 1$ at $(1, 5)$ meets the curve again.

7 Find the equation of the tangent to:

    a $y = \sec 2x$ at $x = \frac{\pi}{3}$ **b** $y = \csc \frac{x}{2}$ at $x = \frac{\pi}{2}$.

8 **a** Find the equation of the normal to $y = e^{2x}$ at the point where $x = a$.

    b Hence find the equation of the normal to $y = e^{2x}$ which passes through the origin.

9 The line $y = 39 - 9x$ is a tangent to $y = \dfrac{k(x + 1)}{x - 2}$. Find the possible values of k, and the point where the tangent meets the curve in each case.

10 Find intervals where $f(x)$ is increasing or decreasing:

    a $f(x) = x^3 - 6x$ **b** $f(x) = e^x(x - 2)$ **c** $f(x) = 2x - \sin x$

11 Find and classify the stationary points of:

    a $f(x) = -x^3 + 2x^2 - x + 3$ **b** $f(x) = \dfrac{x^2}{x + 3}$

12 For each of the following, determine the position and nature of the stationary points on the interval $-\pi \leqslant x \leqslant \pi$, then show them on a graph of the function.

    a $y = \sin \frac{x}{2}$ **b** $y = \cos^2 x$ **c** $y = \cos 2x - 2\sin x$

13 Find intervals where the curve is concave up or concave down:

    a $y = x^3 - 4x^2 + 11$ **b** $y = -\dfrac{x + 1}{x^2}$ **c** $y = \dfrac{x + 2}{x(x + 4)}$

14 Consider the function $f(x) = x + \ln x$.

    a Find the values of x for which $f(x)$ is defined.

    b Draw the sign diagrams of $f'(x)$ and $f''(x)$, and give a geometrical interpretation of each.

    c Sketch the graph of $y = f(x)$.

15 Consider the function $f(x) = e^{x\sqrt{3}} \sin x$.

    a Find $f'(x)$.   **b** Find x on $0 \leqslant x \leqslant 2\pi$ such that $f'(x) = 0$.

    c Draw the sign diagram for $f'(x)$ on $0 \leqslant x \leqslant 2\pi$.

    d Determine the intervals on $0 \leqslant x \leqslant 2\pi$ for which $f(x)$ is:

        i increasing   **ii** decreasing.

16 Consider the function $f(x) = \ln(x^2 + 5)$.

    a Find and classify any turning points.

    b Find and classify any points of inflection.

    c Find intervals where the function is increasing or decreasing.

    d Find intervals where the function is concave up or down.

    e Sketch the function, showing the features you have found.

17 The height of a tree t years after it was planted is given by $H(t) = 60 + 40\ln(2t + 1)$ cm, $t \geqslant 0$.

   a How tall was the tree when it was planted?

   b How long will it take for the tree to reach:

      i 150 cm   **ii** 300 cm?

   c At what rate is the tree's height increasing after:

      i 2 years   **ii** 20 years?

18 The value of a car t years after its purchase is given by $V = 20\,000e^{-0.4t}$ pounds. Calculate:

   a the purchase price of the car

   b the rate at which the value of the car is decreasing 10 years after it was purchased.

19

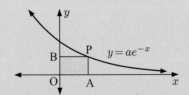

The graph of $y = ae^{-x}$ for $a > 0$ is shown.

P is a moving point on the graph, and A and B lie on the axes as shown so that OAPB is a rectangle.

Find the x-coordinate of P, in terms of a, such that the rectangle OAPB has minimum perimeter.

20 A and B are two houses directly opposite one another and 1 km from a straight road [DC]. MC is 3 km and C is a house at the roadside. A power unit is to be located on [DC] at P such that PA + PB + PC is minimised. This ensures that the cost of trenching and cable will be as small as possible.

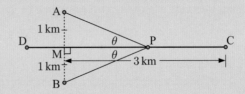

   a What cable length would be required if the power unit P is placed at:

      i M   **ii** C?

   b Show that if $\theta = A\widehat{P}M = B\widehat{P}M$, then the length of cable is given by

$$L(\theta) = 3 + \frac{2 - \cos\theta}{\sin\theta} \text{ km.}$$

   c Show that $\dfrac{dL}{d\theta} = \dfrac{1 - 2\cos\theta}{\sin^2\theta}$, and hence show that the minimum length of cable required is $(3 + \sqrt{3})$ km.

21

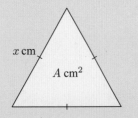

An equilateral triangle has side length x cm and area A cm^2. The side length of the triangle is increasing at 3 cm per minute.

   a State the relationship between A and x.

   b At what rate is the area increasing when the side length is 15 cm?

22 Water exits a conical tank at a constant rate of 0.2 m^3 per minute. The surface of the water has radius r m.

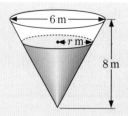

  a Find $V(r)$, the volume of the water remaining.

  b Find the rate at which the surface radius is changing at the instant when the water is 5 m deep.

23 Consider the general quadratic $p(x) = ax^2 + bx + c$.

  a Perform a "shift-expand-shift" on $p(x)$ using these steps:

    (1) Substitute $x = z + r$ into p.

    (2) Expand the brackets.

    (3) Substitute $z = x - r$ into the result, thus writing $p(x)$ in the form
$$p(x) = A(x - r)^2 + B(x - r) + C.$$

  b Prove that for any number r and any quadratic, that
$$p(x) = p(r) + p'(r)(x - r) + \frac{p''(r)}{2}(x - r)^2.$$

  c Explain how this form can be used to find the tangent to $p(x)$ at $x = r$ without using calculus.

Review set 8B

1 Find the equation of the tangent to:

  a $f(x) = x^4 - 2x^2 + 7x - 3$ at $(2, 19)$ **b** $f(x) = \dfrac{1}{\sqrt{x + 7}}$ at $(9, \frac{1}{4})$

  c $f(x) = 3\sin 2x$ at the point where $x = \frac{\pi}{6}$

  d $f(x) = \dfrac{e^x}{2 - x}$ at the point where $x = 0$.

2 Find the equation of the normal to:

  a $y = x\sin x$ at the origin   **b** $y = 3x^e - e^x$ at the point where $x = e$.

3 The graph of $7\sin\frac{\pi x}{2} + 2x^2 = y^2$ is shown alongside.

  a Find $\dfrac{dy}{dx}$.

  b Find the equation of the illustrated normal.

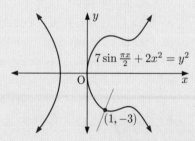

4 Consider the function $f(x) = e^x - x$.

  a Find and classify any stationary points of $y = f(x)$.

  b Discuss what happens to $f(x)$ as $x \to \infty$.

  c Find $f''(x)$ and draw its sign diagram. Give a geometrical interpretation for the sign of $f''(x)$.

  d Sketch the graph of $y = f(x)$.

  e Deduce that $e^x \geqslant x + 1$ for all x.

5 The tangent to $y = x^2\sqrt{1-x}$ at $x = -3$ cuts the axes at points A and B. Determine the area of triangle OAB.

6 Find the point where the normal to $y = x^2 - 4x + 2$ at $x = 3$, meets the curve again.

7 Find the equation of the tangent to:

    a $y = \dfrac{1}{\sin x}$ at the point where $x = \frac{\pi}{3}$ **b** $y = \cos\frac{x}{2}$ at the point where $x = \frac{\pi}{2}$.

8 Show that the curves with equations $y = \sqrt{3x+1}$ and $y = \sqrt{5x - x^2}$ have a common tangent at their point of intersection. Find the equation of this common tangent.

9 The tangent to $y = \dfrac{ax+b}{\sqrt{x}}$ at $x = 1$ has equation $2x - y = 1$. Find a and b.

10 Suppose $f(x) = \dfrac{x+1}{x^2 - 2x - 8}$.

    a Show that $f'(x) = -\dfrac{x^2 + 2x + 6}{(x^2 - 2x - 8)^2}$ and draw its sign diagram.

    b Hence show that $f(x)$ is never increasing.

11 Find and describe the stationary point of $y = \dfrac{x+a}{e^x}$, where a is a constant.

12 $f(x) = \dfrac{\ln(ax)}{bx}$ has a stationary point at $\left(\dfrac{e}{2}, \dfrac{2}{3e}\right)$. Find a and b.

13 Find and classify the inflection points of:

    a $y = x^4 - 3x^3 + 9$ **b** $y = -x^4 + x^3 + 9x^2 + 1$

14 Consider $f(x) = \sqrt{\cos x}$ for $0 \leqslant x \leqslant 2\pi$.

    a For what values of x in this interval is $f(x)$ defined?

    b Find $f'(x)$ and hence find intervals where $f(x)$ is increasing or decreasing.

    c Sketch the graph of $y = f(x)$ on $0 \leqslant x \leqslant 2\pi$.

15 Consider the function $f(x) = x^4 - 4x^3 + 7$.

    a Find $f'(x)$ and $f''(x)$, and draw their sign diagrams.

    b Find and classify any turning points.

    c Find and classify any points of inflection.

    d Find intervals where the function is:

        i increasing   **ii** decreasing   **iii** concave up   **iv** concave down.

    e Sketch the function, showing all important features.

16 For each of the following functions:

    i Find and classify all turning points.

    ii Find and classify all points of inflection.

    iii Sketch the function, showing the features you have found.

    a $f(x) = (x+2)^2(x-2)^2$ **b** $f(x) = \cos^2 x, \; 0 \leqslant x \leqslant 2\pi$

    c $f(x) = \dfrac{4}{\sqrt{x^2+1}}$

17 A cork bobs up and down in a bucket of water. The distance from the centre of the cork to the bottom of the bucket is given by $s(t) = 30 + \cos \pi t$ cm, $t \geqslant 0$ seconds.

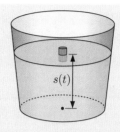

a Find the cork's velocity at times $t = 0$, $\frac{1}{2}$, 1, $1\frac{1}{2}$, and 2 seconds.

b Find the time intervals when the cork is falling.

18 A small population of wasps is observed. After t weeks the population is modelled by $P(t) = \dfrac{50\,000}{1 + 1000e^{-0.5t}}$ wasps, where $0 \leqslant t \leqslant 25$.

a Find the initial population of wasps.

b Find the population after 4 weeks.

c Find $P'(t)$.

d Show that the population is always increasing.

e At what rate is the population increasing after:

    i 2 weeks   **ii** 10 weeks?

19

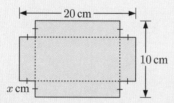

A rectangular sheet of tin-plate is 20 cm by 10 cm. Four squares, each with sides x cm, are cut from its corners. The remainder is bent into the shape of an open rectangular container. Find the value of x which will maximise the capacity of the container.

20 Two runners run in different directions, $60°$ apart. Alex runs at 5 m s^{-1} and Barry runs at 4 m s^{-1}. Barry passes through X 3 seconds after Alex passes through X.

At what rate is the distance between them increasing at the instant when Alex is 20 metres past X?

21 A light bulb hangs from the ceiling at height h metres above the floor, directly above point N. At any point A on the floor which is x metres from the light bulb, the illumination I is given by $I = \dfrac{\sqrt{8} \cos \theta}{x^2}$ units.

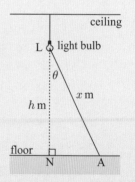

a If $\text{NA} = 1$ metre, show that at A, $I = \sqrt{8} \cos \theta \sin^2 \theta$.

b The light bulb may be lifted or lowered to change the intensity at A. Assuming $\text{NA} = 1$ metre, find the height the bulb should be above the floor to provide the greatest illumination at A.

22

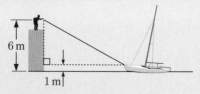

A man on a jetty pulls a boat towards him by hauling a rope at the rate 20 metres per minute. The rope is attached to the boat 1 m above water level, and the man's hands are 6 m above the water level. How fast is the boat approaching the jetty at the instant when it is 15 m from the jetty?

23 Consider the parabola $f(x) = \dfrac{1}{4b}x^2$ where $b > 0$.

   a Show that the tangent to $y = f(x)$ at $P(a, f(a))$ meets the x-axis at the point $P'\left(\dfrac{a}{2}, 0\right)$.

   b   **i** Find the equation of the line perpendicular to this tangent line, and which passes through P'.

       ii Show that this line has y-intercept $F(0, b)$.

       iii Show that the distance FP equals the distance from P to the line $y = -b$.

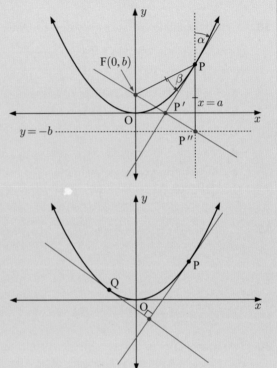

   c The point $F(0, b)$ is *invariant* since it is independent of our choice of a. F is called the *focus* of the parabola. The line $y = -b$ is called the *directrix*.

       i Prove the *reflective* property of the parabola, that any vertical ray will be reflected off the parabola into the focus.
       Hint: Show that $\alpha = \beta$.

       ii Suppose $a \neq 0$ and that $Q(c, f(c))$ is another point on the parabola such that the tangents from P and Q are perpendicular. Show that the intersection of the tangents occurs on the directrix.

9

Integration

Contents:

Opening problem

The function $f(x) = \cos x$ is non-negative for $0 \leqslant x \leqslant \frac{\pi}{2}$.

Things to think about:

a Can you use the sums of areas of rectangles to explain why the shaded area is:

   i less than $\frac{\pi}{2}$ units^2

   ii between $\frac{\pi}{4} \times \frac{1}{\sqrt{2}}$ and $\frac{\pi}{4}\left(1 + \frac{1}{\sqrt{2}}\right)$ units^2

   iii between $\frac{\pi}{6}\left(\frac{\sqrt{3}}{2} + \frac{1}{2}\right)$ and $\frac{\pi}{6}\left(1 + \frac{\sqrt{3}}{2} + \frac{1}{2}\right)$ units^2?

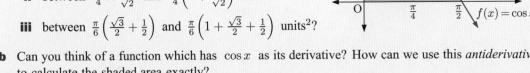

b Can you think of a function which has $\cos x$ as its derivative? How can we use this *antiderivative* to calculate the shaded area exactly?

Historical note

Italian Mathematician **Bonaventura Cavalieri** (1598 - 1647) became Professor of Mathematics at Bologna in 1629. He published tables for many trigonometric and logarithmic functions. However, his best known contribution to mathematics was the invention of **indivisibles**.

In his Method of Indivisibles, Cavalieri considered that a moving point could be used to sketch a curve. The curve could therefore be considered as the set of an infinite number of points, each with no length.

In a similar way, the "indivisibles" that made up a surface were an infinite number of lines. Almost every introduction to integral calculus starts with the division of an area into a number of rectangular strips with finite width.

Bonaventura Cavalieri

Cavalieri's important step was to make the strips narrower and narrower until they were infinitely thin lines. This reduces the "jagged" steps of the strips until they exactly define the curved boundary of the area.

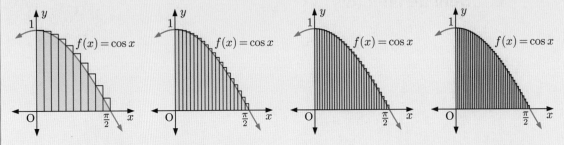

It was not until Englishman **Sir John Wallis** (1616 - 1703) formally introduced the idea of a **limit** in 1656 that Cavalieri's Method of Indivisibles progressed into the foundation for Integral Calculus.

THE DEFINITE INTEGRAL

Consider the lower and upper rectangle sums for a function which is positive and increasing on the interval $a \leqslant x \leqslant b$.

We divide the interval into n subintervals, each of width $w = \dfrac{b-a}{n}$.

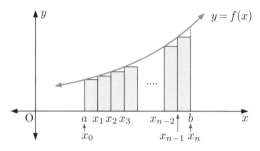 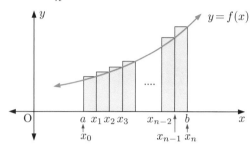

Since the function is increasing:

$$A_L = w\,f(x_0) + w\,f(x_1) + \ldots + w\,f(x_{n-2}) + w\,f(x_{n-1})$$

$$A_U = w\,f(x_1) + w\,f(x_2) + \ldots + w\,f(x_{n-1}) + w\,f(x_n)$$

$$\therefore \quad A_U - A_L = w(f(x_n) - f(x_0))$$

$$= \frac{1}{n}(b-a)(f(b) - f(a))$$

Following Cavalieri's suggestion, we allow there to be infinitely many subintervals, so $n \to \infty$.

In this case $\quad \lim\limits_{n \to \infty} (A_U - A_L) = 0$

$$\therefore \quad \lim_{n \to \infty} A_L = \lim_{n \to \infty} A_U$$

$\therefore \quad$ since $\ A_L < A < A_U \ $ for all values of n, it follows that

$$\lim_{n \to \infty} A_L = A = \lim_{n \to \infty} A_U$$

We can obtain a result like this for every increasing and decreasing interval of a positive function provided the function is *continuous*. This means that the function must have a defined value $f(k)$ for all $\ a \leqslant k \leqslant b$, and that $\lim\limits_{x \to k} f(x) = f(k)$ for all $\ a \leqslant k \leqslant b$.

A formal definition of continuous functions is not required for this course.

The "**definite integral** of $f(x)$ from a to b" is written $\displaystyle\int_a^b f(x)\,dx$.

If $\ f(x) \geqslant 0 \ $ for all $\ a \leqslant x \leqslant b$, then

$$\int_a^b f(x)\,dx \quad \text{is equal to the shaded area } A.$$

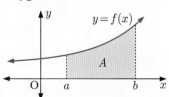

INDEFINITE INTEGRALS

Having seen that a definite integral of $f(x)$ requires a domain such as $a \leqslant x \leqslant b$, we can also consider an *indefinite* integral which does not have a domain specified.

The **indefinite integral** of a function $f(x)$ is $\displaystyle\int f(x)\,dx$.

In this Chapter we consider algebraic techniques for finding definite and indefinite integrals.

A   INTEGRATION

The reverse process of differentiation is **antidifferentiation**. It is the process of finding $f(x)$ given a derivative function $f'(x)$.

> If $F(x)$ is a function where $F'(x) = f(x)$ we say that:
> - the **derivative** of $F(x)$ is $f(x)$ and   • the **antiderivative** of $f(x)$ is $F(x)$.

For example, the derivative of x^3 is $3x^2$, so the antiderivative of $3x^2$ is x^3.

However, any function of the form $x^3 + c$ where c is constant, has derivative $3x^2$. This is because a vertical translation has no effect on the gradient of a function at any point.

We say that the **indefinite integral** or **integral** of $3x^2$ is $x^3 + c$, and write $\displaystyle\int 3x^2\,dx = x^3 + c$.

We read this as "the integral of $3x^2$ with respect to x is $x^3 + c$, where c is a constant".

> If $F'(x) = f(x)$ then $\displaystyle\int f(x)\,dx = F(x) + c$.
>
> The constant c is called the **constant of integration**.

This process is known as **indefinite integration**. It is indefinite because it is not being applied to a particular interval.

DISCOVERING INTEGRALS

Since integration is the reverse process of differentiation, we can sometimes discover integrals by differentiation.

Example 1   ◀◎ **Self Tutor**

Find the derivative of $x^{-\frac{1}{2}}$, and hence find $\displaystyle\int \frac{1}{x\sqrt{x}}\,dx$.

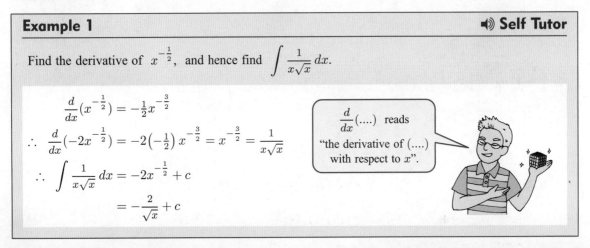

$$\frac{d}{dx}\left(x^{-\frac{1}{2}}\right) = -\tfrac{1}{2}x^{-\frac{3}{2}}$$

$$\therefore \quad \frac{d}{dx}\left(-2x^{-\frac{1}{2}}\right) = -2\left(-\tfrac{1}{2}\right)x^{-\frac{3}{2}} = x^{-\frac{3}{2}} = \frac{1}{x\sqrt{x}}$$

$$\therefore \quad \int \frac{1}{x\sqrt{x}}\,dx = -2x^{-\frac{1}{2}} + c$$

$$= -\frac{2}{\sqrt{x}} + c$$

$\dfrac{d}{dx}(....)$ reads

"the derivative of (....) with respect to x".

EXERCISE 9A

1   **a**   Find the derivative of $x^{\frac{3}{2}}$, and hence find $\displaystyle\int \sqrt{x}\,dx$.

   b   Find the derivative of $x^{-\frac{1}{3}}$, and hence find $\displaystyle\int x^{-\frac{4}{3}}\,dx$.

   c   Find the derivative of x^{n+1}, $n \neq -1$. Hence find $\displaystyle\int x^n\,dx$, $n \neq -1$.

2   **a**   Find the derivative of e^{3x}, and hence find $\displaystyle\int e^{3x}\,dx$.

   b   Find the derivative of $e^{\frac{x}{2}}$, and hence find $\displaystyle\int e^{\frac{x}{2}}\,dx$.

   c   Find the derivative of e^{kx}, $k \neq 0$. Hence find $\displaystyle\int e^{kx}\,dx$, $k \neq 0$.

3   **a**   Find the derivative of $\sin x$, and hence find $\displaystyle\int \cos x\,dx$.

   b   Find the derivative of $\cos x$, and hence find $\displaystyle\int \sin x\,dx$.

4   **a**   Find the derivative of $\ln x$, $x > 0$.   **b**   Find the derivative of $\ln(-x)$, $x < 0$.

   c   Hence find $\displaystyle\int \frac{1}{x}\,dx$, $x \neq 0$.

Example 2
◀)) **Self Tutor**

Find the derivative of $\cos\left(2x + \frac{\pi}{3}\right)$,

and hence find $\displaystyle\int \sin\left(2x + \frac{\pi}{3}\right) dx$.

$$\frac{d}{dx}\left(\cos\left(2x + \tfrac{\pi}{3}\right)\right) = -\sin\left(2x + \tfrac{\pi}{3}\right)(2)$$
$$= -2\sin\left(2x + \tfrac{\pi}{3}\right)$$
$$\therefore \quad \frac{d}{dx}\left(-\tfrac{1}{2}\cos\left(2x + \tfrac{\pi}{3}\right)\right) = \sin\left(2x + \tfrac{\pi}{3}\right)$$
$$\therefore \quad \int \sin\left(2x + \tfrac{\pi}{3}\right) dx = -\tfrac{1}{2}\cos\left(2x + \tfrac{\pi}{3}\right) + c$$

---

5   **a**   Find the derivative of $\sin 3x$, and hence find $\displaystyle\int \cos 3x\,dx$.

   b   Find the derivative of $\cos\left(\frac{\pi}{3} - x\right)$, and hence find $\displaystyle\int \sin\left(\frac{\pi}{3} - x\right) dx$.

   c   Find the derivative of e^{3x+1}, and hence find $\displaystyle\int e^{3x+1}\,dx$.

   d   Find the derivative of $6e^{-2x}$, and hence find $\displaystyle\int e^{-2x}\,dx$.

   e   Find the derivative of $\sqrt{5x-1}$, and hence find $\displaystyle\int \frac{1}{\sqrt{5x-1}}\,dx$.

   f   Find the derivative of $x^4 - x^2$, and hence find $\displaystyle\int (4x^3 - 2x)\,dx$.

   g   Find the derivative of $(2x+1)^4$, and hence find $\displaystyle\int (2x+1)^3\,dx$.

6 Suppose $F(x)$ has the derivative function $f(x)$.

   a Find the derivative of $k\,F(x)$ where k is a constant.

   b Hence show that $\displaystyle\int k\,f(x)\,dx = k\int f(x)\,dx$ where k is a constant.

7 Suppose $F(x)$ and $G(x)$ have the derivative functions $f(x)$ and $g(x)$ respectively.

   a Find the derivative of $F(x)+G(x)$.

   b Show that $\displaystyle\int [f(x)+g(x)]\,dx = \int f(x)\,dx + \int g(x)\,dx$.

B   RULES FOR INTEGRATION

In **Chapter 7** we developed a set of rules to help us differentiate functions more efficiently. These rules or combinations of them can be used to differentiate all of the functions we consider in this course.

However, the task of finding **antiderivatives** is not so easy. Many functions simply do not have antiderivatives which can be expressed easily using standard functions.

Historical note

Robert Henry Risch (1939 -) is an American mathematician. He studied at the University of California, Berkeley.

In his doctorate studies in 1968, Risch devised a method for deciding if a function has an elementary antiderivative, and if it does, finding it. The original summary of his method took over 100 pages. Later developments from this are now used in all computer algebra systems.

After completing his doctorate, he worked at the IBM Thomas Watson Research Centre.

We *can* construct some rules which will allow us to integrate most of the function types we consider in this course.

GENERAL RULES FOR INTEGRATION

In the previous Exercise you should have used differentiation to help prove the following results:

- Any constant within the integral may be written in front of the integral sign.

$$\int k\,f(x)\,dx = k\int f(x)\,dx, \quad k \text{ is a constant}$$

- The integral of a sum is the sum of the separate integrals. This rule enables us to integrate term by term.

$$\int [f(x)+g(x)]\,dx = \int f(x)\,dx + \int g(x)\,dx$$

INTEGRATING BASIC FUNCTION TYPES

For k a constant, $\dfrac{d}{dx}(kx + c) = k$ $\therefore$ $\displaystyle\int k\,dx = kx + c$

If $n \neq -1$, $\dfrac{d}{dx}\left(\dfrac{x^{n+1}}{n+1} + c\right) = \dfrac{(n+1)x^n}{n+1} = x^n$ $\therefore$ $\displaystyle\int x^n\,dx = \dfrac{x^{n+1}}{n+1} + c, \ n \neq -1$

$\dfrac{d}{dx}(e^x + c) = e^x$ $\therefore$ $\displaystyle\int e^x\,dx = e^x + c$

For $x > 0$, $\dfrac{d}{dx}(\ln x + c) = \dfrac{1}{x}$

For $x < 0$, $\dfrac{d}{dx}(\ln(-x) + c) = \dfrac{-1}{-x} = \dfrac{1}{x}$

$\therefore$ $\displaystyle\int \dfrac{1}{x}\,dx = \begin{cases} \ln x & \text{if } x > 0 \\ \ln(-x) & \text{if } x < 0 \end{cases}$

$\therefore$ $\displaystyle\int \dfrac{1}{x}\,dx = \ln|x| + c, \ x \neq 0$

$\dfrac{d}{dx}(\sin x + c) = \cos x$ $\therefore$ $\displaystyle\int \cos x\,dx = \sin x + c$

$\dfrac{d}{dx}(-\cos x + c) = \sin x$ $\therefore$ $\displaystyle\int \sin x\,dx = -\cos x + c$

Function	Integral
k, a constant	$kx + c$
x^n, $n \neq -1$	$\dfrac{x^{n+1}}{n+1} + c$
e^x	$e^x + c$
$\dfrac{1}{x}$	$\ln|x| + c, \ x \neq 0$
$\cos x$	$\sin x + c$
$\sin x$	$-\cos x + c$

c is an arbitrary constant called the **constant of integration** or **integrating constant**.

Example 3

◀) Self Tutor

Find:

a $\displaystyle\int (-2x^3 + 5x - 2)\,dx$ **b** $\displaystyle\int \left(5e^x + \frac{2}{\sqrt{x}}\right) dx$ **c** $\displaystyle\int (2\sin x - \cos x)\,dx$

a $\displaystyle\int (-2x^3 + 5x - 2)\,dx$

$= \dfrac{-2x^4}{4} + \dfrac{5x^2}{2} - 2x + c$

$= -\tfrac{1}{2}x^4 + \tfrac{5}{2}x^2 - 2x + c$

b $\displaystyle\int \left(5e^x + \frac{2}{\sqrt{x}}\right) dx$

$= \displaystyle\int (5e^x + 2x^{-\frac{1}{2}})\,dx$

$= 5e^x + \dfrac{2x^{\frac{1}{2}}}{\frac{1}{2}} + c$

$= 5e^x + 4\sqrt{x} + c$

c $\displaystyle\int (2\sin x - \cos x)\,dx$

$= 2(-\cos x) - \sin x + c$

$= -2\cos x - \sin x + c$

Example 4

◀) Self Tutor

Find:

a $\displaystyle\int \left(x - \frac{1}{x}\right)^3 dx$ **b** $\displaystyle\int \frac{3\sqrt{x} + x^2}{x}\,dx$

a $\displaystyle\int \left(x - \frac{1}{x}\right)^3 dx$

$= \displaystyle\int \left(x^3 + 3x^2\left(-\frac{1}{x}\right) + 3x\left(-\frac{1}{x}\right)^2 + \left(-\frac{1}{x}\right)^3\right) dx$

$= \displaystyle\int (x^3 - 3x + 3x^{-1} - x^{-3})\,dx$

$= \dfrac{x^4}{4} - \dfrac{3x^2}{2} + 3\ln|x| - \dfrac{x^{-2}}{(-2)} + c$

$= \tfrac{1}{4}x^4 - \tfrac{3}{2}x^2 + 3\ln|x| + \dfrac{1}{2x^2} + c$

b $\displaystyle\int \frac{3\sqrt{x} + x^2}{x}\,dx$

$= \displaystyle\int (3x^{-\frac{1}{2}} + x)\,dx$

$= \dfrac{3x^{\frac{1}{2}}}{\frac{1}{2}} + \dfrac{x^2}{2} + c$

$= 6\sqrt{x} + \tfrac{1}{2}x^2 + c$

EXERCISE 9B

1 Find:

a $\displaystyle\int (2x^2 - 3x + 1)\,dx$

b $\displaystyle\int (-x^3 + 4x^2 - 3)\,dx$

c $\displaystyle\int \left(\tfrac{1}{2}x + x^2 + x^3\right) dx$

d $\displaystyle\int \left(4x^2 + \frac{1}{x}\right) dx$

e $\displaystyle\int \left(2\sqrt{x} - \frac{3}{\sqrt{x}}\right) dx$

f $\displaystyle\int \left(\frac{1}{3x} - \frac{2}{x^2}\right) dx$

g $\displaystyle\int (x\sqrt{x} - 9)\,dx$

h $\displaystyle\int (3x^{-\frac{3}{2}} + x^{\frac{1}{4}})\,dx$

Remember that you can check your integration by differentiating the resulting function.

2 Find:

a $\displaystyle\int (2e^x - 3x)\, dx$ 　　　**b** $\displaystyle\int \left(\frac{4}{x} + x^2 - e^x\right) dx$ 　　　**c** $\displaystyle\int \left(5e^x + \tfrac{1}{2}x^2\right) dx$

3 Integrate with respect to x:

a $3\sin x - 2$ 　　　**b** $4x - 2\cos x$ 　　　**c** $\sin x - 2\cos x + e^x$

d $x^2\sqrt{x} - 10\sin x$ 　　　**e** $\dfrac{x(x-1)}{3} + \cos x$ 　　　**f** $-\sin x + 2\sqrt{x}$

4 Find:

a $\displaystyle\int (3x - 2)^2\, dx$ 　　　**b** $\displaystyle\int \left(\sqrt{x} - \frac{1}{\sqrt{x}}\right)^2 dx$ 　　　**c** $\displaystyle\int \frac{3x - x^3}{x^2}\, dx$

d $\displaystyle\int \frac{1 - x^2}{x}\, dx$ 　　　**e** $\displaystyle\int \left(\frac{2}{x} + 1\right)^2 dx$ 　　　**f** $\displaystyle\int \left(x + \frac{1}{x}\right)^3 dx$

g $\displaystyle\int \frac{x^2 - 4}{x\sqrt{x}}\, dx$ 　　　**h** $\displaystyle\int \frac{x^2 - 4x + 10}{x^2}\, dx$ 　　　**i** $\displaystyle\int (x + 1)^4\, dx$

5 Find:

a $\displaystyle\int \left(\sqrt{x} + \tfrac{1}{2}\cos x\right) dx$ 　　　**b** $\displaystyle\int (2e^x - 4\sin x)\, dx$ 　　　**c** $\displaystyle\int (3\cos x - \sin x)\, dx$

6 Find y if:

a $\dfrac{dy}{dx} = 2 - \dfrac{1}{x}$ 　　　**b** $\dfrac{dy}{dx} = \sin x + 2\cos x$ 　　　**c** $\dfrac{dy}{dx} = 2e^x - 5 + x$

7 Find:

a $\displaystyle\int \frac{3x^3 - 2x^2 + 5}{x^2}\, dx$ 　　　　　　**b** $\displaystyle\int \left(x^2 - \frac{1}{x}\right)^2 dx$

c $\displaystyle\int \frac{x^2 - 4x + 2}{\sqrt{x}}\, dx$ 　　　　　　**d** $\displaystyle\int \sqrt{x}(3x - 1)^2\, dx$

Example 5 　　　　　　　　　　　　◀) Self Tutor

Find $\dfrac{d}{dx}(x\sin x)$ and hence determine $\displaystyle\int x\cos x\, dx$.

$\dfrac{d}{dx}(x\sin x) = (1)\sin x + (x)\cos x$ 　　　{product rule}

$\qquad\qquad = \sin x + x\cos x$

$\therefore\ \displaystyle\int (\sin x + x\cos x)\, dx = x\sin x + c$

$\therefore\ \displaystyle\int \sin x\, dx + \int x\cos x\, dx = x\sin x + c$

$\therefore\ -\cos x + \displaystyle\int x\cos x\, dx = x\sin x + c$

$\therefore\ \displaystyle\int x\cos x\, dx = x\sin x + \cos x + c$

8 Find $\dfrac{d}{dx}(e^x \sin x)$ and hence find $\displaystyle\int e^x(\sin x + \cos x)\,dx$.

9 Find $\dfrac{d}{dx}(e^{-x} \sin x)$ and hence find $\displaystyle\int \dfrac{\cos x - \sin x}{e^x}\,dx$.

10 Find $\dfrac{d}{dx}(x \cos x)$ and hence find $\displaystyle\int x \sin x\,dx$.

11 Find $\dfrac{d}{dx}(\ln(3x^2 + 1))$ and hence find $\displaystyle\int \dfrac{2x}{3x^2 + 1}\,dx$.

C　PARTICULAR VALUES

We can find the constant of integration c if we are given a particular value of the function.

Example 6　　　　　◀ﾘ Self Tutor

Find $f(x)$ given that $f'(x) = 2\sin x - \sqrt{x}$ and $f(0) = 4$.

$$f'(x) = 2\sin x - \sqrt{x}$$

$$\therefore\ f(x) = \int \left(2\sin x - x^{\frac{1}{2}}\right) dx$$

$$\therefore\ f(x) = 2(-\cos x) - \frac{x^{\frac{3}{2}}}{\frac{3}{2}} + c$$

$$\therefore\ f(x) = -2\cos x - \tfrac{2}{3}x^{\frac{3}{2}} + c$$

But $f(0) = 4$, so $-2\cos 0 - 0 + c = 4$

$$\therefore\ c = 6$$

Thus $f(x) = -2\cos x - \tfrac{2}{3}x^{\frac{3}{2}} + 6$.

EXERCISE 9C

1 Find $f(x)$ given that:

  a $f'(x) = 3 - x$ and $f(0) = 2$ **b** $f'(x) = 2x^2 - 9$ and $f(1) = -2$

  c $f'(x) = \sqrt{x} - 2$ and $f(4) = 0$ **d** $f'(x) = \dfrac{1}{x}$ and $f(e) = 2$.

2 Find $f(x)$ given that:

  a $f'(x) = x^2 - 4\cos x$ and $f(0) = 3$ **b** $f'(x) = 2\cos x - 3\sin x$ and $f\!\left(\frac{\pi}{4}\right) = \frac{1}{\sqrt{2}}$

  c $f'(x) = \sqrt{x} - 2\sin x$ and $f(0) = -2$ **d** $f'(x) = e^x + 3\cos x$ and $f(\pi) = 0$.

3 A curve has gradient function $\dfrac{dy}{dx} = 1 - e^x$ and passes through $(3,\ e^3)$. Find the equation of the curve.

4 A curve has gradient function $f'(x) = ax^2 + bx$ where a, b are constants. Find $f(x)$ given that $f(-1) = -2$, $f(0) = 1$, and $f(1) = 4$.

5 A function has second derivative $f''(x) = \cos x + 1$. If $f(0) = 1$ and $f(\frac{\pi}{3}) = \frac{1}{4}$, find $f(x)$.

D | INTEGRATING $f(ax + b)$

In this Section we deal with integrals of functions which are composite with the linear function $ax + b$.

Notice that $\quad \dfrac{d}{dx}(e^{ax+b}) = ae^{ax+b}$

$$\therefore \quad \int e^{ax+b} \, dx = \frac{1}{a} e^{ax+b} + c \quad \text{for } a \neq 0$$

Likewise if $n \neq -1$,

$$\frac{d}{dx}((ax+b)^{n+1}) = (n+1)(ax+b)^n \times a$$
$$= a(n+1)(ax+b)^n$$

$$\therefore \quad \int (ax+b)^n \, dx = \frac{1}{a} \frac{(ax+b)^{n+1}}{(n+1)} + c \quad \text{for } n \neq -1$$

We can perform the same process for trigonometric functions:

$$\frac{d}{dx}(\sin(ax+b)) = a\cos(ax+b)$$

So, $$\int \cos(ax+b) \, dx = \frac{1}{a}\sin(ax+b) + c \quad \text{for } a \neq 0.$$

Likewise we can show $$\int \sin(ax+b) \, dx = -\frac{1}{a}\cos(ax+b) + c \quad \text{for } a \neq 0.$$

Finally, $\quad \dfrac{d}{dx}\left(\dfrac{1}{a}\ln(ax+b)\right) = \dfrac{1}{a}\left(\dfrac{a}{ax+b}\right) = \dfrac{1}{ax+b} \quad$ for $\quad ax+b > 0, \quad a \neq 0$

$$\therefore \quad \int \frac{1}{ax+b} \, dx = \frac{1}{a}\ln(ax+b) + c \quad \text{for } ax+b > 0, \quad a \neq 0$$

We can similarly show that $\displaystyle\int \frac{1}{ax+b} \, dx = \frac{1}{a}\ln(-(ax+b)) + c \quad$ for $\quad ax+b < 0, \quad a \neq 0$

$$\therefore \quad \int \frac{1}{ax+b} \, dx = \frac{1}{a}\ln|ax+b| + c \quad \text{for } a \neq 0.$$

For a, b constants with $a \neq 0$, we have:

Function	Integral
e^{ax+b}	$\dfrac{1}{a} e^{ax+b} + c$
$(ax+b)^n, \; n \neq -1$	$\dfrac{1}{a} \dfrac{(ax+b)^{n+1}}{n+1} + c$
$\cos(ax+b)$	$\dfrac{1}{a} \sin(ax+b) + c$
$\sin(ax+b)$	$-\dfrac{1}{a} \cos(ax+b) + c$
$\dfrac{1}{ax+b}$	$\dfrac{1}{a} \ln|ax+b| + c$

Example 7
◀) **Self Tutor**

Find: **a** $\displaystyle \int (2x+3)^4 \, dx$ **b** $\displaystyle \int \frac{1}{\sqrt{1-3x}} \, dx$

a $\displaystyle \int (2x+3)^4 \, dx = \frac{1}{2} \times \frac{(2x+3)^5}{5} + c$

$ = \frac{1}{10}(2x+3)^5 + c$

b $\displaystyle \int \frac{1}{\sqrt{1-3x}} \, dx = \int (1-3x)^{-\frac{1}{2}} \, dx$

$\phantom{\int \frac{1}{\sqrt{1-3x}} \, dx} = \frac{1}{-3} \times \frac{(1-3x)^{\frac{1}{2}}}{\frac{1}{2}} + c$

$\phantom{\int \frac{1}{\sqrt{1-3x}} \, dx} = -\frac{2}{3}\sqrt{1-3x} + c$

EXERCISE 9D

1 Find:

a $\displaystyle \int (2x+5)^3 \, dx$

b $\displaystyle \int \frac{1}{(3-2x)^2} \, dx$

c $\displaystyle \int \frac{4}{(2x-1)^4} \, dx$

d $\displaystyle \int (4x-3)^7 \, dx$

e $\displaystyle \int \sqrt{3x-4} \, dx$

f $\displaystyle \int \frac{10}{\sqrt{1-5x}} \, dx$

g $\displaystyle \int 3(1-x)^4 \, dx$

h $\displaystyle \int \frac{4}{\sqrt{3-4x}} \, dx$

i $\displaystyle \int \frac{5}{(3x-2)^3} \, dx$

2 Find $y = f(x)$ given $\dfrac{dy}{dx} = \sqrt{2x-7}$ and that $f(8) = 11$.

3 The function $f(x)$ has gradient function $f'(x) = \dfrac{4}{\sqrt{1-x}}$, and the curve $y = f(x)$ passes through the point $(-3, -11)$.

Find the point on the graph of $y = f(x)$ with x-coordinate -8.

4 Find:

a $\displaystyle \int 3(2x-1)^2 \, dx$

b $\displaystyle \int (4x-5)^2 \, dx$

c $\displaystyle \int (1-3x)^3 \, dx$

d $\displaystyle \int (2-5x)^2 \, dx$

e $\displaystyle \int 4\sqrt{5-x} \, dx$

f $\displaystyle \int (7x+1)^4 \, dx$

5 Find an expression for y given that $\dfrac{dy}{dx} = x - \dfrac{5}{(1-x)^2}$ and that the curve passes through $(2, 0)$.

Example 8   ◀) **Self Tutor**

Integrate with respect to x:

    a $2e^{2x} - e^{-3x}$ **b** $2\sin 3x + \cos(4x + \pi)$

    a $\displaystyle\int (2e^{2x} - e^{-3x})\,dx$

        $= 2(\tfrac{1}{2})e^{2x} - (\tfrac{1}{-3})e^{-3x} + c$

        $= e^{2x} + \tfrac{1}{3}e^{-3x} + c$

    b $\displaystyle\int (2\sin 3x + \cos(4x + \pi))\,dx$

        $= 2 \times -\tfrac{1}{3}\cos 3x + \tfrac{1}{4}\sin(4x + \pi) + c$

        $= -\tfrac{2}{3}\cos 3x + \tfrac{1}{4}\sin(4x + \pi) + c$

---

6 Integrate with respect to x:

    a $\sin 3x$ **b** $2\cos(-4x) + 1$ **c** $3\cos\frac{x}{2}$

    d $3\sin 2x - e^{-x}$ **e** $2\sin\left(2x + \frac{\pi}{6}\right)$ **f** $-3\cos\left(\frac{\pi}{4} - x\right)$

    g $\cos 2x + \sin 2x$ **h** $2\sin 3x + 5\cos 4x$ **i** $\frac{1}{2}\cos 8x - 3\sin x$

7 Find:

    a $\displaystyle\int (2e^x + 5e^{2x})\,dx$ **b** $\displaystyle\int (3e^{5x-2})\,dx$ **c** $\displaystyle\int (e^{7-3x})\,dx$

    d $\displaystyle\int (e^x + e^{-x})^2\,dx$ **e** $\displaystyle\int (e^{-x} + 2)^2\,dx$ **f** $\displaystyle\int \dfrac{(e^{2x} - 5)^2}{e^x}\,dx$

8 Find an expression for y given that $\dfrac{dy}{dx} = (1 - e^x)^2$, and that the graph has y-intercept 4.

9 Suppose $f'(x) = p\sin\frac{x}{2}$ where p is a constant. $f(0) = 1$ and $f(2\pi) = 0$. Find p and hence $f(x)$.

10 Consider a function g such that $g''(x) = -\sin 2x$.

    Show that the gradients of the tangents to $y = g(x)$ when $x = \pi$ and $x = -\pi$ are equal.

11 Find $f(x)$ given $f'(x) = 2e^{-2x}$ and $f(0) = 3$.

12 A curve has gradient function $\sqrt{x} + \frac{1}{2}e^{-4x}$ and passes through $(1, 0)$. Find the equation of the function.

13 Show that $(\sin x + \cos x)^2 = 1 + \sin 2x$, and hence determine $\displaystyle\int (\sin x + \cos x)^2\,dx$.

14   **a** Rearrange the double angle formulae $\cos 2x = 1 - 2\sin^2 x$ and $\cos 2x = 2\cos^2 x - 1$ to write expressions for $\sin^2 x$ and $\cos^2 x$.

      b Hence find:

         i $\displaystyle\int \sin^2 x\,dx$ **ii** $\displaystyle\int \cos^2 x\,dx$

Example 9
🔊 Self Tutor

Find:

a $\displaystyle\int \frac{1}{5x-3}\,dx$

b $\displaystyle\int \frac{4}{1-2x}\,dx$

a $\displaystyle\int \frac{1}{5x-3}\,dx$

$= \frac{1}{5}\ln|5x-3| + c$

b $\displaystyle\int \frac{4}{1-2x}\,dx$

$= 4\left(\frac{1}{-2}\right)\ln|1-2x| + c$

$= -2\ln|1-2x| + c$

15 Integrate with respect to x:

a $\dfrac{6}{x+4}$

b $1-2x+\dfrac{4}{x-3}$

16 Find:

a $\displaystyle\int \frac{1}{2x-1}\,dx$

b $\displaystyle\int \frac{5}{1-3x}\,dx$

c $\displaystyle\int \left(4+\frac{1}{5x-2}\right)dx$

d $\displaystyle\int \left(e^{-x}-\frac{4}{2x+1}\right)dx$

e $\displaystyle\int \left(\frac{1}{x+2}+\frac{2}{x-3}\right)dx$

f $\displaystyle\int \left(\frac{5}{x-6}-\frac{2}{3x-1}\right)dx$

17 To find $\displaystyle\int \frac{1}{4x}\,dx$, Tracy's answer was $\displaystyle\int \frac{1}{4x}\,dx = \frac{1}{4}\ln|4x| + c$.

Nadine's answer was $\displaystyle\int \frac{1}{4x}\,dx = \frac{1}{4}\int \frac{1}{x}\,dx = \frac{1}{4}\ln|x| + c$.

Which of them has found the correct answer? Prove your statement.

18 Show that $\dfrac{4x+1}{x-1}$ may be written in the form $4+\dfrac{5}{x-1}$. Hence find $\displaystyle\int \frac{4x+1}{x-1}\,dx$.

Example 10
🔊 Self Tutor

Integrate $(2-\sin x)^2$ with respect to x.

$\displaystyle\int (2-\sin x)^2\,dx$

$= \displaystyle\int (4 - 4\sin x + \sin^2 x)\,dx$

$= \displaystyle\int \left(4 - 4\sin x + \frac{1}{2} - \frac{1}{2}\cos 2x\right)dx$

$= \displaystyle\int \left(\frac{9}{2} - 4\sin x - \frac{1}{2}\cos 2x\right)dx$

$= \frac{9}{2}x + 4\cos x - \frac{1}{2}\times\frac{1}{2}\sin 2x + c$

$= \frac{9}{2}x + 4\cos x - \frac{1}{4}\sin 2x + c$

> The identities
> $\cos^2\theta = \frac{1}{2}+\frac{1}{2}\cos 2\theta$
> $\sin^2\theta = \frac{1}{2}-\frac{1}{2}\cos 2\theta$
> are extremely useful!

19 Integrate with respect to x:

- **a** $\cos^2 x + 2$
- **b** $\sin^2 x + 4x$
- **c** $1 + \cos^2 2x$
- **d** $3 - \sin^2 3x$
- **e** $\frac{1}{2}\cos^2 4x$
- **f** $(1 + \cos x)^2$
- **g** $\sin x(2\sin x - 1)$
- **h** $(5 - \sin x)^2$
- **i** $(2\cos x + 1)^2$
- **j** $(1 - 3\sin x)^2$

20
- **a** Show that $\tan^2 x = \sec^2 x - 1$.

- **b** By differentiating $\tan x = \dfrac{\sin x}{\cos x}$, show that

  $$\int \sec^2 x \, dx = \tan x + c.$$

Remember that:

$$\sec x = \frac{1}{\cos x}$$

$$\csc x = \frac{1}{\sin x}$$

$$\cot x = \frac{1}{\tan x}$$

- **c** Use the identity $\tan^2 x = \sec^2 x - 1$ to find:

  i $\displaystyle\int \tan^2 x \, dx$

  ii $\displaystyle\int (\tan x + 3)(\tan x - 3) \, dx$

  iii $\displaystyle\int \left(\tfrac{1}{2}\tan^2 x - 5\right) dx$

  iv $\displaystyle\int (\tan^2 x + \sin^2 x) \, dx$

E   DEFINITE INTEGRALS

At the start of the Chapter we saw how the addition of areas of rectangles can be used to approximate the area under a curve. As the number of rectangles tends to infinity, the approximation becomes exact.

We saw that if $f(x) \geqslant 0$ for all $a \leqslant x \leqslant b$, then the area under the curve between $x = a$ and $x = b$, is

$$A = \int_a^b f(x) \, dx.$$

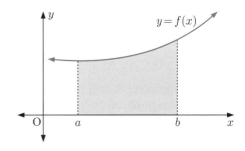

Now consider a function $y = f(t)$ which has antiderivative $F(t)$ and an area function $A(x) = \displaystyle\int_a^x f(t) \, dt$ which is the area from $t = a$ to $t = x$.

$A(x)$ is an increasing function since $f(x) \geqslant 0$, and $A(a) = 0$ .... (1)

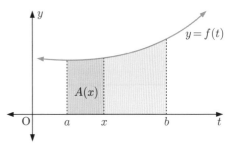

The area of the strip between $t = x$ and $t = x + h$ is $A(x + h) - A(x)$, but we also know it must lie between a lower and upper rectangle on the interval $x \leqslant t \leqslant x + h$ of width h.

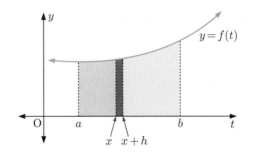

$$\text{area of lower} \atop \text{rectangle} \leqslant A(x + h) - A(x) \leqslant {\text{area of upper} \atop \text{rectangle}}$$

If $f(t)$ is increasing on this interval then

$$hf(x) \leqslant A(x + h) - A(x) \leqslant hf(x + h)$$

$$\therefore \quad f(x) \leqslant \frac{A(x + h) - A(x)}{h} \leqslant f(x + h)$$

Equivalently, if $f(t)$ is decreasing on this interval then

$$f(x + h) \leqslant \frac{A(x + h) - A(x)}{h} \leqslant f(x)$$

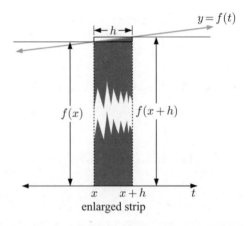

Taking the limit as $h \to 0$ gives

$$f(x) \leqslant A'(x) \leqslant f(x)$$
$$\therefore \quad A'(x) = f(x)$$

So, the area function $A(x)$ must only differ from the antiderivative of $f(x)$ by a constant.

$$\therefore \quad A(x) = F(x) + c$$

Letting $x = a$, $A(a) = F(a) + c$

But from (1), $A(a) = 0$ $\therefore$ $c = -F(a)$

$$\therefore \quad A(x) = F(x) - F(a)$$

and so $\displaystyle\int_a^x f(t)\, dt = F(x) - F(a)$

Letting $x = b$, $\displaystyle\int_a^b f(t)\, dt = F(b) - F(a)$

This result is in fact true for all continuous functions $f(t)$.

However, in situations where a function is negative, the area between the curve and the x-axis is counted as negative. We therefore refer to $A(x)$ as a **signed area function**.

THE FUNDAMENTAL THEOREM OF CALCULUS

From the argument above, the Fundamental Theorem of Calculus can be stated in two forms:

1 For a continuous function $f(t)$, if we define the signed area function from $t = a$ to $t = x$ as

$$A(x) = \int_a^x f(t)\, dt, \quad \text{then} \quad A'(x) = \frac{d}{dx}\left(\int_a^x f(t)\, dt \right) = f(x).$$

or more commonly, exchanging the variable t for the more familiar x:

2 For a continuous function $f(x)$ with antiderivative $F(x)$, $\displaystyle\int_a^b \boldsymbol{f(x)\, dx = F(b) - F(a)}$.

DEFINITE INTEGRALS

The Fundamental Theorem of Calculus allows us to calculate the definite integral $\displaystyle\int_a^b f(x)\,dx$.

$\displaystyle\int_a^b f(x)\,dx$ reads "the integral from $x = a$ to $x = b$ of $f(x)$ with respect to x"
or "the integral from a to b of $f(x)$ with respect to x".

When calculating definite integrals we can omit the constant of integration c as this will always cancel out in the subtraction process.

It is common to write $F(b) - F(a)$ as $[F(x)]_a^b$, and so

$$\int_a^b f(x)\,dx = [F(x)]_a^b = F(b) - F(a)$$

For continuous functions, we can list the following properties of definite integrals:

- $\displaystyle\int_a^b f(x)\,dx = -\int_b^a f(x)\,dx$

- $\displaystyle\int_a^b k\,f(x)\,dx = k\int_a^b f(x)\,dx,$ k is any constant

- $\displaystyle\int_a^b f(x)\,dx + \int_b^c f(x)\,dx = \int_a^c f(x)\,dx$

- $\displaystyle\int_a^b [f(x) + g(x)]\,dx = \int_a^b f(x)\,dx + \int_a^b g(x)\,dx$

Example 11   ◄)) Self Tutor

Find:

a $\displaystyle\int_1^2 (x^3 - 4x + 5)\,dx$ **b** $\displaystyle\int_1^4 \left(2x + \frac{3}{x}\right)dx$ **c** $\displaystyle\int_0^{\frac{\pi}{3}} \sin x\,dx$

a $\displaystyle\int_1^2 (x^3 - 4x + 5)\,dx = \left[\tfrac{1}{4}x^4 - 2x^2 + 5x\right]_1^2$

$\qquad\qquad = \left(\tfrac{2^4}{4} - 2(2)^2 + 5(2)\right) - \left(\tfrac{1^4}{4} - 2(1)^2 + 5(1)\right)$

$\qquad\qquad = 6 - 3\tfrac{1}{4}$

$\qquad\qquad = \tfrac{11}{4}$

b $\displaystyle\int_1^4 \left(2x + \frac{3}{x}\right)dx$ **c** $\displaystyle\int_0^{\frac{\pi}{3}} \sin x\,dx = \left[-\cos x\right]_0^{\frac{\pi}{3}}$

$\quad = \left[x^2 + 3\ln|x|\right]_1^4$ $\qquad = (-\cos\tfrac{\pi}{3}) - (-\cos 0)$

$\quad = (16 + 3\ln 4) - (1 + 3\ln 1)$ $\qquad = -\tfrac{1}{2} + 1$

$\quad = 15 + 6\ln 2$ $\qquad = \tfrac{1}{2}$

EXERCISE 9E

1 Use the Fundamental Theorem of Calculus to prove that for continuous functions:

a $\displaystyle\int_a^b f(x)\,dx = -\int_b^a f(x)\,dx$

b $\displaystyle\int_a^b k\,f(x)\,dx = k\int_a^b f(x)\,dx, \quad k$ is any constant

c $\displaystyle\int_a^b f(x)\,dx + \int_b^c f(x)\,dx = \int_a^c f(x)\,dx$

d $\displaystyle\int_a^b [f(x)+g(x)]\,dx = \int_a^b f(x)\,dx + \int_a^b g(x)\,dx$

2 Find:

a $\displaystyle\int_0^3 (x^2+2x)\,dx$

b $\displaystyle\int_0^2 (3x^2-x+6)\,dx$

c $\displaystyle\int_{-1}^2 (x^3-6x^2+x)\,dx$

d $\displaystyle\int_3^5 \frac{1}{x^2}\,dx$

e $\displaystyle\int_1^4 \left(x^2+\frac{1}{x}\right)dx$

f $\displaystyle\int_1^4 (x+2\sqrt{x})\,dx$

3 Find:

a $\displaystyle\int_0^1 (3x+1)^4\,dx$

b $\displaystyle\int_{-1}^1 (1-2x)^3\,dx$

c $\displaystyle\int_1^5 \sqrt{6x-5}\,dx$

d $\displaystyle\int_{-3}^0 \sqrt{1-x}\,dx$

e $\displaystyle\int_2^3 \frac{1}{(7-4x)^2}\,dx$

f $\displaystyle\int_2^6 \frac{1}{\sqrt{2x-3}}\,dx$

4 Write $\displaystyle\int_3^{12} \frac{1}{x}\,dx$ as a single logarithm.

5 Find:

a $\displaystyle\int_{-6}^{-2} \frac{1}{x}\,dx$

b $\displaystyle\int_{-1}^5 \frac{1}{x+4}\,dx$

c $\displaystyle\int_1^8 \frac{2}{3x+4}\,dx$

d $\displaystyle\int_{-4}^0 \frac{4}{5-2x}\,dx$

6 Evaluate:

a $\displaystyle\int_0^1 e^x\,dx$

b $\displaystyle\int_0^3 (2e^x-3)\,dx$

c $\displaystyle\int_1^4 \left(3\sqrt{x}-\frac{e^x}{2}\right)dx$

d $\displaystyle\int_0^2 e^{3x}\,dx$

e $\displaystyle\int_0^1 e^{1-x}\,dx$

f $\displaystyle\int_0^{\ln 4} e^x(e^x-2)\,dx$

7 Find:

a $\displaystyle\int_1^2 \left(x-\frac{1}{x}\right)^2 dx$

b $\displaystyle\int_1^2 (e^{-x}+1)^2\,dx$

c $\displaystyle\int_0^1 (2x-\sqrt{x})^3\,dx$

d $\displaystyle\int_0^{\ln 2} (e^x+1)^2\,dx$

8 Evaluate:

a $\displaystyle\int_0^\pi \sin x\,dx$

b $\displaystyle\int_0^{\frac{\pi}{6}} \cos x\,dx$

c $\displaystyle\int_{\frac{\pi}{3}}^{\frac{\pi}{2}} \sin x\,dx$

d $\displaystyle\int_0^{\frac{\pi}{6}} \sin 3x\,dx$

e $\displaystyle\int_{\frac{\pi}{6}}^{\frac{\pi}{2}} \cos\left(x-\frac{\pi}{3}\right)dx$

f $\displaystyle\int_{\frac{\pi}{4}}^{\frac{\pi}{2}} \sin\left(2x-\frac{\pi}{4}\right)dx$

9 Evaluate:

a $\displaystyle\int_0^{\frac{\pi}{4}} \cos^2 x \, dx$

b $\displaystyle\int_0^{\frac{\pi}{2}} \sin^2 x \, dx$

c $\displaystyle\int_0^{\frac{\pi}{6}} (\sin 3x - \cos x) \, dx$

d $\displaystyle\int_{\frac{\pi}{3}}^{\frac{2\pi}{3}} (\cos 2x - \sin x)^2 \, dx$

10 Find m such that $\displaystyle\int_m^{-2} \frac{1}{4 - x} \, dx = \ln\left(\frac{3}{2}\right)$.

Review set 9A

1 Find the derivative of $\sin\left(\frac{\pi}{3} - 2x\right)$, and hence find $\displaystyle\int \cos\left(\frac{\pi}{3} - 2x\right) dx$.

2 Find the derivative of $\sqrt{x^2 - 4}$, and hence find $\displaystyle\int \frac{x}{\sqrt{x^2 - 4}} \, dx$.

3 Find:

a $\displaystyle\int (-3x^4 + 6x^2) \, dx$

b $\displaystyle\int \frac{3x^3 - x^2 - 1}{x^2} \, dx$

c $\displaystyle\int (2x - \sqrt{x})^2 \, dx$

d $\displaystyle\int \left(4e^x - \frac{3}{x}\right) dx$

e $\displaystyle\int \sin(4x - 5) \, dx$

f $\displaystyle\int e^{4 - 3x} \, dx$

4 Find y if:

a $\dfrac{dy}{dx} = 3e^{-x} - 2\sin\left(\frac{\pi}{2} - x\right)$

b $\dfrac{dy}{dx} = \cos 4x - \frac{1}{2}x^2$

5 Given that $f'(x) = 3e^{2x}$ and $f(0) = 2$, find $f(x)$.

6 Find $\dfrac{d}{dx}(x^2 \sin x)$ and $\dfrac{d}{dx}(x \cos x)$. Hence find $\displaystyle\int x^2 \cos x \, dx$.

7 Find:

a $\displaystyle\int \frac{x^2 - 7}{x} \, dx$

b $\displaystyle\int \left(e^{2x-3} - \frac{2}{3x - 1}\right) dx$

c $\displaystyle\int (4 - 3x)^3 + \sin(-2x) \, dx$

8 A curve has gradient function $f'(x) = a \cos 3x$ where a is a constant. $f(0) = -1$ and $f(\frac{\pi}{4}) = 1$. Find a and hence $f(x)$.

9 Find m such that $\displaystyle\int_0^m \left(x^2 + \frac{1}{2}x\right) dx = 3$.

10 Find the exact value of:

a $\displaystyle\int_{-5}^{-1} \sqrt{1 - 3x} \, dx$

b $\displaystyle\int_0^{\frac{\pi}{2}} \cos\frac{x}{2} \, dx$

c $\displaystyle\int_2^6 \frac{2}{x} \, dx$

11 If $\displaystyle\int_0^a e^{1-2x} \, dx = \frac{e}{4}$, find the exact value of a.

12 Find $\dfrac{d}{dx}(e^{-2x} \sin x)$ and hence evaluate $\displaystyle\int_0^{\frac{\pi}{2}} \left[e^{-2x}(\cos x - 2\sin x)\right] dx$

Review set 9B

1 Find the derivative of $\ln(2x + 1)$, and hence find $\displaystyle\int \frac{1}{2x+1}\, dx$.

2 By differentiating $(3x^2 + x)^3$, find $\displaystyle\int (3x^2 + x)^2(6x + 1)\, dx$.

3 Find:

   a $\displaystyle\int (2x^3 - 5x + 7)\, dx$ **b** $\displaystyle\int \left(3x - \frac{1}{x}\right) dx$ **c** $\displaystyle\int \left(1 - x^2\right)^3 dx$

   d $\displaystyle\int (2e^{-x} + 3)\, dx$ **e** $\displaystyle\int 4\cos 2x\, dx$ **f** $\displaystyle\int \left(3 + e^{2x-1}\right)^2 dx$

4 Find y if:

   a $\dfrac{dy}{dx} = \left(x - 1 + \dfrac{1}{x}\right)^2$ **b** $\dfrac{dy}{dx} = \cos(3x - \pi) + 2\sin x$

5 Find $f(x)$ given $f'(x) = \dfrac{2}{x} - 1$ and $f(2) = e$.

6 A curve has gradient function $\dfrac{dy}{dx} = ax^2 + b\sqrt{x-1}$ where a, b are constants. Find the equation of the curve given that it contains the points $(1, 4)$, $(2, 4)$, and $(5, 1)$.

7 Find $f(x)$ given $f'(x) = \dfrac{3}{\sqrt{4 - 3x}}$ and $f(-4) = 0$.

8 Show that $(\sin x - \cos x)^2 = 1 - \sin 2x$, and hence determine $\displaystyle\int (\sin x - \cos x)^2\, dx$.

9 Find:

   a $\displaystyle\int \frac{1}{3 - 2x}\, dx$ **b** $\displaystyle\int \frac{4}{5x + 1}\, dx$ **c** $\displaystyle\int (1 - \sin x)^2\, dx$

10 Find the exact value of:

   a $\displaystyle\int_2^3 \frac{1}{\sqrt{3x - 4}}\, dx$ **b** $\displaystyle\int_{2e}^{3e} \frac{4}{x + e}\, dx$ **c** $\displaystyle\int_{\frac{\pi}{4}}^{\frac{\pi}{2}} (2\sin x + 1)^2\, dx$

11 Find the values of b such that $\displaystyle\int_0^b \cos x\, dx = \frac{1}{\sqrt{2}}$, $0 < b < \pi$.

12 **a** Use the identity $\cos^2 \theta = \frac{1}{2} + \frac{1}{2}\cos 2\theta$ to show that $\cos^4 x = \frac{1}{8}\cos 4x + \frac{1}{2}\cos 2x + \frac{3}{8}$.

   b Hence find $\displaystyle\int \cos^4 x\, dx$.

10

Techniques for integration

Contents:

Opening problem

Consider the indefinite integral $\displaystyle\int \frac{2x+1}{x^2+x-6}\,dx$.

Things to think about:

a Can you use any of the rules from **Chapter 9** to find the integral?

b Can you write $\dfrac{2x+1}{x^2+x-6}$ in another form which will enable you to find the integral?

c If we let $u = x^2 + x - 6$, then $\dfrac{du}{dx} = 2x + 1$.

$\therefore \quad \dfrac{2x+1}{x^2+x-6}$ can be written as $\dfrac{1}{u}\dfrac{du}{dx}$.

Can we use this to help find the integral?

In the previous Chapter we integrated some basic function types by applying rules of differentiation in reverse. We now consider some more advanced techniques for integration which will allow us to integrate some other functions of particular forms:

- partial fractions
- integration by substitution
- integration by parts.

A   PARTIAL FRACTIONS

In **Chapter 2** we saw how rational functions of the form $\dfrac{\text{linear}}{\text{quadratic}}$ can be decomposed into the sum of partial fractions. The resulting sum can then be integrated one term at a time.

For example, for the integral in the **Opening Problem**:

$$\int \frac{2x+1}{x^2+x-6}\,dx = \int \left(\frac{1}{x+3} + \frac{1}{x-2}\right) dx$$

$$= \ln|x+3| + \ln|x-2| + c$$

Example 1   ◀)) **Self Tutor**

a Show that $\dfrac{4}{x+5} - \dfrac{1}{x-3} = \dfrac{3x-17}{(x+5)(x-3)}$.

b Hence find $\displaystyle\int \frac{3x-17}{x^2+2x-15}\,dx$.

a $\dfrac{4}{x+5} - \dfrac{1}{x-3} = \dfrac{4(x-3) - 1(x+5)}{(x+5)(x-3)}$

$\qquad\qquad\qquad = \dfrac{3x-17}{(x+5)(x-3)}$

b $\displaystyle\int \frac{3x-17}{x^2+2x-15}\,dx$

$\qquad = \displaystyle\int \frac{3x-17}{(x+5)(x-3)}\,dx$

$\qquad = \displaystyle\int \left(\frac{4}{x+5} - \frac{1}{x-3}\right) dx \qquad \{\text{using } \mathbf{a}\}$

$\qquad = 4\ln|x+5| - \ln|x-3| + c$

EXERCISE 10A

1 a Show that $\dfrac{3}{x+2} - \dfrac{1}{x-2} = \dfrac{2x-8}{(x+2)(x-2)}$.

  b Hence find $\displaystyle\int \dfrac{2x-8}{x^2-4}\,dx$.

2 a Show that $\dfrac{1}{2x-1} - \dfrac{1}{2x+1} = \dfrac{2}{(2x-1)(2x+1)}$.

  b Hence find $\displaystyle\int \dfrac{2}{4x^2-1}\,dx$.

Example 2   ◀) **Self Tutor**

Find $\displaystyle\int \dfrac{7x-1}{2x^2-x-1}\,dx$.

$2x^2 - x - 1 = (2x+1)(x-1)$

Let $\dfrac{7x-1}{2x^2-x-1} = \dfrac{A}{2x+1} + \dfrac{B}{x-1}$

$\therefore\ 7x-1 = A(x-1) + B(2x+1)$

Substituting $x = 1$, $7 - 1 = 3B$
$\therefore\ B = 2$

Substituting $x = -\tfrac{1}{2}$, $7\left(-\tfrac{1}{2}\right) - 1 = A\left(-\tfrac{1}{2} - 1\right)$
$\therefore\ -\tfrac{9}{2} = -\tfrac{3}{2}A$
$\therefore\ A = 3$

$\therefore\ \dfrac{7x-1}{2x^2-x-1} = \dfrac{3}{2x+1} + \dfrac{2}{x-1}$

$\therefore\ \displaystyle\int \dfrac{7x-1}{2x^2-x-1}\,dx = \int \left(\dfrac{3}{2x+1} + \dfrac{2}{x-1} \right) dx$

$\qquad\qquad = \tfrac{3}{2} \ln|2x+1| + 2\ln|x-1| + c$

---

3 a Write $\dfrac{20}{(2x-3)(x+1)}$ as the sum of partial fractions.

  b Hence find $\displaystyle\int \dfrac{20}{2x^2-x-3}\,dx$.

4 a Write $\dfrac{x-9}{x^2-2x-3}$ as the sum of partial fractions.

  b Hence find $\displaystyle\int \dfrac{x-9}{x^2-2x-3}\,dx$.

5 Integrate with respect to x:

  a $\dfrac{3}{(x-2)(x+1)}$ **b** $\dfrac{4x}{x^2+2x-3}$ **c** $\dfrac{1-x}{x^2+2x}$

  d $\dfrac{2x-5}{x^2-x-2}$ **e** $\dfrac{2-x}{2x^2-5x-3}$ **f** $\dfrac{2x}{6x^2-x-15}$

6 Find:

a $\displaystyle\int_0^1 \frac{x+4}{x^2-5x+6}\,dx$

b $\displaystyle\int_{-2}^3 \frac{3x-8}{x^2-x-42}\,dx$

c $\displaystyle\int_1^4 \frac{2x+12}{3x^2+4x-4}\,dx$

d $\displaystyle\int_{-1}^0 \frac{x-3}{8x^2-14x+3}\,dx$

B   INTEGRATION BY SUBSTITUTION

We have already seen how the integral in the **Opening Problem** can be found using partial fractions. We now consider another method which can be used to find this integral, which is available because the expression $\dfrac{2x+1}{x^2+x-6}$ has a particular form. This method is called **integration by substitution**.

Integration by substitution is the reverse process of differentiating using the **chain rule**.

Suppose $F(u)$ is the antiderivative of $f(u)$, so $\dfrac{dF}{du} = f(u)$

$$\therefore \quad \int f(u)\,du = F(u) + c \quad \text{.... (1)}$$

But $\dfrac{dF}{dx} = \dfrac{dF}{du}\dfrac{du}{dx}$ {chain rule}

$$= f(u)\dfrac{du}{dx}$$

$$\therefore \quad \int f(u)\dfrac{du}{dx}\,dx = F(u) + c \quad \text{.... (2)}$$

Comparing (1) and (2), we conclude that: $\displaystyle\int f(u)\frac{du}{dx}\,dx = \int f(u)\,du$

Using the theorem for the **Opening Problem**,

$$\int \frac{2x+1}{x^2+x-6}\,dx$$

$$= \int \frac{1}{u}\frac{du}{dx}\,dx \quad \{\text{letting } u = x^2+x-6, \ \frac{du}{dx} = 2x+1\}$$

$$= \int \frac{1}{u}\,du$$

$$= \ln|u| + c$$

$$= \ln\left|x^2+x-6\right| + c$$

We use u in our solution, but we give our answer in terms of x, since the original integral was with respect to x.

Notice that $\ln\left|x^2+x-6\right| = \ln\left|(x+3)(x-2)\right|$

$$= \ln|x+3| + \ln|x-2|$$

This solution agrees with the one we found using partial fractions.

Example 3

◀) **Self Tutor**

Use substitution to find:

a $\displaystyle\int \sqrt{x^3 + 2x}\,(3x^2 + 2)\,dx$

b $\displaystyle\int xe^{1-x^2}\,dx$

a $\displaystyle\int \sqrt{x^3 + 2x}\,(3x^2 + 2)\,dx$

$= \displaystyle\int \sqrt{u}\,\dfrac{du}{dx}\,dx \qquad \{u = x^3 + 2x,$

$= \displaystyle\int u^{\frac{1}{2}}\,du \qquad\qquad\quad \dfrac{du}{dx} = 3x^2 + 2\}$

$= \dfrac{u^{\frac{3}{2}}}{\frac{3}{2}} + c$

$= \tfrac{2}{3}(x^3 + 2x)^{\frac{3}{2}} + c$

b $\displaystyle\int xe^{1-x^2}\,dx$

$= -\tfrac{1}{2}\displaystyle\int (-2x)e^{1-x^2}\,dx$

$= -\tfrac{1}{2}\displaystyle\int e^u\,\dfrac{du}{dx}\,dx \qquad \{u = 1 - x^2,$

$= -\tfrac{1}{2}\displaystyle\int e^u\,du \qquad\qquad \dfrac{du}{dx} = -2x\}$

$= -\tfrac{1}{2}e^u + c$

$= -\tfrac{1}{2}e^{1-x^2} + c$

Example 4

◀) **Self Tutor**

Integrate with respect to x:

a $\cos^3 x \sin x$

b $\dfrac{\cos x}{\sin x}$

a $\displaystyle\int \cos^3 x \sin x\,dx$

$= \displaystyle\int (\cos x)^3 \sin x\,dx$

$= \displaystyle\int u^3\left(-\dfrac{du}{dx}\right)dx \qquad \{u = \cos x,$

$= -\displaystyle\int u^3\,du \qquad\qquad\quad \dfrac{du}{dx} = -\sin x\}$

$= -\dfrac{u^4}{4} + c$

$= -\tfrac{1}{4}\cos^4 x + c$

b $\displaystyle\int \dfrac{\cos x}{\sin x}\,dx$

$= \displaystyle\int \dfrac{1}{u}\dfrac{du}{dx}\,dx \qquad \{u = \sin x,$

$= \displaystyle\int \dfrac{1}{u}\,du \qquad\qquad \dfrac{du}{dx} = \cos x\}$

$= \ln|u| + c$

$= \ln|\sin x| + c$

The substitutions we make need to be chosen carefully.

For example, in **Example 4** part **b**, if we let $u = \cos x$, then $\dfrac{du}{dx} = -\sin x$ and we obtain

$\displaystyle\int \dfrac{\cos x}{\sin x}\,dx = \int \dfrac{u}{-\frac{du}{dx}}\,dx$. This is not in the correct form to apply our theorem, so we have made the wrong

substitution and we need to try another option.

Example 5   ◀) **Self Tutor**

Find $\displaystyle\int \sin^3 x \, dx$.

$$\int \sin^3 x \, dx$$

$$= \int \sin^2 x \sin x \, dx$$

$$= \int (1 - \cos^2 x) \sin x \, dx$$

$$= \int (1 - u^2)\left(-\frac{du}{dx}\right) dx \qquad \{u = \cos x, \ \frac{du}{dx} = -\sin x\}$$

$$= \int (u^2 - 1) \, du$$

$$= \frac{u^3}{3} - u + c$$

$$= \tfrac{1}{3}\cos^3 x - \cos x + c$$

EXERCISE 10B.1

1 Use the substitution given to perform each integration:

a $\displaystyle\int 3x^2(x^3 + 1)^4 \, dx$ using $u = x^3 + 1$

b $\displaystyle\int x^2 e^{x^3 + 1} \, dx$ using $u = x^3 + 1$

c $\displaystyle\int \sin^4 x \cos x \, dx$ using $u = \sin x$

d $\displaystyle\int 2x \cos(x^2 - 3) \, dx$ using $u = x^2 - 3$

2 Integrate by substitution:

a $4x^3(2 + x^4)^3$

b $\dfrac{2x}{\sqrt{x^2 + 3}}$

c $\dfrac{6x^2}{(2x^3 - 1)^4}$

d $\sqrt{x^3 + x}\,(3x^2 + 1)$

e $(x^3 + 2x + 1)^4(3x^2 + 2)$

f $\dfrac{x}{(1 - x^2)^5}$

g $\dfrac{x^2}{(3x^3 - 1)^4}$

h $\dfrac{x + 2}{(x^2 + 4x - 3)^2}$

i $x^4(x + 1)^4(2x + 1)$

3 Find:

a $\displaystyle\int -2e^{1-2x} \, dx$

b $\displaystyle\int 2xe^{x^2} \, dx$

c $\displaystyle\int \dfrac{e^{\sqrt{x}}}{\sqrt{x}} \, dx$

d $\displaystyle\int (2x - 1)e^{x - x^2} \, dx$

4 Find:

a $\displaystyle\int \dfrac{2x}{x^2 + 1} \, dx$

b $\displaystyle\int \dfrac{x}{2 - x^2} \, dx$

c $\displaystyle\int \dfrac{2x - 3}{x^2 - 3x} \, dx$

d $\displaystyle\int \dfrac{6x^2 - 2}{x^3 - x} \, dx$

e $\displaystyle\int \dfrac{4x - 10}{5x - x^2} \, dx$

f $\displaystyle\int \dfrac{1 - x^2}{x^3 - 3x} \, dx$

5 Integrate with respect to x:

  a $x^2(3 - x^3)^2$

  b $\dfrac{4}{x \ln x}$

  c $x\sqrt{1 - x^2}$

  d xe^{1-x^2}

  e $\dfrac{1 - 3x^2}{x^3 - x}$

  f $\dfrac{(\ln x)^3}{x}$

6 Integrate with respect to x:

  a $\sin^4 x \cos x$

  b $\cos^5 x \sin x$

  c $\dfrac{\sin x}{\sqrt{\cos x}}$

  d $\tan x$

  e $\sqrt{\sin x} \cos x$

  f $\dfrac{\cos x}{(2 + \sin x)^2}$

  g $\dfrac{\sin x}{\cos^3 x}$

  h $\dfrac{\sin x}{1 - \cos x}$

  i $\dfrac{\cos 2x}{\sin 2x - 3}$

7 Find:

  a $\displaystyle\int \cos^3 x \, dx$
  b $\displaystyle\int \sin^5 x \, dx$
  c $\displaystyle\int \sin^4 x \cos^3 x \, dx$
  d $\displaystyle\int \sin^3 2x \cos 2x \, dx$

8 Find $f(x)$ if $f'(x)$ is:

  a $\sin x e^{\cos x}$

  b $x \sin(x^2)$

  c $\dfrac{\sin x + \cos x}{\sin x - \cos x}$

9 Find:

  a $\displaystyle\int \cot 3x \, dx$
  b $\displaystyle\int \sec x \tan x \, dx$
  c $\displaystyle\int \csc x \cot x \, dx$

  d $\displaystyle\int \tan 3x \sec 3x \, dx$
  e $\displaystyle\int \csc \tfrac{x}{2} \cot \tfrac{x}{2} \, dx$
  f $\displaystyle\int \cos x \csc^4 x \, dx$

10 Find:

  a $\displaystyle\int \sqrt{2x + 1} \, dx$ using the substitution $u = 2x + 1$

  b $\displaystyle\int \dfrac{x \ln(x^2 + 7)}{x^2 + 7} \, dx$ using the substitution $u = \ln(x^2 + 7)$

  c $\displaystyle\int x\sqrt{x - 3} \, dx$ using the substitution $u = x - 3$

  d $\displaystyle\int x^2 \sqrt{x - 16} \, dx$ using the substitution $u = x - 16$

  e $\displaystyle\int x^3 \sqrt{3 - x^2} \, dx$ using the substitution $u = 3 - x^2$

  f $\displaystyle\int \dfrac{4 \ln x}{x(1 + [\ln x]^2)} \, dx$ using the substitution $u = \ln x$.

11 Find $\displaystyle\int \dfrac{1}{\sqrt{x} + \sqrt[3]{x}} \, dx$ using the substitution $u = x^{\frac{1}{6}}$.

DEFINITE INTEGRALS INVOLVING SUBSTITUTION

When we evaluate a definite integral using substitution, we need to make sure the endpoints are converted to the new variable.

Example 6   ◀)) **Self Tutor**

Evaluate:

a $\displaystyle\int_2^3 \frac{x}{x^2-1}\,dx$

b $\displaystyle\int_0^1 \frac{6x}{(x^2+1)^3}\,dx$

a Let $u = x^2 - 1$ $\therefore$ $\dfrac{du}{dx} = 2x$

When $x = 2$, $u = 2^2 - 1 = 3$
When $x = 3$, $u = 3^2 - 1 = 8$

$\therefore \displaystyle\int_2^3 \frac{x}{x^2-1}\,dx$

$= \displaystyle\int_2^3 \frac{1}{u}\left(\frac{1}{2}\frac{du}{dx}\right)dx$

$= \dfrac{1}{2}\displaystyle\int_3^8 \frac{1}{u}\,du$

$= \dfrac{1}{2}\Big[\ln|u|\Big]_3^8$

$= \dfrac{1}{2}(\ln 8 - \ln 3)$

$= \dfrac{1}{2}\ln\left(\dfrac{8}{3}\right)$

b Let $u = x^2 + 1$ $\therefore$ $\dfrac{du}{dx} = 2x$

When $x = 0$, $u = 1$
When $x = 1$, $u = 2$

$\therefore \displaystyle\int_0^1 \frac{6x}{(x^2+1)^3}\,dx$

$= \displaystyle\int_0^1 \frac{1}{u^3}\left(3\frac{du}{dx}\right)dx$

$= 3\displaystyle\int_1^2 u^{-3}\,du$

$= 3\left[\dfrac{u^{-2}}{-2}\right]_1^2$

$= 3\left(\dfrac{2^{-2}}{-2} - \dfrac{1^{-2}}{-2}\right)$

$= \dfrac{9}{8}$

EXERCISE 10B.2

1 Evaluate:

a $\displaystyle\int_1^2 2x(x^2-1)^3\,dx$

b $\displaystyle\int_1^2 \frac{x}{(x^2+2)^2}\,dx$

c $\displaystyle\int_0^1 x^2 e^{x^3+1}\,dx$

d $\displaystyle\int_0^3 x\sqrt{x^2+16}\,dx$

e $\displaystyle\int_1^2 xe^{-2x^2}\,dx$

f $\displaystyle\int_2^3 \frac{x}{2-x^2}\,dx$

g $\displaystyle\int_1^2 \frac{\ln x}{x}\,dx$

h $\displaystyle\int_0^1 \frac{1-3x^2}{1-x^3+x}\,dx$

i $\displaystyle\int_2^4 \frac{6x^2-4x+4}{x^3-x^2+2x}\,dx$

Check your answers using technology.

Look for an appropriate substitution.

Example 7

◀) **Self Tutor**

Evaluate: $\displaystyle\int_{\frac{\pi}{6}}^{\frac{\pi}{2}} \sqrt{\sin x}\,\cos x\,dx$

Let $u = \sin x \qquad \therefore \dfrac{du}{dx} = \cos x$

When $x = \frac{\pi}{6}, \quad u = \sin\frac{\pi}{6} = \frac{1}{2}$

When $x = \frac{\pi}{2}, \quad u = \sin\frac{\pi}{2} = 1$

$\therefore \displaystyle\int_{\frac{\pi}{6}}^{\frac{\pi}{2}} \sqrt{\sin x}\,\cos x\,dx = \int_{\frac{\pi}{6}}^{\frac{\pi}{2}} \sqrt{u}\,\frac{du}{dx}\,dx$

$= \displaystyle\int_{\frac{1}{2}}^{1} u^{\frac{1}{2}}\,du$

$= \left[\dfrac{u^{\frac{3}{2}}}{\frac{3}{2}}\right]_{\frac{1}{2}}^{1}$

$= \frac{2}{3}(1)^{\frac{3}{2}} - \frac{2}{3}\left(\frac{1}{2}\right)^{\frac{3}{2}}$

$= \frac{2}{3} - \dfrac{1}{3\sqrt{2}}$

> The endpoints of the integral must always correspond to the variable of integration.

Casio fx-CG20	Casio fx-9860G PLUS	TI-84 Plus CE

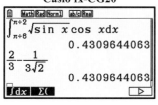 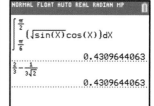

2 Evaluate:

a $\displaystyle\int_{0}^{\frac{\pi}{3}} \dfrac{\sin x}{\sqrt{\cos x}}\,dx$

b $\displaystyle\int_{0}^{\frac{\pi}{6}} \sin^2 x \cos x\,dx$

c $\displaystyle\int_{0}^{\frac{\pi}{4}} \tan x\,dx$

d $\displaystyle\int_{\frac{\pi}{6}}^{\frac{\pi}{2}} \cot x\,dx$

e $\displaystyle\int_{0}^{\frac{\pi}{6}} \dfrac{\cos x}{1 - \sin x}\,dx$

f $\displaystyle\int_{0}^{\frac{\pi}{4}} \sec^2 x \tan^3 x\,dx$

3 Evaluate for $n \in \mathbb{Z}$: $\displaystyle\int_{0}^{1} (x^2 + 2x)^n(x + 1)\,dx$

> Be careful to consider *all* $n \in \mathbb{Z}$

Example 8

◂)) **Self Tutor**

Find the exact value of $\displaystyle\int_{-4}^{6} x\sqrt{x+4}\,dx$.

Let $u = x + 4$ $\quad\therefore\quad \dfrac{du}{dx} = 1$

When $x = -4$, $\quad u = 0$
When $x = 6$, $\quad u = 10$

$\therefore \quad \displaystyle\int_{-4}^{6} x\sqrt{x+4}\,dx$

$= \displaystyle\int_{0}^{10} (u-4)\sqrt{u}\,du$

$= \displaystyle\int_{0}^{10} (u^{\frac{3}{2}} - 4u^{\frac{1}{2}})\,du$

$= \left[\dfrac{2}{5}u^{\frac{5}{2}} - \dfrac{8}{3}u^{\frac{3}{2}}\right]_{0}^{10}$

$= \dfrac{2}{5} \times 10^{\frac{5}{2}} - \dfrac{8}{3} \times 10^{\frac{3}{2}}$

$= 40\sqrt{10} - \dfrac{80}{3}\sqrt{10}$

$= \dfrac{40}{3}\sqrt{10}$

Casio fx-CG20

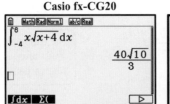

Casio fx-9860G PLUS

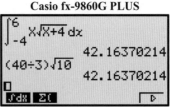

TI-84 Plus CE

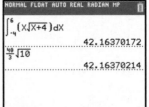

4 Find the exact value of:

a $\displaystyle\int_{3}^{4} x\sqrt{x-1}\,dx$ **b** $\displaystyle\int_{0}^{3} x\sqrt{x+6}\,dx$ **c** $\displaystyle\int_{2}^{5} x^2\sqrt{x-2}\,dx$

Check your answers using technology.

C   INTEGRATION BY PARTS

The method of **integration by parts** comes from the **product rule** of differentiation. In some cases it allows us to integrate a function which is written as a product.

Since $\dfrac{d}{dx}(uv) = u'v + uv',$ $\quad \displaystyle\int (u'v + uv')\,dx = uv$

$\therefore \quad \displaystyle\int u'v\,dx + \int uv'\,dx = uv$

$\therefore \quad \displaystyle\int uv'\,dx = uv - \int u'v\,dx$

So, provided $\displaystyle\int u'v\,dx$ can be found, we can find $\displaystyle\int uv'\,dx$ using

$$\int uv'\,dx = uv - \int u'v\,dx$$

Example 9

◀)) **Self Tutor**

Find:

a $\displaystyle\int xe^{-x}\,dx$ **b** $\displaystyle\int x\cos x\,dx$

a $u = x$ $v' = e^{-x}$
   $u' = 1$ $v = -e^{-x}$

$$\therefore \int xe^{-x}\,dx = -xe^{-x} - \int(-e^{-x})\,dx$$
$$= -xe^{-x} + (-e^{-x}) + c$$
$$= -e^{-x}(x+1) + c$$

Check: $\dfrac{d}{dx}(-e^{-x}(x+1) + c) = e^{-x}(x+1) + -e^{-x}(1)$
$$= xe^{-x} + \cancel{e^{-x}} - \cancel{e^{-x}}$$
$$= xe^{-x} \quad \checkmark$$

b $u = x$ $v' = \cos x$
   $u' = 1$ $v = \sin x$

$$\therefore \int x\cos x\,dx = x\sin x - \int \sin x\,dx$$
$$= x\sin x - (-\cos x) + c$$
$$= x\sin x + \cos x + c$$

Check: $\dfrac{d}{dx}(x\sin x + \cos x + c) = 1 \times \sin x + x\cos x - \sin x$
$$= \cancel{\sin x} + x\cos x - \cancel{\sin x}$$
$$= x\cos x \quad \checkmark$$

EXERCISE 10C

1 Use integration by parts to find integrals of the following
functions with respect to x:

  a xe^x **b** $x\sin x$ **c** $x^2 \ln x$

  d $x\sin 3x$ **e** $x\cos 2x$

> When using integration by
> parts, the function v'
> should be easy to integrate.

2   **a** Find $\displaystyle\int \ln x\,dx$ by first writing $\ln x$ as $1 \times \ln x$.

   b Hence find $\displaystyle\int (\ln x)^2\,dx$.

Example 10   ◀⑨ **Self Tutor**

Find $\displaystyle\int e^x \sin x \, dx$.

$$\int e^x \sin x \, dx$$

$$= e^x(-\cos x) - \int e^x(-\cos x)\,dx \longleftarrow \begin{cases} u = e^x & v' = \sin x \\ u' = e^x & v = -\cos x \end{cases}$$

$$= -e^x \cos x + \int e^x \cos x \, dx$$

$$= -e^x \cos x + e^x \sin x - \int e^x \sin x \, dx \longleftarrow \begin{cases} u = e^x & v' = \cos x \\ u' = e^x & v = \sin x \end{cases}$$

$$\therefore \ 2\int e^x \sin x \, dx = -e^x \cos x + e^x \sin x$$

$$\therefore \ \int e^x \sin x \, dx = \tfrac{1}{2}e^x(\sin x - \cos x) + c$$

For some integrals we need to use integration by parts *twice*.

3 Integrate with respect to x:

    a $x^2 e^{-x}$ **b** $e^x \cos x$ **c** $e^{-x} \sin x$ **d** $x^2 \sin x$

4 **a** Use integration by parts to find $\displaystyle\int u^2 e^u \, du$.

    b Hence find $\displaystyle\int (\ln x)^2 \, dx$ using the substitution $u = \ln x$.

5 Find $\displaystyle\int \sin x \cos x \, dx$ using:

    a a double angle formula   **b** substitution   **c** integration by parts.

6 **a** Use integration by parts to find $\displaystyle\int x \ln(x^2) \, dx$.

    b Hence find $\displaystyle\int_1^e x \ln(x^2) \, dx$.

Review set 10A

1 **a** Show that $\dfrac{4}{x-2} + \dfrac{3}{x+1} = \dfrac{7x-2}{(x-2)(x+1)}$.

    b Hence find $\displaystyle\int_3^4 \dfrac{7x-2}{x^2-x-2}\,dx$.

2 Integrate with respect to x:

    a $\dfrac{x}{x^2 - 2x - 8}$ **b** $\dfrac{3x-4}{2x^2 - 3x - 5}$

3 **a** Find $\int 2x(x^2+1)^3\,dx$ using the substitution $u = x^2 + 1$.

    b Hence evaluate:

        i $\int_0^1 2x(x^2+1)^3\,dx$ **ii** $\int_{-1}^2 -x(1+x^2)^3\,dx$

4 Integrate with respect to x:

    a $\sin^7 x \cos x$ **b** $\tan 2x$ **c** $e^{\sin x} \cos x$ **d** $\sqrt{3x+2}$

5 Use integration by parts to find:

    a $\int x^2 \ln(x^3)\,dx$ **b** $\int_0^\pi e^{2x} \sin x\,dx$

6 Find the following integrals exactly:

    a $\int_0^1 x^2 e^{x+1}\,dx$ **b** $\int_0^{\frac{\pi}{6}} \dfrac{\cos x}{(1+\sin x)^3}\,dx$

7 Find $\int \dfrac{x}{x^2-9}\,dx$ using:

    a the substitution $u = x^2 - 9$

    b partial fractions

    c the substitution $x = 3\sin t$.

Review set 10B

1 **a** Write $\dfrac{x+20}{(x-6)(2x+1)}$ as the sum of partial fractions.

    b Hence find $\int \dfrac{x+20}{2x^2-11x-6}\,dx$.

2 Evaluate $\int_{-2}^2 \dfrac{60}{x^2-16}\,dx$.

3 Find:

    a $\int \dfrac{2x}{\sqrt{x^2-5}}\,dx$ **b** $\int \dfrac{\sin x}{\cos^4 x}\,dx$ **c** $\int 4xe^{-x^2}\,dx$

4 Find the exact value of:

    a $\int_0^1 \dfrac{4x^2}{(x^3+2)^3}\,dx$ **b** $\int_{\frac{\pi}{4}}^{\frac{\pi}{3}} \sin^5 x \cos x\,dx$ **c** $\int_1^2 3x^2 \sqrt{x^3-1}\,dx$

5 Integrate with respect to x:

    a $e^{-x} \cos x$ **b** $x^2 e^x$

6 Find $\int \dfrac{\sin x}{\sqrt{\cos^n x}}\,dx$. Comment on the existence of this integral.

7 Explain what is wrong with the following argument:

Let $I = \displaystyle\int \frac{1}{x} \, dx$

$\therefore \quad I = \displaystyle\int 1 \times \frac{1}{x} \, dx$

$\therefore \quad I = x\left(\dfrac{1}{x}\right) - \displaystyle\int x\left(-\dfrac{1}{x^2}\right) dx \qquad \begin{cases} u = \dfrac{1}{x} & v' = 1 \\ u' = -\dfrac{1}{x^2} & v = x \end{cases}$

$\therefore \quad I = 1 + \displaystyle\int \frac{1}{x} \, dx$

$\therefore \quad I = 1 + I$

$\therefore \quad 0 = 1$

11

Applications of integration

Contents:

Opening problem

On August 8, the intensity of light entering a greenhouse is given by $y = 3\sin\frac{\pi t}{10}$ units per hour if the day is sunny, and $y = \sin\frac{\pi t}{10}$ units per hour if the day is overcast, where t is the number of hours after sunrise, $0 \leqslant t \leqslant 10$.

The graphs of these curves are shown alongside.

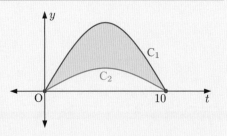

Things to think about:

a Can you identify each curve?

b Can you find the shaded area enclosed by C_1 and C_2 for $0 \leqslant t \leqslant 10$?

c If the intensity of light is regarded as its rate of energy transfer, what does the area in **b** mean about the energy entering the greenhouse?

We have already seen how definite integrals can be related to the areas between functions and the x-axis. In this Chapter we further explore the relationship between integration and area, including interpretation of what the area under a curve represents.

A   THE AREA UNDER A CURVE

We have already established in **Chapter 9** that:

If $f(x)$ is positive and continuous on the interval $a \leqslant x \leqslant b$, then the area bounded by $y = f(x)$, the x-axis, and the vertical lines $x = a$ and $x = b$

is given by $A = \displaystyle\int_a^b f(x)\,dx$ or $\displaystyle\int_a^b y\,dx$.

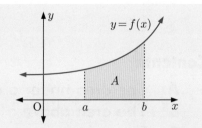

Example 1
◀) **Self Tutor**

Find the area of the region enclosed by $y = 2x$, the x-axis, $x = 0$, and $x = 4$ using:

a a geometric argument

b integration.

a

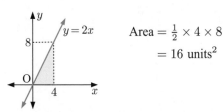

Area $= \frac{1}{2} \times 4 \times 8$

$= 16$ units^2

b Area $= \displaystyle\int_0^4 2x\,dx$

$= \left[x^2\right]_0^4$

$= 4^2 - 0^2$

$= 16$ units^2

Example 2
◀)) **Self Tutor**

Find the area of the region enclosed by $y = x^2 + 1$, the x-axis, $x = 1$, and $x = 2$.

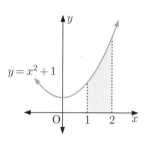

$$\text{Area} = \int_1^2 (x^2 + 1)\, dx$$

$$= \left[\frac{x^3}{3} + x\right]_1^2$$

$$= \left(\frac{8}{3} + 2\right) - \left(\frac{1}{3} + 1\right)$$

$$= 3\tfrac{1}{3} \text{ units}^2$$

It is helpful to sketch the region.

We can check this result using technology.

GRAPHICS
CALCULATOR
INSTRUCTIONS

GRAPHING
PACKAGE

EXERCISE 11A

1 Find the shaded area using:

  a a geometric argument

  b integration.

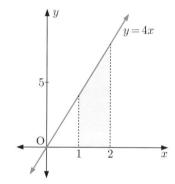

2 Find the area of each region described below using:

    i a geometric argument   **ii** integration

  a the region enclosed by $y = 5$, the x-axis, $x = -6$, and $x = 0$

  b the region enclosed by $y = x$, the x-axis, $x = 4$, and $x = 5$

  c the region enclosed by $y = -3x$, the x-axis, $x = -3$, and $x = 0$

3 Find the exact area of:

  a the blue shaded region

  b the green shaded region.

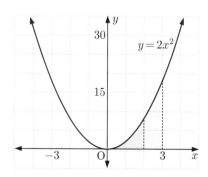

4 Find the area of the region bounded by:

   a $y = x^2$, the x-axis, and $x = 1$

   b $y = x^3$, the x-axis, $x = 1$, and $x = 4$

   c $y = \frac{1}{2}x^2 - 1$, the x-axis, $x = 2$, and $x = 3$.

5 The graph of $y = -x^2 + x + 6$ is shown alongside.

   a Find the coordinates of A and B.

   b Hence find the shaded area.

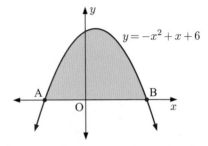

6

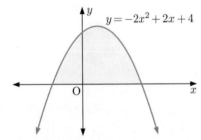

Find the area enclosed by $y = -2x^2 + 2x + 4$ and the x-axis.

Example 3   ◀ᵈ) **Self Tutor**

Find the area enclosed by one arch of the curve $y = \sin 2x$ and the x-axis.

The period of $y = \sin 2x$ is $\frac{2\pi}{2} = \pi$, so the first positive x-intercept is $\frac{\pi}{2}$.

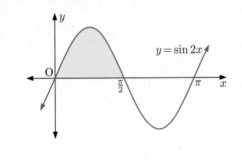

The required area $= \displaystyle\int_0^{\frac{\pi}{2}} \sin 2x \, dx$

$= \left[\tfrac{1}{2}(-\cos 2x) \right]_0^{\frac{\pi}{2}}$

$= -\tfrac{1}{2} \left[\cos 2x \right]_0^{\frac{\pi}{2}}$

$= -\tfrac{1}{2}(\cos \pi - \cos 0)$

$= 1 \text{ unit}^2$

---

7 Show that the area enclosed by $y = \sin x$ and the x-axis from $x = 0$ to $x = \pi$ is 2 units^2.

8 Find the area of the region bounded by $y = \cos x$, the x-axis, $x = 0$, and $x = \frac{\pi}{2}$.

9 Find the area enclosed by one arch of the curve $y = \cos 3x$ and the x-axis.

10

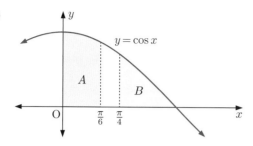

a Which of the shaded regions appears to be larger?

b Calculate the area of each region, and check your answer to **a**.

11 Find the area of the region bounded by:

   a $y = \dfrac{1}{x^2}$, the x-axis, $x = 1$, and $x = 2$

   b $y = e^x$, the axes, and $x = 1$

   c $y = 2 - \dfrac{1}{\sqrt{x}}$, the x-axis, and $x = 4$

   d $y = e^x + e^{-x}$, the x-axis, $x = -1$, and $x = 1$

   e the axes and $y = \sqrt{9 - x}$.

12

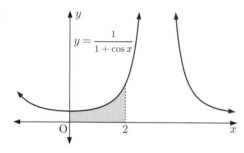

a Show that $\dfrac{d}{dx}\left(\dfrac{\sin x}{1 + \cos x}\right) = \dfrac{1}{1 + \cos x}$.

b Hence find the area of the region bounded by

   $y = \dfrac{1}{1 + \cos x}$, the x-axis, $x = 0$, and $x = 2$.

13 **a** Use integration by parts to show that

$$\int \ln x \, dx = x \ln x - x + c.$$

   b Hence find the shaded area.

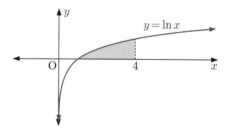

14 **a** Find b, correct to 4 decimal places.

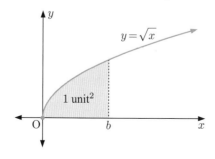

   b Find the exact value of a.

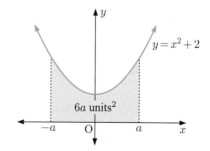

c Find k, correct to 4 decimal places.

d Find k such that area A and area B are equal.

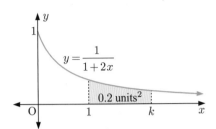

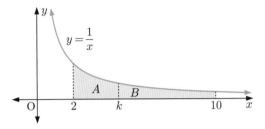

Investigation

$$\int_a^b f(x)\,dx \quad \text{and areas}$$

Can $\displaystyle\int_a^b f(x)\,dx$ always be interpreted as an area?

What to do:

1 **a** Find $\displaystyle\int_0^1 x^3\,dx$ and $\displaystyle\int_{-1}^1 x^3\,dx$.

   b Using a graph, explain why the first integral in **a** gives an area, whereas the second integral does not.

   c Find $\displaystyle\int_{-1}^0 x^3\,dx$ and explain why the answer is negative.

   d Show that $\displaystyle\int_{-1}^0 x^3\,dx + \int_0^1 x^3\,dx = \int_{-1}^1 x^3\,dx$.

   e Find $\displaystyle\int_0^{-1} x^3\,dx$ and interpret its meaning.

2 Suppose $f(x)$ is a function such that $f(x) \leqslant 0$ for all $a \leqslant x \leqslant b$. Write an expression for the area between the function and the x-axis for $a \leqslant x \leqslant b$.

3 Evaluate using area interpretation:

   a $\displaystyle\int_0^3 f(x)\,dx$

   b $\displaystyle\int_3^7 f(x)\,dx$

   c $\displaystyle\int_2^4 f(x)\,dx$

   d $\displaystyle\int_0^7 f(x)\,dx$

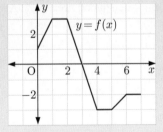

4 Evaluate using area interpretation:

   a $\displaystyle\int_0^4 f(x)\,dx$

   b $\displaystyle\int_4^6 f(x)\,dx$

   c $\displaystyle\int_6^8 f(x)\,dx$

   d $\displaystyle\int_0^8 f(x)\,dx$

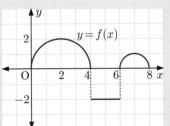

B   THE AREA ABOVE A CURVE

If $f(x)$ is negative and continuous on the interval $a \leqslant x \leqslant b$, then the area bounded by $y = f(x)$, the x-axis, and the vertical lines $x = a$ and $x = b$ is given by

$$A = -\int_a^b f(x)\,dx \quad \text{or} \quad -\int_a^b y\,dx.$$

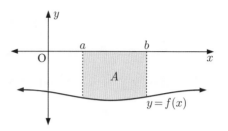

Proof:

The reflection of $y = f(x)$ in the x-axis is $y = -f(x)$.

By symmetry, $-f(x) \geqslant 0$ on the interval $a \leqslant x \leqslant b$, and the area bounded by $y = -f(x)$, the x-axis, $x = a$, and $x = b$ is also A.

$$\therefore \quad A = \int_a^b [-f(x)]\,dx$$

$$= -\int_a^b f(x)\,dx$$

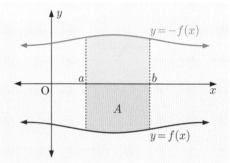

Example 4   ◄⅀ **Self Tutor**

Find the area bounded by the x-axis and $y = x^2 - 2x$.

The curve cuts the x-axis when $y = 0$

$\therefore \quad x^2 - 2x = 0$

$\therefore \quad x(x - 2) = 0$

$\qquad \therefore \quad x = 0$ or 2

$\therefore \quad$ the x-intercepts are 0 and 2.

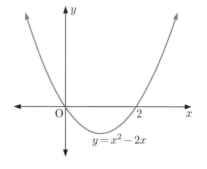

$$\text{Area} = -\int_0^2 (x^2 - 2x)\,dx$$

$$= -\left[\frac{x^3}{3} - x^2\right]_0^2$$

$$= -\left[\left(\tfrac{8}{3} - 4\right) - (0)\right]$$

$$= \tfrac{4}{3}$$

$\therefore \quad$ the area is $\tfrac{4}{3}$ units^2.

EXERCISE 11B

1 Find the shaded area:

a

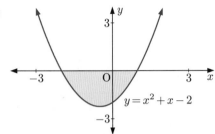

$y = x^2 + x - 2$

b

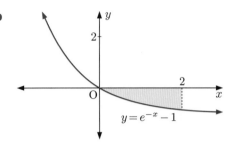

$y = e^{-x} - 1$

c

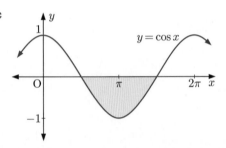

$y = \cos x$

d

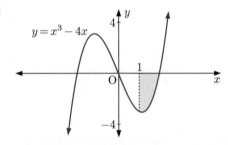

$y = x^3 - 4x$

Example 5   ◀◎ Self Tutor

Find the total area of the regions contained by $y = f(x)$ and the x-axis for $f(x) = x^3 + 2x^2 - 3x$.

$f(x) = x^3 + 2x^2 - 3x$
$ = x(x^2 + 2x - 3)$
$ = x(x - 1)(x + 3)$

$\therefore\ y = f(x)$ cuts the x-axis at 0, 1, and -3.

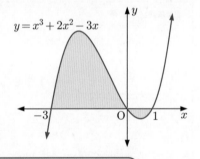

$y = x^3 + 2x^2 - 3x$

Total area

$= \displaystyle\int_{-3}^{0} (x^3 + 2x^2 - 3x)\, dx - \int_{0}^{1} (x^3 + 2x^2 - 3x)\, dx$

$= \left[\dfrac{x^4}{4} + \dfrac{2x^3}{3} - \dfrac{3x^2}{2}\right]_{-3}^{0} - \left[\dfrac{x^4}{4} + \dfrac{2x^3}{3} - \dfrac{3x^2}{2}\right]_{0}^{1}$

$= \left(0 - (-11\tfrac{1}{4})\right) - \left(-\tfrac{7}{12} - 0\right)$

$= 11\tfrac{5}{6}$ units^2

> Graphing the function helps us see where it is above and below the x-axis.

---

2 Find the total area enclosed by the function $y = f(x)$ and the x-axis for:

a $f(x) = x^3 - 9x$ **b** $f(x) = -x(x - 2)(x - 4)$ **c** $f(x) = x^4 - 5x^2 + 4$.

3 a Explain why the total area shaded is *not* equal to

$$\int_1^7 f(x)\,dx.$$

b Write an expression for the total shaded area in terms of integrals.

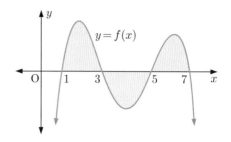

4

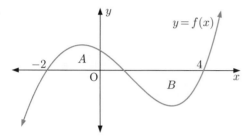

For the graph of $y = f(x)$ alongside,

$$\int_{-2}^4 f(x)\,dx = -6.$$ Which region is larger, A or B?

Explain your answer.

5 The area of the region bounded by $f(x) = -\dfrac{9}{x}$, the x-axis, $x = 3$, and $x = k$, is $9\ln 2$ units^2. Find the value of k.

6 a Sketch the graph of $y = 2\sin x + 1$ for $0 \leqslant x \leqslant 2\pi$.

b Find the area between the x-axis and the part of $y = 2\sin x + 1$ that is below the x-axis on $0 \leqslant x \leqslant 2\pi$.

7

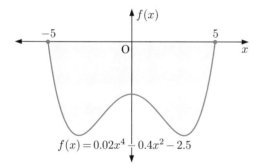

The cross-section of a roof gutter is defined by $f(x) = 0.02x^4 - 0.4x^2 - 2.5$, for $-5 \leqslant x \leqslant 5$ cm.

a Find the cross-sectional area of the gutter.

b The gutter is 20 m long. How much water can it hold in total?

8 a Find $\displaystyle\int x^2 \cos(x^3)\,dx$.

b Show that the red shaded region is twice as large as the blue shaded region.

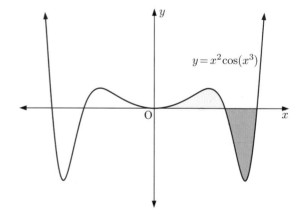

C | THE AREA BETWEEN TWO FUNCTIONS

Consider two functions $f(x)$ and $g(x)$ where $f(x) \geqslant g(x)$ for all $a \leqslant x \leqslant b$.

The area between the two functions on the interval $a \leqslant x \leqslant b$ is given by

$$A = \int_a^b [f(x) - g(x)]\, dx.$$

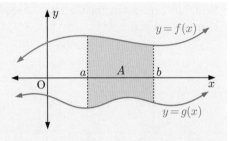

Proof:

If necessary, we translate each curve k units upwards until they are both above the x-axis on the interval $a \leqslant x \leqslant b$.

$\therefore$ $A = (\text{area below } y = f(x) + k) - (\text{area below } y = g(x) + k)$

$$= \int_a^b [f(x) + k]\, dx - \int_a^b [g(x) + k]\, dx$$

$$= \int_a^b [f(x) - g(x)]\, dx$$

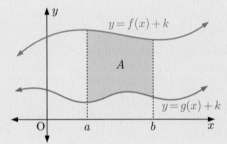

Notice that if $y = g(x)$ is negative on the interval $a \leqslant x \leqslant b$, we can let $f(x) = 0$ and hence derive the formula we saw in the last Section:

$$A = -\int_a^b g(x)\, dx$$

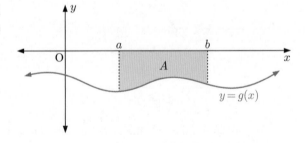

Example 6 ◀)) Self Tutor

Find the area of the region enclosed by $y = x + 2$ and $y = x^2 + x - 2$.

$y = x + 2$ meets $y = x^2 + x - 2$ where $x^2 + x - 2 = x + 2$

$$\therefore \quad x^2 - 4 = 0$$
$$\therefore \quad (x + 2)(x - 2) = 0$$
$$\therefore \quad x = \pm 2$$

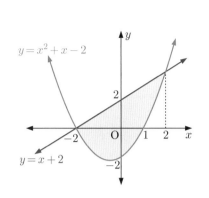

Since $x + 2 \geqslant x^2 + x - 2$ on the interval $-2 \leqslant x \leqslant 2$,

$$\text{area} = \int_{-2}^{2} [(x + 2) - (x^2 + x - 2)]\, dx$$

$$= \int_{-2}^{2} (4 - x^2)\, dx$$

$$= \left[4x - \frac{x^3}{3} \right]_{-2}^{2}$$

$$= \left(8 - \tfrac{8}{3} \right) - \left(-8 + \tfrac{8}{3} \right)$$

$$= 10\tfrac{2}{3} \text{ units}^2$$

$\therefore$ the area is $10\tfrac{2}{3}$ units^2.

EXERCISE 11C

1 **a** Sketch the graphs of $y = x - 3$ and $y = x^2 - 3x$ on the same set of axes.

b Find the coordinates of the points where the graphs meet.

c Find the area of the region enclosed by the two graphs.

2 Find the area of the region enclosed by:

a $y = x^2 - 2x$ and $y = 3$ **b** $y = \sqrt{x}$ and $y = x^2$.

3 Find the area of the region bounded by $y = 2e^x$, $y = e^{2x}$, and $x = 0$.

4 On the same set of axes, sketch $y = 2x$ and $y = 4x^2$.
   Find the area of the region enclosed by these functions.

5 The graphs of $y = \dfrac{5}{2x + 1}$ and $y = 3 - x$ are shown alongside.

a Find the coordinates of A and B.

b Find the exact value of the shaded area.

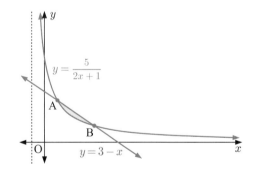

6 A region with $x \geqslant 0$ has boundaries defined by $y = \sin x$, $y = \cos x$, and the y-axis.
   Sketch the region and find its area.

7 The illustrated curves are the graphs of $y = \sin x$ and $y = 4 \sin x$.

a Identify each curve.

b Calculate the shaded area.

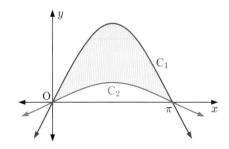

Example 7

◀ **Self Tutor**

Find the total area of the regions enclosed by $y = x^3 - 6x + 3$ and $y = x^2 + 3$.

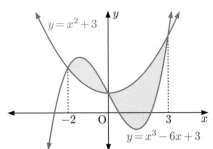

$y = x^3 - 6x + 3$ and $y = x^2 + 3$ meet where

$$x^3 - 6x + 3 = x^2 + 3$$
$$\therefore \ x^3 - x^2 - 6x = 0$$
$$\therefore \ x(x^2 - x - 6) = 0$$
$$\therefore \ x(x + 2)(x - 3) = 0$$
$$\therefore \quad x = 0, \ -2, \text{ or } 3$$

Total area $= \displaystyle\int_{-2}^{0} ((x^3 - 6x + 3) - (x^2 + 3)) \, dx + \int_{0}^{3} ((x^2 + 3) - (x^3 - 6x + 3)) \, dx$

$\displaystyle = \int_{-2}^{0} (x^3 - x^2 - 6x) \, dx + \int_{0}^{3} (-x^3 + x^2 + 6x) \, dx$

$\displaystyle = \left[\frac{x^4}{4} - \frac{x^3}{3} - 3x^2 \right]_{-2}^{0} + \left[-\frac{x^4}{4} + \frac{x^3}{3} + 3x^2 \right]_{0}^{3}$

$\displaystyle = \left(0 - \frac{16}{4} + \frac{-8}{3} + 12 \right) + \left(-\frac{81}{4} + \frac{27}{3} + 27 - 0 \right)$

$= 21\frac{1}{12}$ units^2

8

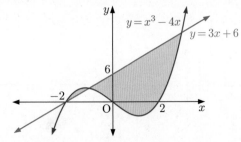

a Write the shaded area as the sum of two definite integrals.

b Find the total shaded area.

9 **a** Sketch the graphs of $y = x^3 - 5x$ and $y = 2x^2 - 6$ on the same set of axes.

    b Find the x-coordinates of their intersection points.

    c Hence find the area enclosed by $y = x^3 - 5x$ and $y = 2x^2 - 6$.

10 Find the area enclosed by:

    a $y = -x^3 + 3x^2 + 6x - 8$ and $y = 5x - 5$

    b $y = 2x^3 - 3x^2 + 18$ and $y = x^3 + 10x - 6$.

11 The shaded area is 2.4 units^2.

Find k, correct to 4 decimal places.

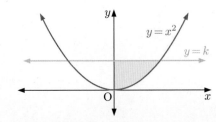

12   **a**   On the same set of axes, graph $y = e^x - 1$ and $y = 2 - 2e^{-x}$. Show all axes intercepts and asymptotes.

  b   Find the points of intersection of $y = e^x - 1$ and $y = 2 - 2e^{-x}$.

  c   Find the area of the region enclosed by the two curves.

13

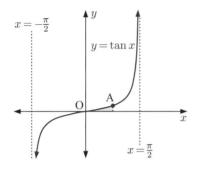

The graph alongside shows $y = \tan x$ for $-\frac{\pi}{2} < x < \frac{\pi}{2}$.

A is the point on the graph with y-coordinate 1.

  a   Find the x-coordinate of A.

  b   Find the shaded area.

14   The illustrated curves are those of $y = \sin x$ and $y = \sin 2x$.

  a   Identify each curve.

  b   Find algebraically the coordinates of A.

  c   Find the total area enclosed by C_1 and C_2 for $0 \leqslant x \leqslant \pi$.

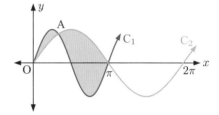

15

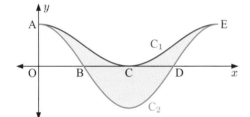

The illustrated curves are $y = \cos 2x$ and $y = \cos^2 x$.

  a   Identify each curve.

  b   Determine the coordinates of A, B, C, D, and E.

  c   Show that the area of the shaded region is $\frac{\pi}{2}$ units^2.

16   **a**   Sketch the graphs of $y = 2xe^x$ and $y = xe^{2x}$ on the same set of axes.

  b   Find the x-coordinates of their points of intersection.

  c   Hence find the area enclosed by the graphs.

D   PROBLEM SOLVING BY INTEGRATION

When we studied differential calculus, we saw how to find the rate of change of a function by differentiation.

Since integration is the reverse process of differentiation, integration can be used to measure the change in a quantity from its rate of change.

When we introduced integration, we separated the area under a curve into rectangles. The area of each rectangle is the rate of change of the quantity times the subinterval width, which tells us the change in the quantity corresponding to that subinterval.

So, the area under a rate function for a particular interval tells us the overall change in the quantity over that interval.

Example 8
◀) **Self Tutor**

The rate of power consumption by the city of Bristol can be modelled by the function

$$E(t) = 0.3 \sin\left(\frac{(t-10)\pi}{12}\right) + 0.1 \cos\left(\frac{(t-6)\pi}{6}\right) + 0.775 \text{ GW}$$

where t is the number of hours after midnight each day, $0 \leqslant t \leqslant 24$.

Find the following quantities and explain what they represent:

a $\displaystyle\int_0^{12} E(t)\, dt$ **b** $\displaystyle\int_{12}^{24} E(t)\, dt$ **c** $\displaystyle\int_0^{24} E(t)\, dt$

$$E(t) = 0.3 \sin\left(\frac{(t-10)\pi}{12}\right) + 0.1 \cos\left(\frac{(t-6)\pi}{6}\right) + 0.775 \text{ GW}$$

$$\therefore \int E(t)\, dt = 0.3\left(-\cos\left(\frac{(t-10)\pi}{12}\right)\right)\left(\frac{12}{\pi}\right) + 0.1 \sin\left(\frac{(t-6)\pi}{6}\right)\left(\frac{6}{\pi}\right) + 0.775t + c$$

$$= -\frac{3.6}{\pi}\cos\left(\frac{(t-10)\pi}{12}\right) + \frac{0.6}{\pi}\sin\left(\frac{(t-6)\pi}{6}\right) + 0.775t + c$$

a $\displaystyle\int_0^{12} E(t)\, dt = \left(-\frac{3.6}{\pi}\cos\frac{\pi}{6} + \frac{0.6}{\pi}\sin\pi + 9.3\right) - \left(-\frac{3.6}{\pi}\cos\left(-\frac{5\pi}{6}\right) + \frac{0.6}{\pi}\sin(-\pi) + 0\right)$

≈ 7.315 GWh

The morning power consumption of Bristol is about 7.32 GWh.

b $\displaystyle\int_{12}^{24} E(t)\, dt = \left(-\frac{3.6}{\pi}\cos\frac{7\pi}{6} + \frac{0.6}{\pi}\sin 3\pi + 18.6\right) - \left(-\frac{3.6}{\pi}\cos\frac{\pi}{6} + \frac{0.6}{\pi}\sin\pi + 9.3\right)$

≈ 11.285 GWh

The afternoon power consumption of Bristol is about 11.29 GWh.

c $\displaystyle\int_0^{24} E(t)\, dt = \int_0^{12} E(t)\, dt + \int_{12}^{24} E(t)\, dt$

≈ 18.6 GWh

The total daily power consumption of Bristol is about 18.6 GWh.

EXERCISE 11D

1 The rate of traffic flow past a pedestrian crossing between 8 am and 8:30 am is given by

$$R(t) = \frac{t^3}{80} - \frac{t^2}{2} + 4t + 40 \text{ cars per minute,}$$

where t is the number of minutes after 8 am, $0 \leqslant t \leqslant 30$. The graph of $R(t)$ against t is shown alongside.

a Find the rate of traffic flow at 8:20 am.

b Use the graph to estimate the time at which the traffic flow was greatest.

c Copy the graph, and shade the region corresponding

to $\displaystyle\int_{10}^{15} R(t)\, dt$. Explain what this region represents.

d How many cars passed the crossing between 8 am and 8:30 am?

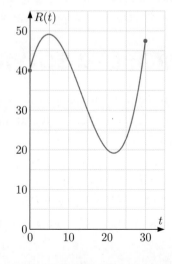

PRINTABLE GRAPH

2 Evan is happily paddling until his kayak strikes a sharp rock. Water begins to leak in at the rate $R_1(t) = 5 - 5e^{-0.2t}$ litres per minute, where t is the time in minutes. Evan tries to bail the water out of the kayak, removing the water at the rate $R_2(t) = 6 - 6e^{-0.1t}$ litres per minute.

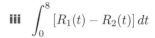

   a After 2 minutes, at what rate is water:

      i leaking into the kayak

      ii being bailed from the kayak?

   b Is the amount of water in the kayak increasing or decreasing after 2 minutes? Explain your answer.

   c Evaluate the following integrals, and interpret their meaning:

$$\textbf{i} \quad \int_0^3 R_1(t)\,dt \qquad\qquad \textbf{ii} \quad \int_2^5 R_2(t)\,dt \qquad\qquad \textbf{iii} \quad \int_0^8 [R_1(t) - R_2(t)]\,dt$$

   d How much water is in the kayak after 10 minutes?

3 Answer the **Opening Problem** on page **252**.

4 The rate of power consumption of the United Kingdom can be modelled by the function

$$E(t) = 13\sin\left(\tfrac{(t+3)\pi}{3}\right) + 70\cos\left(\tfrac{(t-1)\pi}{6}\right) + 196 \ \text{TW}$$

where t is the number of months after January 1st, $0 \leqslant t \leqslant 12$.

   a Use technology to help sketch the function.

   b Find the following quantities and explain what they represent:

$$\textbf{i} \quad \int_3^4 E(t)\,dt \qquad\qquad \textbf{ii} \quad \int_5^8 E(t)\,dt \qquad\qquad \textbf{iii} \quad \int_0^{12} E(t)\,dt$$

Activity   Exponential random variables

The **probability density function** of a continuous random variable X can be used to determine the probability that X lies within a particular interval. If X must lie in the interval $[a,\ b]$ then the probability that X lies in the interval $[c,\ d]$ is given by the area under the probability density function:

$$P(c \leqslant X \leqslant d) = \int_c^d f(x)\,dx$$

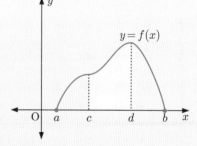

For a function $f(x)$ to be a valid probability density function, it must satisfy the properties:

- $f(x) \geqslant 0$ for all $x \in [a,\ b]$

- $\displaystyle\int_a^b f(x)\,dx = 1$

- Probabilities must be positive.
- The total probability is 1.

The probability density function can also be used to determine other properties of the random variable:

- the **mean** of X is $\mu = \displaystyle\int_a^b x\,f(x)\,dx$

- the **variance** of X is $\displaystyle\int_a^b x^2 f(x)\,dx - \mu^2$.

An **exponential random variable** T has probability density function $f(t) = \lambda e^{-\lambda t}$ on the interval $t \geqslant 0$, where $\lambda > 0$ is a constant.

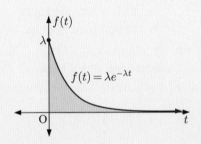

What to do:

1 Explain why $f(t) \geqslant 0$ for all $t \geqslant 0$.

2 **a** Write an expression, in terms of k, for $\displaystyle\int_0^k f(t)\,dt$.

  b By considering $\displaystyle\int_0^k f(t)\,dt$ as $k \to \infty$, show that $\displaystyle\int_0^\infty f(t)\,dt = 1$.

3 **a** Explain why the mean of T is given by $\displaystyle\int_0^\infty \lambda t e^{-\lambda t}\,dt$.

  b Use integration by parts to find an expression for $\displaystyle\int_0^k \lambda t e^{-\lambda t}\,dt$.

  c By considering $\displaystyle\int_0^k \lambda t e^{-\lambda t}\,dt$ as $k \to \infty$, show that T has mean $\dfrac{1}{\lambda}$.

4 **a** Explain why the variance of T is given by $\displaystyle\int_0^\infty \lambda t^2 e^{-\lambda t}\,dt - \dfrac{1}{\lambda^2}$.

  b Use integration by parts twice to find an expression for $\displaystyle\int_0^k \lambda t^2 e^{-\lambda t}\,dt$.

  c By considering $\displaystyle\int_0^k \lambda t^2 e^{-\lambda t}\,dt$ as $k \to \infty$, show that T has variance $\dfrac{1}{\lambda^2}$.

5 Let the exponential random variable T be the time in minutes between customers arriving at a shop. The probability density function for T is $f(t) = 0.5e^{-0.5t}$, $t \geqslant 0$.

  a Draw the graph of $f(t)$.

  b Find the mean and variance of T.

  c Find the probability that the next two customers arrive:

    i less than 1 minute apart   **ii** more than 4 minutes apart.

  d The median of T is the value of m such that $\displaystyle\int_0^m f(t)\,dt = 0.5$. Find the median time between customers.

Review set 11A

1 Find the shaded area:

a
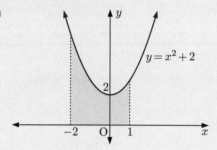

b
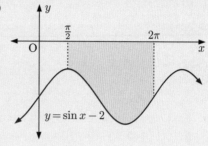

c
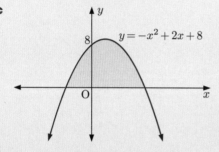

d
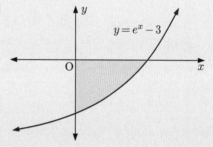

2 Find the area of the region bounded by:

    a $y = x^2$, the x-axis, $x = 2$, and $x = 5$

    b $y = \sqrt{5 - x}$, the x-axis, $x = 1$, and $x = 2$

    c $y = \sin \frac{x}{3}$, the x-axis, $x = \pi$, and $x = 2\pi$.

3 **a** Sketch the graphs of $y = x^2 + 4x + 1$ and $y = 3x + 3$ on the same set of axes.

    b Find the points where the graphs meet.

    c Hence find the area of the region enclosed by $y = x^2 + 4x + 1$ and $y = 3x + 3$.

4 Does $\displaystyle\int_{-1}^{3} f(x)\,dx$ represent the area of the shaded region? Explain your answer.

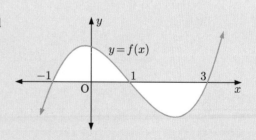

5

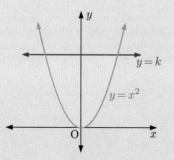

Determine k given that the enclosed region has area $5\frac{1}{3}$ units^2.

6 Find the total area enclosed by $y = \dfrac{x \cos \sqrt{x^2 + 1}}{\sqrt{x^2 + 1}}$,

the x-axis, $x = 2$, and $x = 4$.

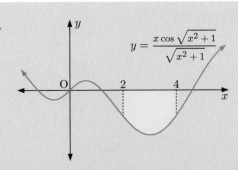

7 **a** Find a given that the area of the region between
  $y = e^x$ and the x-axis from $x = 0$ to $x = a$
  is 2 units^2.

  b Hence determine b such that the area of the region
  from $x = a$ to $x = b$ is also 2 units^2.

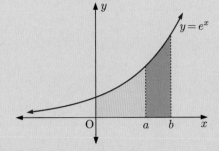

8 Find the area of the region enclosed by $y = x^3 + x^2 + 2x + 6$ and $y = 7x^2 - x - 4$.

9 **a** Sketch the graphs of $y = \sin^2 x$ and $y = \sin x$ on the same set of axes for $0 \leqslant x \leqslant \pi$.

  b Find the exact value of the area enclosed by these curves for $0 \leqslant x \leqslant \frac{\pi}{2}$.

10 Bettina is filling a watering can with water from a tap.
The water enters at the rate $R_1(t) = 6.4$ litres per minute.
Bettina's watering can starts empty, and has capacity
16 litres. Unfortunately it has a hole in the bottom, so it
leaks water at the rate $R_2(t) = 2.5 - 1.25e^{-0.2t}$ litres
per minute, where t is the time in minutes.

  a Evaluate the following integrals, and interpret their
  meaning:

  i $\displaystyle\int_0^{\frac{1}{2}} R_2(t)\,dt$ **ii** $\displaystyle\int_0^1 [R_1(t) - R_2(t)]\,dt$

  b How long will it take for the watering can to be full? Give your answer to the nearest second.

Review set 11B

1 Find the area of the region bounded by:

  a $y = x^3 + 1$, the x-axis, $x = 1$, and $x = 3$

  b $y = \dfrac{1}{x}$, the x-axis, $x = 3$, and $x = 9$

  c $y = e^{2x}$, the x-axis, $x = 0$, and $x = 2$

  d $y = -2 - 2\cos 3x$ and the x-axis from $x = -\frac{\pi}{3}$ to $x = \frac{\pi}{3}$.

2 Find the shaded area.

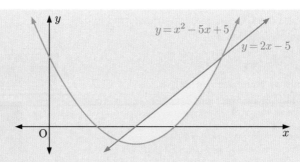

3 Determine the area enclosed by the axes and $y = 4e^x - 1$.

4 **a** Sketch the graphs of $y = x^4 - 12$ and $y = -x^2$ on the same set of axes.

   b Find the points where the graphs meet.

   c Find the exact area of the region enclosed by $y = x^4 - 12$ and $y = -x^2$.

5 OABC is a rectangle and the two shaded regions are equal in area. Find k.

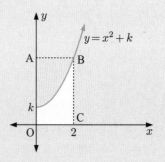

6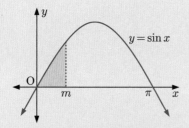

The shaded region has area $\frac{1}{2}$ unit^2.
Find the value of m.

7 Determine the total area enclosed by $y = x^3$ and $y = 7x^2 - 10x$.

8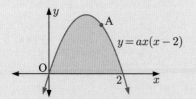

   a Find a given that the shaded area is 4 units^2.

   b Find the x-coordinate of A given [OA] divides the shaded region into equal areas.

9 **a** Find $\displaystyle\int x \sin(x^2)\, dx$.

   b Show that the shaded regions have equal area.

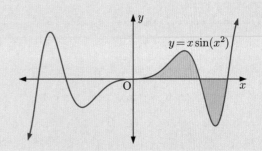

10 Over the course of a day, the solar energy transferred into Callum's solar panels is given by

$$E(t) = 2\sin\left(\tfrac{t-5}{5}\right) + \tfrac{1}{2}\sin\left(\tfrac{t-5}{4}\right) \ \text{kW}$$

where t is the time in hours after midnight, $5 \leqslant t \leqslant 20$. Find the following quantities and explain what they represent:

a $\displaystyle\int_5^{12} E(t)\,dt$

b $\displaystyle\int_{12}^{20} E(t)\,dt$

c $\displaystyle\int_5^{20} E(t)\,dt$

12

Differential equations

Contents:

Opening problem

The population of rodents on an island is currently 500. Its growth rate is expected to be given by $\dfrac{dP}{dt} = 0.1P\left(1 - \dfrac{P}{3000}\right)$, where t is the time in years from now.

Things to think about:

a How would we describe the relationship between the variables P and t?

b What would the graph of P against t look like?

c How can we write P as a function of t?

d Can you find:

   i the expected population after 8 years

   ii the expected time taken for the population to increase to 2000

   iii the maximum population which the island can sustain?

In the **Opening Problem**, the equation $\dfrac{dP}{dt} = 0.1P\left(1 - \dfrac{P}{3000}\right)$ describes the relationship between the function $P(t)$ and its derivative. This is an example of a **differential equation**.

A   DIFFERENTIAL EQUATIONS

A **differential equation** is an equation involving a derivative of a function.

Suppose y is a function of x, so $y = y(x)$. Examples of differential equations for this function are:

- $\dfrac{dy}{dx} = \dfrac{x^2}{y}$
- $\dfrac{dy}{dx} = -0.075y^3$
- $\dfrac{d^2y}{dx^2} - 3\dfrac{dy}{dx} + 4y = 0$

Differential equations not only arise in pure mathematics, but are also used to model and solve problems in applied mathematics, physics, engineering, economics, and other subjects. For example:

A falling object	A parachutist	Object on a spring

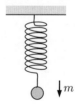

$$\frac{d^2y}{dx^2} = 9.8$$

$$m\frac{dv}{dt} = mg - av^2$$

$$m\frac{d^2y}{dt^2} = -ky$$

Current in an RL Circuit	Water leaking from a tank	A dog pursuing a cat
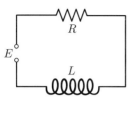		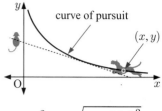
$L \dfrac{dI}{dt} + RI = E$	$\dfrac{dH}{dt} = -a\sqrt{H}$	$x \dfrac{d^2 y}{dx^2} = \sqrt{1 + \left(\dfrac{dy}{dx}\right)^2}$

SOLUTIONS OF DIFFERENTIAL EQUATIONS

A **solution** of a differential equation is a function y in terms of x, which satisfies the differential equation.

For example:

- a solution to the differential equation $\dfrac{dy}{dx} = 2xy$ is $y = e^{x^2}$

- a solution to the differential equation $\dfrac{dy}{dx} = y^2$, is $y = -\dfrac{1}{x}$.

Example 1 ◀)) **Self Tutor**

Verify that $y = e^{x^2}$ is a solution to the differential equation $\dfrac{dy}{dx} = 2xy$.

If $y = e^{x^2}$, then $\dfrac{dy}{dx} = 2x(e^{x^2})$

$\therefore \quad \dfrac{dy}{dx} = 2xy$ as required.

THE GENERAL SOLUTION

In addition to the solution $y = e^{x^2}$, the differential equation $\dfrac{dy}{dx} = 2xy$ has infinitely many other solutions, including $y = 2e^{x^2}$, $y = 3e^{x^2}$, and $y = -7e^{x^2}$.

All of these solutions have the form $y = ce^{x^2}$ where c is a constant, so we say that $y = ce^{x^2}$ is the **general solution** to the differential equation.

If we are given **initial conditions** for the problem, such as a value of y or $\dfrac{dy}{dx}$ for a specific value of x, then we can evaluate c. This gives us a **particular solution** to the problem, which is one particular curve from the family of curves described by the general solution.

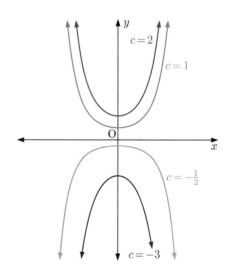

Example 2
◀ᴺ **Self Tutor**

Consider the differential equation $\dfrac{dy}{dx} - 3y = 3$.

a Show that $y = ce^{3x} - 1$ is a solution to the differential equation for any constant c.

b Sketch the solution curves for $c = 0, \pm 1, \pm 2, \pm 3$.

c Find the particular solution which passes through $(0, 2)$.

d Find the equation of the tangent to the particular solution at $(0, 2)$.

a If $y = ce^{3x} - 1$ then $\dfrac{dy}{dx} = 3ce^{3x}$.

$$\therefore \quad \frac{dy}{dx} - 3y = 3ce^{3x} - 3(ce^{3x} - 1)$$
$$= 3ce^{3x} - 3ce^{3x} + 3$$
$$= 3$$

$\therefore$ the differential equation is satisfied for all $x \in \mathbb{R}$ for any constant c.

b The solution curves for $c = 0, \pm 1, \pm 2, \pm 3$ are shown alongside.

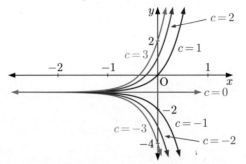

c $y = ce^{3x} - 1$ is a general solution to the differential equation.

The particular solution passes through $(0, 2)$, so $2 = ce^{3 \times 0} - 1$

$$\therefore \quad c = 3$$

$$\therefore \quad \text{the particular solution is} \quad y = 3e^{3x} - 1$$

d $\dfrac{dy}{dx} = 3 + 3y$

$\therefore$ at the point $(0, 2)$, $\dfrac{dy}{dx} = 3 + 3 \times 2 = 9$

$\therefore$ the gradient of the tangent to the particular solution $y = 3e^{3x} - 1$ at $(0, 2)$, is 9.

$\therefore$ the equation of the tangent is

$$y = 9(x - 0) + 2$$
$$\therefore \quad y = 9x + 2$$

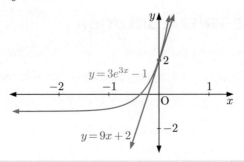

EXERCISE 12A

1 Verify that:

a $y = x^4$ is a solution to $\dfrac{dy}{dx} = 4x^3$

b $y = 5e^{2x}$ is a solution to $\dfrac{dy}{dx} = 2y$

c $y = \sqrt{x^2 + 1}$ is a solution to $\dfrac{dy}{dx} = \dfrac{x}{y}$

d $y = -\dfrac{1}{x}$ is a solution to $\dfrac{dy}{dx} = y^2$

e $y = 3e^{\frac{x^2}{2} + x}$ is a solution to $\dfrac{dy}{dx} - y = xy$.

2   Match each differential equation with the correct solution:

a $\dfrac{dy}{dx} = \dfrac{1}{2y}$ **b** $\dfrac{dy}{dx} = -\dfrac{2y}{x}$ **c** $\dfrac{dy}{dx} = \dfrac{2x}{3y^2}$

A $y = \sqrt[3]{x^2 + 1}$ **B** $y = \sqrt{x + 3}$ **C** $y = \dfrac{1}{x^2}$

3   Verify that:

a $y = x^3 + c$ is the general solution to $\dfrac{dy}{dx} = 3x^2$

b $y = ce^{-x}$ is the general solution to $\dfrac{dy}{dx} = -y$

c $y = -\dfrac{2}{x^2 + c}$ is the general solution to $\dfrac{dy}{dx} = xy^2$.

4   Consider the differential equation $\dfrac{dy}{dx} = 4x$.

a   Show that $y = 2x^2 + c$ is a solution to the differential equation for any constant c.

b   Sketch the solution curves for $c = 0, \pm 1, \pm 2$.

c   Find the particular solution which passes through $(1, \frac{1}{2})$.

d   Find the equation of the tangent to the particular solution at $(1, \frac{1}{2})$.

5   Consider the differential equation $\dfrac{dy}{dx} = 2x - y$.

a   Show that $y = 2x - 2 + ce^{-x}$ is a solution to the differential equation for any constant c.

b   Sketch the solution curves for $c = 0, \pm 1, \pm 2$.

c   Find the particular solution which passes through $(0, 1)$.

d   Find the equation of the tangent to the particular solution at $(0, 1)$.

B   DIFFERENTIAL EQUATIONS OF THE FORM $\dfrac{dy}{dx} = f(x)$

To solve differential equations of the form $\dfrac{dy}{dx} = f(x)$, we simply integrate with respect to x.

Example 3   ◀ Self Tutor

Solve the following differential equations:

a $\dfrac{dy}{dx} = e^{2x}$ **b** $\dfrac{dy}{dx} = \dfrac{2x}{x^2 + 1}$ **c** $\dfrac{dy}{dx} + 1 = \cos x$

a $\dfrac{dy}{dx} = e^{2x}$

$\therefore\; y = \displaystyle\int e^{2x}\, dx$

$\therefore\; y = \frac{1}{2}e^{2x} + c$

b $\dfrac{dy}{dx} = \dfrac{2x}{x^2 + 1}$

$\therefore\; y = \displaystyle\int \dfrac{2x}{x^2 + 1}\, dx$

$\therefore\; y = \ln|x^2 + 1| + c$

c $\dfrac{dy}{dx} + 1 = \cos x$

$\therefore\; \dfrac{dy}{dx} = \cos x - 1$

$\therefore\; y = \displaystyle\int (\cos x - 1)\, dx$

$\therefore\; y = \sin x - x + c$

EXERCISE 12B

1 Solve the following differential equations:

a $\dfrac{dy}{dx} = 4x^3$

b $\dfrac{dy}{dx} = x^2 + 6x$

c $\dfrac{dy}{dx} = e^{3x} + 4$

d $\dfrac{dy}{dx} = \cos x + \sin 2x$

e $\dfrac{dx}{dt} = \cos^2 t$

f $\dfrac{dy}{dx} = \dfrac{1}{x+4} - \dfrac{2}{3x-5}$

g $\dfrac{dM}{dt} = \dfrac{3t^2}{t^3 - 4}$

h $\dfrac{dy}{dx} = \dfrac{x}{\sqrt{25 - x^2}}$

i $f'(t) = te^{-t^2+1} + 2$

j $\dfrac{dS}{dt} = \dfrac{\sqrt{\ln t}}{t}$

k $\dfrac{dy}{dx} + \cos 3x = 1$

l $\dfrac{dy}{dx} + \dfrac{2}{x} = \sqrt{x}$

Example 4
◆)) **Self Tutor**

Solve $\dfrac{dy}{dx} = \sin^2 x$ given that $y(0) = \frac{1}{2}$.

$\dfrac{dy}{dx} = \sin^2 x$

$\therefore \ y = \displaystyle\int \sin^2 x \, dx$

$\therefore \ y = \displaystyle\int \left(\frac{1}{2} - \frac{1}{2}\cos 2x\right) dx$

$\therefore \ y = \frac{1}{2}x - \frac{1}{4}\sin 2x + c$

Now $y(0) = \frac{1}{2}$

$\therefore \ \frac{1}{2} = \frac{1}{2}(0) - \frac{1}{4}\sin 0 + c$

$\therefore \ c = \frac{1}{2}$

So, the solution is $y = \frac{1}{2}x - \frac{1}{4}\sin 2x + \frac{1}{2}$.

---

2 Find the particular solution for each differential equation:

a $\dfrac{dy}{dx} = 3x - 2$ given $y(0) = 5$

b $\dfrac{dy}{dx} = e^{3x} + 1$ given $y(0) = 0$

c $\dfrac{dy}{dx} = \dfrac{1}{x}$ given $y(2) = \ln 12$

3 Find the particular solution to:

a $\dfrac{dy}{dt} = 2e^{2t} - e^{-t}$ given that $y = 2.5$ when $t = \ln 2$

b $\dfrac{dM}{d\alpha} = \cos 2\alpha - 3\sin \alpha$ given that $M = 2$ when $\alpha = \frac{\pi}{2}$

c $\dfrac{dP}{dx} = 2x\cos x$ given that $P = \sqrt{3}$ when $x = \frac{\pi}{6}$.

4 A function $f(x)$ has gradient function $f'(x) = 2x - 5$, and the point $(2, -18)$ lies on the curve. Find the value of $f(-2)$.

5 **a** Find the particular solution to the differential equation $\dfrac{dy}{dx} = e^x - e^{-x}$ given $y(0) = 1$.
  b Sketch the graph of the particular solution.
  c Find the equation of the tangent to the solution curve at the point where $x = \ln 2$.

6 A curve has gradient function $f'(x) = ax + bx^{-2}$ where a and b are constants. It passes through $(-1, -2)$ and has a turning point at $(1, 0)$. Find the function $f(x)$.

7 The curve alongside has gradient function $\dfrac{dy}{dx} = \dfrac{2x}{x^2 + k}$ for some positive constant k. Find the equation of the curve.

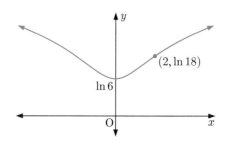

Example 5 ◀) Self Tutor

The marginal cost of producing x urns per week is given by

$\dfrac{dC}{dx} = 2.15 - 0.02x + 0.000\,36x^2$ pounds per urn provided $0 \leqslant x \leqslant 120$.

The initial costs before production starts are £185.

Find the total cost of producing 100 urns per day.

The marginal cost is $\quad \dfrac{dC}{dx} = 2.15 - 0.02x + 0.000\,36x^2$ pounds per urn

$$\therefore \ C(x) = \int (2.15 - 0.02x + 0.000\,36x^2)\,dx$$

$$= 2.15x - 0.02\,\frac{x^2}{2} + 0.000\,36\,\frac{x^3}{3} + c \ \text{ pounds}$$

$$= 2.15x - 0.01x^2 + 0.000\,12x^3 + c \ \text{ pounds}$$

But $\ C(0) = 185$ pounds, so $c = 185$

$$\therefore \ C(x) = 2.15x - 0.01x^2 + 0.000\,12x^3 + 185 \ \text{ pounds}$$

$$\therefore \ C(100) = 2.15(100) - 0.01(100)^2 + 0.000\,12(100)^3 + 185 \ \text{ pounds}$$

$$= 420 \ \text{ pounds}$$

$\therefore \quad$ the total cost is £420.

---

8 The marginal cost per day of producing x gadgets is $C'(x) = 3.15 + 0.004x$ pounds per gadget. Find the total daily production cost for 800 gadgets given that the fixed costs before production commences are £450 per day.

9 The marginal profit for producing x dinner plates per week is given by $P'(x) = 15 - 0.03x$ pounds per plate. If no plates are made then a loss of £650 each week occurs.

   a Find the profit function $P(x)$.

   b What is the maximum profit, and when does it occur?

   c What production levels enable a profit to be made?

Example 6

◀)) Self Tutor

A metal water pipe has an annulus cross-section as shown. The outer radius is 4 cm and the inner radius is 2 cm. Within the pipe, the water temperature is maintained at 100°C. Within the metal, the temperature drops off from inside to outside according to $\dfrac{dT}{dx} = -\dfrac{10}{x}$, where x is the distance from the central axis and $2 \leqslant x \leqslant 4$.

Find the temperature of the outer surface of the pipe.

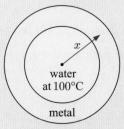

tube cross-section

Since $\dfrac{dT}{dx} = \dfrac{-10}{x}$, $T = \displaystyle\int \dfrac{-10}{x}\, dx$

$\therefore \ T = -10\ln x + c, \quad x > 0$

But when $x = 2$, $T = 100$

$\therefore \ 100 = -10\ln 2 + c$

$\therefore \ c = 100 + 10\ln 2$

Thus $T = -10\ln x + 100 + 10\ln 2$

$\therefore \ T = 100 + 10\ln\left(\dfrac{2}{x}\right)$

When $x = 4$, $T = 100 + 10\ln\left(\dfrac{1}{2}\right) \approx 93.1$

The outer surface temperature is 93.1°C.

---

10 An insulation tube has inner radius 0.02 m and outer radius 0.04 m. Fluid flowing through the tube maintains the temperature on the inner wall at 600°C.

Heat is lost through each metre of tube length according to *Fourier's law* $\dfrac{dT}{dr} = -\dfrac{q}{2\pi kr}$ where $q = 680$ W m^{-1} is the heat transfer rate per metre of length, $k = 0.2$ W m^{-1} °C is the thermal conductivity constant, r is the radius from the centre of the tube, and T is the temperature in °C.

Calculate the external temperature of the tube.

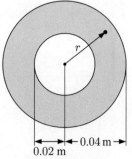

11 Fluid flows through a stainless steel pipe with thermal conductivity $k = 19$ W m^{-1} °C.

The pipe has inner radius $r_1 = 0.14$ m and outer radius $r_2 = 0.20$ m. The inner wall temperature is maintained at 400°C.

To meet occupational health and safety standards, the stainless steel pipe needs to be insulated with urethane foam with $k = 0.018$ W m^{-1} °C so that the temperature on the outside of the foam is not more than 50°C.

The heat loss $q = 60$ W m^{-1} is constant throughout the pipe and the insulation. Use Fourier's law $\dfrac{dT}{dr} = -\dfrac{q}{2\pi kr}$ to find:

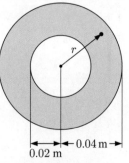

a the temperature on the outer surface of the pipe

b the outer radius r_3 of the insulation

c the thickness of insulation needed.

12 A wooden plank is supported only at its ends O and P, which are 4 m apart. The plank sags under its own weight by y m at the distance x m from O.

The differential equation $\dfrac{d^2y}{dx^2} = 0.01\left(2x - \dfrac{x^2}{2}\right)$ relates

the variables x and y, for $0 \leqslant x \leqslant 4$.

a Find the function $y(x)$ which measures the sag from the horizontal at any point along the plank.

b Find the maximum sag from the horizontal. Does the position of the point of maximum sag seem reasonable?

c Find the sag 1 m away from P.

d Find the angle the plank makes with the horizontal at the point 1 m from P.

C SEPARABLE DIFFERENTIAL EQUATIONS

Differential equations of the form $\dfrac{dy}{dx} = f(x)\,g(y)$ are called **separable differential equations**.

Examples of separable differential equations include:

- $\dfrac{dy}{dx} = xy^2 \qquad \{f(x) = x, \ g(y) = y^2\}$

- $\dfrac{dy}{dx} = y^2 + 3 \quad \{f(x) = 1, \ g(y) = y^2 + 3\}$

- $\dfrac{dy}{dx} = \dfrac{x+2}{y^3} \quad \{f(x) = x+2, \ g(y) = \dfrac{1}{y^3}\}$

> Differential equations of the form $\dfrac{dy}{dx} = g(y)$ are a special case of separable differential equations where $f(x)$ is constant.

These equations are called separable differential equations because we can *separate* the variables so that everything involving y is on one side, and everything involving x is on the other side.

$$\text{Since} \quad \frac{dy}{dx} = f(x)\,g(y),$$

$$\frac{1}{g(y)}\frac{dy}{dx} = f(x)$$

Integrating both sides of this equation with respect to x,

$$\int \frac{1}{g(y)}\frac{dy}{dx}\,dx = \int f(x)\,dx$$

$\therefore$ by the chain rule, $\displaystyle \int \frac{1}{g(y)}\,dy = \int f(x)\,dx.$

The variables are now separated, so we can now find the two integrals separately.

Example 7   ◀) **Self Tutor**

Solve the following differential equations:

a $\dfrac{dy}{dx} = xy$

b $\dfrac{dy}{dx} = \dfrac{1}{y^2}$

a $\qquad \dfrac{dy}{dx} = xy$

$\qquad \therefore \dfrac{1}{y}\dfrac{dy}{dx} = x$

$\qquad \therefore \displaystyle\int \dfrac{1}{y}\dfrac{dy}{dx}\,dx = \int x\,dx$

$\qquad \therefore \displaystyle\int \dfrac{1}{y}\,dy = \int x\,dx$

$\qquad \therefore \ln|y| = \tfrac{1}{2}x^2 + c$

$\qquad \therefore y = \pm e^{\frac{1}{2}x^2 + c}$

$\qquad \therefore y = Ae^{\frac{1}{2}x^2} \qquad \{A = \pm e^c\}$

b $\qquad \dfrac{dy}{dx} = \dfrac{1}{y^2}$

$\qquad \therefore y^2\dfrac{dy}{dx} = 1$

$\qquad \therefore \displaystyle\int y^2\dfrac{dy}{dx}\,dx = \int 1\,dx$

$\qquad \therefore \displaystyle\int y^2\,dy = \int 1\,dx$

$\qquad \therefore \dfrac{y^3}{3} = x + c$

$\qquad \therefore y^3 = 3x + c$

$\qquad \therefore y = \sqrt[3]{3x + c}$

Example 8   ◀) **Self Tutor**

Solve the differential equation $\dfrac{dy}{dx} = e^{2y}\cos 3x$ given that $y(0) = 0$.

$$\dfrac{dy}{dx} = e^{2y}\cos 3x$$

$$\therefore \dfrac{1}{e^{2y}}\dfrac{dy}{dx} = \cos 3x$$

$$\therefore \int \dfrac{1}{e^{2y}}\dfrac{dy}{dx}\,dx = \int \cos 3x\,dx$$

$$\therefore \int e^{-2y}\,dy = \int \cos 3x\,dx$$

$$\therefore -\tfrac{1}{2}e^{-2y} = \tfrac{1}{3}\sin 3x + c$$

$$\therefore e^{-2y} = -\tfrac{2}{3}\sin 3x + c$$

$$\therefore -2y = \ln\left(-\tfrac{2}{3}\sin 3x + c\right)$$

$$\therefore y = -\tfrac{1}{2}\ln\left(-\tfrac{2}{3}\sin 3x + c\right)$$

But $y(0) = 0$, so $\quad 0 = -\tfrac{1}{2}\ln c$

$$\therefore \ln c = 0$$

$$\therefore c = 1$$

The particular solution to the differential equation is $y = -\tfrac{1}{2}\ln\left(-\tfrac{2}{3}\sin 3x + 1\right)$.

EXERCISE 12C

1 Solve the following separable differential equations:

   a $\dfrac{dy}{dx} = \dfrac{x}{y^2}$ **b** $\dfrac{dy}{dx} = \dfrac{2x}{e^y}$ **c** $\dfrac{dy}{dx} = 3xy$

   d $\dfrac{dy}{dx} = 2x\sqrt{y}$ **e** $\dfrac{dy}{dx} = y\sin x$ **f** $\dfrac{dy}{dx} = -x\sqrt{y+1}$

   g $\dfrac{dy}{dx} = \dfrac{y}{x}$ **h** $\dfrac{dy}{dx} = 3x^2 e^y$ **i** $\dfrac{dy}{dx} = \dfrac{y+2}{x-1}$

2 Solve:

   a $\dfrac{dy}{dx} = y$ **b** $\dfrac{dy}{dx} = \dfrac{1}{y}$ **c** $\dfrac{dy}{dt} = y - 4$

   d $\dfrac{dP}{dt} = 3\sqrt{P}$ **e** $\dfrac{dQ}{dt} = 2Q + 3$ **f** $\dfrac{dQ}{dt} = \dfrac{1}{2Q+3}$

3 Solve:

   a $\dfrac{dy}{dx} = \dfrac{y}{3x+1}$ **b** $4 + \dfrac{dy}{dx} = 2y$ **c** $(x^2 + 5)\dfrac{dy}{dx} = \dfrac{2x}{y^2}$

   d $\sqrt{4 - x}\,\dfrac{dy}{dx} = 1 - y$ **e** $\dfrac{dy}{dx} = xy^2 - 2y^2$ **f** $y\dfrac{dy}{dx} = \dfrac{6x\sqrt{y}}{x^2 + 5}$

4 Find the particular solution to:

   a $\dfrac{dy}{dx} = \dfrac{3x}{y^2}$ given that $y(0) = 1$

   b $\dfrac{dy}{dx} = \dfrac{\sqrt{y}}{3}$ given that $y(44) = 9$

   c $\dfrac{dy}{dx} = y + yx^2$ given that $y(0) = 1$

   d $\dfrac{dy}{dx} = \dfrac{3x}{\cos y}$ given that $y(1) = 0$

   e $\dfrac{dy}{dx} = \dfrac{6\cos 2x}{\sqrt{y}}$ given that $y(0) = 3$

   f $e^y(2x^2 + 4x + 1)\dfrac{dy}{dx} = (x + 1)(e^y + 3)$ given that $y(0) = 2$

5   **a** Show that $\dfrac{3 - x}{x^2 - 1} = \dfrac{1}{x - 1} - \dfrac{2}{x + 1}$.

    b Find the particular solution to $\dfrac{dy}{dx} = \dfrac{3y - xy}{x^2 - 1}$ given that $y(0) = 1$.

6   **a** Write $\dfrac{8x + 3}{x^2 - x - 12}$ as a sum of partial fractions.

    b Find the general solution to $\dfrac{dy}{dx} = \dfrac{8x\sqrt{y} + 3\sqrt{y}}{x^2 - x - 12}$.

7 Find the particular solution to $\dfrac{dy}{dx} = \dfrac{5xy^2 + 4y^2}{x^2 + x - 2}$ given that $y(0) = -\tfrac{1}{2}$.

8 Find the general solution of the differential equation $\dfrac{dy}{dx} = \dfrac{x^2 y + y}{x^2 - 1}$.

Example 9

A salmon farm has an initial population of 80. The population grows according to the differential equation $\dfrac{dP}{dt} = \dfrac{1}{5}P$, where t is the time in years.

a Write an expression for P in terms of t.

b Find the salmon population after 10 years.

a
$$\frac{dP}{dt} = \frac{1}{5}P$$

$$\therefore \ \frac{1}{P}\frac{dP}{dt} = \frac{1}{5}$$

$$\therefore \ \int \frac{1}{P}\frac{dP}{dt}\,dt = \int \frac{1}{5}\,dt$$

$$\therefore \ \int \frac{1}{P}\,dP = \int \frac{1}{5}\,dt$$

$$\therefore \ \ln|P| = \frac{1}{5}t + c$$

$$\therefore \ P = e^{\frac{t}{5}+c} \qquad \{\text{since } P \geqslant 0\}$$

$$\therefore \ P = Ae^{\frac{t}{5}} \qquad \{A = e^c\}$$

When $t = 0$, $P = 80$

$$\therefore \ 80 = Ae^0 \quad \text{and so} \quad A = 80$$

$$\therefore \ P = 80e^{\frac{t}{5}}$$

$\dfrac{dP}{dt} = kP$ for some constant k indicates an **exponential relationship** between P and t.

b When $t = 10$, $P = 80e^2$

$$\therefore \ P \approx 591 \qquad \text{After 10 years, there are approximately 591 salmon.}$$

---

9 A population of rabbits in a field grows according to the differential equation $\dfrac{dP}{dt} = \dfrac{1}{2}P$, where t is the time in months. The initial population was 40 rabbits.

   a Write an equation for P in terms of t.

   b Find the rabbit population after 6 months.

10 When a transistor radio is switched off, the current I (in milliamps) falls away according to the differential equation $\dfrac{dI}{dt} = -0.4I$ where t is the time in milliseconds. At the instant the radio is switched off, the current is 350 milliamps.

   a Write an equation for I in terms of t.

   b Find the current after 5 milliseconds.

   c How long will it take for the current to fall to 20 milliamps?

11 Ethylene oxide reacts with water in the presence of the catalyst sulphuric acid to form ethylene glycol. Since water is present in excess, the rate of change in concentration of ethylene oxide (A) will be given by $\dfrac{dC_A}{dt} = -kC_A$ where the reaction rate constant $k = 0.31$ min^{-1}.

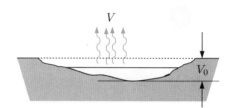

The initial concentration of ethylene oxide is 1 mol L^{-1}. Find the time required for 80% of the ethylene oxide to be used up.

12 In the "inversion" of raw sugar, the rate of change in the weight w kg of raw sugar is directly proportional to the weight w. After 10 hours, 80% of the sugar has been "inverted". What percentage of raw sugar remains after 30 hours?

13 Water evaporates from a lake at a rate proportional to the volume of water remaining.

Suppose V is the total amount of water evaporated after t days, and V_0 is the initial volume of water in the lake.

  a Explain why $\dfrac{dV}{dt} = k(V_0 - V)$.

  b If 50% of the water evaporates in 20 days, find the percentage of the original water remaining after 50 days without rain.

14 *Newton's law of cooling* states that the rate at which an object changes temperature is proportional to the difference between its temperature T and that of the surrounding medium T_m. So, $\dfrac{dT}{dt} \propto (T - T_m)$.

Use Newton's law of cooling to solve these problems:

  a The temperature inside a refrigerator is maintained at 5°C. An object at 100°C is placed in the refrigerator to cool. After 1 minute its temperature drops to 80°C. How long will it take for the temperature to drop to 10°C?

  b At 6 am the temperature of a corpse was 13°C, and 3 hours later had fallen to 9°C. Given that the temperature of a living body is 37°C and the temperature of the surroundings is constant at 5°C, estimate the time of death.

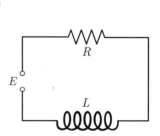

15 In an RL-circuit, the current I amps changes according to the differential equation $L\dfrac{dI}{dt} + RI = E$ where

L is the induction in henrys,
R is the resistance in ohms,
E is the voltage drop in volts, and
t is the time in seconds.

Suppose $L = 0.3$, $R = 10$, and $E = 20$.

  a Find a general solution for $I(t)$.

  b Find a particular solution for $I(t)$ if $I(0) = 0$ amps.

  c By considering I as $t \to \infty$, find the limiting current.

  d Find the time required for the current to reach 99% of its limiting value.

Example 10

◀)) **Self Tutor**

Water flows out of a tap at the bottom of a large cylindrical tank with base radius 2 m. The rate at which the water flows is proportional to the square root of the depth of the water remaining in the tank. Initially the tank is full to a depth of 9 m. After 15 minutes the water is 4 m deep. How long will it take for the tank to empty?

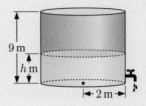

We are given that $\dfrac{dV}{dt} \propto \sqrt{h}$ where h is the depth of the water.

$\therefore \quad \dfrac{dV}{dt} = k\sqrt{h}$ where k is a constant.

$\therefore \quad \dfrac{dV}{dh}\dfrac{dh}{dt} = k\sqrt{h}$ {chain rule}

$\therefore \quad 4\pi\dfrac{dh}{dt} = k\sqrt{h}$ $\{V = \pi r^2 h = 4\pi h \quad \therefore \quad \dfrac{dV}{dh} = 4\pi\}$

$\therefore \quad \dfrac{4\pi}{\sqrt{h}}\dfrac{dh}{dt} = k$

$\therefore \quad \displaystyle\int 4\pi h^{-\frac{1}{2}}\dfrac{dh}{dt}\,dt = \int k\,dt$

$\therefore \quad 4\pi \displaystyle\int h^{-\frac{1}{2}}\,dh = \int k\,dt$

$\therefore \quad 4\pi\dfrac{h^{\frac{1}{2}}}{\frac{1}{2}} = kt + c$

$\therefore \quad 8\pi\sqrt{h} = kt + c$

Now when $t = 0$, $h = 9$

$\therefore \quad 8\pi\sqrt{9} = c$ and so $c = 24\pi$

$\therefore \quad 8\pi\sqrt{h} = kt + 24\pi$

Also, when $t = 15$, $h = 4$

$\therefore \quad 8\pi\sqrt{4} = 15k + 24\pi$

$\therefore \quad 15k = -8\pi$

$\therefore \quad k = -\dfrac{8\pi}{15}$

$\therefore$ the equation connecting the depth of the water and the time t is

$$8\pi\sqrt{h} = -\tfrac{8\pi}{15}t + 24\pi$$

The tank is empty when $h = 0$

$\therefore \quad 0 = -\tfrac{8\pi}{15}t + 24\pi$

$\therefore \quad \tfrac{8\pi}{15}t = 24\pi$

$\therefore \quad t = 45$

The tank empties in 45 minutes.

16 Water flows out of a tap at the bottom of a cylindrical tank of height 4 m and radius 2 m. The tank is initially full, and the water escapes at a rate proportional to the square root of the depth of the water remaining. After 2 hours, the water is 1 m deep. How long will it take for the tank to empty?

17 Water evaporates from a hemispherical bowl with radius r cm at the rate $\dfrac{dV}{dt} = -r^2$, where t is the time in hours.

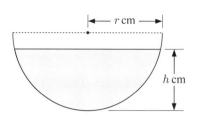

The volume of water of depth h, in a hemispherical bowl of radius r, is given by $V = \frac{1}{3}\pi h^2(3r - h)$.

a Use $\dfrac{dV}{dt} = \dfrac{dV}{dh}\dfrac{dh}{dt}$ to write a differential equation connecting h and t, given that r is a constant.

b Suppose the bowl's radius is 10 cm and that initially the bowl is full of water.

   i Show that $t = \frac{\pi}{300}(h^3 - 30h^2 + 2000)$, and hence find the time taken for the depth of the water to fall to 5 cm.

   ii How long will it take for the bowl to empty?

D LOGISTIC GROWTH

Many populations grow exponentially at first, but then level off to approach a maximum value A due to limited resources such as land or food. This growth model is called **logistic growth**.

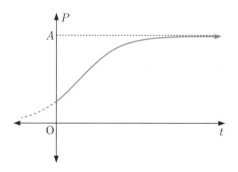

The logistic model can also be used to describe:

- the number of people in a town to be infected by a virus
- the number of "active" molecules in a chemical reaction, whereby a reaction between an "active" and an "inactive" molecule is required to produce two "active" molecules.

Logistic growth is governed by the differential equation

$$\frac{dP}{dt} = \underbrace{kP}\ \underbrace{\left(1 - \frac{P}{A}\right)}$$

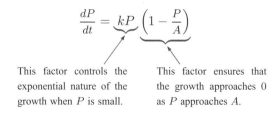

This factor controls the exponential nature of the growth when P is small.

This factor ensures that the growth approaches 0 as P approaches A.

The differential equation for logistic growth is separable, and we also need to use partial fractions.

Example 11

◀)) **Self Tutor**

The population of an island is currently 154. Its expected growth rate is given by
$\dfrac{dP}{dt} = 0.16P\left(1 - \dfrac{P}{500}\right)$, where t is in years.

a Write P as a function of t.

b Find the population after 10 years.

c Find the time taken for the population to increase to 480.

d What is the limiting population size?

a
$$\frac{dP}{dt} = 0.16P\left(1 - \frac{P}{500}\right) = 0.16P\left(\frac{500 - P}{500}\right)$$

$$\therefore \quad \frac{500}{P(500 - P)}\frac{dP}{dt} = 0.16$$

$$\therefore \quad \int \frac{500}{P(500 - P)}\frac{dP}{dt}\, dt = \int 0.16\, dt$$

$$\therefore \quad \int \frac{500}{P(500 - P)}\, dP = \int 0.16\, dt$$

$$\therefore \quad \int \left(\frac{1}{P} + \frac{1}{500 - P}\right) dP = \int 0.16\, dt$$

> We use partial fractions to help with the integration.

$$\therefore \quad \ln|P| + \frac{1}{-1}\ln|500 - P| = 0.16t + c$$

$$\therefore \quad \ln\left|\frac{P}{500 - P}\right| = 0.16t + c$$

$$\therefore \quad \frac{P}{500 - P} = \pm e^{0.16t + c}$$

$$\therefore \quad \frac{500 - P}{P} = be^{-0.16t} \qquad \left\{\text{letting } b = \pm\frac{1}{e^c}\right\}$$

Now when $t = 0$, $P = 154$

$$\therefore \quad \frac{346}{154} = b$$

So, we have $\dfrac{500 - P}{P} = \left(\dfrac{346}{154}\right)e^{-0.16t}$

$$\therefore \quad \frac{500}{P} - 1 = \left(\frac{346}{154}\right)e^{-0.16t}$$

$$\therefore \quad \frac{500}{P} = 1 + \left(\frac{346}{154}\right)e^{-0.16t}$$

$$\therefore \quad P = \frac{500}{1 + \left(\frac{346}{154}\right)e^{-0.16t}}$$

b When $t = 10$, $P = \dfrac{500}{1 + \left(\frac{346}{154}\right)e^{-0.16 \times 10}} \approx 344$

So the population after 10 years is approximately 344 people.

c When $P = 480$, $480 = \dfrac{500}{1 + \left(\frac{346}{154}\right)e^{-0.16t}}$

$\therefore\ 1 + \left(\tfrac{346}{154}\right)e^{-0.16t} = \tfrac{500}{480}$

$\therefore\ e^{-0.16t} = \tfrac{20}{480} \times \tfrac{154}{346} = \tfrac{77}{4152}$

$\therefore\ -0.16t = \ln\left(\tfrac{77}{4152}\right)$

$\therefore\ t = \dfrac{\ln\left(\frac{77}{4152}\right)}{-0.16} \approx 24.9$ years

d as $t \to \infty$, $e^{-0.16t} \to 0$

$\therefore\ P \to \dfrac{500}{1 + 0} = 500$

The limiting population is 500.

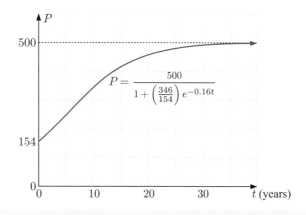

EXERCISE 12D

1 Consider the logistic differential equation $\dfrac{dP}{dt} = 0.2P\left(1 - \dfrac{P}{200}\right)$, $P(0) = 20$.

    a Write P as a function of t.

    b Find the value of P when $t = 10$.

    c Discuss the behaviour of P as $t \to \infty$.

    d Sketch the graph of P against t.

2 Answer the **Opening Problem** on page **272**.

3 A lily in a pond initially has area 20 cm². Its area A grows according

    to the model $\dfrac{dA}{dt} = 0.1A\left(1 - \dfrac{A}{500}\right)$, where t is the time in days.

    a Write A as a function of t.

    b Find the area of the lily after 20 days.

    c Find the limiting size of the lily.

4 In a small country town, rumours spread very fast. At 8 am on Monday, a rumour begins with 2 people. The number of people N who have heard the rumour grows according to the model

$\dfrac{dN}{dt} = 0.8N\left(1 - \dfrac{N}{600}\right)$, where t is the time in hours after 8 am.

    a Write N as a function of t.

    b How many people have heard the rumour by 11 am?

    c How many people do you think live in the town?

    d At what time would 500 people have heard the rumour?

5 There are 10^{30} molecules involved in a chemical reaction. Initially, 200 of the molecules are "active", and any reaction between an "active" and an "inactive" molecule produces two "active" molecules. The number of "active" molecules grows according to the differential

equation $\dfrac{dN}{dt} = kN\left(1 - \dfrac{N}{10^{30}}\right)$, where t is the time in seconds.

  a Solve the differential equation, and hence write N in terms of k and t.

  b Given that 1.5×10^7 molecules were "active" after 10^{-5} seconds, find k.

  c At what time would you expect the reaction to be 99% complete?

6 14 European foxes were released in Australia in 1845 for sport hunting. Spreading rapidly out of control, the fox is now found throughout the mainland, except in the tropical northern regions.

In 1900 there were 30 000 foxes in Australia, and today their population is steady at around 95 000.

  a What features of the growth in the fox population suggest that a logistic model is appropriate?

  b Suppose the population of foxes F grows according to the differential equation

$$\frac{dF}{dt} = kF\left(1 - \frac{F}{A}\right), \quad \text{where } t \text{ is the number of years since 1845.}$$

    i State the value of A.

    ii Solve the differential equation, and use the information provided to write F in terms of t.

  c Estimate the fox population in 1920.

  d Estimate the time at which the fox population was:

    i 15 000   **ii** 65 000.

  e Sketch the graph of F against t.

  f When was the population growth rate a maximum? How does this appear on the graph of F against t?

Historical note   The logistic model

The logistic model was first given its name in 1845 by the Belgian mathematician **Pierre François Verhulst**, in his study of population growth. Verhulst originally described the logistic model as

$\dfrac{dN}{dt} = rN\left(1 - \dfrac{N}{K}\right)$, where r is a measure of the growth rate, and

K is the limiting population.

The model was rediscovered in 1920 by the American scientists Raymond Pearl and Lowell Reed, who promoted its use.

The r/K **selection theory** in ecology derives its name from the symbols in Verhulst's logistic model. The theory hypothesises that species are driven towards either r-*selection* or K-*selection*.

Pierre François Verhulst

r-selected species are characterised by high growth rates and high birth rates, but with individuals having a relatively low chance of surviving to adulthood. Examples of r-selected species include insects and small rodents.

K-selected species operate in dense groups close to their limiting population. They typically produce fewer offspring, but which have high chance of survival. They are characterised by their ability to compete for limited resources. Examples of K-selected species include elephants, humans, and whales.

Review set 12A

1 Verify that $y = 4\ln(x^2 + 3x) + 8$ is a solution of the differential equation $\dfrac{dy}{dx} = \dfrac{8x + 12}{x^2 + 3x}$.

2 Solve the following differential equations:

a $\dfrac{dy}{dx} = \cos 2x - \sin^2 x$

b $\dfrac{dy}{dx} = 3 - e^{-2x}$

c $\dfrac{dy}{dx} = \dfrac{1}{2x + 1}, \ y(0) = 2$

d $\dfrac{dy}{dt} - t = te^{t^2}, \ y(1) = 2e$

3 The marginal profit for producing x vases per day is given by $P'(x) = 20 - \dfrac{x}{4}$ pounds per vase. If no vases are made then a loss of £250 each day occurs.

a Find the profit function $P(x)$.

b What is the maximum profit, and when does it occur?

c What production levels enable a profit to be made?

4 The rate at which spectators entered a stadium before a football match is given by

$$S'(t) = \frac{4000e^{-0.05t}}{(1 + 4e^{-0.05t})^2}$$ spectators per minute, where t is

the time in minutes after noon.

a At what rate were spectators entering the stadium at:

   i noon   **ii** 12:30 pm?

b Find the derivative of $\dfrac{1}{1 + 4e^{-0.05t}}$, and hence find

$$\int S'(t)\, dt.$$

c Find $\displaystyle\int_0^{60} S'(t)\, dt$, and interpret your answer.

d Given that there were 4000 spectators in the stadium at noon, find the number of spectators in the stadium when the match starts at 1:40 pm.

5 Solve:

a $\dfrac{dy}{dx} = 5x^2 y$

b $\dfrac{dy}{dx} = 2xy^2 - y^2$

c $\dfrac{dx}{dt} = \dfrac{3\sin 2t}{4\sqrt{x}}$

6 Solve the differential equation $\dfrac{dy}{dx} = -\dfrac{3}{e^y}$ given that $y = 1$ when $x = 0$.

7 A radioactive substance decays at a rate proportional to the mass M remaining at time t. Suppose the initial mass is M_0.

   a Construct a differential equation involving M, t, and a constant k to model this situation.

   b If the substance is reduced to $\frac{4}{5}$ of its original mass in 30 days, calculate the time required for the substance to decay to half its original mass.

8 The enzyme urease breaks down urea into ammonia and carbon dioxide in the reaction:

$$NH_2CONH_2 + H_2O \xrightarrow{\text{urease}} 2NH_3 + CO_2$$

The rate of change of concentration C of urea after t seconds is given by

$$\frac{dC}{dt} = \frac{-VC}{K+C}$$

$$\text{where} \quad V = 2.66 \times 10^{-4} \text{ mol s}^{-1} \text{ L}^{-1}$$

$$\text{and} \quad K = 0.0266 \text{ mol L}^{-1}$$

   a Given that the initial concentration of urea is 0.1 mol L^{-1}, write t in terms of C.

   b Calculate the time required for 85% of the urea to decompose into ammonia and carbon dioxide.

9 A rare virus is infecting the residents of Smallville. It was introduced initially into the population by a child returning home from boarding school. The number N of people who have been infected after t weeks is given by the differential equation $\dfrac{dN}{dt} = N\left(1 - \dfrac{N}{694}\right)$.

   a Write N in terms of t.

   b Find the number of people who will have had the virus after 4 weeks.

   c How long will it take for 500 people to have been infected?

Review set 12B

1 Verify that $y = 3\sin x - 2\cos x$ is a solution to $3\dfrac{dy}{dx} - 2y = 13\cos x$.

2 Consider the differential equation $\dfrac{dy}{dx} = y - 1$.

   a Show that $y = ce^x + 1$ is a solution to the differential equation for any constant c.

   b Sketch the solution curves for $c = 0, \pm 1, \pm 2$.

   c Find the particular solution which passes through $(0, 4)$.

   d Find the equation of the tangent to the particular solution at $(0, 4)$.

3 Solve the following differential equations:

   a $\dfrac{dy}{dx} = \dfrac{e^x}{e^x - 2}$

   b $\dfrac{dy}{dx} = \frac{1}{2}\cos\left(\frac{\pi}{3} - 2x\right), \quad y\left(\frac{\pi}{2}\right) = 0$

4

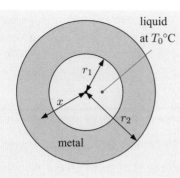

liquid at T_0°C

metal

A metal tube has an annulus cross-section with radii r_1 and r_2 as shown.

Within the tube, a liquid is maintained at temperature T_0 °C.

Within the metal, the temperature drops from inside to outside according to $\dfrac{dT}{dx} = \dfrac{k}{x}$ where k is a negative constant and x is the distance from the central axis.

Show that the outer surface has temperature

$$T_0 + k\ln\!\left(\frac{r_2}{r_1}\right) \text{°C}.$$

5 Solve:

a $\dfrac{dy}{dx} = 2y^4$

b $(t^2 + 1)\dfrac{dP}{dt} = Pt$

6 Find the particular solution to:

a $\dfrac{dy}{dx} = \sqrt{y}$ given that $y(0) = 4$

b $\dfrac{dy}{dx} = y\cos x$ given that $y\!\left(\frac{\pi}{2}\right) = \dfrac{1}{e^2}$.

7 Solve the differential equation $\dfrac{dy}{dx} = \dfrac{xy - 6y}{x^2 - 4}$ given that $y(3) = 1$.

8 Brian has made a cup of tea. The tea's temperature T is initially 85°C, and it cools according to the differential equation $\dfrac{dT}{dt} = -0.1(T - 20)$, where t is the time in minutes.

a Find a formula for T in terms of t.

b Find the temperature of the tea after 4 minutes.

c Describe the behaviour of T as $t \to \infty$.

d Sketch the graph of T against t.

e Brian likes to drink his tea at a temperature between 40°C and 65°C.

   i How long will he have to wait for his tea to cool down?

   ii How much time will he have to drink the tea before it gets too cold?

9 An ecology student is studying the repopulation of wild ostriches in Oudtshoorn after a drought. She notices that the population growth rate is approximately logistic so that $\dfrac{dP}{dt} = kP\!\left(1 - \dfrac{P}{A}\right)$ where $P(t)$ is the population t years after her study begins.

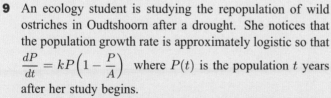

The carrying capacity A is known to be 200 000 ostriches, since this was the population before the drought.

Initially, the student finds that there are 25 000 ostriches.

After two years she finds there are 31 200 ostriches.

a Solve the differential equation, and use the population $P(0)$ and $P(2)$ to find the population $P(t)$ after t years.

b Estimate the ostrich population after 5 years.

c Find the time at which the population growth rate is a maximum.

10 A water tank of height 1 m has a square base 2 m $\times$ 2 m. When a tap at its base is opened, the water flows out at a rate proportional to the square root of the depth of the water at any given time. Suppose the depth of the water is h m, and V is the volume of water remaining in the tank after t minutes.

a Write down a differential equation involving $\dfrac{dV}{dt}$ and h to model this situation.

b Explain why $V = 4h$ m^3 at time t, then use the chain rule to write down a differential equation involving $\dfrac{dh}{dt}$ and h only.

c The tank is initially full. When the tap is opened, the water level drops by 19 cm in 2 minutes. Find the time it takes for the tank to empty.

13

Numerical methods

Contents:

Opening problem

In our introduction to integral calculus, we considered calculating the area under a curve by adding the areas of rectangles.

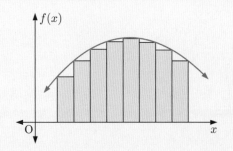

This is a **numerical method** which *approximates* the area. The more rectangles we use on a given interval, the more accurate the estimate will be. Eventually, in the *limit* where we have infinitely many rectangles, the numerical method will give the exact answer.

It is appropriate for us to introduce integral calculus in this way, because it is exactly how the field of integral calculus was historically "discovered" by **Bonaventura Cavalieri**, as described in the **Historical Note** on page **218**.

Things to think about:

a How can the methods of approximating area using "lower rectangles" and "upper rectangles" be improved?

b What can go wrong with numerical methods for integration?

c What are the advantages of using a numerical method rather than performing the integration analytically?

In this Chapter we consider a number of numerical methods, which we can divide into two broad categories:

- methods for solving equations
- methods of integration.

Numerical methods are all **iterative algorithms**, which means they are repetitive processes governed by fixed rules. Each repetition of the process is called an **iteration**.

A   CHANGES OF SIGN METHODS FOR $f(x) = 0$

Suppose we evaluate a function $f(x)$ at $x = a$ and $x = b$, and we find that $f(a)$ and $f(b)$ have opposite signs.

If $f(x)$ is *continuous* on the interval $[a, b]$, then there must be some value $x^* \in [a, b]$ for which $f(x^*) = 0$.

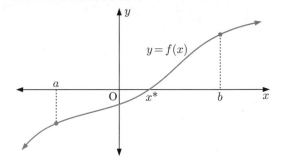

Discussion

- What do you think is meant by *continuous*? Why would it be important?
- What function types have we seen that are not *continuous*?
- Is there necessarily only one value for $x \in [a, b]$ for which $f(x) = 0$?

In this course we do not formally define a continuous function, but an intuitive understanding is useful. Informally:

> A function is **continuous** on a given domain if it is defined for all x in this domain, and its graph never "jumps".

For example, consider the following functions on the interval $[a, b]$:

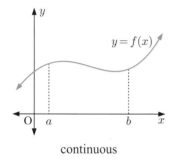

continuous

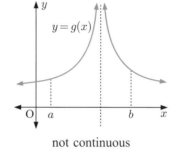

not continuous

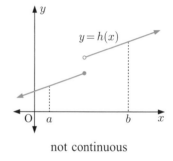

not continuous

In the **changes of sign method**, we use successive iterations to narrow the interval on which we know there is a solution.

Step 1: For the first iteration $n = 1$, we let $a_1 = a$ and $b_1 = b$.

Step 2: For the nth iteration, we choose a value (or values) for x on the interval $[a_n, b_n]$ and evaluate the function for each.

Step 3: If $f(x^*) = 0$ for some particular value x^*, then x^* is a solution, and we stop the algorithm. Alternatively, if we are satisfied that $f(x^*)$ is sufficiently close to zero, then we can accept the approximate solution x^*.

Otherwise, we set new endpoints a_{n+1} and b_{n+1} so that $f(a_{n+1})$ and $f(b_{n+1})$ have opposite signs, and we return to *Step 2*.

There are several ways we can choose points on the interval $[a_n, b_n]$ to test in *Step 2*. We can use:

- a decimal search
- interval bisection
- linear interpolation.

DECIMAL SEARCH

In a decimal search we improve the accuracy of our solution by a decimal point each iteration.

We achieve this by performing 9 calculations equally spaced on the interval.

Example 1

◀)) **Self Tutor**

Approximate $\sqrt{5}$ to 3 decimal places by using a decimal search for the positive solution to $x^2 - 5 = 0$.

Let $f(x) = x^2 - 5$.

We know that $f(x)$ is continuous for all x, and that $f(2) = -1$ and $f(3) = 4$.

∴ $\sqrt{5}$ lies between 2 and 3.

Iteration 1

$a_1 = 2, \ b_1 = 3$

x	$f(x)$
2.0	-1
2.1	-0.59
2.2	-0.16
2.3	0.29
2.4	0.76
2.5	1.25
2.6	1.76
2.7	2.29
2.8	2.84
2.9	3.41
3.0	4

Iteration 2

$a_2 = 2.2, \ b_2 = 2.3$

x	$f(x)$
2.20	-0.1600
2.21	-0.1159
2.22	-0.0716
2.23	-0.0271
2.24	0.0176
2.25	0.0625
2.26	0.1076
2.27	0.1529
2.28	0.1984
2.29	0.2441
2.30	0.2900

Iteration 3

$a_3 = 2.23, \ b_3 = 2.24$

x	$f(x)$
2.230	-0.0271
2.231	$-0.022\,639$
2.232	$-0.018\,176$
2.233	$-0.013\,711$
2.234	$-0.009\,244$
2.235	$-0.004\,775$
2.236	$-0.000\,304$
2.237	$0.004\,169$
2.238	$0.008\,644$
2.239	$0.013\,121$
2.240	0.0176

The solution lies on $[2.236, \ 2.237]$.

Since $f(2.2365) \approx 0.0019 > 0$, we conclude $\sqrt{5} \approx 2.236$ {to 3 d.p.}

DECIMAL SEARCH

EXERCISE 13A.1

1 Approximate $\sqrt{2}$ to 3 decimal places using a decimal search for the positive solution to $x^2 - 2 = 0$.

2 Approximate π to 4 decimal places using a decimal search for the appropriate solution to $\sin x = 0$.

3 Solve $x^3 - 5x^2 + 7x - 1 = 0$ to 3 decimal places using a decimal search.

4 Suppose $f(x) = -x^3 + 4x^2 + x - 2$.

    a Find $f(-5)$ and $f(5)$.

    b Perform one iteration of a decimal search for the solution to $f(x) = 0$ using $a_1 = -5$ and $b_1 = 5$.

    c Explain your result in **b**.

    d Using further decimal searches, find *all* solutions to $f(x) = 0$, correct to 2 decimal places.

Discussion

Will a decimal search always find all solutions to $f(x) = 0$ on the interval $[a, b]$?

INTERVAL BISECTION

In this changes of sign method we only make one calculation each iteration, at the midpoint of the interval.

We therefore test $x^* = \dfrac{a_n + b_n}{2}$.

The interval halves in width each iteration, guaranteeing the rate at which we converge to a solution.

Example 2
◀) **Self Tutor**

Approximate $\sqrt{5}$ to 3 decimal places by using the changes of sign method with interval bisection to find the positive solution to $x^2 - 5 = 0$.

Let $f(x) = x^2 - 5$.

We know that $f(x)$ is continuous for all x, and that $f(2) = -1$ and $f(3) = 4$.

$\therefore$ $\sqrt{5}$ lies between 2 and 3.

Iteration n	a_n	x^*	b_n	$f(a_n)$	$f(x^*)$	$f(b_n)$	
1	2.000 00	2.500 00	3.000 00	−1.000 00	1.250 00	4.000 00	**INTERVAL BISECTION**
2	2.000 00	2.250 00	2.500 00	−1.000 00	0.062 50	1.250 00	
3	2.000 00	2.125 00	2.250 00	−1.000 00	−0.484 38	0.062 50	
4	2.125 00	2.187 50	2.250 00	−0.484 38	−0.214 84	0.062 50	
5	2.187 50	2.218 75	2.250 00	−0.214 84	−0.077 15	0.062 50	
6	2.218 75	2.234 38	2.250 00	−0.077 15	−0.007 57	0.062 50	
7	2.234 38	2.242 19	2.250 00	−0.007 57	0.027 40	0.062 50	
8	2.234 38	2.238 28	2.242 19	−0.007 57	0.009 90	0.027 40	
9	2.234 38	2.236 33	2.238 28	−0.007 57	0.001 16	0.009 90	
10	2.234 38	2.235 35	2.236 33	−0.007 57	−0.003 20	0.001 16	
11	2.235 35	2.235 84	2.236 33	−0.003 20	−0.001 02	0.001 16	
12	2.235 84	2.236 08	2.236 33	−0.001 02	0.000 07	0.001 16	

At iteration 12, the interval is $\approx [2.2358,\ 2.2363]$ so we conclude $\sqrt{5} \approx 2.236$ {to 3 d.p.}

EXERCISE 13A.2

1 Approximate $\sqrt{3}$ to 3 decimal places using the changes of sign method with interval bisection to find the positive solution to $x^2 - 3 = 0$.

2 Approximate e to 5 decimal places using the changes of sign method with interval bisection to find the approximate solution to $\ln x - 1 = 0$.

3 Suppose you know $f(x) = 0$ has a solution which lies in the interval $[a,\ a+1]$. How many iterations of the changes of sign method with interval bisection will you need to find the solution correct to 6 decimal places? Explain your answer.

4 Use the changes of sign method with interval bisection to find, correct to 4 decimal places:

   a the solution to $5 - \sin x - x^2 = 0$ on $[0,\ 4]$

   b the positive solution to $x^2 + 2x - \dfrac{4}{x} = 0$.

5   **a**   Use the changes of sign method with interval bisection to find, correct to 4 decimal places, a solution to $x^3 - 12x - 4 = 0$ on $[-4, 4]$.

  b   Sketch the graph of $y = x^3 - 12x - 4$ on $[-4, 4]$.

  c   Did the method in **a** find *all* of the solutions to $x^3 - 12x - 4 = 0$ on $[-4, 4]$?

6   Consider the general cubic function $f(x) = ax^3 + bx^2 + cx + d$.

  a   Show that if $y = f(x)$ has stationary points, they occur when $x = \dfrac{-b \pm \sqrt{b^2 - 3ac}}{3a}$.

  b   Explain how you can use the changes of sign method with interval bisection to find *all* zeros of a cubic.

LINEAR INTERPOLATION

For a function $f(x)$ which is continuous on $[a, b]$, suppose $(a, f(a))$ and $(b, f(b))$ lie either side of the x-axis.

In this changes of sign method we choose to test the value $x^* \in [a, b]$, where the line through $(a, f(a))$ and $(b, f(b))$ crosses the x-axis.

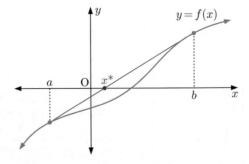

Equating gradients, $\dfrac{f(b) - f(a)}{b - a} = \dfrac{0 - f(a)}{x^* - a}$

$\therefore \ (x^* - a)(f(b) - f(a)) = -f(a)(b - a)$

$\therefore \ x^*(f(b) - f(a)) - a\,f(b) + a\,f(a) = -b\,f(a) + a\,f(a)$

$\therefore \quad x^*(f(b) - f(a)) = a\,f(b) - b\,f(a)$

$\therefore \qquad x^* = \dfrac{a\,f(b) - b\,f(a)}{f(b) - f(a)}$

Example 3   ◀⑨ **Self Tutor**

Approximate $\sqrt{5}$ by using the changes of sign method with linear interpolation to find the positive solution to $x^2 - 5 = 0$.

Let $f(x) = x^2 - 5$.

We know that $f(x)$ is continuous for all x, and that $f(2) = -1$ and $f(3) = 4$.

$\therefore \ \sqrt{5}$ lies between 2 and 3

Iteration n	a_n	x^*	b_n	$f(a_n)$	$f(x^*)$	$f(b_n)$	LINEAR INTERPOLATION
1	2.000 00	2.200 00	3.000 00	−1.000 00	−0.160 00	4.000 00	
2	2.200 00	2.230 77	3.000 00	−0.160 00	−0.023 67	4.000 00	
3	2.230 77	2.235 29	3.000 00	−0.023 67	−0.003 46	4.000 00	
4	2.235 29	2.235 96	3.000 00	−0.003 46	−0.000 50	4.000 00	
5	2.235 96	2.236 05	3.000 00	−0.000 50	−0.000 07	4.000 00	
6	2.236 05	2.236 07	3.000 00	−0.000 07	−0.000 01	4.000 00	
7	2.236 07	2.236 07	3.000 00	−0.000 01	0.000 00	4.000 00	

After iteration 7, $f(x^*) \approx 0$ so we conclude $\sqrt{5} \approx 2.236\,07$

Discussion

- How does the "pattern" of this solution compare to those in **Examples 1** and **2**?
- Are we guaranteed to find a solution using this method?
- Do we know how fast this method will converge to a solution?

EXERCISE 13A.3

1 Approximate $\sqrt{7}$ by using the changes of sign method with linear interpolation to find the positive solution to $x^2 - 7 = 0$.

2 Approximate $\ln 2$ by using the changes of sign method with linear interpolation to find the solution to $e^x - 2 = 0$.

3 Use the changes of sign method with linear interpolation to approximate the positive solution to:

   a $2x - \sqrt{x} = 0$ **b** $\sqrt{3 - 2x^2} = \frac{1}{2}$

4 Use the changes of sign method with linear interpolation to approximate the solution to $x^3 - 5x^2 + 7x - 1 = 0$.

5 **a** Use the changes of sign method with linear interpolation to approximate a solution to $x^3 - 12x - 4 = 0$ on $[-4, 4]$.

   b Did your method in **a** find the same solution as the one in **Exercise 13A.2** question **5 a**? Why do you think this happened?

   c **i** Repeat **a** using the starting interval $[-4, 10]$.

       ii Do you arrive at the same solution?

      iii Do you think linear interpolation is always more efficient than using interval bisection?

B   COBWEB AND STAIRCASE DIAGRAMS FOR $f(x) = 0$

Suppose $x^* \in [a, b]$ is a zero of $f(x)$, and that $f(x)$ is continuous on $[a, b]$.

We can always rearrange the equation $f(x) = 0$ into the form $x = g(x)$, if necessary, by using $g(x) = f(x) + x$.

The solutions to $f(x) = 0$ will be the x-coordinates of any intersection points of $y = g(x)$ and $y = x$.

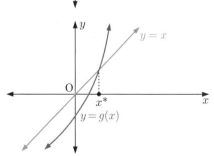

Since $x^* = g(x^*)$, we choose an initial guess x_0 for the solution x^*, and perform an iterative calculation

$$x_{n+1} = g(x_n)$$

We can illustrate the result on a graph:

- In the graph alongside, we *converge* to the solution in a **cobweb**.

  ① For an initial guess x_0, calculate $g(x_0)$.

  ② Move across to $y = x$, to set $x_1 = g(x_0)$.

  ③ Calculate $g(x_1)$.

  ④ Move across to $y = x$, to set $x_2 = g(x_1)$.

  ⑤ Calculate $g(x_2)$.

  ⑥ Move across to $y = x$, to set $x_3 = g(x_2)$.

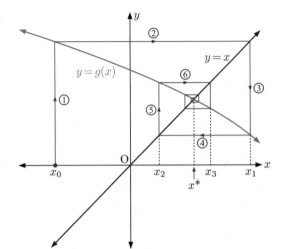

- In the graph alongside, we *converge* to the solution in a **staircase**.

  DEMO

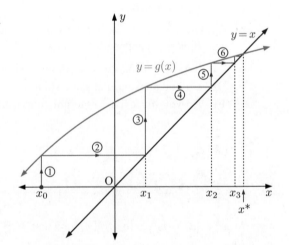

Investigation 1   Cobweb and staircase diagrams

Having seen the algorithm for cobweb and staircase diagrams at work, we now try to understand what scenarios will result in:

- a cobweb   • a staircase   • convergence to a solution to $x = g(x)$.

What to do:

1 Click on the icon and print the cobweb and staircase diagrams.

2 In each case, follow the algorithm to try to find a solution to $x = g(x)$.

3 What feature of $g(x)$ will result in a cobweb, and what cases result in a staircase?

4 **a** What properties of $g(x)$ will *guarantee* convergence to a solution to $x = g(x)$?

  b Does the absence of these properties guarantee the algorithm will *not* converge?

From the **Investigation**, you should have found that when we use the cobweb and staircase algorithm on $x = g(x)$:

- If $g'(x) > 0$ we obtain a staircase. • If $g'(x) < 0$ we obtain a cobweb.
- If $|g'(x)| < 1$ for all x then we are guaranteed to converge to the only solution of $x = g(x)$.
- If $|g'(x)| < 1$ for a sufficient interval including x_0 and the solution x^*, then the algorithm will also converge.
- The algorithm *may* also converge for the case where $|g'(x_0)| \geqslant 1$ if the algorithm takes us into an interval where $|g'(x)| < 1$ and containing x^*, or if we are just plain lucky!

Example 4
◆) **Self Tutor**

Consider the cubic equation $x^3 - 2x^2 + x - 3 = 0$.

a Show that the equation can be rearranged into the form $x = g(x)$ where $g(x) = \sqrt[3]{2x^2 - x + 3}$.

b Apply the algorithm $x_{n+1} = g(x_n)$ with $x_0 = 0$ for 20 iterations to approximate a solution to the equation.

c Use technology to graph $y = x$ and $y = g(x)$ on the same set of axes. Hence draw the cobweb or staircase diagram.

d Explain why $x = \sqrt[3]{2x^2 - x + 3}$ was a suitable rearrangement of the equation, whereas $x = -x^3 + 2x^2 + 3$ would not have converged.

a $x^3 - 2x^2 + x - 3 = 0$

$\therefore \quad x^3 = 2x^2 - x + 3$

$\therefore \quad x = \sqrt[3]{2x^2 - x + 3} = g(x)$

b

n	x_n
0	0
1	1.442 249 57
2	1.788 185 80
3	1.966 701 27
4	2.062 143 00
5	2.113 646 91
6	2.141 528 56
7	2.156 641 49
8	2.164 837 94
9	2.169 284 50
10	2.171 697 09
11	2.173 006 19
12	2.173 716 55
13	2.174 102 03
14	2.174 311 21
15	2.174 424 72
16	2.174 486 32
17	2.174 519 75
18	2.174 537 89
19	2.174 547 73
20	2.174 553 07

$\therefore \quad x \approx 2.1746$

c

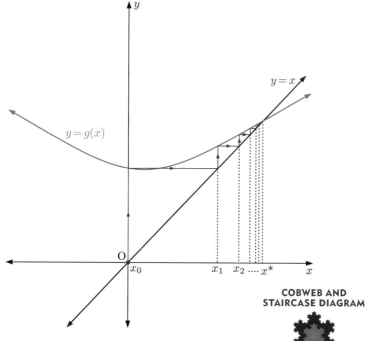

COBWEB AND STAIRCASE DIAGRAM

d $x = g_1(x) = \sqrt[3]{2x^2 - x + 3}$ is a suitable rearrangement since $|g_1'(x)| < 1$ for all x. Hence the algorithm converges.

For the case $x = g_2(x) = -x^3 + 2x^2 + 3$,

$$g_2'(x) = -3x^2 + 4x$$

$\therefore$ $g_2'(2.1746) \approx -5.49$

$\therefore$ $|g_2'(x)| > 1$ in the interval around the solution. The algorithm would therefore not converge.

EXERCISE 13B

1   **a**   Apply the algorithm $x_{n+1} = g(x_n)$ with $x_0 = 0$ and $g(x) = \frac{1}{2}\sqrt{2 - x}$ for 10 iterations.

   b   Hence find an approximate solution to $4x^2 + x - 2 = 0$. Compare your result with the analytic solution.

2   Consider the equation $x - \arctan x = 1$.

   a   Rearrange the equation into the form $x = g(x)$.

   b   Apply the algorithm $x_{n+1} = g(x_n)$ with $x_0 = 1$ to find an approximate solution to the equation.

3   Consider the cubic equation $x^3 - 4x^2 + 2x - 6 = 0$.

   a   Show that the equation can be rearranged into the form $x = g(x)$ where $g(x) = \sqrt[3]{4x^2 - 2x + 6}$.

   b   Apply the algorithm $x_{n+1} = g(x_n)$ with $x_0 = 0$ for 20 iterations to approximate a solution to the equation.

   c   Use technology to graph $y = x$ and $y = g(x)$ on the same set of axes. Hence draw the cobweb or staircase diagram.

4   Consider the equation $x^2 - 5 = 0$.

   a   Suppose the equation is rewritten in the form $x = g_1(x)$ where $g_1(x) = x^2 + x - 5$. Explain why the algorithm $x_{n+1} = g_1(x_n)$ will not converge in this case.

Sometimes we can write an equation in another form in order to get convergence.

   b   Show that the equation can be rearranged into the form $x = g_2(x)$ where $g_2(x) = -\left(\dfrac{x^2}{5} - 1\right) + x$.

   c   Apply the algorithm $x_{n+1} = g_2(x_n)$ with $x_0 = 2$ to approximate $\sqrt{5}$ to 3 decimal places.

C   NEWTON-RAPHSON METHOD FOR $f(x) = 0$

The third numerical method for solving equations which we consider in this course is the **Newton-Raphson method** named after the English mathematicians **Sir Isaac Newton** (1642 - 1726/7) and **Joseph Raphson** (1648 - 1715).

This method also requires the function to be continuous on an interval around the actual solution x^*, and requires an initial guess for the solution x_0 which is sufficiently good.

Suppose the tangent to $y = f(x)$ when $x = x_0$ meets the x-axis at $x = x_1$.

$$\therefore \quad f'(x_0) = \frac{f(x_0) - 0}{x_0 - x_1}$$

$$\therefore \quad f'(x_0)(x_0 - x_1) = f(x_0)$$

$$\therefore \quad x_1 f'(x_0) = x_0 f'(x_0) - f(x_0)$$

$$\therefore \quad x_1 = x_0 - \frac{f(x_0)}{f'(x_0)}$$

We use this value x_1 as our guess for the next iteration.

So, in general, $x_{n+1} = x_n - \dfrac{f(x_n)}{f'(x_n)}$.

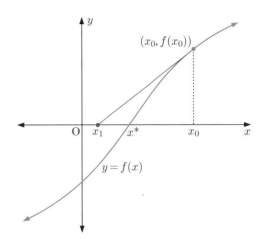

Example 5   ◀)) Self Tutor

Approximate $\sqrt{5}$ to 3 decimal places by using the Newton-Raphson method with $x_0 = 2$ to search for the positive solution to $x^2 - 5 = 0$.

Let $f(x) = x^2 - 5 = 0$
$\therefore \quad f'(x) = 2x$

n	x_n	$f(x_n)$	$f'(x_n)$
0	2	-1	4
1	2.25	0.0625	4.5
2	2.236 11	0.000 19	4.472 22
3	2.236 07	0.000 00	4.472 14
4	2.236 07		

We conclude $\sqrt{5} \approx 2.236$ {to 3 d.p.}

Discussion

- For the equation $x^2 - 5 = 0$ in **Example 5**, what do you think would happen if we used:
  - $x_0 = 3$
  - $x_0 = -2$
  - $x_0 = 0$?

  > Use the software to check your predictions.

- What else do you think could go wrong with this method?

NEWTON-RAPHSON
METHOD

EXERCISE 13C

1 Use the Newton-Raphson method with the given initial guess to find a solution to each equation correct to 3 decimal places:

a $3 - x^2 = 0$, $x_0 = 1$

b $x^3 + 2x + 1 = 0$, $x_0 = 0$

c $2\sqrt{5 - x} = 3$, $x_0 = 4$

d $e^{-(x+1)} = 2$, $x_0 = -2$

2 Consider $f(x) = x + \cos 2x$.

   a Use technology to sketch $y = f(x)$.

   b Use the Newton-Raphson method with $x_0 = 0$ to solve $f(x) = 0$ correct to 4 decimal places.

   c Explain what would happen if the initial guess was $x_0 = 1$.

3 Consider $f(x) = x^3 - 4x^2 + 3x + 1$.

   a Use the Newton-Raphson method with the following initial guesses to find solutions to $f(x) = 0$ correct to 3 decimal places:

      i $x_0 = 0$ **ii** $x_0 = 1$ **iii** $x_0 = 3$

   b Sketch $y = f(x)$ using its axes intercepts.

4 Suppose $f(x) = -x^3 + 5x^2 - 10$.

   a Find the turning points of $y = f(x)$.

   b Hence decide on three initial guesses for x_0 which can be used with the Newton-Raphson method to find the zeros of $f(x)$.

   c Find the zeros of $f(x)$ correct to 3 decimal places.

   d Sketch $y = f(x)$, showing the information you have found.

D   NUMERICAL INTEGRATION

In this Section we consider a number of methods for numerical integration, all with the same theme of partitioning an interval into strips and adding their (signed) areas.

RECTANGLE METHODS

Consider $\displaystyle\int_a^b f(x)\,dx$.

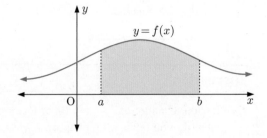

Suppose the interval $[a, b]$ is divided into n rectangular strips of equal width.

Let $x_0 = a$ and the strip width be $h = \dfrac{b - a}{n}$.

$\therefore$ $x_i = a + ih$ is the right endpoint of the ith interval.

We can now choose the (signed) heights of the rectangles according to the value of the function at the left endpoint, the right endpoint, or the midpoint.

- **Left endpoint**

$$\int_a^b f(x)\,dx \approx h\,f(x_0) + h\,f(x_1) + \ldots + h\,f(x_{n-1})$$

$$\approx h\sum_{i=1}^{n} f(x_{i-1})$$

- **Right endpoint**

$$\int_a^b f(x)\,dx \approx h\,f(x_1) + h\,f(x_2) + \ldots + h\,f(x_n)$$

$$\approx h\sum_{i=1}^{n} f(x_i)$$

- **Midpoint or midordinate**

$$\int_a^b f(x)\,dx \approx h\,f\left(\frac{x_0 + x_1}{2}\right) + h\,f\left(\frac{x_1 + x_2}{2}\right) + \ldots + h\,f\left(\frac{x_{n-1} + x_n}{2}\right)$$

$$\approx h\sum_{i=1}^{n} f\left(\frac{x_{i-1} + x_i}{2}\right)$$

Discussion

Which of these rectangle methods do you think will be most accurate? Why do you expect this to be so?

Example 6
◀) **Self Tutor**

Approximate $\displaystyle\int_2^3 (2x^2 + 3)\,dx$ using 4 rectangular strips whose heights correspond to the:

a left endpoint   **b** right endpoint   **c** midpoint.

Compare your answers with the actual value.

$n = 4, \ a = 2, \ b = 3, \ f(x) = 2x^2 + 3$

$h = \dfrac{b - a}{4} = 0.25$

$x_i = 2 + 0.25i$

n	x_i	$f(x_i)$	x_{mid}	$f(x_{mid})$
0	2	11		
1	2.25	13.125	2.125	12.031 25
2	2.5	15.5	2.375	14.281 25
3	2.75	18.125	2.625	16.781 25
4	3	21	2.875	19.531 25

a Using left endpoints, $\displaystyle\int_2^3 (2x^2 + 3)\,dx \approx h\sum_{i=1}^{4} f(x_{i-1})$

$$\approx h(f(x_0) + f(x_1) + f(x_2) + f(x_3))$$

$$\approx 14.4375$$

b Using right endpoints, $\displaystyle\int_2^3 (2x^2 + 3)\,dx \approx h\sum_{i=1}^{4} f(x_i)$

$$\approx h(f(x_1) + f(x_2) + f(x_3) + f(x_4))$$

$$\approx 16.9375$$

c Using midpoints, $\displaystyle\int_2^3 (2x^2 + 3)\,dx \approx h\sum_{i=1}^{4} f\left(\frac{x_{i-1} + x_i}{2}\right)$

GRAPHING PACKAGE

$$\approx 15.656\,25$$

The actual value of $\displaystyle\int_2^3 (2x^2 + 3)\,dx = \left[\tfrac{2}{3}x^3 + 3x\right]_2^3$

$$= \left(\tfrac{2}{3} \times 27 + 3 \times 3\right) - \left(\tfrac{2}{3} \times 8 + 3 \times 2\right)$$

$$= 15\tfrac{2}{3}$$

The approximation using midpoints is much closer to the actual value than the other approximations.

EXERCISE 13D.1

1 Approximate $\displaystyle\int_0^1 (-x^2 + 4x + 1)\, dx$ using 4 rectangular strips whose heights correspond to the:

    a left endpoint   **b** right endpoint   **c** midpoint.

Compare your answers with the actual value.

2 Approximate $\displaystyle\int_{-4}^{-3} (x^2 + 4x - 4)\, dx$ using 4 rectangular strips whose (signed) heights correspond to the:

    a left endpoint   **b** right endpoint   **c** midpoint.

Compare your answers with the actual value.

3 Using rectangular strips with heights corresponding to the midordinate values, approximate $\displaystyle\int_1^2 \sin x\, dx$ using:

    a 3 intervals   **b** 4 intervals   **c** 6 intervals.

Compare your answers with the actual value.

4 Using rectangular strips with heights corresponding to the midordinate values, approximate $\displaystyle\int_2^3 \ln x\, dx$ using:

    a 3 intervals   **b** 4 intervals   **c** 6 intervals.

Compare your answers with the actual value.

5 Using 10 rectangular strips with heights corresponding to the midordinate values, approximate $\displaystyle\int_0^4 5xe^{-x}\, dx$.
Compare your answer with the actual value.

Investigation 2   Improper integrals

An improper integral is a definite integral for which one of the following is true:

- one or both limits is infinite, for example $\displaystyle\int_0^\infty e^{-x}\, dx$

- the function being integrated approaches infinity at least once on the interval of integration, for example $\displaystyle\int_0^1 (-\ln x)\, dx$.

One method for calculating $\displaystyle\int_0^\infty e^{-x}\, dx$ is to use limits.

We calculate the integral for the interval $[0, t]$, then consider the result as t tends to infinity.

> It is not always possible to calculate improper integrals.

$$\int_0^\infty e^{-x}\, dx = \lim_{t\to\infty} \int_0^t e^{-x}\, dx$$

$$= \lim_{t\to\infty} \left[-e^{-x}\right]_0^t$$

$$= \lim_{t\to\infty} (-e^{-t} + 1)$$

$$= 1$$

What to do:

1 Discuss whether you could use a rectangle method of numerical integration to approximate the integral $\displaystyle\int_0^\infty e^{-x}\,dx$.

2　**a** Graph $y = e^{-x}$ and $y = -\ln x$ on the same set of axes. Shade the areas corresponding to the integrals $\displaystyle\int_0^\infty e^{-x}\,dx$ and $\displaystyle\int_0^1 (-\ln x)\,dx$.

　b Comment on the areas you have shaded, and explain this result.

3 Consider the numerical integration of $\displaystyle\int_0^1 (-\ln x)\,dx$.

　a Which of the rectangle methods could you *not* use to approximate the integral? Explain your answer.

　b Which of the rectangle methods would you expect to be more accurate? Explain your answer in the context of *this* integral.

　c Approximate the integral using your method of choice with:

　　i 4 intervals　　　　　　　　**ii** 10 intervals

　　iii 50 intervals　　　　　　　**iv** 200 intervals.

GRAPHING
PACKAGE

　d Comment on your results.

TRAPEZIUM RULE

Using the same set-up as for the rectangle methods, suppose we use the values at either end of the interval to approximate the curve for that interval with a straight line.

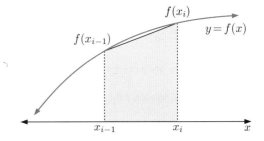

The strip is now a trapezium rather than a rectangle, and

$$\int_a^b f(x)\,dx \approx h\left(\frac{f(x_0) + f(x_1)}{2}\right) + h\left(\frac{f(x_1) + f(x_2)}{2}\right) + \ldots + h\left(\frac{f(x_{n-1}) + f(x_n)}{2}\right)$$

$$\approx \frac{h}{2}\Big(f(x_0) + 2f(x_1) + 2f(x_2) + \ldots + 2f(x_{n-1}) + f(x_n)\Big)$$

$$\approx \frac{h}{2}\left(f(x_0) + 2\sum_{i=1}^{n-1} f(x_i) + f(x_n)\right)$$

You may notice that this is in fact the average of the left endpoint and right endpoint approximations we studied in the previous section.

Example 7

◄ Self Tutor

Use the trapezium rule with 6 intervals to approximate $\int_1^2 \sqrt{6 - x^2}\, dx$.

$n = 6,\ a = 1,\ b = 2,\ f(x) = \sqrt{6 - x^2}$

$h = \dfrac{b - a}{n} = \dfrac{1}{6}$

$x_i = 1 + \frac{1}{6}i$

n	x_i	$f(x_i)$
0	1	2.236 068
1	$1\frac{1}{6}$	2.153 808
2	$1\frac{1}{3}$	2.054 805
3	$1\frac{1}{2}$	1.936 492
4	$1\frac{2}{3}$	1.795 055
5	$1\frac{5}{6}$	1.624 466
6	2	1.414 214

GRAPHING PACKAGE

Using the trapezium rule,

$$\int_1^2 \sqrt{6 - x^2}\, dx \approx \frac{h}{2}\left(f(x_0) + 2f(x_1) + 2f(x_2) + \ldots + 2f(x_5) + f(x_6)\right)$$

$$\approx 1.8983$$

EXERCISE 13D.2

1 Use the trapezium rule with 4 intervals to approximate:

a $\int_2^4 \dfrac{2}{\sqrt{x}}\, dx$

b $\int_1^3 (-x^2 + 6x - 4)\, dx$

Check your answers by performing the integration.

2 Use the trapezium rule with 8 intervals to approximate:

a $\int_0^{\pi^2} \sin(\sqrt{x})\, dx$

b $\int_0^1 \sqrt{x}\, e^{-\pi x}\, dx$

3 **a** Approximate $\int_{-0.6}^1 (x^3 - 2x^2 + 1)\, dx$ using 8 intervals with:

    i the midordinate values

    ii the trapezium rule.

b Calculate $\int_{-0.6}^1 (x^3 - 2x^2 + 1)\, dx$ exactly and decide which numerical approximation was more accurate.

4 Consider the function $f(x) = 5 - x^2 - 2e^{-x}$.

a Use the Newton-Raphson method with initial guesses -2 and 2 to approximate the zeros of $f(x)$.

b Hence estimate the area contained between $y = f(x)$ and the x-axis using the trapezium rule with 8 intervals.

Investigation 3   Simpson's rule

In the trapezium rule we approximated the curve using straight line segments between the endpoints of each interval.

To better approximate a curve, we can use sets of three adjacent points to form a quadratic.

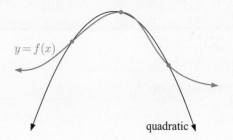

This leads us to **Simpson's rule**, named after the English mathematician **Thomas Simpson** (1710 - 1761), though in fact the German **Johannes Kepler** used the technique over a hundred years earlier.

quadratic

For Simpson's rule, we divide the interval $[a, b]$ into n equal subintervals, choosing n to be even. The ith subinterval is $[x_{i-1}, x_i]$ where $x_i = a + ih$, $h = \dfrac{b-a}{n}$.

We now generate $\dfrac{n}{2}$ quadratics to model the function. Each middle point corresponds to an *odd* value of i.

For the case $n = 2$, there is only one quadratic, and

$$\int_a^b f(x)\,dx \approx \frac{b-a}{6}\left(f(a) + 4f\left(\frac{a+b}{2}\right) + f(b)\right).$$

For the more general case,

$$\int_a^b f(x)\,dx \approx \frac{h}{3}\Big(f(x_0) + 4f(x_1) + 2f(x_2) + 4f(x_3) + 2f(x_4) + \dots + 4f(x_{n-1}) + f(x_n)\Big)$$

$$\approx \frac{h}{3}\left(f(x_0) + 4\underset{\substack{\text{odd } i \\ \text{from 1 to} \\ n-1}}{\sum} f(x_i) + 2\underset{\substack{\text{even } i \\ \text{from 2 to} \\ n-2}}{\sum} f(x_i) + f(x_n)\right)$$

You can find the outline of a proof for Simpson's rule by clicking on the icon.

PROOF OF SIMPSON'S RULE

What to do:

1 Use Simpson's rule with 2 intervals to approximate:

   a $\displaystyle\int_0^1 \sqrt{1-x}\,dx$ **b** $\displaystyle\int_{0.5}^1 (x^3 - 2x)\,dx$

2 Use Simpson's rule with 4 intervals to calculate $\displaystyle\int_0^4 \left(1 + 2x - \tfrac{1}{2}x^2\right) dx$.

   Explain why Simpson's rule gives the exact answer for this integral.

3 Consider the integral $\displaystyle\int_a^b f(x)\,dx$.

   a Let $M \approx \displaystyle\int_a^b f(x)\,dx$ be an approximation found using the midpoint rule with $n = 1$.

   Write an expression for M.

b Let $T \approx \displaystyle\int_a^b f(x)\,dx$ be an approximation found using the trapezium rule with $n = 1$.

Write an expression for T.

c Show that the *weighted average* of the midpoint and trapezium rules $\displaystyle\int_a^b f(x)\,dx \approx \tfrac{2}{3}M + \tfrac{1}{3}T$

gives the Simpson's rule approximation for $n = 2$.

Investigation 4   Euler's method of numerical integration

Euler's method provides a numerical method for approximating the solution curve to

differential equations of the form $\dfrac{dy}{dx} = f(x,\,y)$.

EULER'S METHOD

Click on the icon to obtain an Investigation on Euler's method.

Review set 13A

1 Approximate $\sqrt{6}$ to 3 decimal places using a decimal search for the positive solution to $x^2 - 6 = 0$.

2 Use the changes of sign method with linear interpolation to approximate the solution to $x^3 - 2 = 0$, correct to 4 significant figures.

3 Consider the function $f(x) = x^2 + 1 - \dfrac{3}{x}$.

    a Evaluate $f(-1)$, $f(1)$, and $f(3)$.

    b Use the changes of sign method with interval bisection to find, correct to 2 decimal places, a solution to $f(x) = 0$ on $[1,\,3]$.

    c Explain why we cannot use this method to find a solution to $f(x) = 0$ on $[-1,\,1]$.

4 **a** Apply the iteration $x_{n+1} = g(x_n)$ with $x_0 = 0$ and $g(x) = \tfrac{1}{3}\sqrt{5-x}$ for 10 iterations.

    b Hence find an approximate solution to $9x^2 + x - 5 = 0$. Compare your result with the analytic solution.

5 The equation $\ln(7-x) = x$ has a single solution $x = \alpha$.

    a Show that α lies between 1 and 2.

    b Use the iteration $x_{n+1} = \ln(7 - x_n)$ with $x_0 = 1$ to find x_1 and x_2.

    Copy the graph alongside, and draw a staircase or cobweb diagram, indicating x_1 and x_2 on the x-axis.

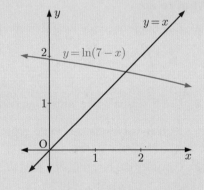

PRINTABLE GRAPH

...vton-Raphson method with the initial guess $x_0 = -1$ to find a solution to ...0.

7 Consider $f(x) = \sin x - x + 1$.

    a Show that there is a solution to $f(x) = 0$ between $x = 0$ and $x = \pi$.

    b Use the Newton-Raphson method with initial guess $x_0 = \pi$ to find the solution, correct to 2 decimal places.

    c Explain why the Newton-Raphson method will fail with the initial guess $x_0 = 0$.

8 Approximate $\displaystyle\int_2^3 (x^2 - 3x + 4)\,dx$ using 4 rectangular strips with heights corresponding to the:

    a left endpoint   **b** right endpoint   **c** midpoint.

    Compare your answers with the actual value of the integral.

9 Use the trapezium rule over 6 intervals to approximate $\displaystyle\int_1^3 (x + \sqrt{x})\,dx$.

10   **a** Approximate $\displaystyle\int_0^{\frac{\pi}{2}} \sin x \cos x \, dx$ with 4 intervals using the trapezium rule.

    b Calculate $\displaystyle\int_0^{\frac{\pi}{2}} \sin x \cos x \, dx$ exactly. Comment on the accuracy of your approximation.

Review set 13B

1 Approximate $\frac{\pi}{2}$ to 3 decimal places, by using the changes of sign method with interval bisection to find the appropriate solution to $\cos x = 0$.

2 Suppose $f(x) = x^3 - 9x^2 + 6x - 1$.

    a Show that $f(x) = 0$ has at least one solution between $x = 0$ and $x = 10$.

    b Perform a decimal search for a solution to $f(x) = 0$ using $a_1 = 0$ and $b_1 = 10$. Give your answer to 2 decimal places.

    c **i** Evaluate $f(0.3)$. What does this result tell you?

        ii Using further decimal searches, find *all* solutions to $f(x) = 0$, correct to 2 decimal places.

3   **a** Show that the equation $x = 5 + \sqrt{3 - \sqrt{x}}$ can be rearranged into $x^2 + \sqrt{x} = 10x - 22$.

    b Use the iteration $x_{n+1} = 5 + \sqrt{3 - \sqrt{x_n}}$ with $x_0 = 5$ to find:

        i x_1 and x_2

        ii a solution to $x^2 + \sqrt{x} = 10x - 22$, correct to 3 decimal places.

4   **a** Sketch the graph of $f(x) = 0.5 \cos x + 3x$ on $[-2, 2]$.

    b Approximate, to 3 decimal places, the solution to $f(x) = 0$ on $[-1, 0]$ using:

        i interval bisection

        ii linear interpolation.

    c Which method obtained the approximate solution more quickly? Explain why this occurred.

5 Consider the equation $5\arccos(x-1) = x$.

   a Show that this equation can be written in the form $x = a + \cos bx$.

   b Hence use the iteration $x_{n+1} = a + \cos bx_n$ with $x_0 = 2$ to solve the equation $5\arccos(x-1) = x$, correct to 3 decimal places.

6 Consider the equation $x^4 - x - 4 = 0$.

   a Use the Newton-Raphson method to show that an iterative formula for solving the equation may be written as $x_{n+1} = \dfrac{3x_n^4 + 4}{4x_n^3 - 1}$.

   b Find the solution to $x^4 - x - 4 = 0$ which lies between $x = 1$ and $x = 2$, correct to 4 significant figures.

7 **a** Use technology to help sketch the graph of $f(x) = e^x - \ln x - 3$.

   b Use the Newton-Raphson method with $x_0 = 1$ to find one of the solutions to $f(x) = 0$, correct to 4 decimal places.

   c Explain why the initial guess $x_0 = 0.5$ would fail to find a solution to $f(x) = 0$.

   d Use a suitable initial guess to approximate the other solution to $f(x) = 0$.

8 Using rectangular strips with heights corresponding to the midordinate values, approximate
$$\int_2^8 \ln x \, dx \quad \text{using:}$$

   a 3 intervals   **b** 6 intervals.

Compare your answer with the actual value.

9 **a** Use the trapezium rule with 6 intervals to approximate $\displaystyle\int_0^2 \sqrt{4 - x^2} \, dx$.

   b Compare your answer with π, and explain this result.

10 The graph of $y = x^4$ is shown alongside.

   a Use the trapezium rule with 8 intervals to estimate $\displaystyle\int_{-2}^{-1} x^4 \, dx$.

   b Use the graph to explain whether your estimate is an over estimate or an under estimate.

   c Check your answer to **b** by performing the integration.

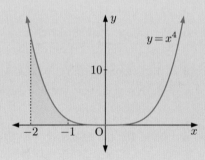

14

Vectors

Contents:

Opening problem

To describe the location of a point on a 2-dimensional plane, we use two coordinates x and y. To describe the location of a point in 3-dimensional space, we need three coordinates X, Y, and Z.

Suppose the point P has coordinates $(3, 3, 2)$.

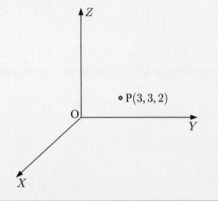

Things to think about:

a What do the coordinates $(3, 3, 2)$ mean? How can we describe how to get to P from the origin O?

b How can we use a vector to represent the position of P?

c How can we use vectors to find:

i the distance from the origin to P

ii the angle between the vector from O to P, and the X-axis?

You should have already seen **2-dimensional** or **plane vectors** on the **Cartesian** or real plane.

You should know that:

- A **scalar** is a quantity which has magnitude but no direction.
- A **vector** is a quantity which has both magnitude and direction.
- The **position vector** of point A is the vector from the origin O to A. We write it as $\overrightarrow{OA}$, **a**, or $\overrightarrow{a}$.
- The **magnitude** or **length** of vector $\overrightarrow{OA}$ is written $|\overrightarrow{OA}|$, OA, $|\mathbf{a}|$, or $|\overrightarrow{a}|$.
- Two vectors are **equal** if they have the same magnitude *and* direction.

For vectors in the plane:

The point $P(x, y)$ has **position vector**

$$\overrightarrow{OP} = \begin{pmatrix} x \\ y \end{pmatrix} = x\mathbf{i} + y\mathbf{j}$$

component form   unit vector form

A **unit vector** has length 1.

where $\mathbf{i} = \begin{pmatrix} 1 \\ 0 \end{pmatrix}$ is the **base unit vector** in the x-direction

and $\mathbf{j} = \begin{pmatrix} 0 \\ 1 \end{pmatrix}$ is the **base unit vector** in the y-direction.

If $\mathbf{a} = \begin{pmatrix} a_1 \\ a_2 \end{pmatrix} = a_1\mathbf{i} + a_2\mathbf{j}$, the **magnitude** or **length** of **a** is $|\mathbf{a}| = \sqrt{a_1^2 + a_2^2}$.

If k is a scalar, $\mathbf{a} = \begin{pmatrix} a_1 \\ a_2 \end{pmatrix}$, and $\mathbf{b} = \begin{pmatrix} b_1 \\ b_2 \end{pmatrix}$ then:

- $-\mathbf{a} = \begin{pmatrix} -a_1 \\ -a_2 \end{pmatrix}$
- $\mathbf{a} + \mathbf{b} = \begin{pmatrix} a_1 + b_1 \\ a_2 + b_2 \end{pmatrix}$
- $\mathbf{a} - \mathbf{b} = \begin{pmatrix} a_1 - b_1 \\ a_2 - b_2 \end{pmatrix}$
- $k\mathbf{a} = \begin{pmatrix} ka_1 \\ ka_2 \end{pmatrix}$

- the **position vector of B relative to A** is $\overrightarrow{AB} = \overrightarrow{OB} - \overrightarrow{OA} = \begin{pmatrix} b_1 - a_1 \\ b_2 - a_2 \end{pmatrix}$

In this Chapter we switch our attention to **3-dimensional** vectors or **vectors in space**. Now that we are dealing with a third dimension, it will become increasingly difficult to visualise the vectors on a 2-dimensional page. For this reason our skills in vector algebra will become much more important.

A   VECTORS IN SPACE

To specify points in **3-dimensional space** we need a point of reference O, called the **origin**.

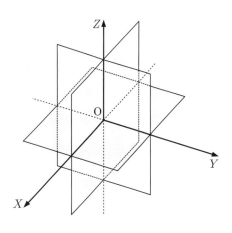

Through O we draw 3 **mutually perpendicular** lines and call them the X, Y, and Z-axes. We often think of the YZ-plane as the plane of the page, with the X-axis coming directly out of the page. However, we cannot of course draw this.

In the diagram alongside, the **coordinate planes** divide space into 8 regions, with each pair of planes intersecting on the axes.

The **positive direction** of each axis is a solid line, while the **negative direction** is "dashed".

Any point P in space can be specified by an **ordered triple** of numbers (x, y, z) where x, y, and z are the **steps** in the X, Y, and Z directions from the origin O, to P.

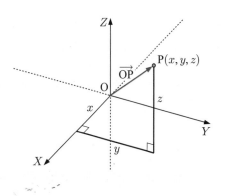

The **position vector** of P is

$$\overrightarrow{OP} = \begin{pmatrix} x \\ y \\ z \end{pmatrix} = x\mathbf{i} + y\mathbf{j} + z\mathbf{k}$$

where $\mathbf{i} = \begin{pmatrix} 1 \\ 0 \\ 0 \end{pmatrix}$, $\mathbf{j} = \begin{pmatrix} 0 \\ 1 \\ 0 \end{pmatrix}$, and $\mathbf{k} = \begin{pmatrix} 0 \\ 0 \\ 1 \end{pmatrix}$

are the **base unit vectors** in the X, Y, and Z directions respectively.

To help us visualise the 3-dimensional position of a point on our 2-dimensional paper, it is useful to complete a rectangular prism or box with the origin O as one vertex, the axes as sides adjacent to it, and P being the vertex opposite O.

3-D POINT PLOTTER

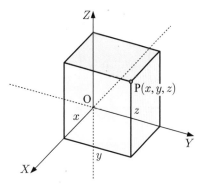

THE MAGNITUDE OF A VECTOR

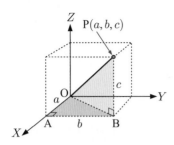

Triangle OAB is right angled at A

$$\therefore \quad OB^2 = a^2 + b^2 \quad .... \text{(1)} \quad \{\text{Pythagoras}\}$$

Triangle OBP is right angled at B

$$\therefore \quad OP^2 = OB^2 + c^2 \qquad \{\text{Pythagoras}\}$$

$$\therefore \quad OP^2 = a^2 + b^2 + c^2 \qquad \{\text{using (1)}\}$$

$$\therefore \quad OP = \sqrt{a^2 + b^2 + c^2}$$

The magnitude or length of the vector $\mathbf{a} = \begin{pmatrix} a_1 \\ a_2 \\ a_3 \end{pmatrix}$ is $|\mathbf{a}| = \sqrt{a_1^2 + a_2^2 + a_3^2}$.

Example 1   ◀) Self Tutor

Illustrate each point and find its distance from the origin O:

  a A(0, 2, 0)   **b** B(3, 0, 2)   **c** C(−1, 2, 3)

a
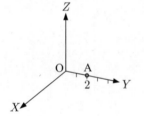

$$|\overrightarrow{OA}|$$
$$= \sqrt{0^2 + 2^2 + 0^2}$$
$$= 2 \text{ units}$$

b
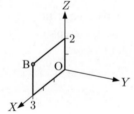

$$|\overrightarrow{OB}|$$
$$= \sqrt{3^2 + 0^2 + 2^2}$$
$$= \sqrt{13} \text{ units}$$

c
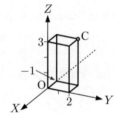

$$|\overrightarrow{OC}|$$
$$= \sqrt{(-1)^2 + 2^2 + 3^2}$$
$$= \sqrt{14} \text{ units}$$

VECTOR EQUALITY

Two vectors are **equal** if they have the same magnitude and direction.

If $\mathbf{a} = \begin{pmatrix} a_1 \\ a_2 \\ a_3 \end{pmatrix}$ and $\mathbf{b} = \begin{pmatrix} b_1 \\ b_2 \\ b_3 \end{pmatrix}$,

then $\mathbf{a} = \mathbf{b} \quad \Leftrightarrow \quad a_1 = b_1, \; a_2 = b_2, \; a_3 = b_3.$

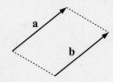

If $\mathbf{a}$ and $\mathbf{b}$ are equal but do not coincide, then they are opposite sides of a parallelogram, and lie in the same plane.

⇔ means
"if and only if".

EXERCISE 14A

1 Consider the point T$(3, -1, 4)$.

    a Draw a diagram to locate the position of T in space.   **b** Find $\overrightarrow{OT}$.

    c Find the distance from O to T.

2 Illustrate P and find its distance from the origin O:

    a P$(0, 0, -3)$ **b** P$(0, -1, 2)$ **c** P$(3, 1, 4)$ **d** P$(-1, -2, 3)$

3 Find a, b, and c if:

    a $\begin{pmatrix} a-4 \\ b-3 \\ c+2 \end{pmatrix} = \begin{pmatrix} 1 \\ 3 \\ -4 \end{pmatrix}$ **b** $\begin{pmatrix} a-5 \\ b-2 \\ c+3 \end{pmatrix} = \begin{pmatrix} 3-a \\ 2-b \\ 5-c \end{pmatrix}$

4 Find k given the unit vector:

    a $\begin{pmatrix} -\frac{1}{2} \\ k \\ \frac{1}{4} \end{pmatrix}$ **b** $\begin{pmatrix} k \\ \frac{2}{3} \\ -\frac{1}{3} \end{pmatrix}$

A unit vector has length 1.

5 Find the possible values of m, given that:

    a $\begin{pmatrix} -2 \\ m \\ 1 \end{pmatrix}$ has length $\sqrt{14}$ units   **b** $\begin{pmatrix} 4 \\ -3 \\ m \end{pmatrix}$ has length 6 units.

B   OPERATIONS WITH VECTORS IN SPACE

For 3-dimensional vectors, it is difficult to illustrate operations. Instead, we perform operations using component form.

The rules for algebra with vectors readily extend from 2 dimensions to 3 dimensions:

If $\mathbf{a} = \begin{pmatrix} a_1 \\ a_2 \\ a_3 \end{pmatrix}$ and $\mathbf{b} = \begin{pmatrix} b_1 \\ b_2 \\ b_3 \end{pmatrix}$ then:

▸ $-\mathbf{a} = \begin{pmatrix} -a_1 \\ -a_2 \\ -a_3 \end{pmatrix}$ ▸ $\mathbf{a} + \mathbf{b} = \begin{pmatrix} a_1 + b_1 \\ a_2 + b_2 \\ a_3 + b_3 \end{pmatrix}$

▸ $\mathbf{a} - \mathbf{b} = \begin{pmatrix} a_1 - b_1 \\ a_2 - b_2 \\ a_3 - b_3 \end{pmatrix}$ ▸ $k\mathbf{a} = \begin{pmatrix} ka_1 \\ ka_2 \\ ka_3 \end{pmatrix}$ for any scalar k.

Example 2
◀)) **Self Tutor**

If $\mathbf{a} = \begin{pmatrix} -1 \\ 3 \\ 2 \end{pmatrix}$ and $\mathbf{b} = \begin{pmatrix} 3 \\ 0 \\ -1 \end{pmatrix}$, find:

a $\mathbf{a} + \mathbf{b}$ **b** $2\mathbf{a} - \mathbf{b}$ **c** $|\mathbf{a}|\mathbf{b}$

a $\mathbf{a} + \mathbf{b}$

$= \begin{pmatrix} -1 \\ 3 \\ 2 \end{pmatrix} + \begin{pmatrix} 3 \\ 0 \\ -1 \end{pmatrix}$

$= \begin{pmatrix} -1+3 \\ 3+0 \\ 2+(-1) \end{pmatrix}$

$= \begin{pmatrix} 2 \\ 3 \\ 1 \end{pmatrix}$

b $2\mathbf{a} - \mathbf{b}$

$= 2\begin{pmatrix} -1 \\ 3 \\ 2 \end{pmatrix} - \begin{pmatrix} 3 \\ 0 \\ -1 \end{pmatrix}$

$= \begin{pmatrix} -2 \\ 6 \\ 4 \end{pmatrix} - \begin{pmatrix} 3 \\ 0 \\ -1 \end{pmatrix}$

$= \begin{pmatrix} -2-3 \\ 6-0 \\ 4-(-1) \end{pmatrix}$

$= \begin{pmatrix} -5 \\ 6 \\ 5 \end{pmatrix}$

c $|\mathbf{a}|\mathbf{b}$

$= \sqrt{(-1)^2 + 3^2 + 2^2}\begin{pmatrix} 3 \\ 0 \\ -1 \end{pmatrix}$

$= \sqrt{14}\begin{pmatrix} 3 \\ 0 \\ -1 \end{pmatrix}$

$= \begin{pmatrix} 3\sqrt{14} \\ 0 \\ -\sqrt{14} \end{pmatrix}$

EXERCISE 14B

1 For $\mathbf{a} = \begin{pmatrix} 2 \\ -1 \\ 1 \end{pmatrix}$, $\mathbf{b} = \begin{pmatrix} 1 \\ 2 \\ -3 \end{pmatrix}$, and $\mathbf{c} = \begin{pmatrix} 0 \\ 1 \\ -3 \end{pmatrix}$, find:

a $\mathbf{a} + \mathbf{b}$ **b** $\mathbf{a} - \mathbf{b}$ **c** $\mathbf{b} + 2\mathbf{c}$

d $\mathbf{c} - \frac{1}{2}\mathbf{a}$ **e** $\mathbf{a} - \mathbf{b} - \mathbf{c}$ **f** $2\mathbf{b} - \mathbf{c} + \mathbf{a}$

2 If $\mathbf{a} = \begin{pmatrix} 1 \\ 0 \\ 3 \end{pmatrix}$ and $\mathbf{b} = \begin{pmatrix} -2 \\ 1 \\ 1 \end{pmatrix}$, find:

a $|\mathbf{a}|$ **b** $|\mathbf{b}|$ **c** $2|\mathbf{a}|$ **d** $|2\mathbf{a}|$

e $-3|\mathbf{b}|$ **f** $|-3\mathbf{b}|$ **g** $|\mathbf{a} + \mathbf{b}|$ **h** $|\mathbf{a} - \mathbf{b}|$

3 If $\mathbf{a} = \begin{pmatrix} -1 \\ 1 \\ 3 \end{pmatrix}$, $\mathbf{b} = \begin{pmatrix} 1 \\ -3 \\ 2 \end{pmatrix}$, and $\mathbf{c} = \begin{pmatrix} -2 \\ 2 \\ 4 \end{pmatrix}$, find:

a $2\mathbf{a} - \mathbf{c}$ **b** $-\mathbf{a} + 3\mathbf{b}$ **c** $|\mathbf{b} + \mathbf{c}|$

d $|\mathbf{a} - \mathbf{c}|$ **e** $|\mathbf{a}|\mathbf{b}$ **f** $\frac{1}{|\mathbf{a}|}\mathbf{a}$

4 Find scalars a, b, and c such that:

a $2\begin{pmatrix} 1 \\ 0 \\ 3a \end{pmatrix} = \begin{pmatrix} b \\ c-1 \\ 2 \end{pmatrix}$

b $a\begin{pmatrix} 1 \\ 1 \\ 0 \end{pmatrix} + b\begin{pmatrix} 2 \\ 0 \\ -1 \end{pmatrix} + c\begin{pmatrix} 0 \\ 1 \\ 1 \end{pmatrix} = \begin{pmatrix} -1 \\ 3 \\ 3 \end{pmatrix}$

c $a\begin{pmatrix} 2 \\ -3 \\ 1 \end{pmatrix} + b\begin{pmatrix} 1 \\ 7 \\ 2 \end{pmatrix} = \begin{pmatrix} 7 \\ c \\ 2 \end{pmatrix}$

d $2\begin{pmatrix} a \\ 1 \\ -2 \end{pmatrix} + \begin{pmatrix} b \\ c \\ 4 \end{pmatrix} = \begin{pmatrix} c \\ -c \\ ab \end{pmatrix}$

C VECTOR ALGEBRA

Investigation 1 **Properties of vectors in space**

There are several properties which are valid for the addition of real numbers. For example, we know that $a + b = b + a$. In this Investigation we look for similar properties of vectors.

What to do:

1 Use general vectors $\mathbf{a} = \begin{pmatrix} a_1 \\ a_2 \\ a_3 \end{pmatrix}$, $\mathbf{b} = \begin{pmatrix} b_1 \\ b_2 \\ b_3 \end{pmatrix}$, and $\mathbf{c} = \begin{pmatrix} c_1 \\ c_2 \\ c_3 \end{pmatrix}$ to find:

    a $\mathbf{a} + \mathbf{b}$ and $\mathbf{b} + \mathbf{a}$ **b** $\mathbf{a} + \mathbf{0}$

    c $\mathbf{a} + (-\mathbf{a})$ and $(-\mathbf{a}) + \mathbf{a}$ **d** $(\mathbf{a} + \mathbf{b}) + \mathbf{c}$ and $\mathbf{a} + (\mathbf{b} + \mathbf{c})$

2 Summarise your observations from **1**. Do they match the rules for real numbers?

3 Prove that for scalar k and vectors $\mathbf{a} = \begin{pmatrix} a_1 \\ a_2 \\ a_3 \end{pmatrix}$ and $\mathbf{b} = \begin{pmatrix} b_1 \\ b_2 \\ b_3 \end{pmatrix}$:

    a $|k\mathbf{a}| = |k||\mathbf{a}|$ **b** $k(\mathbf{a} + \mathbf{b}) = k\mathbf{a} + k\mathbf{b}$

From the **Investigation** you should have found that for vectors **a**, **b**, **c** and scalar $k \in \mathbb{R}$:

-     $\mathbf{a} + \mathbf{b} = \mathbf{b} + \mathbf{a}$ {commutative property}
-     $(\mathbf{a} + \mathbf{b}) + \mathbf{c} = \mathbf{a} + (\mathbf{b} + \mathbf{c})$ {associative property}
-     $\mathbf{a} + \mathbf{0} = \mathbf{0} + \mathbf{a} = \mathbf{a}$ {additive identity}
-     $\mathbf{a} + (-\mathbf{a}) = (-\mathbf{a}) + \mathbf{a} = \mathbf{0}$ {additive inverse}
-     $k(\mathbf{a} + \mathbf{b}) = k\mathbf{a} + k\mathbf{b}$ {distributive property}
-     $\underbrace{|k\mathbf{a}|}_{\text{length of } k\mathbf{a}} = \underbrace{|k|}_{\substack{\text{absolute value of } k}}\ \underbrace{|\mathbf{a}|}_{\text{length of } \mathbf{a}}$ where $k\mathbf{a}$ is parallel to **a**

The rules for solving vector equations are similar to those for solving real number equations. However, there is no such thing as dividing a vector by a scalar. Instead, we multiply by the reciprocal scalar.

For example,   if $2\mathbf{x} = \mathbf{a}$ then $\mathbf{x} = \frac{1}{2}\mathbf{a}$ and *not* $\dfrac{\mathbf{a}}{2}$.

$\dfrac{\mathbf{a}}{2}$ has no meaning in vector algebra.

Two useful rules are:

- If $\mathbf{x} + \mathbf{a} = \mathbf{b}$ then $\mathbf{x} = \mathbf{b} - \mathbf{a}$.

- If $k\mathbf{x} = \mathbf{a},\ k \neq 0$, then $\mathbf{x} = \frac{1}{k}\mathbf{a}$.

Proof:

- If $\mathbf{x} + \mathbf{a} = \mathbf{b}$

 then $\mathbf{x} + \mathbf{a} + (-\mathbf{a}) = \mathbf{b} + (-\mathbf{a})$

 $\therefore\ \mathbf{x} + \mathbf{0} = \mathbf{b} - \mathbf{a}$

 $\therefore\ \mathbf{x} = \mathbf{b} - \mathbf{a}$

- If $k\mathbf{x} = \mathbf{a}$

 then $\frac{1}{k}(k\mathbf{x}) = \frac{1}{k}\mathbf{a}$

 $\therefore\ 1\mathbf{x} = \frac{1}{k}\mathbf{a}$

 $\therefore\ \mathbf{x} = \frac{1}{k}\mathbf{a}$

Example 3   ◀)) Self Tutor

Solve for $\mathbf{x}$:

a $3\mathbf{x} - \mathbf{r} = \mathbf{s}$ **b** $\mathbf{c} - 2\mathbf{x} = \mathbf{d}$

a $3\mathbf{x} - \mathbf{r} = \mathbf{s}$

$\therefore\ 3\mathbf{x} = \mathbf{s} + \mathbf{r}$

$\therefore\ \mathbf{x} = \frac{1}{3}(\mathbf{s} + \mathbf{r})$

b $\mathbf{c} - 2\mathbf{x} = \mathbf{d}$

$\therefore\ \mathbf{c} - \mathbf{d} = 2\mathbf{x}$

$\therefore\ \frac{1}{2}(\mathbf{c} - \mathbf{d}) = \mathbf{x}$

EXERCISE 14C

1 Solve the following vector equations for $\mathbf{x}$:

a $2\mathbf{x} = \mathbf{q}$ **b** $\frac{1}{2}\mathbf{x} = \mathbf{n}$ **c** $-3\mathbf{x} = \mathbf{p}$

d $\mathbf{q} + 2\mathbf{x} = \mathbf{r}$ **e** $4\mathbf{s} - 5\mathbf{x} = \mathbf{t}$ **f** $4\mathbf{m} - \frac{1}{3}\mathbf{x} = \mathbf{n}$

2 Suppose $\mathbf{p} = \begin{pmatrix} 2 \\ -5 \\ 4 \end{pmatrix}$. Find $\mathbf{x}$ such that:

a $2\mathbf{x} = \mathbf{p}$ **b** $\frac{1}{3}\mathbf{x} = \mathbf{p}$ **c** $4\mathbf{x} + \mathbf{p} = \mathbf{0}$

3 Suppose $\mathbf{a} = \begin{pmatrix} -1 \\ 2 \\ 3 \end{pmatrix}$ and $\mathbf{b} = \begin{pmatrix} 2 \\ -2 \\ 1 \end{pmatrix}$. Find $\mathbf{x}$ such that:

a $2\mathbf{a} + \mathbf{x} = \mathbf{b}$ **b** $3\mathbf{x} - \mathbf{a} = 2\mathbf{b}$ **c** $2\mathbf{b} - 2\mathbf{x} = -\mathbf{a}$

D   THE VECTOR BETWEEN TWO POINTS

If $A(a_1,\ a_2,\ a_3)$ and $B(b_1,\ b_2,\ b_3)$ are two points in space then the **position vector of B relative to A** is

$$\overrightarrow{AB} = \overrightarrow{OB} - \overrightarrow{OA} = \begin{pmatrix} b_1 - a_1 \\ b_2 - a_2 \\ b_3 - a_3 \end{pmatrix} \begin{array}{l} \longleftarrow x\text{-step} \\ \longleftarrow y\text{-step} \\ \longleftarrow z\text{-step} \end{array}$$

The magnitude of $\overrightarrow{AB}$ is $AB = \sqrt{(b_1 - a_1)^2 + (b_2 - a_2)^2 + (b_3 - a_3)^2}$ which is the distance between the points A and B.

Example 4   ◀)) **Self Tutor**

If P is $(-3, 1, 2)$ and Q is $(1, -1, 3)$, find:

    a $\overrightarrow{OP}$ **b** $\overrightarrow{PQ}$ **c** $|\overrightarrow{PQ}|$

   a $\overrightarrow{OP} = \begin{pmatrix} -3 \\ 1 \\ 2 \end{pmatrix}$ **b** $\overrightarrow{PQ} = \begin{pmatrix} 1-(-3) \\ -1-1 \\ 3-2 \end{pmatrix} = \begin{pmatrix} 4 \\ -2 \\ 1 \end{pmatrix}$

   c $|\overrightarrow{PQ}| = \sqrt{4^2+(-2)^2+1^2}$
$= \sqrt{21}$ units

EXERCISE 14D

1 Given $A(3, 1, 0)$ and $B(-1, 1, 2)$, find $\overrightarrow{OA}$, $\overrightarrow{OB}$, and $\overrightarrow{AB}$.

2 If $\overrightarrow{OA} = \begin{pmatrix} -2 \\ -1 \\ 1 \end{pmatrix}$ and $\overrightarrow{OB} = \begin{pmatrix} 1 \\ 3 \\ -1 \end{pmatrix}$, find $\overrightarrow{AB}$ and hence the distance from A to B.

3 Given $A(-3, 1, 2)$ and $B(1, 0, -1)$, find:

   a $\overrightarrow{AB}$ and $\overrightarrow{BA}$ **b** the lengths $|\overrightarrow{AB}|$ and $|\overrightarrow{BA}|$.

4 Given $M(4, -2, -1)$ and $N(-1, 2, 0)$, find:

   a the position vector of M relative to N   **b** the position vector of N relative to M

   c the distance between M and N.

5 For $A(-1, 3, -2)$ and $B(3, -2, 1)$ find:

   a $\overrightarrow{AB}$ in terms of **i**, **j**, and **k**   **b** the magnitude of $\overrightarrow{AB}$.

6 Find the shortest distance from $Q(3, 1, -2)$ to:

   a the Y-axis   **b** the origin

   c the YZ-plane.

> The YZ-plane is the plane for which all points have x-coordinate 0.

7 If $\overrightarrow{AB} = \mathbf{i} - \mathbf{j} + \mathbf{k}$ and $\overrightarrow{BC} = -2\mathbf{i} + \mathbf{j} - 3\mathbf{k}$ find $\overrightarrow{AC}$ in terms of **i**, **j**, and **k**.

8   **a** State the coordinates of any general point A on the Y-axis.

     b Hence find the coordinates of two points on the Y-axis which are $\sqrt{14}$ units from $B(-1, -1, 2)$.

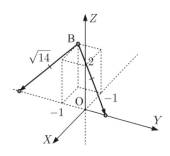

Example 5

◀) **Self Tutor**

Find the coordinates of C and D:

B(-1, -2, 2)   C   D

A(-2, -5, 3)

$$\overrightarrow{AB} = \begin{pmatrix} -1 - (-2) \\ -2 - (-5) \\ 2 - 3 \end{pmatrix} = \begin{pmatrix} 1 \\ 3 \\ -1 \end{pmatrix}$$

$$\overrightarrow{OC} = \overrightarrow{OA} + \overrightarrow{AC}$$
$$= \overrightarrow{OA} + 2\overrightarrow{AB}$$
$$= \begin{pmatrix} -2 \\ -5 \\ 3 \end{pmatrix} + \begin{pmatrix} 2 \\ 6 \\ -2 \end{pmatrix} = \begin{pmatrix} 0 \\ 1 \\ 1 \end{pmatrix}$$

$$\overrightarrow{OD} = \overrightarrow{OA} + \overrightarrow{AD}$$
$$= \overrightarrow{OA} + 3\overrightarrow{AB}$$
$$= \begin{pmatrix} -2 \\ -5 \\ 3 \end{pmatrix} + \begin{pmatrix} 3 \\ 9 \\ -3 \end{pmatrix} = \begin{pmatrix} 1 \\ 4 \\ 0 \end{pmatrix}$$

∴ C is (0, 1, 1)   ∴ D is (1, 4, 0)

9 Consider the points A(2, 1, −2), B(0, 3, −4), C(1, −2, 1), and D(−2, −3, 2).
Deduce that $\overrightarrow{BD} = 2\overrightarrow{AC}$.

10 Find the coordinates of C, D, and E.

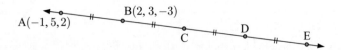

A(−1, 5, 2)   B(2, 3, −3)
                         C   D   E

11 A sphere has centre C(−1, 2, 4) and diameter [AB] where A is (−2, 1, 3).
Find the coordinates of B and the radius of the sphere.

12 Consider A(−1, 2, 5), B(2, 0, 3), and C(−3, 1, 0).

    a Find the position vector $\overrightarrow{AB}$ and its length AB.

    b Find the position vector $\overrightarrow{AC}$ and its length AC.

    c Find the position vector $\overrightarrow{CB}$ and its length CB.

    d Hence classify triangle ABC.

13 Show that P(0, 4, 4), Q(2, 6, 5), and R(1, 4, 3) are vertices of an isosceles triangle.

14 Use side lengths to classify triangle ABC given the coordinates:

    a A(0, 0, 3), B(2, 8, 1), and C(−9, 6, 18)

    b A(1, 0, −3), B(2, 2, 0), and C(4, 6, 6).

15 The vertices of triangle ABC are A(5, 6, −2), B(6, 12, 9), and C(2, 4, 2).

    a Use distances to show that the triangle is right angled.

    b Hence find the area of the triangle.

Example 6

ABCD is a parallelogram. A is $(-1, 2, 1)$, B is $(2, 0, -1)$, and D is $(3, 1, 4)$.
Find the coordinates of C.

Let C be (a, b, c).

[AB] is parallel to [DC], and they have the same
length, so $\overrightarrow{DC} = \overrightarrow{AB}$

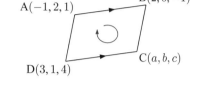

$$\therefore \begin{pmatrix} a - 3 \\ b - 1 \\ c - 4 \end{pmatrix} = \begin{pmatrix} 2 - (-1) \\ 0 - 2 \\ -1 - 1 \end{pmatrix} = \begin{pmatrix} 3 \\ -2 \\ -2 \end{pmatrix}$$

$\therefore \ a - 3 = 3, \quad b - 1 = -2, \quad$ and $\quad c - 4 = -2$
$\quad \therefore \ a = 6, \qquad \quad b = -1, \quad$ and $\qquad \ c = 2$

So, C is $(6, -1, 2)$.

Check: The midpoint of [DB] is $\left(\dfrac{3 + 2}{2}, \dfrac{1 + 0}{2}, \dfrac{4 + -1}{2} \right)$ which is $\left(\frac{5}{2}, \frac{1}{2}, \frac{3}{2} \right)$.

The midpoint of [AC] is $\left(\dfrac{-1 + 6}{2}, \dfrac{2 + -1}{2}, \dfrac{1 + 2}{2} \right)$ which is $\left(\frac{5}{2}, \frac{1}{2}, \frac{3}{2} \right)$.

The midpoints are the same, so the diagonals of the parallelogram bisect.   ✓

16 $A(-1, 3, 4)$, $B(2, 5, -1)$, $C(-1, 2, -2)$, and $D(r, s, t)$ are four points in space.
Find r, s, and t if:

    a $\overrightarrow{AC} = \overrightarrow{BD}$ **b** $\overrightarrow{AB} = \overrightarrow{DC}$

17 A quadrilateral has vertices $A(1, 2, 3)$, $B(3, -3, 2)$, $C(7, -4, 5)$, and $D(5, 1, 6)$.

    a Find $\overrightarrow{AB}$ and $\overrightarrow{DC}$.

    b What can be deduced about the quadrilateral ABCD?

18 PQRS is a parallelogram. P is $(-1, 2, 3)$, Q is $(1, -2, 5)$, and R is $(0, 4, -1)$.

    a Use vectors to find the coordinates of S.

    b Use the midpoints of the diagonals to check your answer.

19 Use vectors to determine whether ABCD is a parallelogram:

    a $A(5, 0, 3)$, $B(-1, 2, 4)$, $C(4, -3, 6)$, and $D(10, -5, 5)$

    b $A(2, -3, 2)$, $B(1, 4, -1)$, $C(-2, 6, -2)$, and $D(-1, -1, 2)$.

20 Use vector methods to find the remaining vertex of:

    a

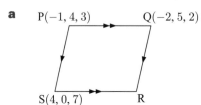

    b

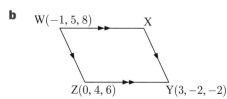

21   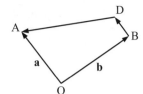 In the given figure [BD] is parallel to [OA] and half its length. Find, in terms of **a** and **b**, vector expressions for:

a $\overrightarrow{BD}$	**b** $\overrightarrow{AB}$
d $\overrightarrow{OD}$	**e** $\overrightarrow{AD}$

c $\overrightarrow{BA}$

f $\overrightarrow{DA}$

22 If $\overrightarrow{AB} = \begin{pmatrix} -1 \\ 3 \\ 2 \end{pmatrix}$, $\overrightarrow{AC} = \begin{pmatrix} 2 \\ -1 \\ 4 \end{pmatrix}$, and $\overrightarrow{BD} = \begin{pmatrix} 0 \\ 2 \\ -3 \end{pmatrix}$, find: **a** $\overrightarrow{AD}$ **b** $\overrightarrow{CB}$ **c** $\overrightarrow{CD}$

E   PARALLELISM

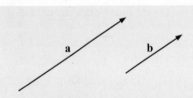 $\frac{1}{2}\mathbf{a}$ are parallel vectors of different length.

Two non-zero vectors are **parallel** if and only if one is a scalar multiple of the other.

Given any non-zero vector **v** and non-zero scalar k, the vector $k\mathbf{v}$ is parallel to **v**.

- If **a** is parallel to **b**, then there exists a scalar k such that $\mathbf{a} = k\mathbf{b}$.
- If $\mathbf{a} = k\mathbf{b}$ for some scalar k, then
  - **a** is parallel to **b**,   and
  - $|\mathbf{a}| = |k||\mathbf{b}|$.

Example 7   ◀) Self Tutor

Find r and s given that $\mathbf{a} = \begin{pmatrix} 2 \\ -1 \\ r \end{pmatrix}$ is parallel to $\mathbf{b} = \begin{pmatrix} s \\ 2 \\ -3 \end{pmatrix}$.

Since **a** and **b** are parallel, $\mathbf{a} = k\mathbf{b}$ for some scalar k.

$$\therefore \quad \begin{pmatrix} 2 \\ -1 \\ r \end{pmatrix} = k \begin{pmatrix} s \\ 2 \\ -3 \end{pmatrix}$$

$\therefore \; 2 = ks, \; -1 = 2k, \; \text{and} \; r = -3k$

Consequently, $k = -\frac{1}{2}$

$\therefore \quad 2 = -\frac{1}{2}s \; \text{ and } \; r = -3(-\frac{1}{2})$

$\therefore \quad s = -4 \; \text{ and } \; r = \frac{3}{2}$

UNIT VECTORS

A **unit vector** is a vector of length 1.

A unit vector in the direction of **v** is $\dfrac{1}{|\mathbf{v}|}\mathbf{v}$.

A vector **b** of length k in the same direction as **a** is $\mathbf{b} = \dfrac{k}{|\mathbf{a}|}\mathbf{a}$.

A vector **b** of length k which is *parallel to* **a** could be $\mathbf{b} = \pm\dfrac{k}{|\mathbf{a}|}\mathbf{a}$.

Example 8
◀ೃ Self Tutor

Find a vector **b** of length 7 units in the opposite direction to $\mathbf{a} = \begin{pmatrix} 2 \\ -1 \\ 1 \end{pmatrix}$.

The unit vector in the direction of **a** is

$$\frac{1}{|\mathbf{a}|}\mathbf{a} = \frac{1}{\sqrt{4+1+1}}\begin{pmatrix} 2 \\ -1 \\ 1 \end{pmatrix} = \frac{1}{\sqrt{6}}\begin{pmatrix} 2 \\ -1 \\ 1 \end{pmatrix}.$$

We multiply the unit vector by -7. The negative reverses the direction. The 7 gives the required length.

$$\therefore \quad \mathbf{b} = -\frac{7}{\sqrt{6}}\begin{pmatrix} 2 \\ -1 \\ 1 \end{pmatrix}.$$

COLLINEAR POINTS

Three or more points are said to be **collinear** if they lie on the same straight line.

A, B, and C are **collinear** if $\overrightarrow{AB} = k\overrightarrow{BC}$ for some scalar k.

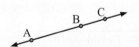

Example 9
◀ೃ Self Tutor

Prove that $A(-1, 2, 3)$, $B(4, 0, -1)$, and $C(14, -4, -9)$ are collinear.

$$\overrightarrow{AB} = \begin{pmatrix} 5 \\ -2 \\ -4 \end{pmatrix} \quad \text{and} \quad \overrightarrow{BC} = \begin{pmatrix} 10 \\ -4 \\ -8 \end{pmatrix} = 2\begin{pmatrix} 5 \\ -2 \\ -4 \end{pmatrix}$$

$\therefore \quad \overrightarrow{BC} = 2\overrightarrow{AB}$

$\therefore \quad$ [BC] is parallel to [AB].

Since B is common to both, A, B, and C are collinear.

EXERCISE 14E

1 Find r and s given that **a** and **b** are parallel:

a $\mathbf{a} = \begin{pmatrix} 2 \\ -1 \\ 3 \end{pmatrix}$ and $\mathbf{b} = \begin{pmatrix} -6 \\ r \\ s \end{pmatrix}$
b $\mathbf{a} = \begin{pmatrix} 3 \\ -1 \\ 2 \end{pmatrix}$ and $\mathbf{b} = \begin{pmatrix} r \\ 2 \\ s \end{pmatrix}$

2 What can be deduced from the following?

a $\overrightarrow{PQ} = 4\overrightarrow{RS}$ **b** $\overrightarrow{AB} = -\frac{2}{3}\overrightarrow{CD}$ **c** $\overrightarrow{LN} = 3\overrightarrow{LM}$

3 The position vectors of P, Q, R, and S are $\begin{pmatrix} 3 \\ 2 \\ -1 \end{pmatrix}$, $\begin{pmatrix} 1 \\ 4 \\ -3 \end{pmatrix}$, $\begin{pmatrix} 2 \\ -1 \\ 2 \end{pmatrix}$, and $\begin{pmatrix} -1 \\ -2 \\ 3 \end{pmatrix}$

respectively.

Deduce that [PR] and [QS] are parallel, and state the ratio PR : QS.

4 Find the unit vector in the direction of:

a $2\mathbf{i} + 3\mathbf{k}$ **b** $-\mathbf{i} + 2\mathbf{j} - \mathbf{k}$ **c** $2\mathbf{i} - 2\mathbf{j} + \mathbf{k}$

5 **a** Find vectors of length 1 unit which are parallel to $\mathbf{a} = \begin{pmatrix} 2 \\ -1 \\ -2 \end{pmatrix}$.

b Find vectors of length 2 units which are parallel to $\mathbf{b} = \begin{pmatrix} -2 \\ -1 \\ 2 \end{pmatrix}$.

6 Find a vector **b** in:

a the same direction as $\begin{pmatrix} -1 \\ 4 \\ 1 \end{pmatrix}$ and with length 6 units

b the opposite direction to $\begin{pmatrix} -1 \\ -2 \\ -2 \end{pmatrix}$ and with length 5 units.

7 Find the coordinates of the point:

a 6 units from $(1, 3, -2)$ in the direction $\begin{pmatrix} -3 \\ 0 \\ 4 \end{pmatrix}$

b 4 units from $(2, -1, 4)$ in the direction $\begin{pmatrix} 1 \\ -1 \\ 2 \end{pmatrix}$.

8 **a** Prove that $A(-2, 1, 4)$, $B(4, 3, 0)$, and $C(19, 8, -10)$ are collinear.

b Prove that $P(2, 1, 1)$, $Q(5, -5, -2)$, and $R(-1, 7, 4)$ are collinear.

c $A(2, -3, 4)$, $B(11, -9, 7)$, and $C(-13, a, b)$ are collinear. Find a and b.

d $K(1, -1, 0)$, $L(4, -3, 7)$, and $M(a, 2, b)$ are collinear. Find a and b.

Investigation 2 — The equation of a sphere

A **sphere** is the set of all points in space which are a fixed distance from a fixed point called its centre. The distance from the centre to any point on the sphere is called the **radius**.

In this Investigation we derive the Cartesian equation for generating a sphere.

What to do:

1 Consider a sphere of radius 1 unit with centre the origin.

Let P(x, y, z) be a point on the sphere.

   a Write down the vector $\overrightarrow{OP}$.

   b Use the distance formula for OP to write a Cartesian equation connecting x, y, and z.

2 Let P(x, y, z) be a point on a sphere with radius r units which is centred at the origin. Use the distance formula for OP to write a Cartesian equation for the sphere.

3 Let P(x, y, z) be a point on a sphere with radius r units and centre C(X, Y, Z).

   a Write down the vector $\overrightarrow{CP}$.

   b Use the distance formula for CP to write a Cartesian equation for the sphere.

F   PROOF USING VECTOR GEOMETRY

We can use vectors to prove important geometric facts. The use of vectors in such proofs provides us with an alternative to using deductive or coordinate arguments, and can often simplify the proofs.

Example 10 ◀) Self Tutor

Prove that if the diagonals of a quadrilateral bisect each other, then the quadrilateral is a parallelogram.

Consider a quadrilateral ABCD in which the diagonals bisect each other at P.

Let $\overrightarrow{AP} = \mathbf{a}$ and $\overrightarrow{BP} = \mathbf{b}$

$\therefore$ $\overrightarrow{PC} = \mathbf{a}$ and $\overrightarrow{PD} = \mathbf{b}$

Now $\overrightarrow{AB} = \overrightarrow{AP} + \overrightarrow{PB} = \mathbf{a} - \mathbf{b}$

and $\overrightarrow{DC} = \overrightarrow{DP} + \overrightarrow{PC} = -\mathbf{b} + \mathbf{a}$

$\therefore$ $\overrightarrow{AB} = \overrightarrow{DC}$

Also, $\overrightarrow{AD} = \overrightarrow{AP} + \overrightarrow{PD} = \mathbf{a} + \mathbf{b}$

and $\overrightarrow{BC} = \overrightarrow{BP} + \overrightarrow{PC} = \mathbf{b} + \mathbf{a}$

$\therefore$ $\overrightarrow{AD} = \overrightarrow{BC}$

Since the opposite sides of the quadrilateral are parallel and equal in length, the quadrilateral is a parallelogram.

EXERCISE 14F

1 **a** and **b** are the position vectors of two distinct points A and B, neither of which is the origin.
Show that if $|\mathbf{a} + \mathbf{b}| = |\mathbf{a} - \mathbf{b}|$ then **a** is perpendicular to **b**.

2 Consider a quadrilateral ABCD where the midpoints of the
sides are M, N, P, and Q as shown.
Prove that MNPQ is a parallelogram.

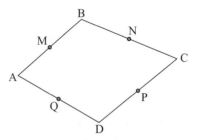

3 Prove that the diagonals of a rhombus bisect the angle of the rhombus.

4 Consider a triangle OAB where the midpoints of the sides are
X, Y, and Z as shown. [OX], [AY], and [BZ] are called the
medians of the triangle.

Let $\overrightarrow{OA} = \mathbf{a}$ and $\overrightarrow{OB} = \mathbf{b}$.

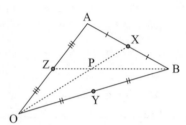

  a Write $\overrightarrow{OX}$ and $\overrightarrow{BZ}$ in terms of **a** and **b**.

  b Suppose [OX] and [BZ] intersect at P. Write $\overrightarrow{OP}$ in terms
of **a** and **b**.

  c Write $\overrightarrow{YP}$ and $\overrightarrow{PA}$ in terms of **a** and **b**.

  d Show that the medians of a triangle meet at a common point.

G   LINES IN 2 AND 3 DIMENSIONS

In both 2-dimensional and 3-dimensional geometry we can determine the **equation of a line** using its
direction and any **fixed point** on the line.

In 2 dimensions we are dealing with a **line in a plane**. In 3 dimensions we are dealing with a **line in space**.

Suppose a line passes through a fixed point A with
position vector **a**, and that the line is parallel to the
vector **b**.

Consider a point R on the line with position vector **r**.

By vector addition, $\mathbf{r} = \mathbf{a} + \overrightarrow{AR}$.

Since $\overrightarrow{AR}$ is parallel to **b**,

$$\overrightarrow{AR} = \lambda\mathbf{b} \quad \text{for some scalar } \lambda \in \mathbb{R}$$
$$\therefore \quad \mathbf{r} = \mathbf{a} + \lambda\mathbf{b}$$

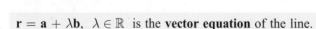

So, $\mathbf{r} = \mathbf{a} + \lambda\mathbf{b}, \ \lambda \in \mathbb{R}$ is the **vector equation** of the line.

Suppose   R(x, y, z) is any point on the line,
  A(a_1, a_2, a_3) is the known or fixed point on the line,

$$\mathbf{b} = \begin{pmatrix} b_1 \\ b_2 \\ b_3 \end{pmatrix}$$

and $\mathbf{b} = \begin{pmatrix} b_1 \\ b_2 \\ b_3 \end{pmatrix}$ is the **direction vector** of the line.

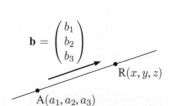

- $$\begin{pmatrix} x \\ y \\ z \end{pmatrix} = \begin{pmatrix} a_1 \\ a_2 \\ a_3 \end{pmatrix} + \lambda \begin{pmatrix} b_1 \\ b_2 \\ b_3 \end{pmatrix} \quad \text{is the } \textbf{vector equation} \\ \text{of the line.}$$

- The **parametric equations** of the line are:

$$\begin{cases} x = a_1 + \lambda b_1 \\ y = a_2 + \lambda b_2 \qquad \text{where } \lambda \in \mathbb{R} \text{ is the } \textbf{parameter.} \\ z = a_3 + \lambda b_3 \end{cases}$$

In 2 dimensions we simply leave out the z components.

- By equating λ values, we obtain the **Cartesian equation**

of the line $\quad \dfrac{x - a_1}{b_1} = \dfrac{y - a_2}{b_2} = \dfrac{z - a_3}{b_3}.$

Discussion

In 2 dimensions we talk about the gradient of a line.

- If a line in 2 dimensions has direction vector $\mathbf{b} = \begin{pmatrix} b_1 \\ b_2 \end{pmatrix}$, what is its gradient?

- Does it make sense to talk about the gradient of a line in space?

Example 11
◀ **Self Tutor**

A line in space passes through $(1, -2, 3)$ in the direction $4\mathbf{i} + 5\mathbf{j} - 6\mathbf{k}$.

Describe the line using:

a a vector equation **b** parametric equations **c** a Cartesian equation.

a The vector equation is $\mathbf{r} = \mathbf{a} + \lambda \mathbf{b}$

$$\therefore \quad \begin{pmatrix} x \\ y \\ z \end{pmatrix} = \begin{pmatrix} 1 \\ -2 \\ 3 \end{pmatrix} + \lambda \begin{pmatrix} 4 \\ 5 \\ -6 \end{pmatrix}, \quad \lambda \in \mathbb{R}.$$

b The parametric equations are: $x = 1 + 4\lambda, \quad y = -2 + 5\lambda, \quad z = 3 - 6\lambda, \quad \lambda \in \mathbb{R}.$

c Equating λ values, the Cartesian equations are: $\dfrac{x - 1}{4} = \dfrac{y + 2}{5} = \dfrac{z - 3}{-6}.$

NON-UNIQUENESS OF THE VECTOR EQUATION OF A LINE

Consider the line passing through $(5, 4)$ and $(7, 3)$. When writing the equation of the line, we could use either point to give the position vector $\mathbf{a}$.

Similarly, we could use the direction vector $\begin{pmatrix} 2 \\ -1 \end{pmatrix}$, but we could

also use $\begin{pmatrix} -2 \\ 1 \end{pmatrix}$ or any non-zero scalar multiple of these vectors.

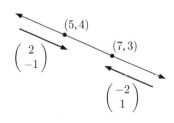

We could thus write the equation of the line as

$$\mathbf{r} = \begin{pmatrix} 5 \\ 4 \end{pmatrix} + \lambda \begin{pmatrix} 2 \\ -1 \end{pmatrix}, \ \lambda \in \mathbb{R} \quad \text{or} \quad \mathbf{r} = \begin{pmatrix} 7 \\ 3 \end{pmatrix} + \mu \begin{pmatrix} -2 \\ 1 \end{pmatrix}, \ \mu \in \mathbb{R} \quad \text{and so on.}$$

Notice how we use different parameters λ and μ when we write these equations. This is because the parameters are clearly not the same: when $\lambda = 0$, we have the point $(5, 4)$

when $\mu = 0$, we have the point $(7, 3)$.

In fact, the parameters are related by $\mu = 1 - \lambda$.

Example 12   ◄ϑ **Self Tutor**

Find parametric equations for the line through A$(2, -1, 4)$ and B$(-1, 0, 2)$.

We require a direction vector for the line, either $\overrightarrow{AB}$ or $\overrightarrow{BA}$.

$$\overrightarrow{AB} = \begin{pmatrix} -1 - 2 \\ 0 - (-1) \\ 2 - 4 \end{pmatrix} = \begin{pmatrix} -3 \\ 1 \\ -2 \end{pmatrix}$$

Using point A, the equations are:

$x = 2 - 3\lambda, \ y = -1 + \lambda, \ z = 4 - 2\lambda, \ \lambda \in \mathbb{R}$

or   using point B, the equations are:

$x = -1 - 3\mu, \ y = \mu, \ z = 2 - 2\mu, \ \mu \in \mathbb{R}.$

EXERCISE 14G

1 Describe each of the following lines using:

  i a vector equation   **ii** parametric equations   **iii** a Cartesian equation.

  a a line with direction $\begin{pmatrix} -1 \\ 2 \end{pmatrix}$ which passes through $(1, 4)$

  b a line passing through $(5, 2)$ which is perpendicular to $\begin{pmatrix} 5 \\ 2 \end{pmatrix}$

  c a line parallel to $3\mathbf{i} + 7\mathbf{j}$ which cuts the x-axis at -6

  d a line perpendicular to $3\mathbf{i} - \mathbf{j}$ which cuts the y-axis at 2

  e a line passing through $(3, 0)$ and $(7, 2)$

  f a line passing through $(-2, 5)$ and $(4, -6)$.

2 A line passes through $(4, -3)$ with direction vector $\begin{pmatrix} -1 \\ 2 \end{pmatrix}$.

  a Write parametric equations for the line using the parameter λ.

  b Find the points on the line for which $\lambda = 0, 1, 2, -1$, and -3.

3 Line L has vector equation $\mathbf{r} = \begin{pmatrix} 1 \\ 5 \end{pmatrix} + \lambda \begin{pmatrix} -1 \\ 3 \end{pmatrix}$, $\lambda \in \mathbb{R}$.

  a Locate the point on the line corresponding to $\lambda = 1$.

  b Explain why the direction of the line could also be described by $\begin{pmatrix} 1 \\ -3 \end{pmatrix}$.

  c Use your answers to **a** and **b** to write an alternative vector equation for line L.

4 Describe each of the following lines using:

    i a vector equation   **ii** parametric equations   **iii** a Cartesian equation.

    a a line parallel to $\begin{pmatrix} 2 \\ 1 \\ 3 \end{pmatrix}$ which passes through $(1,\, 3,\, -7)$

    b a line which passes through $(0,\, 1,\, 2)$ with direction vector $\mathbf{i} + \mathbf{j} - 2\mathbf{k}$

    c a line parallel to the X-axis which passes through $(-2,\, 2,\, 1)$

    d a line parallel to $2\mathbf{i} - \mathbf{j} + 3\mathbf{k}$ which passes through $(0,\, 2,\, -1)$

    e a line perpendicular to the XY-plane which passes through $(3,\, 2,\, -1)$.

5 Find the vector equation of the line which passes through:

    a $A(1,\, 2,\, 1)$ and $B(-1,\, 3,\, 2)$ **b** $C(0,\, 1,\, 3)$ and $D(3,\, 1,\, -1)$

    c $E(1,\, 2,\, 5)$ and $F(1,\, -1,\, 5)$ **d** $G(0,\, 1,\, -1)$ and $H(5,\, -1,\, 3)$.

6 Find the direction vector of the line:

    a $\begin{pmatrix} 4 \\ -1 \\ 1 \end{pmatrix} + \lambda \begin{pmatrix} -2 \\ 0 \\ 3 \end{pmatrix}, \quad \lambda \in \mathbb{R}$ **b** $x = 5 - t, \quad y = 1 + t, \quad z = 2 - 3t, \quad t \in \mathbb{R}$

    c $\dfrac{x-2}{3} = \dfrac{y+1}{2} = z - 1$ **d** $\dfrac{1-x}{2} = \dfrac{y}{4} = \dfrac{z-3}{3}$

7 Find the coordinates of the point where the line with parametric equations $x = 1 - \lambda$, $y = 3 + \lambda$, and $z = 3 - 2\lambda$, $\lambda \in \mathbb{R}$ meets:

    a the XY-plane

    b the YZ-plane

    c the XZ-plane.

The XY-plane has equation $z = 0$.

8 The parametric equations of a line are $x = x_0 + \lambda l$, $y = y_0 + \lambda m$, $z = z_0 + \lambda n$ for $\lambda \in \mathbb{R}$.

    a Find the coordinates of the point on the line corresponding to $\lambda = 0$.

    b Find the direction vector of the line.

    c Write the Cartesian equation for the line.

9 Find points on the line with parametric equations $x = 2 - \lambda$, $y = 3 + 2\lambda$, and $z = 1 + \lambda$, $\lambda \in \mathbb{R}$ which are $5\sqrt{3}$ units from the point $(1,\, 0,\, -2)$.

10 The perpendicular from a point to a line minimises the distance from the point to that line. Use quadratic theory to find the coordinates of the point where the perpendicular:

    a from $(1,\, 1,\, 2)$ meets the line with parametric equations $x = 1 + \lambda$, $y = 2 - \lambda$, $z = 3 + \lambda$, $\lambda \in \mathbb{R}$

    b from $(2,\, 1,\, 3)$ meets the line with vector equation $\begin{pmatrix} x \\ y \\ z \end{pmatrix} = \begin{pmatrix} 1 \\ 2 \\ 0 \end{pmatrix} + \mu \begin{pmatrix} 1 \\ -1 \\ 2 \end{pmatrix}, \quad \mu \in \mathbb{R}$.

H ▌ CONSTANT VELOCITY PROBLEMS

An object moving with a constant velocity will travel in a straight line. To model the position using vectors:

- the **velocity vector** of the motion gives the direction vector of the line
- **time** is the parameter
- the **initial position** of the object gives a fixed point on the line.

If an object has initial position vector **a** and moves with constant velocity **b**, its position at time t is given by

$$\mathbf{r} = \mathbf{a} + t\mathbf{b} \quad \text{for } t \geq 0.$$

The **speed** of the object is $|\mathbf{b}|$.

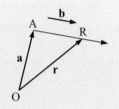

Make sure the units of distance, time, and velocity correspond.

Example 13 ◀)) Self Tutor

A mermaid is initially at $(15, \ 8, \ -1)$. She swims with velocity vector $3\mathbf{i} - 5\mathbf{j} - \frac{9}{2}\mathbf{k}$ metres per second.

Find:

a the position of the mermaid after t seconds

b the speed of the mermaid

c the time when the mermaid reaches depth 28 m.

a
$$\mathbf{r} = \mathbf{a} + t\mathbf{b}$$

$$\therefore \quad \begin{pmatrix} x \\ y \\ z \end{pmatrix} = \begin{pmatrix} 15 \\ 8 \\ -1 \end{pmatrix} + t \begin{pmatrix} 3 \\ -5 \\ -\frac{9}{2} \end{pmatrix}, \quad t \in \mathbb{R}$$

$$\therefore \quad \begin{pmatrix} x \\ y \\ z \end{pmatrix} = \begin{pmatrix} 15 + 3t \\ 8 - 5t \\ -1 - \frac{9}{2}t \end{pmatrix}$$

After t seconds the mermaid is at $(15 + 3t, \ 8 - 5t, \ -1 - \frac{9}{2}t)$.

b The speed of the mermaid

$$= \sqrt{3^2 + (-5)^2 + (-\tfrac{9}{2})^2}$$

$$= \sqrt{54\tfrac{1}{4}}$$

$$\approx 7.37 \text{ m s}^{-1}$$

c The mermaid reaches depth 28 m when

$$z(t) = -1 - \tfrac{9}{2}t = -28$$

$$\therefore \quad \tfrac{9}{2}t = 27$$

$$\therefore \quad t = 6 \text{ seconds}$$

EXERCISE 14H

1 Each of the following vector equations represents the path of a moving object. t is measured in seconds, $t \geqslant 0$. Distances are measured in metres. In each case, find:

  i the initial position   **ii** the velocity vector   **iii** the speed of the object.

  a $\begin{pmatrix} x \\ y \\ z \end{pmatrix} = \begin{pmatrix} -4 \\ 3 \\ 0 \end{pmatrix} + t \begin{pmatrix} 12 \\ 5 \\ 6 \end{pmatrix}$ **b** $x = 3 + 2t, \quad y = -t, \quad z = 4 - 2t$

2 Find the velocity vector of:

  a a hot air balloon moving parallel to $4\mathbf{i} + 7\mathbf{j} + \mathbf{k}$ with speed 33 km h^{-1}

  b a swooping eagle moving in the direction $-2\mathbf{i} + 5\mathbf{j} - 14\mathbf{k}$ with speed 90 km h^{-1}.

3 A helicopter at A$(6, 9, 3)$ moves with constant velocity. 10 minutes later it is at B$(3, 10, 2.5)$. Distances are in kilometres.

  a Find $\overrightarrow{AB}$.   **b** Find the helicopter's speed.

  c Determine the vector equation which represents the path of the helicopter.

  d The helicopter is travelling directly towards its helipad, which has Z-coordinate 0. Find the total time taken for the helicopter to land.

I   PARAMETRIC EQUATIONS

We have already seen how the equation of a line can be written as a **Cartesian equation** which directly relates the coordinates, or as a **parametric equation** in which each coordinate is written in terms of a parameter.

Parametric equations can be useful for representing physical situations, especially motion where the parameter is time. In such cases the equations give the position of an object at a particular time.

Parametric equations are also useful for describing relations which are not functions.

Investigation 3   Parametric equations

Parametric curves can be drawn using technology. You can use a graphics calculator or the software provided.

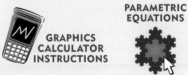

GRAPHICS CALCULATOR INSTRUCTIONS

PARAMETRIC EQUATIONS

What to do:

1 **a** Plot $x = t^2 - 1$, $y = t^3$ for $-2 \leqslant t \leqslant 2$.

  b Show that this curve has Cartesian equation $y^2 = (x + 1)^3$.

Make sure your calculator is set to radians.

2 **a** Plot $x = \cos t$, $y = \sin t$ for $0 \leqslant t \leqslant 2\pi$, using the same scale on both axes.

  b Describe the resulting graph. Is it the graph of a function?

  c Evaluate $x^2 + y^2$. Hence determine the Cartesian equation of this relation.

3   **a**   Use your graphics calculator to plot $x = 2\cos t$, $y = \sin 2t$ for $0 \leqslant t \leqslant 2\pi$.

   b   Show that this curve has Cartesian equation
   $$x^4 - 4x^2 + 4y^2 = 0.$$

Example 14
◀)) **Self Tutor**

Find the Cartesian equation of the curve which has parametric equations:

a $x = 1 - t$, $y = 2t + 3$ **b** $x = 2t + 1$, $y = t^2 - t$

a $x = 1 - t$

  $\therefore\ t = 1 - x$

Substituting into the second equation, we have

$$y = 2(1 - x) + 3$$

$$\therefore\ y = 5 - 2x$$

b $x = 2t + 1$

  $\therefore\ t = \dfrac{x-1}{2}$

Substituting into the second equation, we have

$$y = \left(\frac{x-1}{2}\right)^2 - \left(\frac{x-1}{2}\right)$$

$$\therefore\ y = \left(\frac{x-1}{2}\right)\left(\frac{x-1}{2} - 1\right)$$

$$\therefore\ y = \left(\frac{x-1}{2}\right)\left(\frac{x-3}{2}\right)$$

$$\therefore\ y = \frac{(x-1)(x-3)}{4}$$

EXERCISE 14I.1

1   Find the Cartesian equation of the curve with parametric equations:

  a $x = t$, $y = 1 - 5t$ **b** $x = 1 + 2t$, $y = 3 - t$ **c** $x = t^2$, $y = t^3$

  d $x = t^2$, $y = 4t$ **e** $x = 1 + \dfrac{2}{t}$, $y = 2 - 3t$ **f** $x = t^3 - 2$, $y = t^2 + 3$

2   Consider the curve with parametric equations $x = -t^2 + 6t - 8$, $y = 3t - 9$.

  a   Use technology to help sketch the curve for $0 \leqslant t \leqslant 5$.

  b   Find the point(s) at which the curve crosses the:

    i x-axis   **ii** y-axis.

  c   Find the Cartesian equation of the curve.

3   Determine the Cartesian equation for the following curves, and describe each curve:

  a $x = 3\cos t$, $y = 3\sin t$, $0 \leqslant t \leqslant 2\pi$ **b** $x = -1 + 2\cos t$, $y = 2 + 2\sin t$, $0 \leqslant t \leqslant 2\pi$

  c $x = \sin t$, $y = \cos t$, $0 \leqslant t \leqslant 2\pi$ **d** $x = \cos t$, $y = \sin t$, $\frac{\pi}{2} \leqslant t \leqslant \frac{3\pi}{2}$

4   For each of the following curves:

    i   Use technology to help sketch the curve.

    ii   Describe the curve.

    iii   Find the Cartesian equation of the curve.

  a $x = 2\cos t$, $y = \sin t$, $0 \leqslant t \leqslant 2\pi$ **b** $x = 4 + \cos t$, $y = 3\sin t$, $0 \leqslant t \leqslant 2\pi$

  c $x = \cos t$, $y = \cos^2 t - 1$, $0 \leqslant t \leqslant 2\pi$

5   Use technology to help sketch the curve, then find its Cartesian equation:

   a $x = 2\cos t, \ y = 2\sin 3t, \ 0 \leqslant t \leqslant 2\pi$ **b** $x = 2\cos t, \ y = \cos t - \sin t, \ 0 \leqslant t \leqslant 2\pi$

   c $x = \cos^2 t + \sin 2t, \ y = \cos t, \ 0 \leqslant t \leqslant 2\pi$ **d** $x = \cos^3 t, \ y = \sin t, \ 0 \leqslant t \leqslant 2\pi$

6   What curve do you think would be represented by $x = \cos t, \ y = \sin t, \ z = t, \ t \in \mathbb{R}$? Explain your answer.

Discussion

What *region* do you think would be represented by:

   • $x = \cos t, \ y = \sin t, \ z = s, \ s, t \in \mathbb{R}$ • $x = s\cos t, \ y = s\sin t, \ z = s, \ s, t \in \mathbb{R}, \ s \geqslant 0$

   • $x = \cos t, \ y = \sin t \cos s, \ z = \sin t \sin s, \ 0 \leqslant t \leqslant \pi, \ 0 \leqslant s \leqslant 2\pi$?

FINDING TANGENTS TO CURVES IN PARAMETRIC FORM

When the variables x and y are expressed in terms of a parameter such as t, we can use the chain rule to find $\dfrac{dy}{dx}$.

$$\frac{dy}{dx} = \frac{dy}{dt} \times \frac{dt}{dx} \qquad \{\text{chain rule}\}$$

$$\therefore \quad \frac{dy}{dx} = \frac{dy}{dt} \times \left(1 \div \frac{dx}{dt}\right)$$

$$\therefore \quad \frac{dy}{dx} = \frac{dy}{dt} \div \frac{dx}{dt}$$

Example 15

◀》 **Self Tutor**

A curve is defined by the parametric equations $x = 3 + 2t, \ y = t^3 - 1$. Find the equation of the tangent to the curve at the point where the curve crosses the x-axis.

The curve crosses the x-axis when $y = 0$

$$\therefore \ t^3 - 1 = 0$$

$$\therefore \ t = 1 \quad \text{and} \quad x = 3 + 2(1) = 5$$

$\therefore$ the curve crosses the x-axis at $(5, \, 0)$.

Now $\dfrac{dy}{dx} = \dfrac{dy}{dt} \div \dfrac{dx}{dt}$

$\phantom{Now \ \dfrac{dy}{dx}} = 3t^2 \div 2$

$\phantom{Now \ \dfrac{dy}{dx}} = \tfrac{3}{2}t^2$

When $t = 1$, $\dfrac{dy}{dx} = \dfrac{3}{2}$

$\therefore$ the tangent at $(5, \, 0)$ has gradient $\tfrac{3}{2}$

$\therefore$ the tangent has equation $y = \tfrac{3}{2}(x - 5) - 0$

$$\therefore \ y = \tfrac{3}{2}x - \tfrac{15}{2}$$

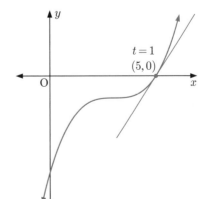

EXERCISE 14I.2

1 A curve is defined by the parametric equations $x = t^3 - 3, \ y = t^2 + t.$

    a Find $\dfrac{dy}{dx}$ in terms of t.

    b The point P on the curve corresponds to $t = 1$.

       i Find the coordinates of P.

      ii Find the equation of the tangent to the curve at P.

    c Find the equation of the tangent to the curve at the point on the curve with x-coordinate 5.

2 A curve has parametric equations $x = \cos\theta, \ y = 2\sin\theta, \ 0 \leqslant \theta \leqslant 2\pi.$

    a Find $\dfrac{dy}{dx}$ in terms of θ.

    b Find the equation of the tangent to the curve at the point where $\theta = \frac{\pi}{6}$.

    c Suppose this tangent cuts the x-axis at A and the y-axis at B. Find the area of triangle AOB. Give your answer in the form $k\sqrt{3}$ units^2, where $k \in \mathbb{Q}$.

3 The curve with parametric equations $x = 2t^2 - 7t + 3,$ $y = t^2 - 1$ is shown alongside. The curve cuts the y-axis at P and Q.

    a Find the coordinates of P and Q.

    b Find the point of intersection of the tangents to the curve at P and Q.

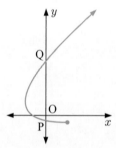

4 The parametric equations of a curve are $x = \cos\theta, \ y = \sin 2\theta$ for $0 \leqslant \theta \leqslant 2\pi.$

    a Find $\dfrac{dy}{dx}$ in terms of θ.

    b The tangents to the curve at the points A and B pass through the point $\left(-\frac{1}{4}, 0\right)$.

       i Show that the values of θ corresponding to points A and B satisfy $4\cos^3\theta + 2\cos^2\theta - 1 = 0$.

      ii Given that $2X - 1$ is a factor of $4X^3 + 2X^2 - 1$, solve $4\cos^3\theta + 2\cos^2\theta - 1 = 0$.

     iii Hence find the coordinates of A and B.

5 The *lemniscate* shown alongside has parametric equations

$$x = \frac{\cos t}{1 + \sin^2 t}, \ \ y = \frac{\sin t \cos t}{1 + \sin^2 t}, \ \ 0 \leqslant t \leqslant 2\pi.$$

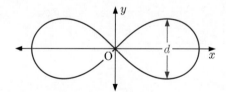

    a Find $\dfrac{dy}{dx}$ in terms of t.

    b **i** Find the values of t at which the curve passes through the origin.

      ii For each of these values, find the equation of the tangent to the curve.

    c **i** Find the value of $\sin^2 t$ such that $\dfrac{dy}{dx} = 0$.

      ii Hence find the distance d between the highest and lowest points of the lemniscate.

Activity   Lissajous curves

Lissajous curves are named after the French physicist **Jules Antoine Lissajous** (1822 - 1880).

Lissajous curves have parametric equations of the form

$$x = a\sin(\omega t + \delta), \quad y = b\sin t.$$

Some examples of Lissajous curves are shown below.

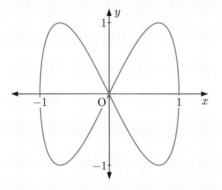

$$x = \sin\left(\tfrac{t}{2} + \tfrac{\pi}{2}\right), \quad y = \sin t$$

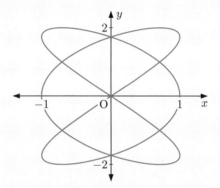

$$x = \sin\left(\tfrac{3t}{2} + \tfrac{\pi}{2}\right), \quad y = 2\sin t$$

What to do:

1 Using technology to help, sketch the following Lissajous curves:

   a $x = \sin(2t + \pi), \quad y = 2\sin t$

   b $x = -\sin\tfrac{3t}{4}, \quad y = \sin t$

   c $x = \sin\left(\sqrt{2}t + \tfrac{\pi}{2}\right), \quad y = -\sin t$

PARAMETRIC
PLOTTER

2 A curve is **closed** if it forms a closed loop with no endpoints. Under what conditions is the Lissajous curve $x = a\sin(\omega t + \delta), \ y = b\sin t$ closed?

3 The logo for the Australian Broadcasting Corporation is a Lissajous curve with parametric equations $x = \sin\left(\tfrac{t}{3} + \tfrac{\pi}{2}\right), \ y = \sin t$.

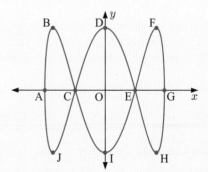

   a   Find $\dfrac{dy}{dx}$ in terms of t.

   b   Find the exact coordinates of the labelled points.

4 Using technology, experiment with different values of a, b, ω, and δ, to create other Lissajous curves.

5 Determine the conditions under which the Lissajous curve $x = a\sin(\omega t + \delta), \ y = b\sin t$ is:

   a   a circle   **b**   an ellipse   **c**   a straight line.

6 Research the use of Lissajous curves in physics, astronomy, and engineering.

J ▌ BÉZIER CURVES

Curves are very important in the design of virtually every manufactured object, including cars, aeroplanes, and toys. It is much harder to mathematically describe a curve between two points than it is to describe the straight line segment between them. One method to describe a curve between two points is to construct a **Bézier curve**.

CONSTRUCTION OF BÉZIER CURVES

Suppose we wish to draw a curve starting at S and ending at E.

We introduce two control points C_1 and C_2. These are not points which the curve needs to pass through but rather they give the initial direction of departure from S and arrival to E. They also control how fast the path curves.

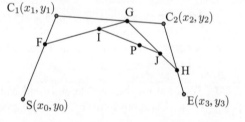

We also introduce a parameter t where $0 \leqslant t \leqslant 1$. When $t = 0$, we are at S, and when $t = 1$, we are at E.

To locate $P(x(t), y(t))$ on the Bézier curve for a particular value of t:

Step 1: Locate F, G, and H on $[SC_1]$, $[C_1C_2]$, and $[C_2E]$ respectively such that F, G, and H divide these line segments in the ratio $t : 1 - t$.

Step 2: Join [FG] and [GH], and likewise locate I and J so they divide [FG] and [GH] in the same ratio $t : 1 - t$.

Step 3: Find P on [IJ] so that P divides [IJ] in the ratio $t : 1 - t$. This point P lies on the Bézier curve.

As t takes all values between 0 and 1, $P(x(t), y(t))$ traces out a curve between S and E. This curve is the Bézier curve.

Click on the icon to construct Bézier curves of your choosing between any two points S and E. The control points can be dragged to different positions to create different curves.

BÉZIER CURVES

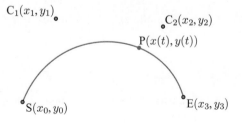

PARAMETRIC EQUATIONS OF BÉZIER CURVES

The parametric equations $x(t)$ and $y(t)$ for a Bézier curve are polynomials with highest order 3.

Given the starting point (x_0, y_0) and finishing point (x_3, y_3), and control points (x_1, y_1) and (x_2, y_2), the Bézier curve has parametric equations

$$x(t) = a_x t^3 + b_x t^2 + c_x t + d_x \quad \text{and}$$
$$y(t) = a_y t^3 + b_y t^2 + c_y t + d_y, \quad 0 \leqslant t \leqslant 1, \quad \text{where}$$

$$\begin{cases} a_x = x_3 - 3x_2 + 3x_1 - x_0 \\ b_x = 3x_2 - 6x_1 + 3x_0 \\ c_x = 3x_1 - 3x_0 \\ d_x = x_0 \end{cases} \quad \text{and} \quad \begin{cases} a_y = y_3 - 3y_2 + 3y_1 - y_0 \\ b_y = 3y_2 - 6y_1 + 3y_0 \\ c_y = 3y_1 - 3y_0 \\ d_y = y_0 \end{cases}$$

Click on the icon to see how these results are established.

BÉZIER CURVE EQUATIONS

Example 16

◆)) **Self Tutor**

A Bézier curve from $S(0, 1)$ to $E(3, 0)$ has control points $C_1(1, 3)$ and $C_2(2, 4)$.

a Find the parametric equations of the curve.

b Find the highest and lowest points on the curve.

a $x_0 = 0, \quad x_1 = 1, \quad x_2 = 2, \quad x_3 = 3$

$y_0 = 1, \quad y_1 = 3, \quad y_2 = 4, \quad y_3 = 0$

$$\begin{aligned} a_x &= 3 - 3(2) + 3(1) - 0 = 0 & a_y &= 0 - 3(4) + 3(3) - 1 = -4 \\ b_x &= 3(2) - 6(1) + 3(0) = 0 & b_y &= 3(4) - 6(3) + 3(1) = -3 \\ c_x &= 3(1) - 3(0) = 3 & c_y &= 3(3) - 3(1) = 6 \\ d_x &= 0 & d_y &= 1 \end{aligned}$$

$\therefore \quad x(t) = 3t, \quad 0 \leqslant t \leqslant 1 \qquad \therefore \quad y(t) = -4t^3 - 3t^2 + 6t + 1, \quad 0 \leqslant t \leqslant 1$

Check: $\underbrace{x(0) = 0, \quad y(0) = 1}_{S(0,\,1)} \checkmark \qquad \underbrace{x(1) = 3, \quad y(1) = -4 - 3 + 6 + 1 = 0}_{E(3,\,0)} \checkmark$

b The highest and lowest points occur either when $y'(t) = 0$, or when $t = 0$ or 1.

$$y'(t) = -12t^2 - 6t + 6$$

$\therefore \quad y'(t) = 0$ when $-12t^2 - 6t + 6 = 0$

$\therefore \quad -6(2t^2 + t - 1) = 0$

$\therefore \quad -6(2t - 1)(t + 1) = 0$

$\therefore \quad t = \frac{1}{2} \qquad \{\text{as } 0 \leqslant t \leqslant 1\}$

Now $y(0) = 1, \quad y\left(\frac{1}{2}\right) = -4\left(\frac{1}{8}\right) - 3\left(\frac{1}{4}\right) + 6\left(\frac{1}{2}\right) + 1 = 2\frac{3}{4}, \quad y(1) = 0$

$\therefore$ the highest point on the curve is $\left(\frac{3}{2}, 2\frac{3}{4}\right) \qquad \{\text{when } t = \frac{1}{2}\}$

and the lowest point on the curve is $(3, 0) \qquad \{\text{when } t = 1\}$.

EXERCISE 14J

1 The Bézier curve alongside starts at $(-3, -1)$ and ends at $(3, 2)$. The control points are $(0, 3)$ and $(2, 4)$.

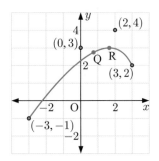

a Find the parametric equations of the Bézier curve.

b Verify that $t = 0$ corresponds to $(-3, -1)$ and $t = 1$ corresponds to $(3, 2)$.

c The point Q on the curve corresponds to $t = \frac{1}{2}$. Find the coordinates of Q.

d R is the highest point on the curve. Find the coordinates of R.

2 Find the parametric equations for the Bézier curve defined by each set of points:

	Starting point	Ending point	Control points	
	S	E	C_1	C_2
a	$(4, 1)$	$(3, -2)$	$(1, 3)$	$(2, 0)$
b	$(2, 4)$	$(3, 6)$	$(-1, -1)$	$(-2, 5)$
c	$(-1, 2)$	$(-3, 8)$	$(0, 0)$	$(2, 3)$

BÉZIER CURVES

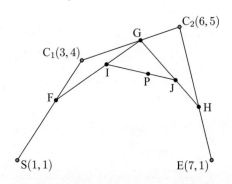

Use the software to check your answers.

3 A Bézier curve is defined by $x(t) = -5t^3 + 15t - 5$, $y(t) = -9t^3 - 18t^2 + 15t - 1$, $0 \leqslant t \leqslant 1$.

a Find the starting and ending points of the curve.

b Use technology to help sketch the curve.

c Find the lowest and highest points on the curve.

d Find the left-most and right-most points on the curve.

4 A Bézier curve is defined by $x(t) = \frac{2}{3}t^3 - \frac{1}{2}t^2 + 1$, $y(t) = 2t^3 - \frac{7}{2}t^2 + 2t + 4$ where $0 \leqslant t \leqslant 1$.

a Find the starting and ending points on the curve.

b Find the left-most and right-most points on the curve.

c Find the highest and lowest points on the curve.

5 A Bézier curve goes from $S(1, 2)$ to $E(-2, 4)$, and has control points $C_1(0, 4)$ and $C_2(-4, -2)$.

a Find the position equations for the position of any point P on the curve.

b Use technology to determine the:

  i highest and lowest points on the curve

  ii left-most and right-most points on the curve.

6 A Bézier curve starts at $S(1, 1)$ and ends at $E(7, 1)$. $C_1(3, 4)$ and $C_2(6, 5)$ are its control points. When $t = 0.6$, $SF : FC_1 = 0.6 : 0.4$, $C_1G : GC_2 = 0.6 : 0.4$, and so on.

a Use ratio of division to find the coordinates of F, G, H, I, J, and P.

b Find the parametric equations of the Bézier curve.

c Show that P lies on the Bézier curve.

d Show that [IJ] is a tangent to the Bézier curve at P.

7 We can construct a 3-dimensional Bézier curve by defining a parametric equation for the z-coordinate in the same way as for the x- and y-coordinates.

Suppose a Bézier curve starts at $(-2, -3, 1)$, ends at $(4, 2, 3)$, and has control points $(0, 2, 0)$ and $(-1, 0, 2)$.

   a Find the parametric equations for the curve.

   b Verify that $t = 0$ corresponds to $(-2, -3, 1)$ and $t = 1$ corresponds to $(4, 2, 3)$.

   c Find the point on the curve corresponding to $t = \frac{1}{2}$.

Discussion

What advantages are there to defining a curve mathematically, as opposed to simply drawing the curve freehand?

Review set 14A

1 Illustrate P and find its distance from the origin O:

   a P$(1, 3, 2)$ **b** P$(2, -1, 3)$

2 Find k given the unit vector:

   a $\begin{pmatrix} \frac{3}{4} \\ -\frac{1}{2} \\ k \end{pmatrix}$ **b** $\begin{pmatrix} 3k \\ k \\ \frac{3}{8} \end{pmatrix}$

3 Find scalars a, b, and c such that $3 \begin{pmatrix} 1 \\ -2 \\ -1 \end{pmatrix} + a \begin{pmatrix} 2 \\ 0 \\ 6 \end{pmatrix} - \begin{pmatrix} b \\ -1 \\ -4 \end{pmatrix} = \begin{pmatrix} 4 \\ c \\ 7 \end{pmatrix}$.

4 Suppose $\mathbf{a} = \begin{pmatrix} 2 \\ -3 \\ 1 \end{pmatrix}$, $\mathbf{b} = \begin{pmatrix} -1 \\ 2 \\ 3 \end{pmatrix}$, and $\mathbf{a} - 3\mathbf{x} = \mathbf{b}$. Find $\mathbf{x}$.

5 Given P$(2, 3, -1)$ and Q$(-4, 4, 2)$, find:

   a $\overrightarrow{PQ}$ **b** the distance between P and Q   **c** the midpoint of [PQ].

6 Show that A$(-2, -1, 3)$, B$(4, 0, -1)$, and C$(-2, 1, -4)$ are vertices of an isosceles triangle.

7 If A$(4, 2, -1)$, B$(-1, 5, 2)$, C$(3, -3, c)$ are vertices of triangle ABC which is right angled at B, find the value of c.

8 The points A, B, and C have position vectors $2\mathbf{i} + \mathbf{j} - 4\mathbf{k}$, $5\mathbf{i} - \mathbf{k}$, and $-\mathbf{i} - 4\mathbf{j} + \mathbf{k}$ respectively.

   a M is the midpoint of BC. Find the magnitude of $\overrightarrow{AM}$.

   b Point D is such that $\overrightarrow{AC} = \overrightarrow{BD}$. Find the position vector of D.

9 Find m and n if $\begin{pmatrix} 3 \\ m \\ n \end{pmatrix}$ and $\begin{pmatrix} -12 \\ -20 \\ 2 \end{pmatrix}$ are parallel vectors.

10 Prove that P$(-6, 8, 2)$, Q$(4, 6, 8)$, and R$(19, 3, 17)$ are collinear.

11   Find the vector which is 5 units long and has the opposite direction to $\begin{pmatrix} 3 \\ 2 \\ -1 \end{pmatrix}$.

12   Suppose that two opposite sides of a quadrilateral are parallel and equal in length. Prove that the quadrilateral is a parallelogram.

13   For the line that passes through $(-6, 3)$ with direction $\begin{pmatrix} 4 \\ -3 \end{pmatrix}$, write down the corresponding:

    a vector equation   **b** parametric equations   **c** Cartesian equation.

14 $(-3, m)$ lies on the line with vector equation $\begin{pmatrix} x \\ y \end{pmatrix} = \begin{pmatrix} 18 \\ -2 \end{pmatrix} + t \begin{pmatrix} -7 \\ 4 \end{pmatrix}$, $t \in \mathbb{R}$. Find m.

15   Consider $A(2, -1, 3)$ and $B(0, 1, -1)$.

    a Find the vector equation of the line through A and B.

    b Hence find the coordinates of C on (AB) which is 2 units from A.

16   Consider $A(4, 2, -1)$, $B(2, 1, 5)$, and $C(9, 4, 1)$. Find parametric equations for the line through:

    a A and B   **b** A and C.

17   Find the velocity vector of a drone moving parallel to $3\mathbf{i} - 4\mathbf{j} + 2\mathbf{k}$ with speed 72 km h^{-1}.

18   Find the Cartesian equation of the curve $x = 3\cos t$, $y = \sin 2t$, $0 \leqslant t \leqslant 2\pi$. Use technology to help sketch the curve.

19   A curve is defined by the parametric equations $x = \sqrt{t - 3}$, $y = 2t - 8$. The curve cuts the x-axis at P, and includes the point Q with x-coordinate 2. Find:

    a the coordinates of P and Q

    b the gradient of the tangent to the curve at P

    c the equation of the tangent to the curve at Q

    d the Cartesian equation of the curve.

20   A Bézier curve is defined by $x(t) = t^3 - t + 1$, $y(t) = 3 - t^2 - t^3$ where $0 \leqslant t \leqslant 1$ and t is in seconds.

    a Find the starting and ending points for the curve.

    b Find the point on the curve corresponding to $t = 0.5$.

    c Show that the curve passes through only one point with y-coordinate 1.

Review set 14B

1   Find a, b, and c such that $\begin{pmatrix} a - 1 \\ b^2 \\ c + 7 \end{pmatrix} = \begin{pmatrix} 2a \\ 4 - b \\ 2c - 3 \end{pmatrix}$.

2 For $\mathbf{m} = \begin{pmatrix} 6 \\ -3 \\ 1 \end{pmatrix}$, $\mathbf{n} = \begin{pmatrix} 2 \\ 3 \\ -4 \end{pmatrix}$, and $\mathbf{p} = \begin{pmatrix} -1 \\ 3 \\ 6 \end{pmatrix}$, find:

    a $\mathbf{m} - \mathbf{n} + \mathbf{p}$ **b** $2\mathbf{n} - 3\mathbf{p}$ **c** $|\mathbf{m} + \mathbf{p}|$

3 Suppose $\mathbf{p} = \begin{pmatrix} -3 \\ 1 \\ 2 \end{pmatrix}$, $\mathbf{q} = \begin{pmatrix} 2 \\ -4 \\ 1 \end{pmatrix}$, and $\mathbf{r} = \begin{pmatrix} 3 \\ 2 \\ 0 \end{pmatrix}$. Find $\mathbf{x}$ such that:

    a $\mathbf{p} - 3\mathbf{x} = 0$ **b** $2\mathbf{q} - \mathbf{x} = \mathbf{r}$

4 If $\overrightarrow{AB} = \begin{pmatrix} 2 \\ -7 \\ 4 \end{pmatrix}$ and $\overrightarrow{AC} = \begin{pmatrix} -6 \\ 1 \\ -3 \end{pmatrix}$, find $\overrightarrow{CB}$.

5 Given $P(2, -5, 6)$ and $Q(-1, 7, 9)$, find:

    a the position vector of Q relative to P   **b** the distance from P to Q

    c the distance from P to the X-axis.

6 Find k given that $\begin{pmatrix} k \\ \frac{1}{\sqrt{2}} \\ -k \end{pmatrix}$ is a unit vector.

7 Find two points on the Z-axis which are 6 units from $P(-4, 2, 5)$.

8 Show that $K(4, 3, -1)$, $L(-3, 4, 2)$, and $M(2, 1, -2)$ are vertices of a right angled triangle.

9 Find a and b if $J(-4, 1, 3)$, $K(2, -2, 0)$, and $L(a, b, 2)$ are collinear.

10 Find *two* vectors of length 4 units which are parallel to $3\mathbf{i} - 2\mathbf{j} + \mathbf{k}$.

11 In the diagram alongside, $AD = kBC$, $k \in \mathbb{R}$.

Let $\overrightarrow{AB} = \mathbf{a}$ and $\overrightarrow{BC} = \mathbf{b}$.

    a Write $\overrightarrow{CD}$ in terms of $\mathbf{a}$, $\mathbf{b}$, and k.

    b Prove that $\overrightarrow{ED} = k\overrightarrow{BE}$.

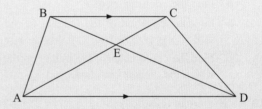

12 The points A, B, and C have position vectors $\mathbf{i} + 2\mathbf{j} - 4\mathbf{k}$, $4\mathbf{i} + 5\mathbf{k}$, and $-5\mathbf{i} - \mathbf{j} + 3\mathbf{k}$ respectively.

    a Show that triangle ABC is isosceles.

    b Point D is such that ABCD is a parallelogram. Find the coordinates of D.

13 Line L has equation $r = \begin{pmatrix} 3 \\ -3 \end{pmatrix} + t\begin{pmatrix} 2 \\ 5 \end{pmatrix}$, $t \in \mathbb{R}$.

    a Locate the point on the line corresponding to $t = 1$.

    b Explain why the direction of the line could also be described by $\begin{pmatrix} 4 \\ 10 \end{pmatrix}$.

    c Use your answers to **a** and **b** to write an alternative vector equation for line L.

14 Find the vector equation of the line which cuts the y-axis at $(0, 8)$ and has direction $5\mathbf{i} + 4\mathbf{j}$.

15 Write down   **i** a vector equation   **ii** parametric equations   for the line passing through:

a $(2, -3, 1)$ with direction $\begin{pmatrix} 4 \\ 2 \\ -1 \end{pmatrix}$
   b $(-1, 6, 3)$ and $(5, -2, 0)$.

16 The vector equation $\begin{pmatrix} x \\ y \\ z \end{pmatrix} = \begin{pmatrix} -10 \\ 5 \\ 12 \end{pmatrix} + t \begin{pmatrix} 6 \\ 14 \\ -0.4 \end{pmatrix}$

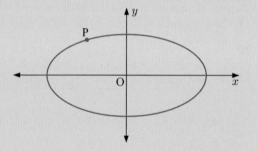

represents the path of a person zip-lining through a forest. t is measured in seconds, $t \geqslant 0$. Distances are measured in metres. Find:

a the initial position   **b** the velocity vector

c the speed of the zip-liner

d the time taken to reach the end of the line at $z = 0$

e the location of the endpoint of the line.

17 For each of the following curves:

   i Use technology to help sketch the curve.

   ii Find the Cartesian equation of the curve.

a $x = -2\cos t, \ y = 2\sin t, \ 0 \leqslant t \leqslant 2\pi$

b $x = 3\sin t - 1, \ y = 2\cos t + 5, \ 0 \leqslant t \leqslant 2\pi$

c $x = t^2 - 1, \ y = 3 - 2t, \ t \in \mathbb{R}$

18 The curve alongside has parametric equations $x = 2\cos\theta, \ y = \sin\theta, \ 0 \leqslant \theta \leqslant 2\pi$. The point P corresponds to $\theta = \frac{2\pi}{3}$.

a Find the coordinates of P.

b Find the equation of the tangent to the curve at P.

c Find the coordinates of the points on the curve where the tangent has gradient 1.

19 A Bézier curve is defined by $x(t) = 2t^2 - 3t + 1, \ y(t) = 4t^3 + 3t^2 + 4, \ 0 \leqslant t \leqslant 1$.

a Find the starting and ending points of the curve.

b Find the left-most point of the curve.

20 A Bézier curve from $S(3, -1)$ to $E(-4, 2)$ has control points $C_1(4, -4)$ and $C_2(4, -3)$.

a Find the parametric equations of the curve.

b Find the highest and lowest points on the curve.

c Find the left-most and right-most points on the curve.

15

Kinematics

Contents:

Opening problem

A trebuchet is a siege engine used to launch projectiles at enemy castles from a considerable distance away. They were mainly used in the Middle Ages, but continued to be used long after the invention of gun powder and cannon.

The Trebuchet at Warwick Castle claims to be the largest working siege machine in the world. It is constructed mainly of oak, with the long throwing arm made of ash, a more flexible wood. It is 8.7 m high, increasing to 18 m high with the arm fully extended. Using a counterweight of approximately 6 tonnes, it is capable of throwing an 18 kg projectile a horizontal distance of up to 242 m.

Things to think about:

a Can you describe how the projectile moves?

b Suppose an 18 kg projectile is fired from the origin. For the projectile to reach its maximum range of 242 m:

    i   at what angle of trajectory should it be thrown

    ii what initial velocity is needed

    iii how high will it reach

    iv for how long will it stay in the air?

Kinematics is the study of motion.

In this Chapter we revise motion in a straight line, and the relationships between the vector quantities displacement, velocity, and acceleration.

We then use vectors to consider the two-dimensional motion of projectiles fired under the influence of constant gravity.

A   MOTION IN A STRAIGHT LINE

DISPLACEMENT

Suppose an object P moves along a straight line. The position of P relative to an origin O is called the **displacement** of P.

We can define a **displacement function** $s(t)$ which gives the displacement of the object at any time $t \geqslant 0$ as it moves along the line.

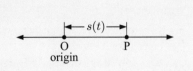

When $s(t) > 0$, P is to the right of the origin.

When $s(t) < 0$, P is to the left of the origin.

For example, consider $s(t) = 3\sin \pi t + t$ cm for the first two seconds of motion.

t	$s(t)$
0	0
$\frac{1}{2}$	$3\frac{1}{2}$
1	1
$\frac{3}{2}$	$-\frac{3}{2}$
2	2

To appreciate the motion of P, we draw a **motion diagram**.

DEMO

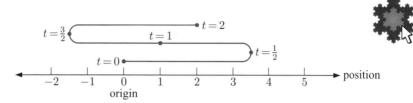

VELOCITY

The **velocity** of an object is its rate of change of displacement.

If $s(t)$ is the displacement function of an object moving in a straight line, then the **instantaneous velocity** or **velocity function** of the object at time t is $v(t) = s'(t)$.

When $v(t) > 0$, the object is moving to the right.

When $v(t) < 0$, the object is moving to the left.

When $v(t) = 0$, the object is instantaneously at rest. A change in the sign of $v(t)$ at this time indicates that the object has changed direction.

ACCELERATION

The **acceleration** of an object is its rate of change of velocity.

The **acceleration function** $a(t)$ of an object is given by $a(t) = v'(t) = s''(t)$.

SPEED

The **speed** of an object at any instant is the magnitude of the object's velocity.

Speed cannot be negative.

For example, an object with velocity -8 m s^{-1} is moving to the left with speed 8 m s^{-1}.

To determine when the speed of an object P with displacement $s(t)$ is increasing or decreasing, we use a **sign test**.

- If the signs of $v(t)$ and $a(t)$ are the same (both positive or both negative), then the speed of P is increasing.
- If the signs of $v(t)$ and $a(t)$ are opposite, then the speed of P is decreasing.

We can use **sign diagrams** to interpret:

- where the object is located relative to O
- the direction of motion and where a change of direction occurs
- when the object's speed is increasing or decreasing.

Example 1 　　　　　　　　　　　　　　　　　　　　　　　　◀ﹾ **Self Tutor**

A particle moves in a straight line with position relative to O given by $s(t) = t^3 - 6t^2 + 9t - 1$ cm, where t is the time in seconds, $t \geqslant 0$.

a Find expressions for the particle's velocity and acceleration, and draw sign diagrams for each of them.

b Find the initial conditions and hence describe the motion at this instant.

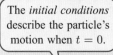

The *initial conditions* describe the particle's motion when $t = 0$.

c Find the position of the particle when it changes direction.

d Draw a motion diagram for the particle.

e For what time interval is the particle's speed increasing?

f Find the total distance travelled in the first 3 seconds.

a 　　　$s(t) = t^3 - 6t^2 + 9t - 1$ cm

$\therefore \ v(t) = 3t^2 - 12t + 9 \qquad \{v(t) = s'(t)\}$

$\qquad\quad = 3(t^2 - 4t + 3)$

$\qquad\quad = 3(t-1)(t-3)$ cm s^{-1}

which has sign diagram:

and 　$a(t) = 6t - 12$ cm s$^{-2} \qquad \{a(t) = v'(t)\}$

which has sign diagram:

b When $t = 0$, $s(0) = -1$ cm

$\qquad\qquad\qquad v(0) = 9$ cm s^{-1}

$\qquad\qquad\qquad a(0) = -12$ cm s^{-2}

$\therefore$ the particle is 1 cm to the left of O, moving to the right with speed 9 cm s^{-1}, and has acceleration -12 cm s^{-2}.

c Since $v(t)$ changes sign when $t = 1$ and $t = 3$, a change of direction occurs at these instants.

$s(1) = 1 - 6 + 9 - 1 = 3$, and $s(3) = 27 - 54 + 27 - 1 = -1$.

So, the particle changes direction when it is 3 cm to the right of O, and when it is 1 cm to the left of O.

d

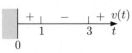

e The speed is increasing when $v(t)$ and $a(t)$ have the same sign. This occurs for $1 \leqslant t \leqslant 2$ and $t \geqslant 3$.

f The total distance travelled $= 4 + 4 = 8$ cm.

EXERCISE 15A.1

1 An object moves in a straight line with position from O given by $s(t) = t^2 - 6t + 7$ m, where t is in seconds, $t \geqslant 0$.

   a Find expressions for the object's velocity and acceleration, and show sign diagrams for each of them.

   b Find the initial conditions and hence describe the motion at this instant.

   c Find the position of the object when it changes direction.

   d Draw a motion diagram for the object.

   e During which time interval is the object's speed decreasing?

2 A particle moves in a straight line with displacement function $s(t) = 12t - 2t^3 - 1$ cm where t is in seconds, $t \geqslant 0$.

   a Find velocity and acceleration functions for the particle's motion.

   b Find the initial conditions, and interpret their meaning.

   c Find the times and positions when the particle reverses direction.

   d At what times is the particle's:

      i speed increasing   **ii** velocity increasing?

   e Draw a motion diagram for the particle.

3 A shell is accidentally fired vertically from a mortar at ground level. Its height above the ground after t seconds is given by $s(t) = bt - 4.9t^2$ metres where b is constant.

   a Show that the initial velocity of the shell is b m s^{-1} upwards.

   b The shell reaches its maximum height after 7.1 seconds.

      i Find the initial velocity of the shell.

      ii Find the maximum height reached by the shell.

4 A particle has displacement function $s(t) = 4 - \sqrt{t + 1}$ m, where t is in seconds, $t \geqslant 0$.

   a Find the velocity and acceleration functions for the particle's motion, and draw sign diagrams for each of them.

   b Find the initial conditions and interpret their meaning.

   c Describe the motion of the particle after 3 seconds.

   d Describe what is happening to the speed of the particle.

5 A flotation device is thrown from a jetty into the water below. It takes k seconds for the device to reach the water. The height of the device above sea level is given by $s(t) = -4.9t^2 + 4.9t + 8$ m, where t is in seconds, $0 \leqslant t \leqslant k$ seconds.

   a Find the value of k.

   b Draw sign diagrams for the device's velocity and acceleration functions.

   c Was the speed of the device increasing or decreasing after:

      i 0.2 seconds   **ii** 1 second?

6 A particle P moves in a straight line with displacement function $s(t) = 100t + 200e^{-\frac{t}{5}}$ cm, where t is the time in seconds, $t \geqslant 0$.

   a Find the velocity and acceleration functions.

   b Find the initial position, velocity, and acceleration of P.

   c Discuss the velocity of P as $t \to \infty$.

   d Sketch the graph of the velocity function.

   e Find the time at which the velocity of P is 80 cm s^{-1}.

7 An object has displacement function $s(t) = t - \ln(2t + 1)$ cm, where t is in seconds, $t \geqslant 0$.

   a Show that the object is initially at the origin.

   b Find the velocity function.

   c Over what time interval is the object moving:

     i to the right   **ii** to the left?

   d Show that the object's acceleration is positive for all $t \geqslant 0$.

   e Find the distance travelled by the object in the first 2 seconds.

8 A particle P moves along the x-axis with position given by $x(t) = 1 - 2\cos t$ cm, $t \geqslant 0$ seconds.

   a State the initial position, velocity, and acceleration of P.

   b Describe the motion of P when $t = \frac{\pi}{4}$ seconds.

   c Find the times when the particle reverses direction on $0 < t < 2\pi$, and find the position of the particle at these instants.

   d When is the particle's speed increasing on $0 \leqslant t \leqslant 2\pi$?

9 A dog paces back and forth along a fence, guarding its owner's property. The dog's horizontal displacement relative to its kennel is given by $s(t) = 8\sin\frac{t}{2}$ m, $t \geqslant 0$ seconds.

   a Is the dog to the left or the right of its kennel after:

     i 3 seconds   **ii** 7 seconds?

   b Find the velocity function $v(t)$.

   c Is the dog moving to the left or the right after:

     i 4 seconds   **ii** 10 seconds?

   d Show that the dog's speed is maximised when it is moving past its kennel.

10 The velocity of an object after t seconds is given by $v(t) = 25te^{-2t}$ cm s^{-1}, $t \geqslant 0$.

   a Use technology to help sketch the velocity function.

   b Show that the object's acceleration at time t is given by $a(t) = 25(1 - 2t)e^{-2t}$ cm s^{-2}, $t \geqslant 0$.

   c When is the velocity of the object increasing?

   d When is the speed of the object decreasing?

FINDING DISPLACEMENT FROM VELOCITY AND ACCELERATION

Since the velocity and acceleration functions can be obtained by differentiation of the displacement function, we can use integration to reverse the process.

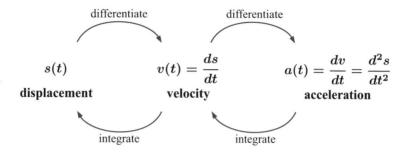

CHANGE IN DISPLACEMENT

We can determine the change in displacement in a time interval $t_1 \leqslant t \leqslant t_2$ using the integral:

$$\textbf{Change in displacement} = s(t_2) - s(t_1) = \int_{t_1}^{t_2} v(t)\, dt$$

DISTANCE TRAVELLED

For a velocity-time function $v(t)$ where $v(t) \geqslant 0$ on the interval $t_1 \leqslant t \leqslant t_2$,

$$\textbf{distance travelled} = \int_{t_1}^{t_2} v(t)\, dt.$$

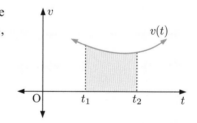

If $v(t) \geqslant 0$, the distance travelled is the area under the velocity curve.

If we have a change of direction within the time interval then the velocity will change sign. We therefore need to add the components of area above and below the t-axis to find the total distance travelled.

Example 2
◀) **Self Tutor**

When a car does a U-turn, its velocity function is $v(t) = \pi \cos \frac{\pi t}{8}$ m s^{-1} for $0 \leqslant t \leqslant 10$ seconds.

a How far does the car travel in the first 10 seconds?

b Find the change in displacement of the car after 10 seconds.

a $v(t) = \pi \cos \frac{\pi t}{8}$

∴ the sign diagram of v for the first 10 seconds is:

The car changes direction after 4 seconds.

Now $s(t) = \displaystyle\int v(t)\,dt = \int \pi \cos\frac{\pi t}{8}\,dt = 8\sin\frac{\pi t}{8} + c$

Hence $s(0) = 0 + c = c$

$s(4) = 8 + c$

$s(10) = 8\left(-\frac{1}{\sqrt{2}}\right) + c = -4\sqrt{2} + c$

Motion diagram:

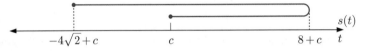

$\therefore$ total distance travelled $= 8 + (8 + 4\sqrt{2})$

≈ 21.7 m

b Displacement $=$ final position $-$ original position

$= s(10) - s(0)$

$= -4\sqrt{2} + c - c$

$= -4\sqrt{2}$ m

After 10 seconds, the car is $4\sqrt{2}$ metres behind its original position.

EXERCISE 15A.2

1 A particle has velocity function $v(t) = t^2 + t - 6$ m s^{-1}.

  a Find the total distance travelled in the first 3 seconds of motion.

  b Find the change in displacement of the particle after 3 seconds.

2 Particle P is initially at the origin O. It moves with velocity function $v(t) = t^2 + 3t + 2$ cm s^{-1}.

  a Write a formula for the displacement function $s(t)$.

  b Find the total distance travelled in the first 4 seconds of motion.

  c Find the particle's change in displacement from its starting position after 4 seconds.

3 A football is kicked from ground level. Its velocity after t seconds is
$32.4 - 9.8t$ m s^{-1}.

  a Find the displacement function $s(t)$.

  b Find the maximum height reached by the ball.

4 An object has velocity function $v(t) = \cos 2t$ m s^{-1}.

  a Show that the particle oscillates between two points, and find the distance between them.

  b If $s(\frac{\pi}{4}) = 1$ m, determine $s(\frac{\pi}{3})$ exactly.

5 A cyclist is travelling along a series of hills. His velocity after t hours is given by
$v(t) = 20 + 5\sin 10t$ miles per hour.

  a Sketch the graph of $v(t)$ against t for $0 \leqslant t \leqslant 2$.

  b Find the cyclist's velocity after:

    i 0.5 hours   **ii** 75 minutes.

  c Find the total distance travelled by the cyclist after 2 hours.

6 A lion is chasing a zebra. The zebra notices the lion when the lion is 40 m away. From this time, the lion approaches with speed $v_1(t) = 15e^{-0.1t}$ m s^{-1}, and the zebra runs away with speed $v_2(t) = 20 - 20e^{-0.1t}$ m s^{-1}.

a Find the speed of each animal after 1 second.

b Show that the lion's speed decreases over time, whereas the zebra's speed increases over time.

c Find $\displaystyle\int_0^3 v_1(t)\, dt$, and interpret your answer.

d Find $\displaystyle\int_0^3 [v_1(t) - v_2(t)]\, dt$, and interpret your answer.

e Explain why the lion will be closest to the zebra when $v_1(t) = v_2(t)$.

f Find the time at which $v_1(t) = v_2(t)$.

g Did the lion catch the zebra? If not, how close did the lion get to the zebra?

7 An object moves with velocity $v(t) = \cos^2 t$ m s^{-1}, where $t \geqslant 0$, t in seconds.

a Find the velocity of the object:

   i initially   **ii** after $\frac{\pi}{3}$ seconds.

b Show that the object never changes direction.

c Given that the object is initially at the origin, find the displacement function $s(t)$.

d Find the position of the object after:

   i $\frac{\pi}{2}$ seconds   **ii** $\frac{7\pi}{6}$ seconds.

e Find the acceleration function $a(t)$.

f Find the times during the first 3π seconds when the object's speed is:

   i increasing   **ii** decreasing.

g Find the total distance travelled by the object from $t = \frac{\pi}{6}$ to $t = \frac{2\pi}{3}$.

8 The velocity of a moving object is given by $v(t) = -4 + \sqrt{t}$ m s^{-1}. Suppose the object is initially at the origin. Find:

a the displacement function

b the time when the object changes direction

c the change in displacement after the first 30 seconds

d the total distance travelled in the first 30 seconds.

9 The velocity of a particle travelling in a straight line is given by $v(t) = 50 - 10e^{-0.5t}$ m s^{-1}, where $t \geqslant 0$, t in seconds.

a State the initial velocity of the particle.

b Find the velocity of the particle after 3 seconds.

c How long will it take for the particle's velocity to reach 45 m s^{-1}?

d Discuss $v(t)$ as $t \to \infty$.

e Show that the particle's acceleration is always positive.

f Draw the graph of $v(t)$ against t.

g Find the total distance travelled by the particle in the first 3 seconds of motion.

Example 3
◀) **Self Tutor**

A train is initially at rest at a station. It begins to accelerate according to the function $a(t) = \frac{1}{2}e^{-\frac{t}{100}}$ m s^{-2}.

a Find the velocity function of the train, and sketch its graph.

b How long will it take for the train to reach the speed 40 m s^{-1}?

c How far will the train travel in this time?

a $a(t) = \frac{1}{2}e^{-\frac{t}{100}}$ m s^{-2}

$\therefore\ v(t) = \int \frac{1}{2}e^{-\frac{t}{100}}\, dt$

$\qquad = -50e^{-\frac{t}{100}} + c$ m s^{-1}

The train was initially stationary.

$\therefore\ -50e^{0} + c = 0$

$\therefore\ c = 50$

$\therefore\ v(t) = 50 - 50e^{-\frac{t}{100}}$ m s^{-1}

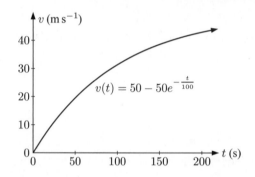

b   If $v = 40$ then

$50 - 50e^{-\frac{t}{100}} = 40$

$\therefore\ 50e^{-\frac{t}{100}} = 10$

$\therefore\ e^{-\frac{t}{100}} = \frac{1}{5}$

$\therefore\ -\frac{t}{100} = \ln\left(\frac{1}{5}\right) = -\ln 5$

$\therefore\ t = 100\ln 5 \approx 160.9$ s

c   The train does not change direction.

$\therefore$ distance travelled

$= \displaystyle\int_{0}^{100\ln 5} \left(50 - 50e^{-\frac{t}{100}}\right) dt$

$= \left[50t + 5000e^{-\frac{t}{100}}\right]_{0}^{100\ln 5}$

$= 50(100\ln 5) + 5000e^{-\frac{100\ln 5}{100}} - 5000e^{0}$

$= 5000\ln 5 + \frac{5000}{5} - 5000$

$= 5000\ln 5 - 4000$

≈ 4047 m

10   A particle is initially stationary at the origin. It accelerates according to the function $a(t) = \dfrac{2}{(t+1)^3}$ m s^{-2}.

   a   Find the velocity function $v(t)$ for the particle.

   b   Find the displacement function $s(t)$ for the particle.

   c   Describe the motion of the particle at the time $t = 2$ seconds.

11 An object has initial velocity 20 m s^{-1} as it moves in a straight line with acceleration function $a(t) = 4e^{-\frac{t}{20}} \text{ m s}^{-2}$.

    a Show that as t increases, the object approaches a limiting velocity.

    b Find the total distance travelled in the first 10 seconds of motion.

12 A droplet of water is flowing down a window with acceleration $a(t) = \dfrac{5}{\sqrt{t+5}} \text{ cm s}^{-2}$. Its initial velocity is $5\sqrt{5} \text{ cm s}^{-1}$.

    a Determine the velocity function $v(t)$.

    b Evaluate $\displaystyle\int_0^4 v(t)\,dt$, and interpret your result.

13 When a tyre hits a bump in the road, the end point P of the car's suspension is 10 cm from its top T.

    The velocity of P relative to T is given by
$$v(t) = 10e^{-2t}\sin 5t - 5e^{-2t}\cos 5t \text{ cm s}^{-1}$$
    where t is the time in seconds.

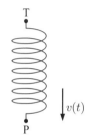

    a Find the velocity of P relative to T after 0.3 seconds.

    b Find the acceleration of P relative to T after 0.3 seconds.

    c Use integration by parts to find $s(t)$.

    d Hence find the distance by which the suspension has contracted after 0.1 seconds.

B   PROJECTILE MOTION

A **projectile** is an object which is released into the air, and then moves under the force of gravity only. Examples of projectiles include a ball which is thrown into the air, or a bullet which is fired from a gun.

In this Section we consider the motion of a projectile after it has been released.

The path the projectile takes is called its trajectory.

MODELLING ASSUMPTIONS

To create a mathematical model for a projectile we need to make a number of assumptions:

- The projectile is a "particle", which means we assume it has no size.
- There is no air resistance.
- When released, the only force acting on the particle is weight, the force due to gravity.
- The projectile remains relatively close to the surface of the Earth so that gravity remains constant and acts vertically downwards.
- We assume the surface of the Earth is well approximated as a flat plane, rather than a circular arc.
- The mass of the projectile remains constant.

MODELLING LIMITATIONS

- Air resistance has a very important practical effect on most projectiles, so this assumption seriously affects the accuracy of our models. We still make this assumption, otherwise the equations become significantly more difficult to solve.

- Many objects generate their own thrust, for example rockets and aeroplanes. Projectiles do not create thrust, and so the only forces are external.

- Large rigid objects can rotate or spin, and this can significantly affect the motion of the object.

- Projectiles can have very large trajectories. While no projectile has been fired into orbit (rocket motors are needed) some naval artillery shells are fired very long distances, well over the visible horizon.

For example, during the Second World War the British battleship HMS Warspite was engaged with the Italian Royal Navy in the Mediterranean Sea. On 9 July 1940, Warspite achieved one of the longest range hits from a moving ship to a moving target, when a shell hit the Giulio Cesare at a range of approximately 24 km. To actually see the target vessel from this distance at sea, the curvature of the Earth would require you to be 45 m above sea level.

COMPONENTS

The projectile motion we consider in this course takes place in a two-dimensional plane.

The only force acting on the projectile is weight. By Newton's second law, this force is $-mg$ where m is the mass of the projectile and $g \approx 9.8 \text{ m s}^{-2}$ is the acceleration due to gravity. We can therefore conclude that:

- the horizontal component of the projectile motion has zero acceleration

- the vertical component of the projectile motion has constant acceleration which is $-g$.

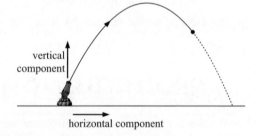

MODELLING PROJECTILE MOTION USING THE *suvat* EQUATIONS

In previous years you should have seen the *suvat* equations which can be applied to motion in a straight line with constant acceleration.

> Suppose an object moves displacement s in time t under constant acceleration a. If the initial velocity is u and the final velocity is v, then:
>
> - $v = u + at$
> - $s = \dfrac{u + v}{2}t$
> - $s = ut + \frac{1}{2}at^2$
> - $v^2 = u^2 + 2as$

Since the horizontal and vertical components of projectile motion are perpendicular, we can apply the *suvat* equations to each component separately.

Example 4

◀) **Self Tutor**

A diver jumps with vertical velocity 3.2 m s^{-1} from a diving platform 10 m above a pool.

a How long does it take for the diver to reach the top of her jump?

b How high does she jump?

c What is her velocity when she enters the water?

a The diver is at the top of her jump when her vertical velocity is zero.

The upwards direction is taken to be positive, so the acceleration due to gravity is negative.

Now $v = u + at$

$\therefore \; 0 = 3.2 + (-9.8)t$

$\therefore \; 9.8t = 3.2$

$\therefore \; t \approx 0.327 \text{ s}$

It takes about 0.327 seconds for the diver to reach the top of her jump.

b Displacement $s = ut + \frac{1}{2}at^2$

$\therefore \; s \approx 3.2 \times 0.327 + \frac{1}{2} \times (-9.8) \times 0.327^2$

$\approx 0.522 \text{ m}$ The diver jumps about 52.2 cm above the platform.

c The diver has vertical velocity zero at height 10.522 m above the pool.

Now $v^2 = u^2 + 2as$

$\therefore \; v^2 \approx 0^2 + 2(-9.8)(-10.522)$

$\therefore \; v^2 \approx 206.23$

$\therefore \; v \approx -14.36 \text{ m s}^{-1}$ {since $v < 0$}

The diver's vertical velocity is about 14.4 m s^{-1} downwards when she enters the water.

Example 5

◀) **Self Tutor**

A C-130 Hercules transport aircraft is travelling horizontally at 125 mph or 56 m s^{-1}. The pilot wishes to make a humanitarian supply drop in a remote area from a height of only 200 m. His GPS will tell him the horizontal distance from the intended target drop zone.

a At what horizontal distance from the target should the team drop the supplies?

b The average reaction time for humans is 0.25 seconds. If the team misses the drop by 0.25 seconds, how far will the supplies land from the target?

a We assume the supplies are dropped from a height of 200 m with initial horizontal velocity 56 m s^{-1} and initial vertical velocity 0 m s^{-1}.

First, consider the vertical motion.

$$s = ut + \frac{1}{2}at^2$$

$\therefore \; -200 = 0t + \frac{1}{2}(-9.8)t^2$

$\therefore \; 4.9t^2 = 200$

$\therefore \; t^2 \approx 40.82$

$\therefore \; t \approx 6.39 \text{ s}$ {since $t > 0$}

The drop will take ≈ 6.39 s to reach the ground.

In this time, the horizontal distance travelled is

$$s \approx 56 \times 6.39 \qquad \{\text{since horizontal velocity is constant}\}$$
$$\therefore \quad s \approx 358 \text{ m}$$

$\therefore$ the team should drop the supplies 358 m before the target.

b In the 0.25 s reaction time, the horizontal distance travelled is $s = 56 \times 0.25 = 14$ m

$\therefore$ the supplies will land 14 m past the target.

EXERCISE 15B.1

1 When a bouncy ball rebounds off the ground, it moves directly upwards for 3.5 seconds.

   a How high does it bounce?

   b What will its velocity be when it hits the ground again?

2 A particle is projected from a height h with horizontal velocity v_x. In each of the following cases, find:

   i the time taken to hit the ground

   ii the horizontal distance travelled in that time.

   a $h = 100$ m, $v_x = 20$ m s^{-1} **b** $h = 5$ m, $v_x = 0.7$ m s^{-1}

   c $h = 490$ m, $v_x = 2.5$ m s^{-1}

3 Jane kicks a stone off a cliff horizontally with velocity 1.7 m s^{-1}. The cliff is 58 m high. How far horizontally will the stone land from the base of the cliff?

4 A ball is thrown horizontally from a third floor balcony 9.4 m above the ground. It lands on the road 26 m horizontally from its point of release.

   a How long will it take for the ball to reach the ground?

   b Find the initial horizontal velocity of the ball.

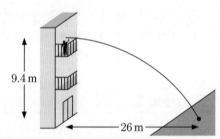

5 A particle is projected horizontally from a height h with velocity v_x. Find the value of v_x needed for the particle to hit a target s m away, given:

   a $h = 11.5$ m, $s = 3.0$ m   **b** $h = 0.5$ m, $s = 2.0$ m   **c** $h = 75$ m, $s = 0.5$ m

6 A stone is thrown horizontally from a bridge into a river. The bridge is 15 m high, and the stone lands 38 m horizontally from the bridge. Find the horizontal velocity.

7 Two friends stand on a bridge 10 m above a river, each holding a small stone. At the same time, one drops his stone while the other throws his stone horizontally as fast as possible.

   a Explain why the stones will hit the river at the same time.

   b Explain why the vertical velocity of each stone will always be the same.

   c Will the overall velocity of the stones be the same? Explain your answer.

Investigation   Projectile motion equations

Hubert is a professor of medieval history. He is taking part in a documentary on medieval weapons of war. He is about to film a catapult in action, but he wants to calculate where his projectile will land, to make sure he does not accidentally damage nearby houses.

Suppose Hubert's catapult will fire a projectile from the origin, to land on level ground. The initial horizontal velocity is u_x, and the initial vertical velocity is u_y. The acceleration due to gravity is $-g$.

What to do:

1 Consider the vertical component of the projectile motion.

a Explain why the height of the projectile at time t is given by $y(t) = u_y t - \frac{1}{2}gt^2$.

b Hence show that:

i the maximum height is reached when $t = \dfrac{u_y}{g}$

ii the maximum height reached is $y_{\max} = \dfrac{u_y^2}{2g}$

iii the projectile has total flight time $\dfrac{2u_y}{g}$.

c Explain why the vertical velocity at time t is given by $v_y(t) = u_y - gt$.

2 Consider the horizontal component of the projectile motion. Explain why:

a the horizontal velocity at time t is given by $v_x(t) = u_x$

b the horizontal displacement at time t is given by $x(t) = u_x t$

c the projectile will land distance $x_{\max} = \dfrac{2u_x u_y}{g}$ from the catapult.

3 Explain why:

a the initial speed of the projectile is given by $\sqrt{u_x^2 + u_y^2}$

b the initial angle of trajectory of the projectile is given by $\arctan \frac{u_y}{u_x}$.

VECTOR MODEL FOR PROJECTILE MOTION

We have seen how projectile motion can be separated into horizontal and vertical components. This leads us to consider a **vector model** for the motion.

We are free to choose the origin of the coordinate system, and we choose this to be the point at which the particle is released. We choose the x-axis to be horizontal in the direction of travel, and the y-axis to be vertically upwards.

Suppose a projectile P is fired from the origin at time 0.

The position, velocity, and acceleration of P can all be described using vectors.

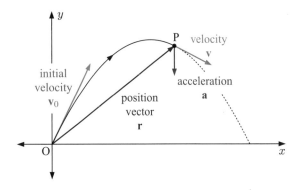

These vectors can all depend on time, so we can write them as **vector functions**:

- The **position vector** is $r(t) = \begin{pmatrix} x(t) \\ y(t) \end{pmatrix}$ —— horizontal component
  —— vertical component

DEMO

- The **velocity vector** is $v(t) = \begin{pmatrix} v_x(t) \\ v_y(t) \end{pmatrix}$ —— horizontal component
  —— vertical component

  and we define the **initial velocity vector** $v_0 = \begin{pmatrix} u_x \\ u_y \end{pmatrix} = \begin{pmatrix} v_x(0) \\ v_y(0) \end{pmatrix}$

- The **acceleration vector** is $a(t) = \begin{pmatrix} 0 \\ -g \end{pmatrix}$ which is a
  constant vector for projectile motion. There is no horizontal acceleration, and the vertical acceleration due to gravity is $-g$.

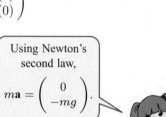

Using Newton's second law,

$$ma = \begin{pmatrix} 0 \\ -mg \end{pmatrix}.$$

In the **Investigation** on page **359**, you should have used the *suvat* equations to find formulae for the components of the displacement and velocity vectors:

> For a projectile P fired from the origin with initial velocity $v_0 = \begin{pmatrix} u_x \\ u_y \end{pmatrix}$:
>
> - the position vector is $r(t) = \begin{pmatrix} x(t) \\ y(t) \end{pmatrix} = \begin{pmatrix} u_x t \\ u_y t - \frac{1}{2}gt^2 \end{pmatrix}$
>
> - the velocity vector is $v(t) = \begin{pmatrix} v_x(t) \\ v_y(t) \end{pmatrix} = \begin{pmatrix} u_x \\ u_y - gt \end{pmatrix}$
>
> - the acceleration vector is $a = \begin{pmatrix} 0 \\ -g \end{pmatrix}$.

We can prove these results by integrating the components of the acceleration and applying the initial conditions.

EXERCISE 15B.2

1 Begin with the horizontal acceleration 0 m s^{-2}.
   By integrating twice and using the initial horizontal velocity u_x and displacement 0, prove that $v_x(t) = u_x$ and $x(t) = u_x t$.

2 Begin with the vertical acceleration $-g$ m s^{-2}.
   By integrating twice and using the initial vertical velocity u_y and displacement 0, prove that $v_y(t) = u_y - gt$ and $y(t) = u_y t - \frac{1}{2}gt^2$.

INITIAL SPEED AND ANGLE OF TRAJECTORY

It is quite common to be given the **initial speed** $s = |\mathbf{v}_0|$ and the **angle of trajectory** θ at which the projectile is launched.

Using right-angled triangle trigonometry, we find

$$\mathbf{v}_0 = \begin{pmatrix} u_x \\ u_y \end{pmatrix} = \begin{pmatrix} s\cos\theta \\ s\sin\theta \end{pmatrix}.$$

On the other hand, if we are given $\mathbf{v}_0 = \begin{pmatrix} u_x \\ u_y \end{pmatrix}$ we can calculate

$$s = |\mathbf{v}_0| = \sqrt{u_x^2 + u_y^2} \quad \text{and} \quad \tan\theta = \frac{u_y}{u_x}.$$

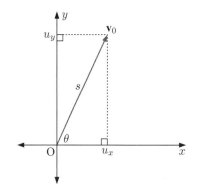

The process of splitting a vector into perpendicular components is called *resolving* the vector.

Example 6

◀) **Self Tutor**

A sniper fires a bullet from ground level at a rooftop target.

The initial velocity vector $\mathbf{v}_0 = \begin{pmatrix} 240 \\ 180 \end{pmatrix}$ m s^{-1}.

a Find the initial speed and angle of trajectory of the bullet.

b Find the maximum height the bullet will reach if it misses its target.

c The above initial velocity is quite realistic for a rifle. Do you think the bullet would actually reach the maximum height in **b**? Explain your answer.

a Initial speed $s = \sqrt{u_x^2 + u_y^2}$ Initial angle of trajectory $\theta = \arctan\frac{u_y}{u_x}$

$\qquad\qquad = \sqrt{240^2 + 180^2}$ $\qquad\qquad\qquad\qquad\qquad = \arctan\frac{180}{240}$

$\qquad\qquad = 300$ m s^{-1} $\qquad\qquad\qquad\qquad\qquad\qquad \approx 36.87°$

b The vertical velocity $v_y(t) = u_y - gt$

$\qquad\qquad\qquad\qquad = 180 - 9.8t$

The maximum height occurs when the vertical velocity is zero. This is when $t \approx 18.37$ s.

The height $y(t) = u_y t - \frac{1}{2}gt^2$

$\qquad\qquad\qquad = 180t - \frac{1}{2}(9.8)t^2$

$\therefore \quad y(18.37) \approx 180(18.37) - 4.9(18.37)^2$

$\qquad\qquad\quad \approx 1653$ m

The bullet would reach maximum height ≈ 1653 m.

c In reality, the bullet would be significantly slowed by air resistance, and would therefore have a much lower maximum height.

Example 7

◄)) **Self Tutor**

A projectile is fired from a cliff 21 m above the sea, at angle $30°$ to the horizontal and with speed 16 m s^{-1}. Assuming $g \approx 10 \text{ m s}^{-2}$, find:

a the initial velocity vector

b the time the projectile remains in the air

c how far the projectile travels horizontally

d the speed of the projectile when it hits the water.

a $\mathbf{v}_0 = \begin{pmatrix} s\cos\theta \\ s\sin\theta \end{pmatrix} = \begin{pmatrix} 16\cos 30° \\ 16\sin 30° \end{pmatrix} = \begin{pmatrix} 8\sqrt{3} \\ 8 \end{pmatrix} \text{ m s}^{-1}$

b The projectile remains in the air until it lands in the sea 21 m below the starting point.

$$y(t) = u_y t - \tfrac{1}{2}gt^2$$
$$\therefore \quad -21 \approx 8t - \tfrac{1}{2}(10)t^2$$
$$\therefore \quad 5t^2 - 8t - 21 \approx 0$$
$$\therefore \quad (5t+7)(t-3) \approx 0$$
$$\therefore \quad t \approx 3 \text{ s} \quad \{\text{since } t > 0\}$$

The projectile will spend 3 seconds in the air.

c $x(t) = u_x t$
$\therefore \quad x(3) = 8\sqrt{3} \times 3 \approx 41.57 \text{ m}$

The projectile will travel about 41.6 m horizontally.

d $v_x(t) = u_x = 8\sqrt{3} \text{ m s}^{-1}$
$v_y(t) = u_y - gt$
$\therefore \quad v_y(3) \approx 8 - (10)(3) \approx -22 \text{ m s}^{-1}$

$$\text{Speed} = \sqrt{v_x^2 + v_y^2}$$
$$\approx \sqrt{(8\sqrt{3})^2 + (-22)^2}$$
$$\approx 26 \text{ m s}^{-1}$$

EXERCISE 15B.3

1 A golf ball is hit across a level field. Its initial velocity vector is $\begin{pmatrix} 60 \\ 18 \end{pmatrix} \text{ m s}^{-1}$.

   a Find the initial speed and angle of trajectory of the golf ball.

   b Find the maximum height the golf ball will reach.

   c Do you think that in reality, the golf ball would reach this height? Explain your answer.

2 A particle is projected with initial velocity 120 m s^{-1} at an angle of $30°$.

   a Find the initial velocity vector.

   b Find the time at which the particle will reach its maximum height.

   c Find the maximum height reached by the particle.

3 A particle is projected from a point on a horizontal plane with initial velocity 11 m s^{-1} at an angle of $45°$. Assuming $g \approx 10 \text{ m s}^{-2}$, find the maximum height reached by the particle above the plane.

4 A projectile is fired from a cliff 6 m above the sea, at angle $30°$ to the horizontal and with speed of 14 m s^{-1}. Assuming $g \approx 10 \text{ m s}^{-2}$, find:

   a the time the projectile remains in the air   **b** how far the projectile travels horizontally

   c the speed of the projectile when it hits the water.

5 A rock is hurled from a castle battlement 30 m above the field below, at an angle $45°$ to the horizontal and with speed $5\sqrt{2}$ m s^{-1}. Assuming $g \approx 10$ m s^{-2}, how far from the castle will the projectile land?

6 A projectile is fired across level ground from the origin, with initial velocity $\mathbf{v}_0 = \begin{pmatrix} u_x \\ u_y \end{pmatrix}$.

Write an expression for:

   a the time when the maximum height is reached

   b the maximum height reached

   c the flight time until the projectile returns to the ground

   d the horizontal distance the projectile will travel.

7 A cannon can fire a cannonball across level ground with initial speed s and angle of trajectory θ.

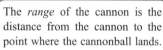

> The *range* of the cannon is the distance from the cannon to the point where the cannonball lands.

   a Write an expression for the horizontal distance the cannonball will travel.

   b Hence find the value of θ which will maximise the *range* of the cannon.

8 A football is kicked at 27 m s^{-1}. What is the maximum range it can travel?

9 Answer the **Opening Problem** on page **346**.

10   **a** Find the minimum muzzle velocity needed to send a projectile a horizontal distance of 24 km. Convert this to miles per hour.

     b What height will be reached by this projectile?

THE CARTESIAN EQUATION OF THE TRAJECTORY

For each time t, the position vector of the particle is given by $\mathbf{r}(t) = \begin{pmatrix} x \\ y \end{pmatrix} = \begin{pmatrix} u_x t \\ u_y t - \frac{1}{2}gt^2 \end{pmatrix}$.

If $u_x = 0$, the particle travels vertically along the line $x = 0$. We cannot write y in terms of x.

If $u_x \neq 0$, then $x = u_x t$ .... (1)

$\qquad\qquad\qquad y = u_y t - \frac{1}{2}gt^2$ .... (2)

Rearranging (1), $t = \dfrac{x}{u_x}$.

Substituting into (2), we find $y = \dfrac{u_y}{u_x} x - \dfrac{g}{2u_x^2} x^2$.

This is a quadratic equation in x, showing that the trajectory is a parabola.

EXERCISE 15B.4

1 A particle is projected with initial velocity $\mathbf{v}_0 = \begin{pmatrix} 0.5 \\ 2.5 \end{pmatrix}$.

Show that the trajectory of the projectile is $y = 5x - 2gx^2$.

2 A projectile is fired with initial speed s and angle of trajectory θ.

Prove that the Cartesian equation of its trajectory is $y = x \tan\theta - \dfrac{g}{2s^2}(1 + \tan^2\theta)x^2$.

3   **a** Prove that the motion of a projectile is symmetrical about a vertical line through the highest point.

     b Hence find the velocity with which a projectile, having been fired with initial velocity $\begin{pmatrix} u_x \\ u_y \end{pmatrix}$ across level ground, will impact with the ground.

Game   Castle crush

Click on the icon to practise your skills with projectiles!

Puzzle   Projectiles on a slope

When projectiles are used in war, it is rarely on level ground.

What to do:

1 Prove that if a projectile is fired up an inclined plane, the range is maximum when the initial angle of trajectory bisects the angle between the plane and the vertical.

2 Suppose a projectile is fired on an inclined plane. For a given initial speed of projection, there are two angles of trajectory such that a particular range is achieved.

Prove that these directions make equal angles with the bisector between the plane and the vertical.

Review set 15A

1 A particle P moves in a straight line with position from O given by $s(t) = 15t - \dfrac{60}{(t+1)^2}$ cm, where t is the time in seconds, $t \geqslant 0$.

  a Find velocity and acceleration functions for P's motion.

  b Describe the motion of P at $t = 3$ seconds.

  c For what values of t is the particle's speed increasing?

2 A particle moves in a straight line along the x-axis with position given by $x(t) = 3 + \sin 2t$ cm after t seconds.

  a Find the initial position, velocity, and acceleration of the particle.

  b Find the times when the particle changes direction during $0 \leqslant t \leqslant \pi$ seconds.

  c Find the total distance travelled by the particle in the first π seconds.

3 A particle has velocity function $v(t) = 2\cos 4t$ m s^{-1}.

  a Show that the particle oscillates between two points, and find the distance between them.

  b Given that $s\left(\frac{\pi}{12}\right) = 6$ m, determine $s\left(\frac{\pi}{6}\right)$.

  c Find the total distance travelled by the particle in the first π seconds.

4 The graph alongside shows the path of a projectile under gravity. Ignore air resistance. At each point A, B, and C, draw arrows to indicate the direction of the forces acting on the projectile.

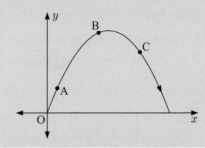

5 A boat travelling in a straight line has its engine turned off at time $t = 0$. Its velocity at time t seconds thereafter is given by $v(t) = \dfrac{100}{(t+2)^2}$ $\mathrm{m\,s}^{-1}$.

  a Find the initial velocity of the boat, and its velocity after 3 seconds.

  b Discuss $v(t)$ as $t \to \infty$.

  c Sketch the graph of $v(t)$ against t.

  d Find $\displaystyle\int_0^2 v(t)\,dt$, and interpret your answer.

  e From when the engine is turned off, how long will it take for the boat to travel 30 metres?

6 A particle is projected horizontally from a height h with velocity v_x. Using the *suvat* equations, find the value of v_x needed for the particle to hit a target s m away, given:

  a $h = 37$ m, $s = 12$ m
  b $h = 44.1$ m, $s = 27$ m

7 A stone is tossed horizontally from a bridge into a river. The bridge is 20 m high, and the stone lands 7 m horizontally from the bridge. Find the horizontal velocity.

8 A projectile is projected from a horizontal plane at an angle of $60°$ and an initial speed of 27 $\mathrm{m\,s}^{-1}$. Find:

  a the initial velocity vector
  b the time it remains in the air
  c the distance it travels horizontally
  d the maximum height it reaches.

9 A projectile is fired with initial velocity 75 $\mathrm{m\,s}^{-1}$ and angle of trajectory $50°$ to the horizontal. Find the maximum height reached by the particle.

10 A particle is projected with initial velocity $\mathbf{v_0} = \begin{pmatrix} 2 \\ 6 \end{pmatrix}$.

Show that the trajectory of the projectile is $y = 3x - \dfrac{g}{8}x^2$.

11 A projectile is fired across a level field with initial speed s and angle of trajectory θ to the horizontal. For what angle θ is the maximum height a half of the range?

Review set 15B

1 A particle P moves in a straight line with position given by $s(t) = 80e^{-\frac{t}{10}} - 40t$ m, $t \geqslant 0$ seconds.

  a Find the velocity and acceleration functions.

  b Find the initial position, velocity, and acceleration of P.

  c Sketch the graph of the velocity function.

  d Find the exact time when the velocity is -44 $\mathrm{m\,s}^{-1}$.

2 A cork bobs up and down in a bucket of water. The distance from the centre of the cork to the bottom of the bucket is given by $s(t) = 30 + \cos \pi t$ cm, $t \geqslant 0$ seconds.

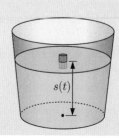

  a Find the cork's velocity at times $t = 0, \frac{1}{2}, 1, 1\frac{1}{2}$, and 2 seconds.

  b Find the time intervals when the cork is falling.

3 A feather is falling with velocity function $v(t) = -\frac{1}{24}t^3 - \frac{1}{12}t$ ms^{-1}. The feather is initially 2 metres above the ground.

   a Find the displacement function $s(t)$.

   b Find the time taken for the feather to reach the ground.

4 Tyson and Maurice are competing in a 100 m sprint. Tyson runs with speed $v_1(t) = 10(1 - e^{-1.25t})$ ms^{-1}, and Maurice runs with speed $v_2(t) = 10.5(1-e^{-t})$ ms^{-1}, for $t \geqslant 0$ seconds.

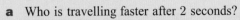

   a Who is travelling faster after 2 seconds?

   b Find $\displaystyle\int_0^5 v_1(t)\, dt$, and interpret your answer.

   c Find displacement functions $s_1(t)$ and $s_2(t)$ for each sprinter.

   d Who is winning the race after 3 seconds?

   e Show that Tyson completes the 100 m in approximately 10.8 seconds.

   f Who will win the race?

5 A particle is projected from a height h with horizontal velocity v_x. In each of the following cases, use the *suvat* equations to find:

   i the time taken to hit the ground

   ii the horizontal distance travelled in that time.

   a $h = 83$ m, $v_x = 1.5$ ms^{-1} **b** $h = 207$ m, $v_x = 0.83$ ms^{-1}

6 A particle is projected from a point on a horizontal plane with initial velocity 9 ms^{-1} at an angle of 60°. Find:

   a the initial velocity vector

   b the time at which the particle will reach its maximum height

   c the greatest height reached by the particle.

7 An aircraft is travelling horizontally at 250 kmh^{-1}. At a 3 km horizontal distance from a target, it releases a parcel of mail and supplies. The parcel hits the target. Assuming there is no air resistance, find the altitude at which the aircraft is flying.

8 A projectile is fired across a horizontal plane at an angle of 50° and an initial speed of 35 ms^{-1}. Find:

   a the length of time it remains in the air   **b** the distance it travels horizontally.

9 A particle is projected with initial velocity $\mathbf{v}_0 = \begin{pmatrix} 2 \\ 5 \end{pmatrix}$.

Show that the trajectory of the projectile is $y = \frac{5}{2}x - \frac{g}{8}x^2$.

10 Prove that if a projectile fired at 45° has range r, then the maximum height reached is $\frac{r}{4}$.

11 A particle is projected from a horizontal surface with initial speed s and angle of trajectory θ to the horizontal. Prove that the range of the projectile is $\dfrac{s^2}{g} \sin 2\theta$.

16

Mechanics: Forces and moments

Contents:

Opening problem

An aeroplane is travelling with constant speed and constant height, trailing an advertising banner behind it.

Things to think about:

a What forces are acting on the banner?

b What "holds up" the banner?

In previous years you should have studied **Newton's Laws of Motion**.

- **Newton's First Law** states:

  A body will remain at rest, or will continue to move with a constant velocity, unless an external force acts upon it.

  The banner remains in its state of constant velocity because there is no overall force acting upon it.

- **Newton's Second Law** tells us what happens when there is an overall resultant force acting on a body:

  For a body with constant mass m, a resultant force F causes an acceleration a which is proportional to the force.
  $$F = ma.$$

Forces are vector quantities, having magnitude and direction. Newton's Second Law can therefore be written more completely

as the **vector equation** $\mathbf{F} = m\mathbf{a}$.

The units of force are the Newton, abbreviated N.

- **Newton's Third Law** states:

  Every action has an equal and opposite reaction.

  This means that for any force the aeroplane exerts on the banner, the banner will exert a force with the same magnitude but the opposite direction on the aeroplane.

We say that the force $\mathbf{F}$ has magnitude $|\mathbf{F}|$ or F.

In this Chapter we consider how the forces on an object can be *resolved* into components which act perpendicular to one another.

A   FORCES AT A POINT

The **resultant** of two forces is a single force which has the same effect as the sum of the two component forces. The resultant can be calculated by vector addition.

Example 1   ◀)) **Self Tutor**

Find the resultant force when two forces **a** and **b** act on the particle P alongside. State its magnitude and direction.

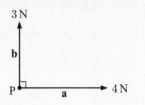

We assume **a** is horizontal, so **b** is vertical.

$$\therefore \quad \mathbf{a} = 4\mathbf{i} = \begin{pmatrix} 4 \\ 0 \end{pmatrix} \quad \text{and} \quad \mathbf{b} = 3\mathbf{j} = \begin{pmatrix} 0 \\ 3 \end{pmatrix}$$

$\therefore$ the resultant force $\mathbf{u} = \mathbf{a} + \mathbf{b} = (4\mathbf{i} + 3\mathbf{j})$ N   or $\begin{pmatrix} 4 \\ 3 \end{pmatrix}$ N.

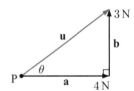

The resultant force has magnitude $|\mathbf{u}| = \sqrt{4^2 + 3^2} = 5$ N   and direction $\theta = \tan^{-1}\left(\frac{3}{4}\right) \approx 36.9°$ to the horizontal.

When more than two forces act on a point P, the resultant of the forces is found by repeated vector addition.

Example 2   ◀)) **Self Tutor**

Find the resultant when the forces $\mathbf{a} = (\mathbf{i} + 2\mathbf{j})$ N, $\mathbf{b} = (3\mathbf{i} - \mathbf{j})$ N,   and $\mathbf{c} = -2\mathbf{i}$ N   act on a point P.

There are three forces acting on the point P.

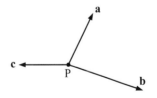

The resultant force is $\mathbf{u} = \mathbf{a} + \mathbf{b} + \mathbf{c}$
$$= (\mathbf{i} + 2\mathbf{j}) + (3\mathbf{i} - \mathbf{j}) + (-2\mathbf{i})$$
$$= (1 + 3 - 2)\mathbf{i} + (2 - 1)\mathbf{j}$$
$$= (2\mathbf{i} + \mathbf{j}) \text{ N}$$

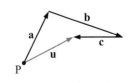

EXERCISE 16A

1 Find the resultant force **u** when the forces **a** and **b** act on the particle P. State its magnitude and direction.

a

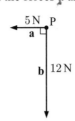

b

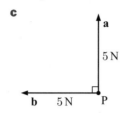

c

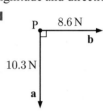

d

e $a = -2i$ N, $b = 3j$ N

f $a = -1.8j$ N, $b = -2.4i$ N

2 Find the resultant force when the following forces act on the particle P:

a $a = (i - 2j)$ N, $b = 3j$ N

b $a = (-i + 3j)$ N, $b = (2i + j)$ N

c $p = (3i - j)$ N, $q = (-2i + j)$ N

d $u = 4i$ N, $v = (-i + 2j)$ N, $w = 3j$ N

e $a = (2i - 5j)$ N, $b = (-2i + 2j)$ N, $c = (-i + 3j)$ N

Example 3   ◄)) **Self Tutor**

Two forces **a** and **b** act on a particle P as shown. Calculate the resultant force.

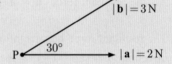

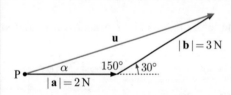

Using the cosine rule,

$$|u|^2 = |a|^2 + |b|^2 - 2|a||b|\cos 150°$$
$$= 2^2 + 3^2 - 2(2)(3)\left(-\frac{\sqrt{3}}{2}\right)$$
$$= 13 + 6\sqrt{3}$$
$$\therefore \ |u| \approx 4.84 \text{ N}$$

Using the sine rule, $\dfrac{\sin \alpha}{3} \approx \dfrac{\sin 150°}{4.84}$

$$\therefore \ \sin \alpha \approx \frac{3 \times \frac{1}{2}}{4.84} \approx 0.310$$

$$\therefore \ \alpha \approx 18.1°$$

The resultant force is 4.84 N in the direction 18.1° to the left of **a**.

---

3 Two forces **a** and **b** act on a particle P as shown. Calculate the resultant force.

a

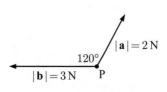

b

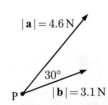

c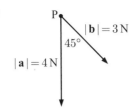

d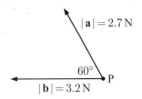

4 Two forces **a** and **b** act on a particle P with magnitudes $|\mathbf{a}| = 133$ N and $|\mathbf{b}| = 156$ N. Find the angle between **a** and **b** so that the resultant satisfies $|\mathbf{u}| = 205$ N.

5 Calculate the resultant force acting on the point P:

a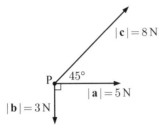

b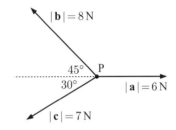

B EQUILIBRIUM

A particle is in **equilibrium** if there is no resultant force acting upon it.

For a particle P to be in equilibrium, the forces acting upon it must sum to zero.

For example, for the forces   we can draw the closed triangle   .

Being in "equilibrium" does not mean that a particle is not moving. Zero resultant force means zero acceleration, and therefore constant velocity.

Example 4   ◄‹) **Self Tutor**

The forces $\mathbf{a} = (\mathbf{i} + 2\mathbf{j})$ N, $\mathbf{b} = (2\mathbf{i} - \mathbf{j})$ N, and $\mathbf{c} = (-3\mathbf{i} - \mathbf{j})$ N act on a point P. Show that P is in equilibrium.

The resultant force is $\mathbf{u} = \mathbf{a} + \mathbf{b} + \mathbf{c}$
$$= (\mathbf{i} + 2\mathbf{j}) + (2\mathbf{i} - \mathbf{j}) + (-3\mathbf{i} - \mathbf{j})$$
$$= (1 + 2 - 3)\mathbf{i} + (2 - 1 - 1)\mathbf{j}$$
$$= \mathbf{0} \text{ N}$$

The overall force is zero, so P is in equilibrium.

Example 5

🔊 **Self Tutor**

Three forces **a**, **b**, and **c** act on a particle which is moving with constant velocity.

$$\mathbf{a} = \begin{pmatrix} 3 \\ -1 \end{pmatrix}, \quad \mathbf{b} = \begin{pmatrix} -2 \\ b \end{pmatrix}, \quad \text{and} \quad \mathbf{c} = c\begin{pmatrix} 1 \\ -1 \end{pmatrix}.$$

Find the values of b and c.

Since the particle is in equilibrium, $\mathbf{a} + \mathbf{b} + \mathbf{c} = \mathbf{0}$

$$\therefore \quad \begin{pmatrix} 3 \\ -1 \end{pmatrix} + \begin{pmatrix} -2 \\ b \end{pmatrix} + c\begin{pmatrix} 1 \\ -1 \end{pmatrix} = \begin{pmatrix} 0 \\ 0 \end{pmatrix}$$

$$\therefore \quad \begin{pmatrix} 3 - 2 + c \\ -1 + b - c \end{pmatrix} = \begin{pmatrix} 0 \\ 0 \end{pmatrix}$$

$$\therefore \quad c + 1 = 0 \quad \text{and} \quad b - 1 - c = 0$$
$$\therefore \quad c = -1 \quad \text{and} \quad b - 1 + 1 = 0$$
$$\therefore \quad b = 0$$

EXERCISE 16B

1 Forces $\mathbf{a} = (\mathbf{i} + 3\mathbf{j})$, $\mathbf{b} = (-2\mathbf{i} - 4\mathbf{j})$, and $\mathbf{c} = (\mathbf{i} + \mathbf{j})$ act on a particle. Show that the particle is in equilibrium.

2 **a** Two forces **a** and **b** act on an object. Write an expression for **b** such that the object travels with constant velocity.

    b Three forces **a**, **b**, and **c** act on an object. Write an expression for **c** such that the object travels with constant velocity.

3 The forces $\mathbf{a} = \begin{pmatrix} 3 \\ k^2 \end{pmatrix}$, $\mathbf{b} = \begin{pmatrix} -1 \\ -k \end{pmatrix}$, and $\mathbf{c} = \begin{pmatrix} -2 \\ -6 \end{pmatrix}$ acting on a point have resultant force zero.

Find the possible values of k. Illustrate your answer.

4 Three forces acting in 3-dimensional space are $\mathbf{a} = 2\mathbf{i} + \mathbf{j} + z\mathbf{k}$, $\mathbf{b} = x\mathbf{i} + 2\mathbf{j} + \mathbf{k}$, and $\mathbf{c} = 3\mathbf{i} + y\mathbf{j} + \mathbf{k}$. Find x, y, and z so that the three forces are in equilibrium.

5 Three forces act on a particle in equilibrium:

$$\mathbf{a} = \begin{pmatrix} 1 \\ 4 \end{pmatrix}, \quad \mathbf{b} = b\begin{pmatrix} 1 \\ -3 \end{pmatrix}, \quad \text{and} \quad \mathbf{c} = \begin{pmatrix} c \\ 2 \end{pmatrix}$$

Find the values of b and c.

6 A particle has three forces acting on it: $\mathbf{a} = \begin{pmatrix} 3 \\ 1 \end{pmatrix}$, $\mathbf{b} = \begin{pmatrix} 2 \\ -2 \end{pmatrix}$, and **c**.

What force must **c** be, given that the particle:

    a is stationary   **b** moves with constant velocity?

7 Forces $\mathbf{a} = \begin{pmatrix} 2x \\ y \end{pmatrix}$, $\mathbf{b} = \begin{pmatrix} 0 \\ x \end{pmatrix}$, and $\mathbf{c} = \begin{pmatrix} y \\ 2 \end{pmatrix}$ are in equilibrium. Find x and y.

8 An object is being pulled by three ropes. One rope exerts the force $\mathbf{a} = (2\mathbf{i} + 2\mathbf{j})$, while another exerts the force $\mathbf{b} = (-2\mathbf{i} + 2\mathbf{j})$. What force must the third rope exert so that the object remains in equilibrium?

Example 6

◀ﬔ Self Tutor

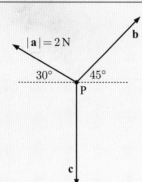

Find $|\mathbf{b}|$ and $|\mathbf{c}|$ such that P is in equilibrium.

Since P is in equilibrium, $\mathbf{a} + \mathbf{b} + \mathbf{c} = \mathbf{0}$.

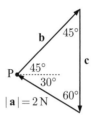

Using the sine rule,

$$\frac{|\mathbf{a}|}{\sin 45°} = \frac{|\mathbf{b}|}{\sin 60°} = \frac{|\mathbf{c}|}{\sin 75°}$$

$$\therefore \quad |\mathbf{b}| = \sin 60° \times \frac{|\mathbf{a}|}{\sin 45°} = \left(\frac{\sqrt{3}}{2}\right) \times \frac{2}{\left(\frac{1}{\sqrt{2}}\right)} = \sqrt{6} \text{ N}$$

and $\quad |\mathbf{c}| = \sin 75° \times \frac{|\mathbf{a}|}{\sin 45°} = \sin 75° \times \frac{2}{\left(\frac{1}{\sqrt{2}}\right)} \approx 2.73 \text{ N}$

9 Three forces act on a particle as shown in the diagram.

Lami's Theorem states that the particle is in equilibrium if:

$$\frac{|\mathbf{a}|}{\sin \alpha} = \frac{|\mathbf{b}|}{\sin \beta} = \frac{|\mathbf{c}|}{\sin \gamma}$$

Prove Lami's Theorem.

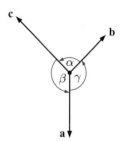

10 Find $|\mathbf{b}|$ and $|\mathbf{c}|$ such that P is in equilibrium:

a

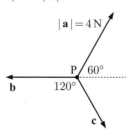

b

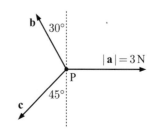

11 A boat is held at anchor against the tide exerting force 600 N on the bearing 120°, and the wind exerting force 250 N on the bearing 210°.

Find the magnitude and direction of the force **F** exerted by the anchor on the boat.

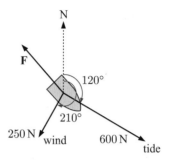

C RESOLVING FORCES

We can *resolve* forces into perpendicular components in the same way as we did for displacement and velocity in kinematics.

HORIZONTAL AND VERTICAL DIRECTIONS

It is most common for us to resolve forces into horizontal and vertical directions. In this case we use **i** for the unit vector in the horizontal direction, and **j** for the unit vector in the vertical direction.

> For a force **F** acting at angle θ measured anticlockwise from the positive horizontal direction:
>
> • the horizontal component of **F** is
> $\mathbf{F}_x = |\mathbf{F}| \cos \theta \, \mathbf{i}$
> • the vertical component of **F** is
> $\mathbf{F}_y = |\mathbf{F}| \sin \theta \, \mathbf{j}$.
>
> The components can be recombined using $\mathbf{F} = \mathbf{F}_x + \mathbf{F}_y$.

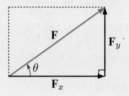

Example 7   ◀)) **Self Tutor**

A mass of 5 kg is suspended from an inextensible string. What horizontal force would maintain equilibrium with the string at an angle of 45° to the horizontal?

The object has weight $5g$ N acting vertically downwards.

Let **T** be the tension in the string and let **F** be the horizontal force.

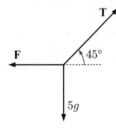

> *An inextensible string does not stretch.*

Together these forces are in equilibrium.

Now **T** has horizontal component $|\mathbf{T}| \cos 45° \mathbf{i} = \frac{1}{\sqrt{2}} |\mathbf{T}| \, \mathbf{i}$

and vertical component $|\mathbf{T}| \sin 45° \mathbf{j} = \frac{1}{\sqrt{2}} |\mathbf{T}| \, \mathbf{j}$

Using the horizontal components, $\frac{1}{\sqrt{2}}|\mathbf{T}| - |\mathbf{F}| = 0$ .... (1)

Using the vertical components, $\frac{1}{\sqrt{2}}|\mathbf{T}| - 5g = 0$ .... (2)

Comparing (1) and (2), $|\mathbf{F}| = 5g$.

EXERCISE 16C.1

1 A mass sitting on a table is connected to a string under tension. The string is pulling on the mass to the right and upwards, as shown alongside.

Write down the horizontal and vertical components of the tension.

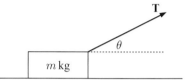

2 Resolve each vector into the form $a\mathbf{i} + b\mathbf{j}$:

a

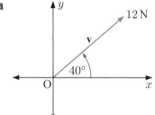

b

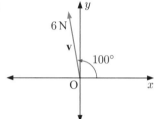

c

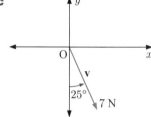

3 A mass m is suspended from an inextensible string.

a What horizontal force would maintain equilibrium with the string at an angle of θ to the horizontal? Give your answer in terms of m and θ.

b What force is needed:

i when $\theta = 90°$ **ii** as $\theta \to 0°$?

COMPONENTS PERPENDICULAR AND PARALLEL TO A PLANE

We do not need to always resolve vectors horizontally and vertically. We can choose any two *perpendicular* directions.

For example, if an object is sitting on an inclined plane, it may be useful to resolve forces parallel to the slope and perpendicular to the slope.

Consider an object with mass m kg which is sitting on a plane at angle θ to the horizontal.

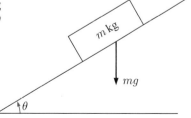

The weight mg N downwards can be resolved into the components $\mathbf{F}_{\text{parallel}}$ and $\mathbf{F}_{\text{perpendicular}}$, where:

- $|\mathbf{F}_{\text{parallel}}| = mg \sin \theta$
- $|\mathbf{F}_{\text{perpendicular}}| = mg \cos \theta$

Example 8   ◀◎ Self Tutor

A block of mass 5 kg sits on a smooth slope angled 30° to the horizontal. A string parallel to the slope stops the block from sliding and maintains equilibrium. Find:

a the contact force **R** between the block and the slope

b the tension **T** in the string.

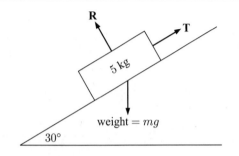

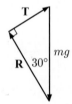

Contact forces are always perpendicular to the surface.

Since the block is in equilibrium, the forces must sum to zero.

a $|\mathbf{R}| = mg \cos 30°$

$\qquad = 5 \times 9.8 \times \frac{\sqrt{3}}{2}$

$\qquad \approx 42.4 \text{ N}$

∴   the contact force is 42.4 N perpendicular to the surface.

b $|\mathbf{T}| = mg \sin 30°$

$\qquad = 5 \times 9.8 \times \frac{1}{2}$

$\qquad = 24.5 \text{ N}$

∴   the tension is 24.5 N up the slope.

EXERCISE 16C.2

1 In each situation, find the component of weight which is parallel and which is perpendicular to the surface:

a

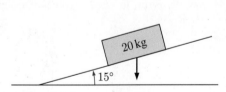

b

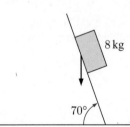

2

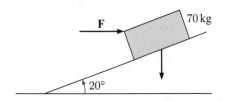

A 70 kg block sits on a smooth slope, which is inclined 20° to the horizontal. To prevent it from slipping, a horizontal force **F** is applied.

Find the magnitude of **F** necessary to maintain the equilibrium.

3 A block of mass 8.2 kg sits on a smooth slope angled 45° to the horizontal. A string parallel to the slope stops the block from sliding and maintains equilibrium. Find:

    a the contact force **R** between the block and the slope

    b the tension **T** in the string.

4 A 10 kg mass hangs from the end of a light rigid beam. The beam is supported by a rope which makes an angle of 20° to the horizontal.

Find the tension **T** in the rope.

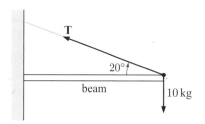

5

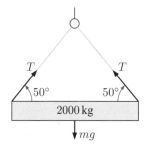

A steel beam is supported by rope at each end. Each rope makes an angle of 50° to the horizontal beam, as shown. Find the value of T, the magnitude of the tension in each rope.

6 On 8 June 2017, British wingsuit pilot Fraser Corsan jumped from a plane at an altitude of 10 832 m. His flight lasted about nine minutes and he covered a horizontal distance of 30.6 km. His mass, including all equipment, was 95 kg.

As he fell, his speed increased, but so did the air resistance forces. He eventually stopped accelerating, even though gravity was still acting downwards.

The diagram below shows the forces acting on Corsan when he reached his *terminal velocity*.

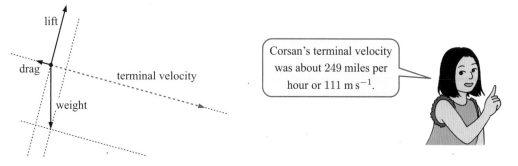

Corsan's terminal velocity was about 249 miles per hour or 111 m s^{-1}.

Assume his glide angle is constant throughout the flight, and that for most of the time he maintained a constant velocity.

    a Calculate the glide angle.

    b Calculate the force of drag, which is parallel to the direction of travel.

    c Calculate the lift force generated by his wingsuit.

7 A rock climber of mass 70 kg ties himself to the rock using two symmetrical anchors at A and B. The angle between these is θ.

He falls off and is hanging in equilibrium, putting his full weight on P where the anchors join. Suppose the tension in each rope has magnitude T.

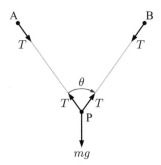

  a Find T as a function of the angle θ.

  b For what values of θ does the tension exceed 4900 N?

  c Comment on your answer to **b** by comparing it with the static weight generated by a mass of 500 kg.

Example 9   ◀ Self Tutor

A block of mass 6.3 kg is released from rest on a smooth slope angled 40° to the horizontal. Find the acceleration of the block.

Suppose **F** is the resultant force. It must be parallel to the surface, since this is the direction the block must move in.

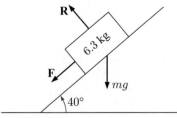

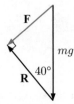

$|\mathbf{F}| = mg \sin 40°$

By Newton's Second Law, $|\mathbf{F}| = m\,|\mathbf{a}|$

$$\therefore \ |\mathbf{a}| = \frac{|\mathbf{F}|}{m} = g \sin 40° \approx 6.299 \text{ m s}^{-2}$$

The block accelerates down the slope at approximately 6.30 m s^{-2}.

---

8 A 9.1 kg mass is released from rest on a smooth slope angled 35° to the horizontal. Find the acceleration of the block.

9 A block of mass m kg is released from rest on a smooth slope at angle θ to the horizontal. If the resulting acceleration is 1 m s^{-2}, find the value of θ.

D CONNECTED PARTICLES

In the **Opening Problem**, the aeroplane is connected to the banner. To make the situation easier to understand, we consider the forces on each object separately.

We make the following assumptions:

- The plane P and banner B are modelled as points with masses m_P and m_B respectively.
- The string connecting them has no mass and does not stretch.
- The plane and banner move horizontally with constant velocity.

The forces on the plane are:

- weight due to gravity vertically downwards
- thrust from the engines which we choose to be horizontal to the right
- lift from the wings perpendicular to the thrust, so vertically upwards
- drag (air resistance) which opposes the thrust, so horizontally to the left
- tension in the string along the line of the string.

The forces on the banner are:

- weight due to gravity vertically downwards   • drag (air resistance) horizontally to the left
- tension in the string along the line of the string.

We illustrate all of these forces on the diagram alongside. We assume the tension in the string has magnitude T.

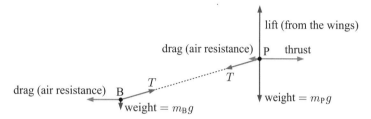

Since the plane and banner move with constant velocity, the system is in equilibrium and all forces balance.

Consider the forces on the plane:

- The weight and vertical component of the tension must be equal and opposite to the lift.
- The horizontal component of the tension and the drag on P must be equal and opposite to the thrust from the engine.

Consider the forces on the banner:

- The banner is "held up" by the vertical component of tension. This must be equal and opposite to the weight.
- The drag on B must be equal to the horizontal component of tension.

Example 10   ◀》 Self Tutor

A 15 kg mass A lies on a smooth horizontal table. A light inextensible string attached to the mass passes over a smooth pulley and carries a mass B of 15 kg suspended freely at its other end. The string between the mass on the table and the pulley makes an angle of $35°$ to the horizontal. The system is kept in equilibrium by a horizontal force applied to the mass on the table. Find the magnitude of this force, and of the other forces involved.

Suppose the tension in the string has magnitude T, the contact force has magnitude R, and the force holding A has magnitude F.

Resolving vertically on B, $T = 15g = 147$ N.

Resolving horizontally on A, $F = T \cos 35° \approx 120$ N.

Resolving vertically on A, $R + T \sin 35° = 15g$

$$\therefore \quad R = 147 - 147 \sin 35° \approx 63 \text{ N}$$

EXERCISE 16D

1 The windsock at an airport is tied to a post with light inextensible strings.
A steady breeze keeps the windsock in equilibrium.

   a Draw a diagram showing the forces acting on the windsock.

   b Describe each of these forces, and explain how the windsock is "held up".

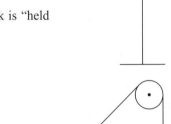

2 A 12 kg mass A lies on a smooth horizontal table. A light inextensible string attached to the mass passes over a smooth pulley and carries a mass B of 12 kg suspended freely at its other end. The string between the mass on the table and the pulley makes an angle of $45°$ to the horizontal. The system is kept in equilibrium by a horizontal force **F** applied to the mass on the table. Find the magnitude of this force, and of the other forces involved.

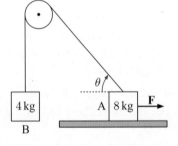

3 An 8 kg mass A lies on a smooth horizontal table. A light inextensible string attached to the mass passes over a smooth pulley, and carries a 4 kg mass B suspended freely at its other end. A force **F** of 20 N is applied horizontal to mass A as shown.

The string between the mass on the table and the pulley makes an angle θ to the horizontal.

   a Draw a diagram showing the forces acting on A and B.

   b Suppose $\theta = 20°$. Find the magnitude and direction of the force acting on A. Explain your answer.

   c Find the value of θ so that the system will be in equilibrium.

4

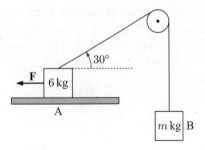

A force **F** of 40 N is applied horizontal to mass A, to keep the system in equilibrium.

Find the value of m.

Activity   Belaying

In rock climbing, **belaying** is a process where one climber uses a rope under tension to stop their climbing partner from falling.

When climbing at a climbing wall, or a small outdoor crag, the *belayer* is the person standing at the bottom who is responsible for the safety of the climber. The rope passes over a pulley above the climber to protect the climber if they fall off, and to safely lower them back to the ground at the end of the climb.

Suppose Bob is belaying the climber Claire. Claire, who has mass 70 kg, falls off and hangs in equilibrium. Bob is standing away from the wall, and is able to brace himself against a solid object to stop him moving towards the wall.

We can represent this situation using a mathematical diagram with forces labelled. The rope passes from Bob (B) over a solid (frictionless) anchor at P and supports Claire (C). Both B and C are in equilibrium.

What to do:

1 What causes the horizontal force **c** acting on Bob, and why does it act to the left?

2 Bob only weighs 60 kg, so he is lighter than Claire. Why can Bob hold Claire up without being lifted off the ground?

3 Find the largest angle θ for which Bob remains on the ground.

4 To increase his margin of safety, should Bob move towards or away from the wall? Explain your answer.

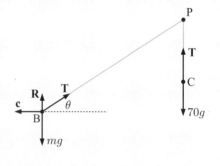

E ▌ FRICTION

If you slide a book across a table, its velocity will decrease until the book comes to rest. The acceleration which brings the book to rest is caused by a force, and this force is called *friction*.

Coulomb's Law of Friction says that friction is always parallel to the line of contact and opposite to the direction of motion.

The force of friction **F** is limited to μR where R is the magnitude of the **normal reaction force R** which is the contact force perpendicular to the surface.

If **F** has magnitude F then $F \leqslant \mu R$.

The number $\mu \geqslant 0$ is called the **coefficient of friction** and is determined by the nature of the two surfaces.

- If $\mu = 0$ then there is no friction and we describe the surfaces as **smooth**.
- If $\mu > 0$ then friction occurs and we describe the surfaces as **rough**.

Imagine a mass sitting on a rough horizontal surface with coefficient of friction $\mu > 0$. It is connected to a horizontal string under tension.

The vertical forces on the mass are its weight mg due to gravity acting vertically downwards, and a reaction force **R** acting on the mass. The mass is in equilibrium, so **R** is equal and opposite to the weight. Its magnitude $R = mg$.

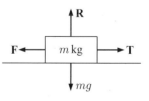

The horizontal forces on the mass are the tension **T** with magnitude T, and the friction **F** with magnitude F.

- If $T < \mu R = \mu mg$ then $F = T$ and the mass remains in equilibrium.
- If $T = \mu R = \mu mg$ then $F = T$ and the mass remains in limiting equilibrium.
- If $T > \mu R = \mu mg$ then $F = \mu R$ and the mass experiences a resulting force $T - \mu R > 0$ acting to the right. The mass will accelerate as a result of the resultant force.

Example 11

◀⫶) **Self Tutor**

A mathematics book with mass 1.4 kg rests on a horizontal table. Calculate the maximum frictional force between the book and the surface if the coefficient of friction $\mu = 0.2$.

Coulomb's Law of Friction states that $F \leqslant \mu R$ where $R = mg$.

$\therefore$ the maximum frictional force is

$$\mu mg = 0.2 \times 1.4 \times 9.8$$
$$= 2.744 \text{ N}$$

F and R are both measured in N, so μ is dimensionless.

EXERCISE 16E

1 A block with mass 7 kg rests on a horizontal surface. Calculate the maximum frictional force between the surfaces for:

    a $\mu = 0.1$ **b** $\mu = 4$

2 In 1992-93, Mike Stroud and Sir Ranulph Fiennes made the first unsupported crossing of the Antarctic continent. Each man pulled a sledge weighing about 225 kg containing food and other supplies needed for the 1500 mile journey.

    If the coefficient of friction between snow and the sledge was 0.1, calculate the force needed to move one of the sledges.

3 A block of mass 50 kg sits on a rough horizontal surface. The coefficient of friction is 0.2, and the block is subjected to a horizontal force of 250 N. Calculate the resulting acceleration.

Example 12

◀⫶) **Self Tutor**

A 2 kg mass is sitting on a rough surface with coefficient of friction $\mu = 0.55$. It is connected to a string under tension 12 N which pulls the mass to the right and upwards at an angle $40°$ to the horizontal. Find the maximum friction force.

Let the maximum friction force **F** have magnitude F, the reaction force **R** have magnitude R, and the tension **T** have magnitude $T = 12$ N.

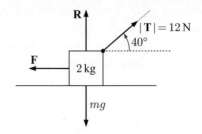

Resolving forces vertically, $R + T \sin 40° = mg$
$$\therefore\ R = mg - T \sin 40°$$

The maximum friction force is

$$F = \mu R$$
$$= \mu(mg - T\sin 40°)$$
$$= 0.55(2 \times 9.8 - 12 \sin 40°)$$
$$\approx 6.54 \text{ N}$$

The friction force is reduced by the vertical component of the tension.

4 A 3.3 kg mass is sitting on a rough surface with coefficient of friction $\mu = 0.18$. It is connected to a string under tension 6.8 N which pulls the mass to the right and upwards at an angle 16° to the horizontal. Find the maximum friction force.

5 A 1.14 kg mass is sitting on a table with coefficient of friction $\mu = 0.37$. It is connnected to a string which pulls the mass to the left and upwards at an angle 50° to the horizontal. Find the tension necessary to move the mass.

6 An object of mass 6 kg is pulled by a string along a horizontal rough surface. The string is at an angle of 25° to the horizontal, and the tension in the string is 15 N.

   a Given that the object moves at a constant speed, show that the coefficient of friction between the object and the surface is $\mu \approx 0.26$.

   b Suppose the tension in the string is increased to 20 N. Find the resulting acceleration of the object.

   c Find the tension required to produce an acceleration of 1 m s^{-2}.

7 A mass of m kg is sitting on a table with coefficient of friction μ. It is connnected to a string which pulls the mass to the left and upwards at an angle $\theta°$ to the horizontal.

   a Find the tension necessary to move the mass.

   b Show that the tension needed to move the mass is minimised when $\theta = \arctan \mu$.

   c For a 100 kg mass and $\mu = 0.4$, find the minimum tension needed to move the mass.

Example 13   ◀) **Self Tutor**

A rough slope with $\mu = 0.1$ is angled at 50° to the horizontal. A block of mass 2 kg sits on the slope and is released from rest.

Will the friction keep the block in equilibrium? If not, find the resulting acceleration.

Let F be the magnitude of friction **F**, which is always parallel to the surface.

Let R be the magnitude of the normal reaction force **R** between the block and the slope.

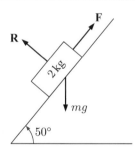

Contact forces are always perpendicular to the surface.

The reaction force balances the component of weight perpendicular to the slope.

∴ $R = mg \cos 50° \approx 12.6$ N

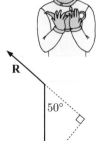

Friction will oppose the motion of the body, so in this case friction will act up the slope. The maximum value of friction is $\mu R \approx 0.1 \times 12.6 \approx 1.26$ N.

The component of weight parallel to the slope is
$mg \sin 50° = 2 \times 9.8 \times \sin 50° \approx 15.0$ N.

Since 15.0 N > 1.26 N the block will not remain in equilibrium under friction.

Instead, the overall force on the block is 15.0 N − 1.26 N ≈ 13.8 N down the slope.

Using Newton's Second Law, the block will accelerate at $\dfrac{13.8}{2} \approx 6.9$ m s^{-2}.

8 A rough slope with $\mu = 0.45$ is angled at $48°$ to the horizontal. A block of mass 1.6 kg sits on the slope and is released from rest.

Will the friction keep the block in equilibrium? If not, find the resulting acceleration.

9 A rough slope is angled at $30°$ to the horizontal. A block of mass 5 kg sits on the slope. Friction stops the block sliding and maintains equilibrium.

Find the minimum value of μ .

10

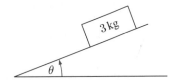

A rough slope has $\mu = 0.5$. A block of mass 3 kg sits on the slope, and is released from rest.

Find the minimum angle θ of the slope that would cause the block to slide down the slope.

11 An object of mass 5 kg is on a rough surface inclined $35°$ to the horizontal. The coefficient of friction of the surface is 0.2. The object is connected to a string which pulls the object with tension **T** parallel to the slope.

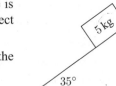

  a Suppose there is no tension on the string. Find the acceleration of the object as it slides down the slope.

  b Find the magnitude of the tension required so that:

   i   the object is in equilibrium, but is at the point of sliding *down* the slope

   ii the object is in equilibrium, but is at the point of being pulled *up* the slope.

  c Suppose the string has tension 60 N.

   i   Find the acceleration of the object as it is pulled up the slope.

   ii Find the distance the object has travelled after 3 seconds.

12 A rough slope with coefficient of friction μ is angled θ to the horizontal. A block of mass m kg sits on the slope and horizontal force **P** is applied to push the block up the slope.

Find the minimum magnitude of **P** required for the block to move, giving your answer in terms of m, μ, and θ.

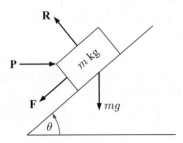

F | MOMENTS

So far we have assumed that bodies are points and that all forces act through the point.

We now consider **rigid bodies** which have size and shape. Rigid bodies do not change shape when forces are applied, but they can **rotate**.

Rigid bodies rotate about an **axis**. In this course all our examples are two dimensional, so the axis of rotation is perpendicular to the plane of the page. In this situation it makes sense to think of rotation about a point.

We measure a rotation using an **angle**.

- An anticlockwise rotation is given a positive angle.

- A clockwise rotation is given a negative angle.

Discussion

A rigid object is held in place at a point P, but it is free to rotate about this point.

A force **F** is applied to a point X on the edge of the object.

What do you expect to happen if the force is:

- $\mathbf{F}_1$ pointing directly at P
- $\mathbf{F}_2$ pointing perpendicular to XP
- $\mathbf{F}_3$ at an angle θ to XP?

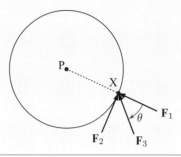

When a force **F** is applied to a particular point on a rigid body, the force acts along a line given by its direction.

The **moment** of **F** about any point P on a rigid body is a measure of the rotational effect about point P.

The **size of the moment of F about P** is the magnitude of **F** multiplied by the perpendicular distance from the line of the force to P.

$$|\text{moment of } \mathbf{F} \text{ about P}| = |\mathbf{F}| \times l$$

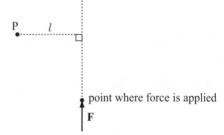

point where force is applied

The **sign of the moment of F about P** is:

- positive if the resulting rotation is anticlockwise
- negative if the resulting rotation is clockwise.

For the situation in the previous **Discussion**, $l = \text{XP} \sin \theta$

$\therefore$ the moment of **F** about P $= \text{XP} \sin \theta \, |\mathbf{F}|$

For the case $\mathbf{F}_1$, $\theta = 0°$ and the moment of $\mathbf{F}_1$ about P is 0.

For the case $\mathbf{F}_2$, $\theta = 90°$ and the moment of $\mathbf{F}_2$ about P takes its maximum value $\text{XP} \, |\mathbf{F}|$.

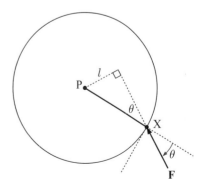

Notice that if a force acts on a point, there can be no moment about that point, since the perpendicular distance is always zero.

Example 14

◀⟩ **Self Tutor**

Consider a light rigid rod ABC.

A force of 6 N is applied to the point A as shown.

Find the moment about the point:

a A   b B   c C

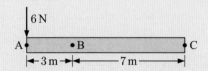

|Moment| = force × perpendicular distance

a Moment = 6 N × 0 m = 0 Nm

b Moment = 6 N × 3 m = 18 Nm

c Moment = 6 N × 10 m = 60 Nm

The rotation will be anticlockwise, so the moments are positive.

Example 15

◀⟩ **Self Tutor**

Consider the forces acting on the light rigid rod AB shown alongside.

Find the total moment about the point P.

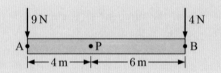

The force at A operates anticlockwise

∴ it has moment about P = 9 N × 4 m

$$= 36 \text{ Nm}$$

The force at B operates clockwise

∴ it has moment about P = −4 N × 6 m

$$= -24 \text{ Nm}$$

The total moment about P = 36 Nm − 24 Nm

$$= 12 \text{ Nm}$$

The total moment is positive, resulting in an anticlockwise rotation.

EXERCISE 16F

1 Consider a light rigid rod ABC. A force of 5 N is applied to the point A as shown.

Find the moment at:

a A   b B   c C.

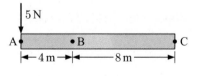

2 Consider a light rigid rod PQR. A force of 8 N upwards is applied to the point P as shown.

Find the moment at:

a P   b Q   c R.

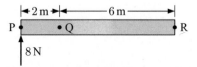

3 Consider the forces acting on a light rigid rod AB.
Find the total moment about the point P.

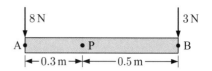

4 The diagram below shows two forces acting on a rigid rod AB. Find the total moment about the point P.

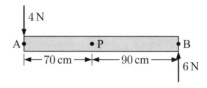

Example 16

◀) **Self Tutor**

A rectangular block ABCD stands on a rough table. A force is applied to point D as shown.

Find the moment about the point B.

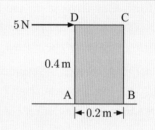

The force is applied to point D, but the perpendicular distance from the line of the force to point B is CB = 0.4 m.

The force will generate clockwise rotation

∴ the moment about B = −| force | × perpendicular distance

$$= -5 \text{ N} \times 0.4 \text{ m}$$

$$= -2 \text{ N m}$$

---

5 A rectangular block ABCD stands on a rough surface. A force is applied to point D as shown.
Find the moment about point B.

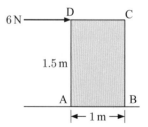

6 A triangular block ABC stands on a rough surface. A force of 6 N is applied to point C as shown.

   a Find the moment at point A.

   b Explain why the moment at any point on the base will be the same as at point A.

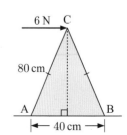

G   COUPLES

Consider a light rigid rod AB with length l and midpoint P.

- Suppose identical forces are applied at each end.

  The overall force on the rod is 2**F**.

  Taking moments about P, each force contributes $\dfrac{l}{2}|\mathbf{F}|$ Nm,

  but the directions of rotation are opposite.

  $\therefore$ the total moment is zero.

  We have an overall translational force but no turning effect.

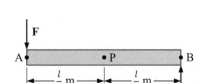

- Suppose forces with equal size but opposite sign are applied at each end.

  The overall force on the rod is **F** − **F** = **0**.

  $\therefore$ there is no *translational* force on the rod.

  Taking moments about P, each force contributes $\dfrac{l}{2}|\mathbf{F}|$ Nm,

  and the directions of rotation are both anticlockwise.

  $\therefore$ the total moment about P is $l|\mathbf{F}|$ Nm.

  We have a turning effect but no translational force. We call this a *couple*.

> A **couple** is when the forces applied to a rigid body result in a turning effect
> but no translational effect.

EQUILIBRIUM

> A rigid body is in equilibrium if:
> - there is no overall translational force
> - there is no couple about any point.

Investigation   Equilibrium

Following the definition of equilibrium for a rigid body above, it appears that we might need to take
moments about *every* point in order to establish equilibrium.

What to do:

1 Consider two equal but opposite forces acting on a rod AB of length l as shown below.
   Suppose P is the point x m from A.

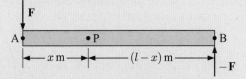

   Show that the total moment about point P is $l|\mathbf{F}|$.

2   Now suppose the rod is extended so that P is the other side of B.

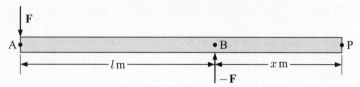

Find the total moment about point P.

3   For a rigid body in equilibrium, does the total moment about P depend on the location of P? Explain your answer.

If a rigid body is in equilibrium, the total moment about any point is zero.

Example 17
◀) **Self Tutor**

Three given forces act on a rod as shown below. Apply a single additional force **u**, perpendicular to the rod and a distance x m from A, to maintain equilibrium.

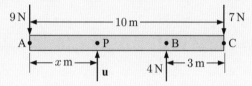

At equilibrium there must be no overall translational force.

Resolving forces vertically with upwards being positive, we find $|\mathbf{u}| + 4 - 9 - 7 = 0$
$$\therefore \quad |\mathbf{u}| = 12 \text{ N}$$

At equilibrium there is also no moment.

Taking moments about A, we find
$$0 + |\mathbf{u}| \times x + 4 \times 7 - 7 \times 10 = 0$$
$$\therefore \quad |\mathbf{u}| \times x = 42 \text{ N m}$$
$$\therefore \quad x = \tfrac{42}{12} = 3.5 \text{ m}$$

> Since the rod is in equilibrium, the total moment about *any* point is zero.

A force of 12 N upwards should be applied at the point 3.5 m from A.

EXERCISE 16G

1   Given the solution to **Example 17** above, evaluate the total moment about each of the points P, B, and C. Hence confirm our choice to use point A in the solution was as valid as any other.

2   Two given forces act on a rod as shown below. Apply a single additional force **F**, perpendicular to the rod and a distance x m from A, to maintain equilibrium.

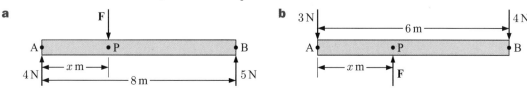

3 Three given forces act on a rod as shown below. Apply a single additional force **F**, perpendicular to the rod and a distance x m from A, to maintain equilibrium.

a

4 N — 10 m — 3 N
A • B • P • C
2.6 m → 2 N   F
x m

b

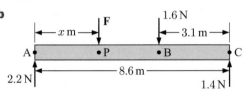

c

— 8 m —
3 m → 6 N   2 N
A • B • P • C
1 N — x m — F

d

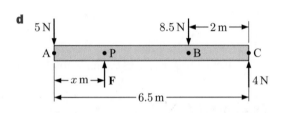

4 Each rod is in equilibrium. Find $|\mathbf{F}|$ and x.

a

— 3 m — F
4 N — x m — 6 N

b

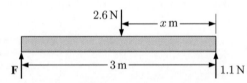

c

F
1 N — x m — 1 N 1 m — x m — 1.5 N

d

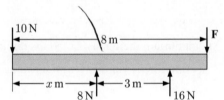

5 Forces act on a rod in equilibrium. If the forces act in the locations and directions shown, which of the following scenarios are possible? Explain your answers.

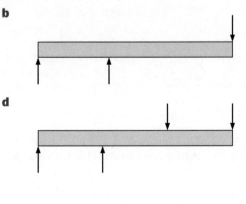

6 A man has mass 78 kg. He stands 3 m from one end of a bridge which is 18 m long and in equilibrium. Find the forces **a** and **b** needed to support each end of the bridge.

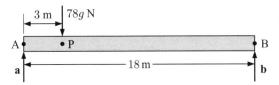

7 A uniform rod AB has length 5 m and mass 8 kg. The rod is placed on a support at S, which is 2 m from A. An object of mass m kg is placed at A to balance the rod in equilibrium.

> If a uniform rod has mass, the force of its weight is applied at the *centre* of the rod.

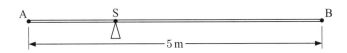

a Find the moment of the weight of the rod about S.

b Find the value of m.

c Find the magnitude of the reaction force exerted by the support on the rod.

Historical note

The Greek mathematician **Archimedes of Syracuse** is supposed to have said "Give me a place to stand and a lever long enough and I will move the Earth".

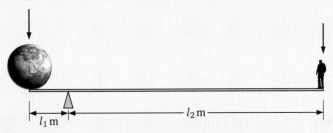

Suppose Archimedes stood at the end of a very long, strong seesaw, and that the Earth was at the other end.

The concept of weight due to the gravity of the Earth is meaningless here, so we suppose there is some other gravitational influence with coefficient g^*.

The mass of the Earth is approximately 5.9×10^{26} kg. We suppose Archimedes had mass 80 kg, and that he managed to position the Earth at distance $l_1 = 0.1$ mm from the pivot of the seesaw.

To "move the Earth", $l_2 \times 80 \times g^* > l_1 \times 5.9 \times 10^{26} \times g^*$

$$\therefore \ l_2 > \frac{1 \times 10^{-4} \times 5.9 \times 10^{26}}{80}$$

$$\therefore \ l_2 > 7.375 \times 10^{20} \text{ m}$$

$$\therefore \ l_2 > 77\,955 \text{ light years}$$

Archimedes would have required a very long lever indeed!

H   SYNOPTIC PROBLEMS

In this Section we consider situations which bring together forces, resolving forces, friction, and moments. These provide a very useful collection of tools for solving a wide variety of problems.

Example 18   ◀) Self Tutor

A light uniform ladder AB of length l rests in equilibrium with end B at an angle θ to a rough horizontal floor, and with end A against a smooth vertical wall. Suppose a man of mass m kg climbs up the ladder to a point P which is x m from B. Find the minimum coefficient of friction necessary to maintain equilibrium. Leave your answer in terms of θ and other constants.

Since the vertical wall is smooth, there is no vertical frictional force at A. The only force at A is a reaction force with magnitude R_A perpendicular to the wall.

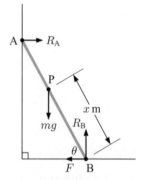

The man at P has weight mg N downwards.

We suppose the frictional force at B has magnitude F and that the reaction force at B has magnitude R_B.

Resolving forces horizontally we find $R_A = F$.

Resolving forces vertically we find $R_B = mg$.

Taking moments about B,

$$-R_A \times l \sin \theta + mg \times x \cos \theta = 0 \qquad \text{\{the system is in equilibrium\}}$$

$$\therefore \quad R_A = \frac{mgx \cos \theta}{l \sin \theta}$$

$$\therefore \quad F = \frac{mgx}{l \tan \theta} \qquad \{F = R_A\}$$

If μ is the coefficient of friction then $F \leqslant \mu R_B$. Hence $\mu \geqslant \dfrac{F}{R_B} = \dfrac{mgx}{l \tan \theta} \times \dfrac{1}{mg} = \dfrac{x}{l \tan \theta}$.

EXERCISE 16H

1 A ladder of length 5 m leans against a wall as shown alongside. A person of mass m kg stands on the middle of the ladder. In addition to the weight of the person, there are four other forces with magnitudes R_A, R_B, F_A, and F_B. The system is in equilibrium.

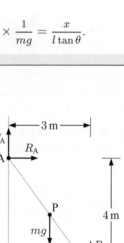

   a Describe the four forces other than weight.

   b Discuss whether the position of the person on the ladder affects:

      i translational force

      ii moments.

   c Generate a system of four equations in the four variables R_A, R_B, F_A, F_B by:

      i resolving horizontal forces   **ii** resolving vertical forces

      iii taking moments about A.   **iv** taking moments about B.

   d Show that this system of equations does not have a unique solution.

e Suppose the wall is smooth, so $F_A = 0$.

    i Find the other forces involved.

    ii Show that the horizontal coefficient of friction must satisfy $\mu \geqslant \frac{3}{8}$.

f Now suppose the floor is smooth, so $F_B = 0$. Can the ladder balance against the wall?

2 A castle wall is inclined at $10°$ to the vertical. An invading army uses a light ladder of length l to cross the moat.

A person with mass m kg climbs up the ladder, and is at point P which is distance x along the ladder.

The system is in equilibrium. Assume the castle wall is smooth.

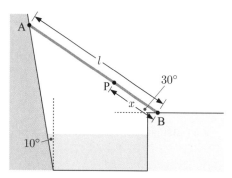

  a Draw a diagram labelling the forces in this problem, and give a description of each force.

  b Generate a system of four equations by resolving horizontal and vertical forces, and taking moments about A and B.

  c Find the minimum coefficient of friction at B necessary to maintain equilibrium.

  d Does the position of the person on the ladder affect whether the ladder will slip? Explain your answer.

3

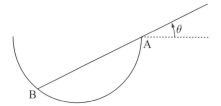

A uniform smooth rod of mass m and length $2l$ rests partly inside and partly outside a fixed smooth hemispherical bowl of radius a. The rod is inclined at angle θ to the horizontal. Assume there is no friction.

  a Show that $2a \cos 2\theta = l \cos \theta$.

  b Hence find the reaction force at A, the contact point with the rim, in terms of m, g, l, and a.

4 A light square table with side length 2 m sits on horizontal ground. It has four identical vertical legs A, B, C, and D at the corners. We choose a coordinate system so that looking from above, A$(-1, -1)$, B$(-1, 1)$, C$(1, 1)$, and D$(1, -1)$.

A mass m is placed on the top of the table at M(x, y).

Let the vertical forces supporting the table have magnitudes a, b, c, and d at A, B, C, and D respectively.

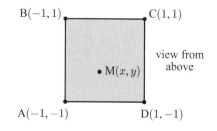

  a Explain why $a + b + c + d = mg$ .... (1).

  b By taking moments on the table about the line $y = -1$, explain why $2b + 2c = (1 + y)mg$ .... (2).

  c By taking moments on the table about the line $x = -1$, explain why $2c + 2d = (1 + x)mg$ .... (3).

  d Can the system (1) – (3) be uniquely solved to find the forces through each leg?

  e Show that taking moments on the table about the lines $y = 1$ and $x = 1$ produces equations which are consistent with the system (1) – (3).

Since the problem is 3-dimensional, we take moments about a *line*, the axis of rotation, rather than a point.

Review set 16A

1 Find the resultant force **u** when the forces **a** and **b** act on the particle P. State its magnitude and direction.

a

a ← 1.7 N — P

0.9 N

↓ **b**

b

↑ **b**

7 N

a ← 6 N — P

2 Calculate the resultant force acting on the point P.

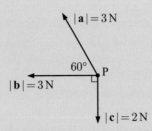

$|\mathbf{a}| = 3$ N

60°   P

$|\mathbf{b}| = 3$ N

$|\mathbf{c}| = 2$ N

3 The forces $\mathbf{a} = \begin{pmatrix} 1 \\ 3 \end{pmatrix}$, $\mathbf{b} = \begin{pmatrix} 2 \\ -2 \end{pmatrix}$, and $\mathbf{c} = \begin{pmatrix} -3 \\ -1 \end{pmatrix}$ act on a point P.

Show that P is in equilibrium.

4 A particle travels with constant velocity under the forces $\begin{pmatrix} -1 \\ 4 \end{pmatrix}$, $\begin{pmatrix} 3 \\ -2 \end{pmatrix}$, and **F**.

Find the force **F**.

5

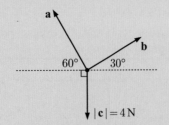

$|\mathbf{c}| = 4$ N

Three forces acting on a particle are in equilibrium. Find the magnitude of forces **a** and **b**.

6 **a** Find the component of weight which is:
     i parallel to the plane
     ii perpendicular to the plane.

  b Hence find the contact force **R** which the slope exerts on the mass.

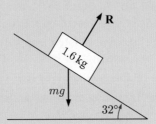

7 A 12.3 kg mass is released from rest on a smooth slope angled 19° to the horizontal. Find the acceleration of the block.

8 A block of mass 6.2 kg sits on a rough horizontal surface with coefficient of friction 0.18. The block is subjected to a horizontal force of 16 N to the right. Calculate the resulting acceleration.

9 A 5 kg mass A lies on a smooth horizontal table. A light inextensible string attached to the mass passes over a smooth pulley, and carries a 3 kg mass B suspended freely at its other end. A force **F** of 12 N is applied horizontal to mass A as shown.

The string between the mass on the table and the pulley makes an angle θ to the horizontal.

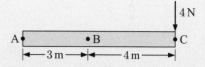

   a Draw a diagram showing the forces acting on A and B.

   b Suppose $\theta = 16°$. Find the magnitude and direction of the force acting on A. Explain your answer.

   c Find the value of θ so that the system will be in equilibrium.

10 A 5.2 kg mass is sitting on a rough surface with coefficient of friction $\mu = 0.4$. It is connected to a string under tension 4 N which pulls the mass to the right and upwards at an angle $9°$ to the horizontal. Find the maximum friction force.

11 Consider a light rigid rod ABC. A force of 4 N is applied to point C as shown.

Find the moment at:

   a A   **b** B   **c** C.

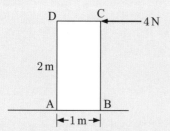

12

A rectangular block ABCD stands on a rough surface. A force of 4 N is applied to the rectangular block at C as shown. Find the moment about the point A.

13 Two forces act on a rod at points A and B. Apply a single additional force **F**, perpendicular to the rod, so that the forces are in equilibrium. Find the magnitude of **F** and the position where it needs to be applied.

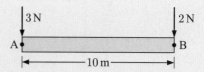

14 A ladder of length 2.5 m leans against a smooth wall as shown alongside. A child with mass 40 kg is standing at P, two thirds of the way up the ladder.

   a Describe the forces whose magnitudes are marked as R_A, R_B, and F_B.

   b Find R_A, R_B, and F_B.

   c Find the minimum coefficient of friction needed to keep the system in equilibrium.

   d Do you think it is reasonable to assume the wall is smooth?

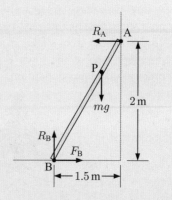

15 The forces acting on the rod shown below are in equilibrium. Find $|\mathbf{F}|$ and x.

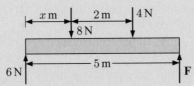

Review set 16B

1 Find the resultant force when the following forces act on the particle P:

   a $\mathbf{a} = (3\mathbf{i} + 5\mathbf{j})$ N, $\mathbf{b} = (-4\mathbf{i} + \mathbf{j})$ N

   b $\mathbf{x} = (-2\mathbf{i} + \mathbf{j})$ N, $\mathbf{y} = 3\mathbf{j}$ N, $\mathbf{z} = (\mathbf{i} - 5\mathbf{j})$ N

2 Two forces $\mathbf{a}$ and $\mathbf{b}$ act on a particle P. Calculate the resultant force.

   a 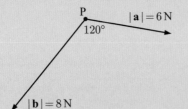   **b**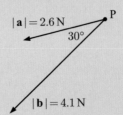

3 Find the values of a and b given that the forces $\mathbf{a} = \begin{pmatrix} 2a \\ -1 \end{pmatrix}$, $\mathbf{b} = \begin{pmatrix} b \\ a \end{pmatrix}$, and $\mathbf{c} = \begin{pmatrix} 3 \\ -4 \end{pmatrix}$ are in equilibrium.

4 Resolve each vector into the form $a\mathbf{i} + b\mathbf{j}$:

   a 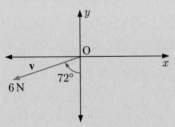   **b**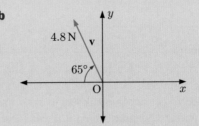

5 A skier with mass 95 kg (including equipment) is resting on a slippery slope inclined 18° to the horizontal. To prevent her from sliding, she holds on to a tree. Find the magnitude of the force $\mathbf{F}$ the tree exerts on her in order to maintain equilibrium.

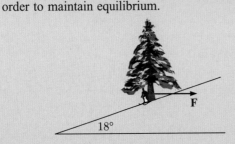

6 A cargo ship is pulled with constant velocity up a river with current
$\mathbf{c} = \begin{pmatrix} 0 \\ -4 \end{pmatrix}$. The two tug boats pull with forces $\mathbf{T_1} = T\begin{pmatrix} -1 \\ 1 \end{pmatrix}$
and $\mathbf{T_2} = T\begin{pmatrix} 1 \\ 1 \end{pmatrix}$. Find the value of T.

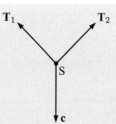

7 A block of mass m kg is released from rest on a smooth slope at angle $\theta°$ to the horizontal. If the resulting acceleration is 3 m s^{-2}, find the value of θ.

8

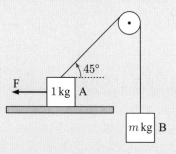

A force **F** of 20 N is applied horizontal to mass A, to keep the system in equilibrium. Find the value of m.

9 A box with mass 2.6 kg rests on a horizontal floor. Calculate the maximum frictional force between the surfaces for:

   a $\mu = 0.2$ **b** $\mu = 1.6$

10 A rough slope is angled at $18°$ to the horizontal. A block of mass 4.2 kg sits on the slope. Friction stops the block sliding and maintains equilibrium.
Find the minimum value of μ.

11 Two objects P and Q of mass m kg and $5m$ kg are connected by a taut string as shown. P lies on a slope with coefficient of friction 0.2, and Q lies on a smooth slope. P is initially held at rest.
Show that, once P is released, it will accelerate towards
the pulley at $\frac{g}{60}(24\sqrt{3} - 5) \text{ m s}^{-2}$.

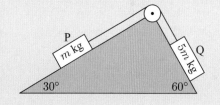

12 Consider the forces acting on a rigid rod shown below.

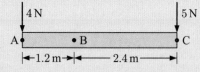

Find the total moment at point B.

13 Two forces act on a rod at points A and B. Apply a single additional force **F**, perpendicular to the rod, so that the forces are in equilibrium. Find the magnitude of **F** and the position where it needs to be applied.

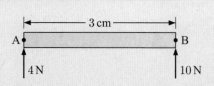

14 The forces acting on the rod shown below are in equilibrium. Find $|\mathbf{F}|$ and x.

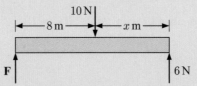

15 Two people are sitting on a bench 2 m long. Person A has mass 81 kg and sits 0.5 m from the left end. Person B has mass 62 kg and sits 0.7 m from the right end. Find the forces acting on the legs at each end of the bench.

17

Probability

Contents:

Opening problem

Sarah runs a performing arts school which offers piano lessons and acting classes. She has a total of 35 students, of which 15 take her acting class, and 27 learn piano.

Things to think about:

a How can we represent this scenario in:
  i a Venn diagram   **ii** a two-way table?

b Suppose one of Sarah's students is randomly chosen. What is the probability that the selected student:
  i learns piano given that they take the acting class
  ii does not take the acting class given that they learn piano?

From previous years, you should be familiar with the following concepts in **probability**:

Outcome:	a possible result from one trial of a chance experiment
Sample space:	the set $\mathcal{E}$ of all possible outcomes of an experiment, it is sometimes referred to as the **universal set**
Event:	a set of outcomes with a particular property
Complementary events:	Two events are complementary if *exactly one* of the events *must* occur. If A is an event, then A' is its complement and $P(A) + P(A') = 1$.
Independent events:	Events are independent if the occurrence of each of them does not affect the probability that the others occur. If two events A and B are independent, then $P(A \cap B) = P(\text{both } A \text{ and } B) = P(A) \times P(B)$.
Addition law:	For any two events A and B: $P(A \cup B) = P(\text{either } A \text{ or } B \text{ or both})$ $= P(A) + P(B) - P(A \cap B)$
Mutually exclusive events:	Two events A and B are mutually exclusive if $P(A \cap B) = 0$. For mutually exclusive events, $P(A \cup B) = P(A) + P(B)$.

You should be familiar with using diagrams to help calculate probabilities involving independent and mutually exclusive events, including:

- **2-dimensional grids**
- **tree diagrams**
- **Venn diagrams**.

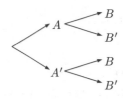

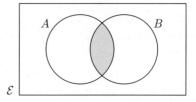

In this Chapter, we will extend these ideas to events which may not necessarily be independent.

A | TWO-WAY TABLES

Two-way tables are tables which compare **two** categorical variables.

For example, the two-way table alongside summarises the numbers of boys and girls in two classes, A and B.

In this case, the variables are *class* and *gender*.

	Boys	Girls
Class A	14	12
Class B	11	15

We can use two-way tables to estimate probabilities.

there are 11 boys in class B

Example 1

◀ **Self Tutor**

People exiting a new ride at a theme park were asked whether they liked or disliked the ride. The results are shown in the two-way table alongside.

	Child	Adult
Liked the ride	55	28
Disliked the ride	17	30

Use this table to estimate the probability that a randomly selected person who went on the ride:

a liked the ride
b is a child *and* disliked the ride
c is an adult *or* disliked the ride.

We extend the table to include totals for each row and column.

	Child	Adult	Total
Liked the ride	55	28	83
Disliked the ride	17	30	47
Total	72	58	130

a 83 out of the 130 people surveyed liked the ride.

$\therefore$ P(liked the ride) $\approx \frac{83}{130} \approx 0.638$

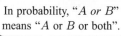

In probability, "*A or B*" means "*A or B or both*".

b 17 of the 130 people surveyed are children who disliked the ride.

$\therefore$ P(child *and* disliked the ride) $\approx \frac{17}{130} \approx 0.131$

c $28 + 30 + 17 = 75$ of the 130 people are adults or people who disliked the ride.

$\therefore$ P(adults *or* disliked the ride) $\approx \frac{75}{130} \approx 0.577$

EXERCISE 17A

1 The types of ticket used to gain access to a basketball game were recorded as people entered the stadium. The results are shown alongside.

	Adult	Child
Season ticket holder	1824	779
Not a season ticket holder	3247	1660

a What was the total attendance for the match?

b One person is randomly selected to sit on the home team's bench. Find the probability that the person selected is:

i a child   **ii** not a season ticket holder   **iii** an adult season ticket holder.

2 A sample of adults in a suburb were surveyed about their current employment status and their level of education. The results are summarised in the table below.

	Employed	Unemployed
Attended university	225	7
Did not attend university	197	18

Estimate the probability that the next randomly chosen adult:

a attended university
b did not attend university and is currently employed
c is unemployed
d is unemployed or attended university.

3 Students at a school were asked whether they played a sport.

- In the junior school, 131 students played a sport and 28 did not.
- In the middle school, 164 students played a sport and 81 did not.
- In the senior school, 141 students played a sport and 176 did not.

a Copy and complete this table.

	Junior	Middle	Senior	Total
Sport				
No sport				
Total				

b Find the probability that a randomly selected student:
  i plays sport
  ii plays sport and is in the junior school
  iii does not play sport and is in middle school or higher.

4 A small hotel in London has kept a record of all room bookings made for the year. The results are summarised in the two-way table.

Find the probability that a randomly selected booking was:

	Single	Double	Family
Peak season	225	420	98
Off-peak season	148	292	52

a in the peak season
b a single room in the off-peak season
c a single or a double room
d during the peak season or a family room.

B   CONDITIONAL PROBABILITY

Suppose a class has 30 students. 22 students study Italian (I), 15 students study French (F), and 8 students study both languages. We can represent this information on a **Venn diagram** as follows:

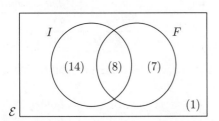

The universal set $\mathcal{E}$ is represented by a rectangle. Events are represented by circles.

Suppose a student is randomly selected from the class and it is found that the student studies French.

We can determine the probability that this student also studies Italian. We call this a **conditional probability** because it is the probability of I occurring on the *condition* that F has occurred.

$\text{P}(I \text{ given that } F \text{ has occurred}) = \frac{8}{15}$ ← number of students that study Italian and French
    ← number of students that study French

For events A and B, we use the notation "$A \mid B$" to represent the event "A *given that B has occurred*".

$$\text{P}(A \mid B) = \frac{n(A \cap B)}{n(B)}$$

"∩" means "and".

Example 2   ◀) **Self Tutor**

The two-way table alongside summarises the types and colours of cars made by a car manufacturer.

One car is randomly selected for quality inspection. Find the probability that the selected car is:

a white given that it is a sedan

b a hatchback given that it is white.

		Car colour	
		White	Other
Vehicle type	Sedan	15	32
	Hatchback	20	41

a Of the 47 sedans, 15 are white.

∴ $\text{P}(\text{white} \mid \text{sedan}) = \frac{15}{47} \approx 0.319$

b Of the 35 white cars, 20 are hatchbacks.

∴ $\text{P}(\text{hatchback} \mid \text{white}) = \frac{20}{35} \approx 0.571$

		Car colour		
		White	Other	Total
Vehicle type	Sedan	15	32	47
	Hatchback	20	41	61
	Total	35	73	108

If the outcomes in each of the events are equally likely, notice that:

$$\frac{n(A \cap B)}{n(B)} = \frac{\frac{n(A \cap B)}{n(\mathcal{E})}}{\frac{n(B)}{n(\mathcal{E})}} = \frac{\text{P}(A \cap B)}{\text{P}(B)}$$

This gives us the **conditional probability formula**: $\text{P}(A \mid B) = \dfrac{\textbf{P}(A \cap B)}{\textbf{P}(B)}$

EXERCISE 17B

1 Find $\text{P}(A \mid B)$ if:

    a $\text{P}(A \cap B) = 0.1$ and $\text{P}(B) = 0.4$ **b** $\text{P}(A) = 0.3$, $\text{P}(B) = 0.4$, and $\text{P}(A \cup B) = 0.5$

    c A and B are mutually exclusive.

2 Suppose A and B are independent events.

    a Write an expression for $\text{P}(A \cap B)$.   **b** Hence write an expression for $\text{P}(A \mid B)$.

    c Interpret your answer to **b**.

3 The probability that it is cloudy on a particular day is 0.4. The probability that it is cloudy *and* rainy on a particular day is 0.2. Find the probability that it will be rainy on a day when it is cloudy.

4 In a group of 50 students, 40 study Mathematics, 32 study Physics, and each student studies at least one of these subjects.

    a Use a Venn diagram to find how many students study both subjects.

    b If a student from this group is randomly selected, find the probability that he or she:

      i studies Mathematics but not Physics

      ii studies Physics given that he or she studies Mathematics.

5 Out of 40 boys, 23 have dark hair, 18 have brown eyes, and 26 have dark hair, brown eyes, or both.

    a Copy and complete the two-way table.

    b One of the boys is selected at random. Determine the probability that he has:

      i dark hair and brown eyes

      ii brown eyes given that he has dark hair.

Eye colour

		Brown	Other	Total
Hair colour	Dark			
	Fair			
	Total			

6 50 hikers participated in an orienteering event during summer. 23 were sunburnt, 22 were bitten by ants, and 5 were both sunburnt and bitten by ants.

    a Construct a two-way table to display this information.

    b Determine the probability that a randomly selected hiker:

      i escaped being bitten

      ii was bitten or sunburnt (or both)

      iii was bitten given that he or she was sunburnt

      iv was sunburnt given that he or she was not bitten.

Example 3   ◀) Self Tutor

In a town, 25% of the residents own a cat, 55% own a dog, and 30% do not own either animal.

    a Draw a Venn diagram to describe the situation.

    b Find the probability that a randomly selected resident:

      i owns a cat given that they own a dog

      ii does not own a dog given that they own a cat.

a
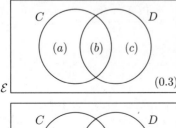

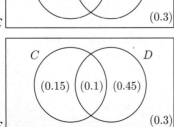

Let C represent residents who own a cat and D represent residents who own a dog.

We are given:

$$a + b = 0.25 \quad \{\text{cat owners}\}$$
$$b + c = 0.55 \quad \{\text{dog owners}\}$$
$$a + b + c = 0.7 \quad \{\text{own at least one}\}$$

$\therefore \quad c = 0.7 - 0.25 = 0.45$

and so $b = 0.1$ and $a = 0.15$

b   **i** $P(C \mid D) = \dfrac{P(C \cap D)}{P(D)}$

$= \dfrac{0.1}{0.55}$

≈ 0.182

ii $P(D' \mid C) = \dfrac{P(D' \cap C)}{P(C)}$

$= \dfrac{0.15}{0.25}$

$= 0.6$

7 400 families were surveyed. It was found that 90% had a TV set and 80% had a computer. Every family had at least one of these items. One of the families is randomly selected, and it is found that they have a computer. Find the probability that they also have a TV set.

8 In a certain town three newspapers are published. 20% of the population read A, 16% read B, 14% read C, 8% read A and B, 5% read A and C, 4% read B and C, and 2% read all 3 newspapers. A person is selected at random. Use a Venn diagram to help determine the probability that the person reads:

a none of the papers

b at least one of the papers

c exactly one of the papers

d A or B (or both)

e A, given that the person reads at least one paper

f C, given that the person reads either A or B or both.

Discussion

Consider events A and B. Is it meaningful to talk about $P(A \mid B)$ when B is:

- a certain event
- an impossible event?

C   THE PRODUCT RULE OF PROBABILITY

By rearranging the conditional probability formula, we obtain: $\mathbf{P(A \cap B) = P(B)P(A \mid B)}$

In words, we can interpret this as "P(B **then** A) = P(B) $\times$ P(A given that B has occurred)".

This is known as the **product rule of probability**, and it can be extended to any number of events.

For example:

For **any** events A, B, and C:

$\text{P}(A \text{ then } B \text{ then } C) = \text{P}(A \cap B \cap C)$

$= \text{P}(A) \times \text{P}(B \mid A) \times \text{P}(C \mid A \cap B)$

> The probability of each successive event *depends* on the events that came before it.

Example 4
◀) **Self Tutor**

A box contains 4 red and 2 yellow tickets. Two tickets are randomly selected from the box one by one *without* replacement. Find the probability that:

a both are red   **b** the first is red and the second is yellow.

Let A be the event that the first ticket is red, and B be the event that the second ticket is red.

∴ $P(A) = \frac{4}{6} = \frac{2}{3}$

a $P(B \mid A) = \frac{3}{5}$ ◀—— number of red tickets remaining
                           ◀—— total number of tickets remaining

∴ $P(A \cap B) = P(A) \times P(B \mid A)$

           $= \frac{2}{3} \times \frac{3}{5} = \frac{2}{5}$

The probability that both tickets are red is $\frac{2}{5}$.

b $P(B' \mid A) = \frac{2}{5}$ ◀—— number of yellow tickets remaining
                            ◀—— total number of tickets remaining

∴ $P(A \cap B') = P(A) \times P(B' \mid A)$

           $= \frac{2}{3} \times \frac{2}{5} = \frac{4}{15}$

The probability that the first ticket is red and the second ticket is yellow is $\frac{4}{15}$.

Example 5
◀) **Self Tutor**

A hat contains 20 tickets numbered 1, 2, 3, ..., 20. If three tickets are drawn from the hat without replacement, determine the probability that they are all prime numbers.

Let P_k be the event that the kth number drawn is prime, $k = 1, 2, 3$.

Now $\{2, 3, 5, 7, 11, 13, 17, 19\}$ are the primes on tickets.

∴ 8 of the 20 numbers are primes.

P(3 primes)

$= P(P_1 \cap P_2 \cap P_3)$

$= P(P_1) \times P(P_2 \mid P_1) \times P(P_3 \mid P_1 \cap P_2)$

$= \frac{8}{20}$ ◀———— 8 primes out of 20 numbers

   $\times \frac{7}{19}$ ◀———— 7 primes out of 19 numbers after a successful first draw

      $\times \frac{6}{18}$ ◀— 6 primes out of 18 numbers after two successful draws

≈ 0.0491

$P_3 \mid P_1 \cap P_2$ is the event that the 3rd number is prime given that the first two numbers were prime.

In each fraction, the numerator is the number of outcomes in the event, and the denominator is the total number of possible outcomes.

EXERCISE 17C

1 A box contains 7 red and 3 green balls. Two balls are drawn one after another from the box without replacement. Find the probability that:

    a both are red
                 b the first is green and the second is red.

2 A bin contains 12 identically shaped chocolates of which 8 are strawberry creams. Three chocolates are selected simultaneously from the bin. Find the probability that:

Drawing three chocolates *simultaneously* implies there is no replacement.

    a they are all strawberry creams

    b none of them are strawberry creams.

3 A lottery has 100 tickets which are placed in a barrel. Three tickets are drawn at random from the barrel, without replacement, to decide 3 prizes. If John has 3 tickets in the lottery, determine his probability of winning:

    a first prize   **b** first and second prize   **c** all 3 prizes   **d** none of the prizes.

4 A hat contains 7 names of players in a tennis squad including the captain and the vice captain. A team of three is chosen at random by drawing the names from the hat. Find the probability that it does *not* contain:

    a the captain   **b** the captain or the vice captain.

5 Two students are chosen at random from a group of two girls and five boys, all of different ages. Find the probability that the two students chosen will be:

    a two boys   **b** the eldest two students.

6 Suppose A, B, and C are events. Use the conditional probability formula to prove that

$P(A \cap B \cap C) = P(A) \times P(B \mid A) \times P(C \mid A \cap B)$.

Hint: $P(A \cap B \cap C) = P(A \cap (B \cap C))$.

Activity 1   Pólya's urn

Pólya's urn is a statistical model which uses the following sampling scheme:

Imagine an urn containing n black balls and m white balls. When a ball is drawn from the urn, the colour of the ball is observed. The ball is then returned to the urn and an additional ball of the same colour is added.

What to do:

1 Consider an urn with 3 black balls and 4 white balls. If Pólya's urn scheme is used, find the probability that:

    a 2 black balls are drawn in succession   **b** 3 black balls are drawn in succession.

2 Click on the icon to access a computer simulation of Pólya's urn.   **DEMO**

    a Set $n = 5$ and $m = 5$. Let the total number of balls drawn $= 1000$.
        Run the simulation several times and comment on your results. What happens to the proportion of black balls in the urn as more balls are drawn?

    b Vary the values of n and m. What effects do changing these values have on your answers in **a**? Explain your answers.

D | TREE DIAGRAMS

In previous years, you should have used **tree diagrams** to help illustrate sample spaces and calculate probabilities involving independent events. Tree diagrams are particularly useful in calculations involving conditional probabilities.

Example 6   ◀) Self Tutor

The top shelf in a cupboard contains 3 cans of pumpkin soup and 2 cans of chicken soup. The bottom shelf contains 4 cans of pumpkin soup and 1 can of chicken soup. Lukas is twice as likely to take a can from the bottom shelf as he is from the top shelf.

Suppose Lukas takes one can of soup without looking at the label. Find the probability that it:

a is chicken   **b** was taken from the top shelf given that it is chicken.

Let T represent the top shelf, B represent the bottom shelf, P represent the pumpkin soup, and C represent the chicken soup.

shelf	soup

a $P(C)$

$= \underbrace{P(T \cap C)}_{\text{branch ①}} + \underbrace{P(B \cap C)}_{\text{branch ②}}$

$= \frac{1}{3} \times \frac{2}{5} + \frac{2}{3} \times \frac{1}{5}$

$= \frac{4}{15}$

b $P(T \mid C)$

$= \frac{P(T \cap C)}{P(C)}$

$= \dfrac{\frac{1}{3} \times \frac{2}{5}}{\frac{4}{15}}$ ⟵ branch ① from **a**

$= \frac{1}{2}$

EXERCISE 17D

1 Urn A contains 2 red and 3 blue marbles, and urn B contains 4 red and 1 blue marble. Peter selects an urn by tossing a coin, and takes a marble from that urn.

  a Draw a tree diagram to illustrate the situation.

  b Determine the probability that the marble is red.

  c Given that the marble is red, what is the probability that it came from urn B?

2 When Greta's mother goes shopping, the probability that she takes Greta with her is $\frac{2}{5}$. When Greta goes shopping with her mother she gets an ice cream 70% of the time. When Greta does not go shopping with her mother she gets an ice cream 30% of the time.

Determine the probability that:

  a Greta's mother buys her an ice cream when shopping

  b Greta went shopping with her mother, given that her mother buys her an ice cream.

3 On a given day, machine A has a 10% chance of malfunctioning and machine B has a 7% chance of the same. Given that at least one of the machines malfunctioned today, what is the chance that machine B malfunctioned?

4 On any day, the probability that a boy eats his prepared lunch is 0.5. The probability that his sister eats her lunch is 0.6. The probability that the girl eats her lunch given that the boy eats his lunch is 0.9.

    a Find the probability that the girl eats her lunch but the boy does not.

    b Draw a tree diagram to illustrate the situation.

    c Find the probability that at least one of the children eats their lunch.

5 A test to detect cancer is not always reliable. It gives a positive result 95% of the time if the person does have cancer, and it gives a positive result 3% of the time if the person does not.

The probability that a randomly selected person has cancer is 0.02.

    a Find the probability that the test on a randomly selected person gives a positive result.

    b Given that a test on a randomly selected person is positive, find the probability that he or she does have cancer.

6 A double-headed, a double-tailed, and a fair coin are placed in a tin can. One of the coins is randomly chosen without identifying it. The coin is tossed and falls "heads". Determine the probability that the coin is the "double-header".

Example 7   ◀) **Self Tutor**

A bag contains 10 balls. 6 balls are black and 4 balls are white. Three balls are drawn from the bag without replacement. Determine the probability that 1 black ball is drawn.

Let B represent drawing a black ball and W represent drawing a white ball.

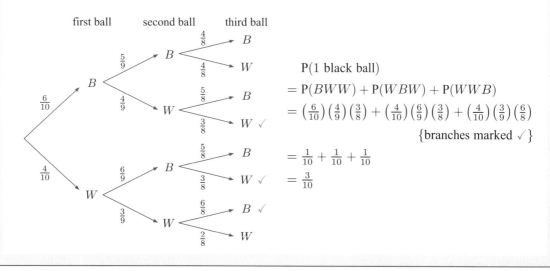

$P(1 \text{ black ball})$
$= P(BWW) + P(WBW) + P(WWB)$
$= \left(\frac{6}{10}\right)\left(\frac{4}{9}\right)\left(\frac{3}{8}\right) + \left(\frac{4}{10}\right)\left(\frac{6}{9}\right)\left(\frac{3}{8}\right) + \left(\frac{4}{10}\right)\left(\frac{3}{9}\right)\left(\frac{6}{8}\right)$
                             {branches marked ✓}
$= \frac{1}{10} + \frac{1}{10} + \frac{1}{10}$
$= \frac{3}{10}$

---

7 In a class of 25 students, 11 students participate in extra-curricular activities. Suppose 3 students are randomly selected to be on the student representative council. Find the probability that at least two students selected for the council also participate in extra-curricular activities.

8 A standard deck of playing cards contains 52 cards. Four cards are drawn from a well-shuffled deck without replacement. Find the probability that:

    a two red cards are drawn
                                      b at least one black card is drawn.

Activity 2   The Monty Hall problem

The Monty Hall problem is a mathematical paradox first posed by **Steve Selvin** to the *American Statistician* in 1975. It became famous after its publication in *Parade* magazine in 1990. The problem is named after the original host of the American television game show *Let's Make a Deal*, on which the problem is loosely based.

The problem as posed in *Parade* reads:

> *Suppose you're on a game show, and you're given the choice of three doors: Behind one door is a car; behind the others, goats. You pick a door, say No. 1, and the host, who knows what's behind the doors, opens another door, say No. 3, which has a goat. He then says to you, "Do you want to switch your choice to door No. 2?" Is it to your advantage to switch your choice?*

What to do:

1 Draw a tree diagram to represent the problem. Let C represent the event that a contestant's choice is correct, and C' represent the event that the choice is incorrect.

2 Find the probability that:

  a the contestant's first choice has the car

  b the contestant's second choice has the car *given* they decide to change their guess.

3 Suppose this game is being played in front of a live studio audience. One of the audience members arrives late, so when they enter the room, they see two closed doors and the third (incorrect) door open. They do not know the contestant's original choice.

  a If this audience member is asked to choose a door, what is the probability they will choose the one with the car?

  b Explain why the contestant has an advantage over this audience member.

Historical note

A **Markov process** is one example of a **stochastic process** or **stochastic model**. Stochastic processes are systems in which we cannot predict what can happen with absolute certainty, but we can model the random variation to get an idea of what we can expect.

The word "stochastic" comes from the Greek word *stokhastikós* which means "acting based on guesswork".

Stochastic processes were the main research focus of the Russian mathematician **Andrey Markov** (1856 - 1922), after whom Markov processes were named.

Today, Markov processes have many applications in genetics, economics, and telecommunications.

Andrey Markov

Activity 3 Markov processes

In springtime in a particular seaside town:

- Tomorrow's weather *only* depends on today's weather.
- The probability that it will be sunny tomorrow given that it was sunny today is 0.8.
- The probability that it will be rainy tomorrow given that it was rainy today is 0.5.

This is an example of a **Markov process**, since the state of the system tomorrow is only dependent on the state of the system today.

What to do:

1 Suppose it was sunny in this town on Monday.

a Copy and complete this tree diagram.

b Find the probability that on Wednesday it will be:

    i sunny   **ii** rainy.

c Extend the tree diagram in **a** to include the weather for Thursday.

d Hence find the probability that it will be sunny on Thursday.

e Penelope is hoping that it will be sunny on Friday because she is planning to host a beach party. Find the probability that Penelope will have to postpone her party.

2 Click on the icon to access a simulation of the weather in this town.

p_{sunny} is the probability that it will be sunny tomorrow given that it was sunny today.

p_{rainy} is the probability that it will be rainy tomorrow given that it was rainy today.

SIMULATION

a Run the simulation for:

    i a month   **ii** 100 days   **iii** 1000 days.

b Describe how the *proportion* of sunny days changes during the simulation.

c Discuss what happens to the proportion of rainy days throughout the simulation.

d Run the simulation again for different values of p_{sunny} and p_{rainy}. Comment on your results.

3 How reasonable do you think this model is in describing the weather? Discuss the assumptions that have been made in formulating the model.

Activity 4 Hidden Markov processes

Markov processes are not always directly observable. A Markov process will sometimes generate a sequence of events that we can see, but we know nothing about the underlying process itself. In this case we call it a **hidden Markov process**.

Consider again the weather in the seaside town in **Activity 3**. Suppose you are a resident in the town and arc bed-ridden for several days, so you cannot observe what the weather is like outside. Your friend comes to visit you once a day to prepare your meals and do any chores.

Your friend does not explicitly comment on the weather, but they do bring an umbrella depending on what the weather is like outside.

- The probability that your friend brings an umbrella given that it is sunny is 0.1.
- The probability that your friend brings an umbrella given that it is rainy is 0.7.

What to do:

1 On the 1st day of your sickness, your friend brings an umbrella. You do not remember what the weather was like the day before, so you guess that each type of weather was *equally likely*.

a Copy and complete the tree diagram alongside.

b Use the tree diagram to find:

    i $P(S \cap U)$ **ii** $P(R \cap U)$ **iii** $P(U)$.

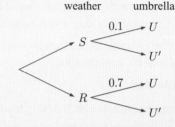

c Hence find:

    i $P(S \mid U)$ **ii** $P(R \mid U)$.

d Hence explain why, on the basis that your friend brought an umbrella, it is *more likely* to have rained on the 1st day.

e Explain why we could have reached the same conclusion by comparing **b i** and **b ii** instead.

2 a By extending your tree diagram, find the probability that:

    i it rains on the first day, and on the second day your friend brings an umbrella and it is sunny

    ii it rains on the first day, and on the second day your friend brings an umbrella and it is rainy

    iii it is sunny on the first day, and on the second day your friend brings an umbrella and it is sunny

    iv it is sunny on the first day, and on the second day your friend brings an umbrella and it is rainy.

b Hence decide which of the following sequences of weather is most likely:

    A sunny, sunny   **B** sunny, rainy   **C** rainy, sunny   **D** rainy, rainy

c How do you think your answer in **2 b** relates to your answer in **1 d**?

3 The steps outlined in **1** and **2** are the first few steps in the **Viterbi algorithm** which is used to find the *most probable* sequence for the underlying Markov process. Click on the icon to see the Viterbi algorithm in action.

SIMULATION

Review set 17A

1 T and M are events such that $n(\mathcal{E}) = 30$, $n(T) = 10$, $n(M) = 17$, and $n((T \cup M)') = 5$.

a Draw a Venn diagram to display this information.

b Hence find:

    i $P(T \cap M)$ **ii** $P((T \cap M) \mid M)$

2 A survey of 200 people included 90 females. It found that 60 people smoked, 40 of whom were male.

a Use the given information to complete the two-way table.

	Female	Male	Total
Smoker			
Non-smoker			
Total			

b A person is selected at random. Find the probability that this person is:

    i a female non-smoker

    ii a male given the person was a non-smoker.

c If two people from the survey are selected at random, calculate the probability that:

    i both of them are non-smoking females

    ii one is a smoker and the other is a non-smoker.

3 An urn contains 3 red balls and 6 blue balls.

    a A ball is drawn at random and found to be blue. Find the probability that a second draw with no replacement will also produce a blue ball.

    b Two balls are drawn without replacement and the second is found to be red. What is the probability that the first ball was also red?

4 A team of three is randomly chosen from six doctors and four dentists. Determine the probability that it consists of:

    a all doctors   **b** at least two doctors.

5 Answer the **Opening Problem** on page **400**.

6 The probability that a particular salesman will leave his sunglasses behind in any store is $\frac{1}{5}$. Suppose the salesman visits two stores in succession and leaves his sunglasses behind in one of them. What is the probability that the salesman left his sunglasses in the first store?

7 Two different numbers were chosen at random from the digits 1 to 9 inclusive. It was observed that their sum was even. Determine the probability that both numbers were odd.

Review set 17B

1 A survey of 50 men and 50 women was conducted to see how many people prefer coffee or tea. It was found that 15 men and 24 women prefer tea.

    a Display this information in a two-way table.

    b Let C represent the people who prefer coffee and M represent the men. Hence complete the Venn diagram.

    c Calculate $P(M \mid C)$.

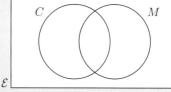

2 A and B are independent events where $P(A) = 0.8$ and $P(B) = 0.65$. Determine:

    a $P(A \cup B)$ **b** $P(A \mid B)$ **c** $P(A' \mid B')$ **d** $P(B \mid A)$.

3 If I buy 4 tickets in a 500 ticket lottery and the prizes are drawn without replacement, determine the probability that I will win:

    a the first 3 prizes   **b** at least one of the first 3 prizes.

4 The students in a school are all vaccinated against measles. 48% of the students are males, of whom 16% have an allergic reaction to the vaccine. 35% of the females also have an allergic reaction. A student is randomly chosen from the school. Find the probability that the student:

    a has an allergic reaction

    b is female given that a reaction occurs.

5 Let C be the event that "a person has a cat" and D be the event that "a person has a dog".
$P(C) = \frac{3}{7}$, $P(D \mid C') = \frac{2}{5}$, and $P(D' \mid C) = \frac{3}{4}$.

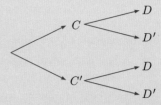

  a Copy and complete the tree diagram by marking a probability on each branch.

  b If a person is chosen at random, find the probability that the person has:

    i a cat and a dog

    ii at least one cat or dog.

6 Jon goes cycling on three random mornings of each week. When he goes cycling he has eggs for breakfast 70% of the time. When he does not go cycling he has eggs for breakfast 25% of the time. Determine the probability that Jon:

  a has eggs for breakfast   **b** goes cycling given that he has eggs for breakfast.

7 With each pregnancy, a particular woman will give birth to either a single baby or twins. There is a 15% chance of having twins during each pregnancy. Suppose that after 2 pregnancies she has given birth to 3 children. Find the probability that she had twins first.

18

The normal distribution

Contents:

Opening problem

A salmon breeder is interested in the distribution of the *weight of female adult salmon, w*.

He catches hundreds of female adult fish and records their weights in a frequency table with class intervals $3 \leqslant w < 3.1$ kg, $3.1 \leqslant w < 3.2$ kg, $3.2 \leqslant w < 3.3$ kg, and so on.

The mean weight is 4.73 kg, and the standard deviation is 0.53 kg.

Things to think about:

a Which of these do you think is the most likely distribution for the weights of the female adult salmon?

A
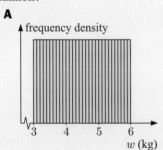

B
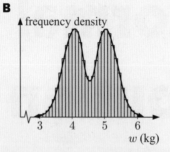

C
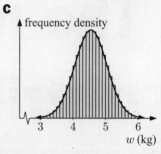

b How can we use the mean and standard deviation to estimate the proportion of salmon that weigh:

   i more than 6 kg   ii between 4 kg and 6 kg?

c How can we find the weight which:

   i 90% of salmon weigh less than   ii 25% of salmon weigh more than?

CONTINUOUS RANDOM VARIABLES

A **random variable** uses numbers to describe the possible outcomes which could result from a random experiment.

A **continuous random variable** can take any value within some interval on the number line.

In previous years you would have used a **histogram** to display data for a continuous variable. These graphs show how the data is distributed when organised into **class intervals**.

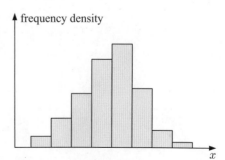

Suppose we have a very large number of data values. If we decrease the widths of the class intervals, the bars of the histogram start to form a smooth curve. Click on the icon to see this process in action.

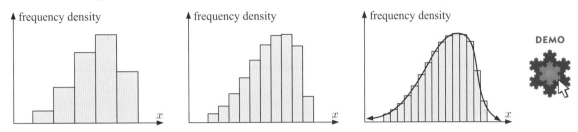

For this reason, we model the shape of a continuous variable's **distribution** using a function called a distribution curve.

A   THE NORMAL DISTRIBUTION

In this Chapter, we consider variables with **symmetrical**, **bell-shaped** distribution curves. These variables are said to have a **normal distribution**, which is one of the most important distributions in statistics.

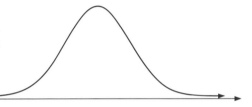

The normal distribution arises in nature when many different factors affect the value of the variable.

For example, consider the apples harvested from an apple orchard. They do not all have the same weight. This variation may be due to genetic factors, the soil, the amount of sunlight reaching the leaves and fruit, weather conditions, and so on.

The result is that most of the fruit will have weights centred about the mean weight, and there will be fewer apples that are much heavier or much lighter than this mean.

Some examples of quantities that may be normally distributed or approximately normally distributed are:

- the heights of 16 year old boys
- the lengths of adult sharks
- the yields of corn or wheat

- the volumes of liquid in soft drink cans
- the weights of peaches in a harvest
- the life times of batteries

EXERCISE 18A.1

1 Which of the following appear to be normal distribution curves?

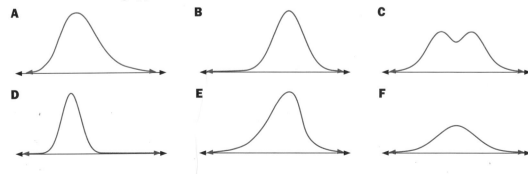

2 Explain why it is likely that the following variables will be normally distributed:

    a the diameter of wooden rods cut using a lathe

    b scores for tests taken by a large population

    c the amount of time a student takes to walk to school each day.

3 Discuss whether the following variables are likely to be normally distributed. Sketch a graph to illustrate the possible distribution of each variable.

    a the ages of people at a football match   **b** the distances recorded by a long jumper

    c the numbers drawn in a lottery   **d** the lengths of carrots in a supermarket

    e the amounts of time passengers spend waiting in a queue at an airport

    f the numbers of brown eggs in a sample of cartons which each contain a dozen eggs

    g the numbers of children in families living in Cardiff

    h the heights of buildings in a city.

THE NORMAL DISTRIBUTION CURVE

Although all normal distributions have the same general bell-shaped curve, the exact location and shape of the curve is determined by:

- the **mean** μ which measures the **centre** of the distribution
- the **standard deviation** σ which measures the **spread** of the distribution.

If X is a normally distributed random variable with mean μ and standard deviation σ, we write $X \sim N(\mu, \sigma^2)$.

> $\sim$ is read "is distributed as".

We say that μ and σ are the **parameters** of the distribution.

The distribution curve of X is defined by the function:

$$f(x) = \frac{1}{\sigma\sqrt{2\pi}}\, e^{-\frac{1}{2}\left(\frac{x-\mu}{\sigma}\right)^2}, \quad \text{for } x \in \mathbb{R}$$

> You are not required to remember this formula.

We refer to this curve as the **normal distribution curve** or just **normal curve**.

Investigation 1   Properties of the normal curve

In this Investigation, we will look at some interesting properties of the normal distribution curve

$f(x) = \dfrac{1}{\sigma\sqrt{2\pi}}\, e^{-\frac{1}{2}\left(\frac{x-\mu}{\sigma}\right)^2}$ with the help of **graphing software** and **calculus**.

What to do:

1 Click on the icon to explore the normal distribution curve and how it changes when μ and σ are altered.

DEMO

    a What effects do variations in μ and σ have on the curve? How do these relate to what μ and σ represent?

    b Does the curve have a line of symmetry? If so, what is it?

    c Is the function ever negative? Why is this important?

d Discuss the behaviour of the normal curve as $x \to \pm\infty$.

e What do you think happens to the *area* under the curve as you change μ and σ?

2 For a general normal curve, use calculus to find and classify the:

a stationary point   **b** inflection points.

3 Sketch a normal distribution curve to illustrate your answers to **2**.

From the **Investigation**, you should have found that:

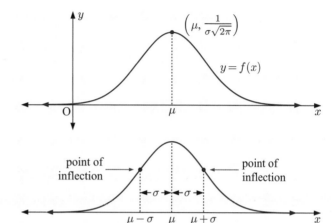

- The normal curve is symmetrical about the vertical line $x = \mu$.
- $f(x) > 0$ for all x.
- The x-axis is a horizontal asymptote.
- The maximum occurs at $x = \mu$.

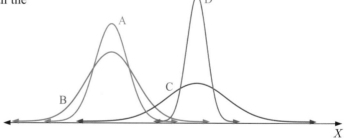

- The points of inflection occur at $x = \mu \pm \sigma$, so the standard deviation is the horizontal distance from the line of symmetry $x = \mu$ to a point of inflection.

EXERCISE 18A.2

1 Match each pair of parameters with the correct normal distribution curve:

a $\mu = 5, \ \sigma = 2$

b $\mu = 15, \ \sigma = 0.5$

c $\mu = 5, \ \sigma = 1$

d $\mu = 15, \ \sigma = 3$

2 Sketch the following normal distributions on the same set of axes.

Distribution	Mean (mL)	Standard deviation (mL)
A	25	5
B	30	2
C	21	10

3 Suppose X_1 is a normally distributed random variable with mean 4 and standard deviation 3. Let $X_2 = X_1 + 2$.

a Write down the formula for the normal distribution curve of X_1 in terms of x_1.

b Write down the formula for the normal distribution curve of X_2 by letting $x_2 = x_1 + 2$ and substituting for x_1 in your formula in **a**.

GRAPHING PACKAGE

c Hence describe the distribution of X_2.

d Sketch the distribution curves for X_1 and X_2 on the same set of axes.

Historical note

The normal distribution was first characterised by **Carl Friedrich Gauss** in 1809 as a way to rationalise his **method of least squares** for linear regression.

In fact, the normal distribution curve is a special case of the **Gaussian function** which has the form:

$$f(x) = ae^{-\frac{1}{2}\left(\frac{x-b}{c}\right)^2} \quad \text{where } a, b, \text{ and } c \text{ are constants.}$$

Since the normal distribution has such strong ties to Gauss, it is sometimes called the **Gaussian distribution**.

Carl Friedrich Gauss

B   CALCULATING PROBABILITIES

You should have seen previously that in a histogram, the **area of a bar** gives the *frequency* for its corresponding class interval.

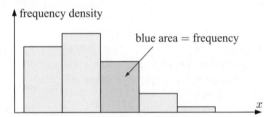

By dividing the frequency density of each interval by the total number of data values, the area of the bars now gives us the *proportion* of data values in an interval.

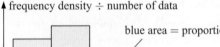

Suppose we make the class intervals for this new histogram narrower and narrower until the tops of the bars produce a smooth curve. This curve defines the continuous distribution, and the **area under the curve** on a particular interval is the proportion of the population that lies in that interval.

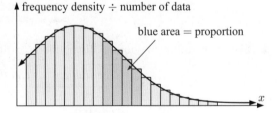

If $f(x)$ describes the *distribution* of a continuous random variable X over a domain $[a, b]$, then

$$P(c \leqslant X \leqslant d) = \int_c^d f(x)\, dx$$

$$= \text{the area under } y = f(x)$$
$$\text{between } x = c \text{ and } x = d$$

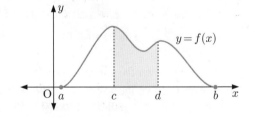

$f(x)$ must satisfy:

- $f(x) \geqslant 0$ for all $a \leqslant x \leqslant b$

  {as probabilities cannot be negative}

- $\displaystyle\int_a^b f(x)\,dx = 1$ {as the total probability must be 1}

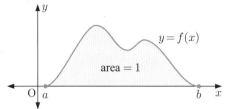

For the normal distribution curve $f(x) = \dfrac{1}{\sigma\sqrt{2\pi}}\, e^{-\frac{1}{2}\left(\frac{x-\mu}{\sigma}\right)^2}$, the coefficient $\dfrac{1}{\sigma\sqrt{2\pi}}$ ensures the total area under the curve is always 1, irrespective of the values of μ and σ.

Investigation 2   Proportions from a normal distribution

In this Investigation we find the proportions of normally distributed data which lie within σ, 2σ, and 3σ of the mean.

What to do:

DEMO

1 Click on the icon to run a demonstration which randomly generates 1000 data values from a normal distribution with mean μ and standard deviation σ. Set $\mu = 0$ and $\sigma = 1$.

2 Find:

   a $\mu - \sigma$ and $\mu + \sigma$ **b** $\mu - 2\sigma$ and $\mu + 2\sigma$ **c** $\mu - 3\sigma$ and $\mu + 3\sigma$

3 Use the frequency table provided to find the proportion of data values which lie between:

   a $\mu - \sigma$ and $\mu + \sigma$ **b** $\mu - 2\sigma$ and $\mu + 2\sigma$ **c** $\mu - 3\sigma$ and $\mu + 3\sigma$

4 Repeat **2** and **3** for values of μ and σ of your choosing. Summarise your answers in a table like the one below.

μ	σ	$\mu - \sigma$ to $\mu + \sigma$		$\mu - 2\sigma$ to $\mu + 2\sigma$		$\mu - 3\sigma$ to $\mu + 3\sigma$	
		Interval	*Proportion*	*Interval*	*Proportion*	*Interval*	*Proportion*
0	1	$[-1, 1]$					

5 What do you notice about the proportion of data values in each interval?

6 To more accurately calculate proportions from a normal distribution, we can integrate the normal distribution curve. However, we cannot easily write an antiderivative for

   $f(x) = \dfrac{1}{\sigma\sqrt{2\pi}}\, e^{-\frac{1}{2}\left(\frac{x-\mu}{\sigma}\right)^2}$, so we calculate definite integrals numerically using technology.

   a Suppose X is normally distributed with $\mu = 0$ and $\sigma = 1$. Use the **graphing package** to help you estimate the following probabilities with an appropriate definite integral:

   GRAPHING PACKAGE

   i $P(\mu - \sigma \leqslant X \leqslant \mu + \sigma)$ **ii** $P(\mu - 2\sigma \leqslant X \leqslant \mu + 2\sigma)$
   iii $P(\mu - 3\sigma \leqslant X \leqslant \mu + 3\sigma)$

   b Repeat **a** for each of the values of μ and σ that you chose in **4**. How do the definite integrals compare to your proportions from the simulation?

   c Does changing the mean and standard deviation of a normal distribution change the proportion of the population that lies within 1, 2, or 3 standard deviations of the mean?

From the **Investigation**, you should have found that:

For any population that is normally distributed with mean μ and standard deviation σ:
- approximately **0.68** or **68%** of the population will lie between $\mu - \sigma$ and $\mu + \sigma$
- approximately **0.95** or **95%** of the population will lie between $\mu - 2\sigma$ and $\mu + 2\sigma$
- approximately **0.997** or **99.7%** of the population will lie between $\mu - 3\sigma$ and $\mu + 3\sigma$.

The proportion of data values that lie within different ranges relative to the mean are:

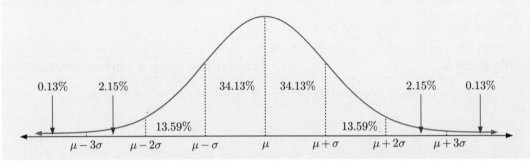

For any variable that is normally distributed, we can use the mean and standard deviation to estimate the proportion of data that will lie in a given interval. This proportion tells us the probability that a randomly selected member of the population will be in that interval.

Example 1   ◀)) Self Tutor

A sample of cans of peaches was taken from a warehouse, and the contents of each can was weighed. The sample mean was 486 g with standard deviation 6 g.

State the proportion of cans that weigh:

a between 480 g and 486 g   **b** more than 492 g.

For a manufacturing process such as this, the distribution of weights is approximately normal.

a

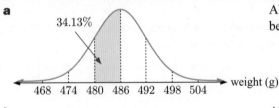

About 34.13% of the cans are expected to weigh between 480 g and 486 g.

b

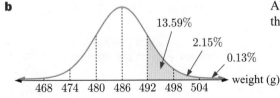

About $13.59\% + 2.15\% + 0.13\% = 15.87\%$ of the cans are expected to weigh more than 492 g.

> The probability of randomly selecting a can which weighs more than 492 g is approximately 0.1587.

EXERCISE 18B.1

1 Suppose X is normally distributed with mean 30 and standard deviation 5.

  a State the value which is:

    i 2 standard deviations above the mean   **ii** 1 standard deviation below the mean.

  b Describe the following values in terms of the number of standard deviations above or below the mean:

    i 35   **ii** 20   **iii** 45

  c Draw a curve to illustrate the distribution of X.

  d What proportion of values of X are between 25 and 30?

  e Find the probability that a randomly selected member of the population will measure between 35 and 40.

2 Suppose the variable X is normally distributed according to the curve shown.

  a State the mean and standard deviation of X.

  b Find the proportion of values of X which are:

    i between 20 and 24

    ii between 12 and 16

    iii greater than 28.

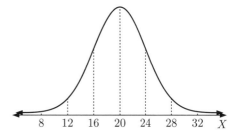

3 A school's Grade 12 students sat for a Mathematics examination. Their marks were approximately normally distributed with mean 75 and standard deviation 8.

  a Copy and complete this bell-shaped curve, assigning scores to the markings on the horizontal axis.

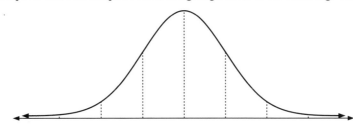

  b What proportion of students would you expect to have scored:

    i more than 83   **ii** less than 59   **iii** between 67 and 91?

4 State the probability that a randomly selected, normally distributed value:

  a lies within one standard deviation either side of the mean

  b is more than two standard deviations above the mean.

Example 2

◀)) **Self Tutor**

The chest measurements of 18 year old male rugby players are normally distributed with mean 95 cm and standard deviation 8 cm.

a From a group of 200 18 year old male rugby players, how many would you expect to have a chest measurement between 87 cm and 111 cm?

b Find the value of k such that approximately 16% of chest measurements are below k cm.

a About $34.13\% + 34.13\% + 13.59\%$
$= 81.85\%$ of the rugby players have a chest measurement between 87 cm and 111 cm. So, we would expect 81.85% of $200 \approx 164$ of the rugby players to have a chest measurement between 87 cm and 111 cm.

b Approximately 16% of data lies more than one standard deviation below the mean.

∴ k is σ below the mean μ

∴ $k = 95 - 8$
$= 87$

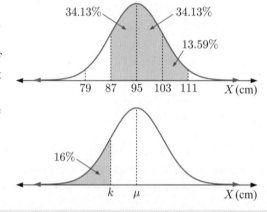

---

5 The height of female students at a university is normally distributed with mean 170 cm and standard deviation 8 cm.

   a Find the percentage of female students whose height is:

   i between 162 cm and 170 cm   **ii** between 170 cm and 186 cm.

   b Find the probability that a randomly chosen female student has a height:

   i less than 154 cm   **ii** greater than 162 cm.

   c From a group of 500 female university students, how many would you expect to be between 178 cm and 186 cm tall?

   d Estimate the value of k such that 16% of the female students are taller than k cm.

6 The weights of the 545 babies born at a maternity hospital last year were normally distributed with mean 3.0 kg and standard deviation 200 grams. Estimate the number that weighed:

   a less than 3.2 kg   **b** between 2.8 kg and 3.4 kg.

7 An industrial machine fills an average of 20 000 bottles each day with standard deviation 2000 bottles. Assuming that production is normally distributed and the year comprises 260 working days, estimate the number of working days on which:

   a under 18 000 bottles are filled

   b over 16 000 bottles are filled

   c between 18 000 and 24 000 bottles are filled.

8 Two hundred lifesavers competed in a swimming race. Their times were normally distributed with mean 10 minutes 30 seconds and standard deviation 15 seconds. Estimate the number of competitors who completed the race in a time:

   a longer than 11 minutes
   b less than 10 minutes 15 seconds
   c between 10 minutes 15 seconds and 10 minutes 45 seconds.

9 The weights of Jason's oranges are normally distributed. 84% of the crop weighs more than 152 grams and 16% weighs more than 200 grams.

   a Find μ and σ for the crop.
   b What percentage of the oranges weigh between 152 grams and 224 grams?

10 When a particular variety of radish is grown without fertiliser, the weights of the radishes produced are normally distributed with mean 40 g and standard deviation 10 g.
   When these radishes are grown in the same conditions but with fertiliser added, their weights are also normally distributed, but with mean 140 g and standard deviation 40 g.

   a Determine the proportion of radishes grown:
      i without fertiliser which weigh less than 50 grams
      ii with fertiliser which weigh less than 60 grams.

   b Find the probability that a randomly selected radish weighs between 20 g and 60 g, if it is grown:
      i with fertiliser
      ii without fertiliser.

   c One radish grown with fertiliser and one radish grown without fertiliser are selected at random. Find the probability that *both* radishes weigh more than 60 g.

USING TECHNOLOGY

When calculating normal distribution probabilities, we have so far only considered numbers that are a whole number of standard deviations from the mean.

To calculate other probabilities, we could use definite integrals of the normal curve. However, since the normal distribution occurs so often in statistics, your **graphics calculator** has built-in functions to calculate these integrals.

GRAPHICS CALCULATOR INSTRUCTIONS

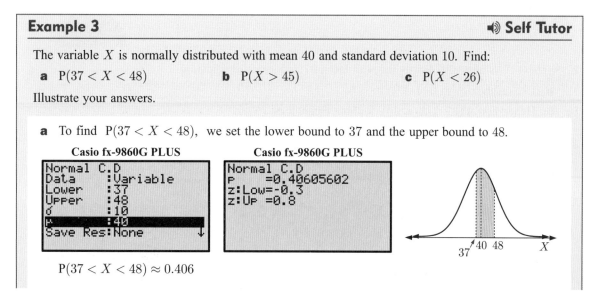

Example 3   ◀) **Self Tutor**

The variable X is normally distributed with mean 40 and standard deviation 10. Find:

   a $P(37 < X < 48)$
   b $P(X > 45)$
   c $P(X < 26)$

Illustrate your answers.

   a To find $P(37 < X < 48)$, we set the lower bound to 37 and the upper bound to 48.

$$P(37 < X < 48) \approx 0.406$$

b To find $P(X > 45)$, we use a very high value such as 10^{99} to represent the upper bound.

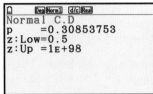

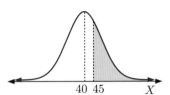

$P(X > 45) \approx 0.309$

c To find $P(X < 26)$, we use a very low value such as -10^{99} to represent the lower bound.

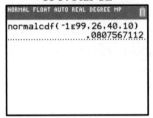

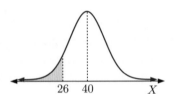

$P(X < 26) \approx 0.081$

EXERCISE 18B.2

1 Suppose X is normally distributed with mean 60 and standard deviation 5.   Find:

a $P(60 \leqslant X \leqslant 65)$

b $P(62 \leqslant X \leqslant 67)$

c $P(X \geqslant 64)$

d $P(X \leqslant 68)$

e $P(X \leqslant 61)$

f $P(57.5 \leqslant X \leqslant 62.5)$

Illustrate your answers.

A continuous random variable X can never be *exactly* 64, so $P(X \geqslant 64) = P(X > 64)$.

2 Suppose X is normally distributed with mean 37 and standard deviation 7.

a Use technology to find $P(X > 40)$.

b Hence find $P(37 \leqslant X \leqslant 40)$ without technology.

3 A machine produces metal bolts. The lengths of these bolts have a normal distribution with mean 19.8 cm and standard deviation 0.3 cm.

If a bolt is selected at random from the machine, find the probability that it will have a length between 19.7 cm and 20 cm.

4 The speed of cars passing a supermarket is normally distributed with mean 46.3 $km\,h^{-1}$ and standard deviation 7.4 $km\,h^{-1}$. Find the probability that a randomly selected car is travelling:

   a between 50 and 65 $km\,h^{-1}$

   b slower than 60 $km\,h^{-1}$

   c faster than 50 $km\,h^{-1}$.

5 Eels are washed onto a beach after a storm. Their lengths have a normal distribution with mean 41 cm and standard deviation 5.5 cm.

   a If an eel is randomly selected, find the probability that it is at least 50 cm long.

   b Find the percentage of eels measuring between 40 cm and 50 cm long.

   c How many eels from a sample of 200 would you expect to measure at least 45 cm in length?

6 Max's customers put money for charity into a collection box on the front counter of his shop. The weekly collection is approximately normally distributed with mean £40 and standard deviation £6.

   a On what percentage of weeks would Max expect to collect:

     i between £30 and £50   **ii** at least £50?

   b How much money would you expect Max to collect in two years?

7 The amount of petrol bought by customers at a petrol station is normally distributed with mean 36 L and standard deviation 7 L.

   a What percentage of customers buy:

     i less than 28 L of petrol

     ii between 30 L and 40 L of petrol?

   b On a particular day, the petrol station has 600 customers.

     i How much petrol would you expect the petrol station to sell on this day?

     ii How many customers would you expect to buy at least 44 L of petrol?

8 The times Enrique and Damien spend working out at the gym each day are both normally distributed with mean 45 minutes. The standard deviation of Enrique's times is 9 minutes, and the standard deviation of Damien's times is 6 minutes.

   a On what percentage of days does:

     i Enrique spend between 32 and 40 minutes at the gym

     ii Damien spend less than 55 minutes at the gym?

   b Tomorrow, who do you think is more likely to spend:

     i at least 1 hour at the gym

     ii between 40 minutes and 50 minutes at the gym?

   Explain your answers.

   c Perform calculations to check your answers to **b**.

Example 4

◀ Self Tutor

The times taken by students to complete a puzzle are normally distributed with mean 28.3 minutes and standard deviation 3.6 minutes. Calculate the probability that:

a a randomly selected student took at least 30 minutes to complete the puzzle

b out of 10 randomly selected students, 5 or fewer of them took at least 30 minutes to complete the puzzle.

a Let X denote the time for a student to complete the puzzle.

$X \sim N(28.3, \ 3.6^2)$

$\therefore \ P(X \geqslant 30) \approx 0.318\,38$

≈ 0.318

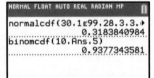

TI-84 Plus CE

NORMAL FLOAT AUTO REAL RADIAN MP

normalcdf(30,1ε99,28.3,3.◢
        0.3183840984

b Let Y denote the number of students who took at least 30 minutes to complete the puzzle.

$Y \sim B(10, \ 0.318\,38)$

$\therefore \ P(Y \leqslant 5) \approx 0.938$

$B(n, \ p)$ is the binomial distribution with n independent trials, each with probability of success p.

TI-84 Plus CE

NORMAL FLOAT AUTO REAL RADIAN MP

normalcdf(30,1ε99,28.3,3.◢
        0.3183840984
binomcdf(10,Ans,5)
        0.9377343581

9 Apples from a grower's crop were normally distributed with mean 173 grams and standard deviation 34 grams. Apples weighing less than 130 grams were too small to sell.

a Find the percentage of apples from this crop which were too small to sell.

b Find the probability that in a picker's basket of 100 apples, more than 10 apples were too small to sell.

10 People found to have high blood pressure are prescribed a course of tablets. They have their blood pressure checked at the end of 4 weeks. The drop in blood pressure over the period is normally distributed with mean 5.9 units and standard deviation 1.9 units.

a Find the proportion of people who show a drop of more than 4 units.

b Eight people taking the course of tablets are selected at random. Find the probability that at least six of them will show a drop in blood pressure of more than 4 units.

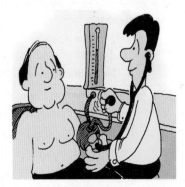

C   THE STANDARD NORMAL DISTRIBUTION

Suppose a random variable X is normally distributed with mean μ and standard deviation σ.

For each value of x we can calculate a **z-score** using the algebraic transformation $z = \dfrac{x - \mu}{\sigma}$.

This algebraic transformation is known as the **Z-transformation**.

Investigation 3   z-scores

In this Investigation we consider how the z-scores for a distribution are themselves distributed.

What to do:

1 Consider the x-values: 1, 2, 2, 3, 3, 3, 3, 4, 4, 4, 4, 4, 5, 5, 5, 5, 6, 6, 7.

   a Draw a histogram of the x-values to check that the distribution is approximately normal.

   b Find the mean μ and standard deviation σ of the x-values.

   c Calculate the z-score for each x-value.

   d Find the mean and standard deviation of the z-scores.

2 Click on the icon to access a demo which randomly generates data values from a normal distribution with given mean and standard deviation. The z-score of each data value is calculated, and histograms of the original data and the z-scores are also shown.

   DEMO

   a Generate samples using various values of μ and σ of your choosing.

   b Record the mean and standard deviation of the z-scores in a table like the one below.

	x-values		z-scores
Mean	Standard deviation	Mean	Standard deviation

   c How does the histogram of the z-scores generally compare with the histogram of the x-values?

   d What conclusions can you make about the *distribution* of the z-scores?

From the **Investigation**, you should have found that the z-scores were normally distributed with mean 0 and standard deviation 1.

$$\text{If } X \sim N(\mu, \sigma^2) \text{ and } Z = \frac{X - \mu}{\sigma} \text{ then } Z \sim N(0, 1^2).$$

No matter what the parameters μ and σ of the original X-distribution are, we always end up with the same Z-distribution, $Z \sim N(0, 1^2)$.

THE
Z-TRANSFORMATION

Click on the icon for a proof of this result.

The distribution $Z \sim N(0, 1^2)$ is called the **standard normal distribution** or **Z-distribution**.

The diagram below shows how z-scores are related to a general normal curve:

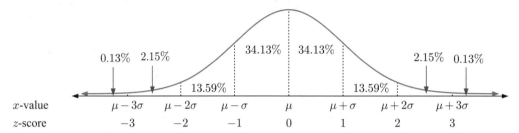

Notice that the value of the z-score corresponds to the coefficient of σ in the x-value.

> The **z-score** of x is the number of standard deviations x is from the mean.

For example:

- if $z = 1.84$, then x is 1.84 standard deviations to the *right* of the mean
- if $z = -0.273$, then x is 0.273 standard deviations to the *left* of the mean.

z-scores are particularly useful when comparing two populations with different μ and σ. However, these comparisons will only be reasonable if both distributions are approximately normal.

Example 5
◄)) **Self Tutor**

Kelly scored 73% in History, where the class mean was 68% and the standard deviation was 10.2%. In Mathematics she scored 66%, where the class mean was 62% and the standard deviation was 6.8%.

In which subject did Kelly perform better compared with the rest of her class?

Assume the scores for both subjects were normally distributed.

Kelly's z-score for History $= \dfrac{73 - 68}{10.2} \approx 0.490$

Kelly's z-score for Mathematics $= \dfrac{66 - 62}{6.8} \approx 0.588$

Kelly's result in Mathematics was 0.588 standard deviations above the mean, whereas her result in History was 0.490 standard deviations above the mean.

$\therefore$ Kelly's result in Mathematics was better compared to her class, even though her percentage was lower.

EXERCISE 18C.1

1 In Emma's classes, the exam results for each subject are normally distributed with the mean μ and standard deviation σ shown in the table.

   a Find the z-score for each of Emma's scores.

   b Arrange Emma's subjects from best to worst in terms of the z-scores.

   c Explain why the z-scores are a reasonable way to compare Emma's performances with the rest of her class.

Subject	Emma's score	μ	σ
English	48	40	4.4
Mandarin	81	60	9
Geography	84	55	18
Biology	68	50	20
Mathematics	84	50	15

2 The table alongside shows Sergio's results in his final examinations, along with the class means and standard deviations.

    a Find Sergio's z-score for each subject.

    b Arrange Sergio's performances in each subject from best to worst.

Subject	Sergio's score	μ	σ
Physics	73%	78%	10.8%
Chemistry	77%	72%	11.6%
Mathematics	76%	74%	10.1%
German	91%	86%	9.6%
Biology	58%	62%	5.2%

3 At a swimming competition, Frederick competed in the 50 m freestyle, 100 m backstroke, 200 m breaststroke, and 100 m butterfly events. His times are summarised in the table below, along with the event means and standard deviations.

Event	Time (seconds)	μ (seconds)	σ (seconds)
50 m freestyle	32.1	27.8	2.2
100 m backstroke	53.5	58.1	4.3
200 m breaststroke	140.0	143.7	6.4
100 m butterfly	59.6	57.7	5.5

    a Calculate the z-scores for each of Frederick's times.

    b Explain why in this case a lower z-score indicates a better performance.

    c Hence arrange Frederick's performances in each event from best to worst.

CALCULATING PROBABILITIES USING THE Z-DISTRIBUTION

Consider the variables $X \sim N(\mu, \sigma^2)$ and $Z \sim N(0, 1^2)$.

For any $x_1, x_2 \in \mathbb{R}$, $x_1 < x_2$, with corresponding z-scores $z_1 = \dfrac{x_1 - \mu}{\sigma}$ and $z_2 = \dfrac{x_2 - \mu}{\sigma}$:

- $P(X \geqslant x_1) = P(Z \geqslant z_1)$

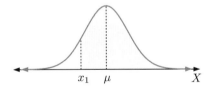
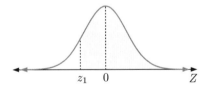

- $P(X \leqslant x_1) = P(Z \leqslant z_1)$

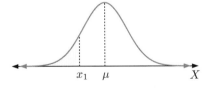
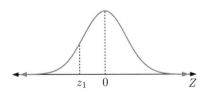

- $P(x_1 \leqslant X \leqslant x_2) = P(z_1 \leqslant Z \leqslant z_2)$

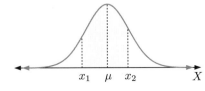
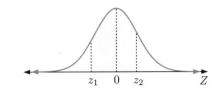

Example 6

◀)) **Self Tutor**

Use technology to illustrate and calculate:

a $P(-0.41 \leqslant Z \leqslant 0.67)$ **b** $P(Z \leqslant 1.5)$ **c** $P(Z > 0.84)$

$Z \sim N(0, 1^2)$

a

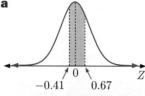

b

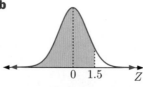

c

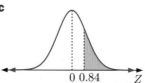

$P(-0.41 \leqslant Z \leqslant 0.67)$
≈ 0.408

$P(Z \leqslant 1.5) \approx 0.933$

$P(Z > 0.84) \approx 0.200$

EXERCISE 18C.2

1 Consider the normal distribution curve below.

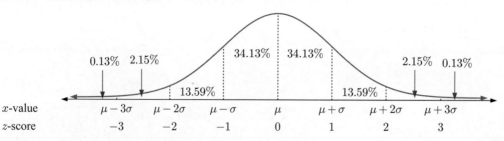

Use the diagram to calculate the following probabilities. In each case sketch the Z-distribution and shade in the region of interest.

PRINTABLE CURVES

a $P(-1 < Z < 1)$ **b** $P(-1 \leqslant Z \leqslant 3)$ **c** $P(-1 < Z < 0)$

d $P(Z < 2)$ **e** $P(-1 < Z)$ **f** $P(Z \geqslant 1)$

2 Given $X \sim N(\mu, \sigma^2)$ and $Z \sim N(0, 1^2)$, determine the values of a and b such that:

a $P(\mu - \sigma < X < \mu + 2\sigma) = P(a < Z < b)$

b $P(\mu - 0.5\sigma < X < \mu) = P(a < Z < b)$

c $P(0 \leqslant Z \leqslant 3) = P(\mu - a\sigma \leqslant X \leqslant \mu + b\sigma)$

3 If $Z \sim N(0, 1^2)$, find the following probabilities using technology. In each case sketch the region under consideration.

   a $P(0.5 \leqslant Z \leqslant 1)$ **b** $P(-0.86 \leqslant Z \leqslant 0.32)$ **c** $P(-2.3 \leqslant Z \leqslant 1.5)$

   d $P(Z \leqslant 1.2)$ **e** $P(Z \leqslant -0.53)$ **f** $P(Z \geqslant 1.3)$

   g $P(Z \geqslant -1.4)$ **h** $P(Z > 4)$ **i** $P(-0.5 < Z < 0.5)$

   j $P(-1.960 \leqslant Z \leqslant 1.960)$ **k** $P(-1.645 \leqslant Z \leqslant 1.645)$ **l** $P(|Z| > 1.645)$

4 **a** Suppose X is normally distributed with mean μ and standard deviation σ.

     i Explain why $P(\mu - 3\sigma < X < \mu + 2\sigma) = P(-3 < Z < 2)$.

     ii Hence find $P(\mu - 3\sigma < X < \mu + 2\sigma)$.

  b For a random variable $X \sim N(\mu, \sigma^2)$, find:

     i $P(\mu - 2\sigma < X < \mu + 1.5\sigma)$ **ii** $P(\mu - 2.5\sigma < X < \mu - 0.5\sigma)$

5 Suppose X is normally distributed with mean $\mu = 58.3$ and standard deviation $\sigma = 8.96$.

  a Let the z-score of $x_1 = 50.6$ be z_1 and the z-score of $x_2 = 68.9$ be z_2.

     i Calculate z_1 and z_2.   **ii** Find $P(z_1 \leqslant Z \leqslant z_2)$.

  b Check your answer by calculating $P(50.6 \leqslant X \leqslant 68.9)$ directly using technology.

Historical note

The normal distribution has two parameters μ and σ, whereas the standard normal distribution has no parameters. This means that a unique table of probabilities can be constructed for the standard normal distribution.

Before graphics calculators and computer packages, it was impossible to calculate probabilities for a general normal distribution $N(\mu, \sigma^2)$ directly.

Instead, all data was transformed using the Z-transformation, and the standard normal distribution table was consulted for the required probabilities.

D   QUANTILES

Consider a population of crabs where the length of a shell, X mm, is normally distributed with mean 70 mm and standard deviation 10 mm.

A biologist wants to protect the population by allowing only the largest 5% of crabs to be harvested. He therefore wants to know what length corresponds to the 95th percentile of crabs.

To answer this question we need to find k such that $P(X \leqslant k) = 0.95$.

The number k is known as a **quantile**. In this case it is the 95% quantile.

When finding quantiles, we are given a probability and are asked to calculate the corresponding measurement. This is the *inverse* of finding probabilities, so we use the **inverse normal function** on our calculator.

GRAPHICS CALCULATOR INSTRUCTIONS

Example 7

◀)) **Self Tutor**

A population of crabs has shell length X mm. X is normally distributed with mean 70 and standard deviation 10. Find k for which $P(X < k) = 0.95$.

Casio fx-9860G PLUS	Casio fx-CG20	TI-84 Plus CE
	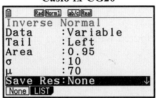	

If $P(X < k) = 0.95$ then
$$k \approx 86.45$$

The 95% quantile corresponds to a shell width of 86.45 mm.

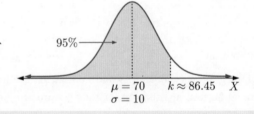

When using the **TI-84 Plus CE** calculator, we must always use the area to the *left* of k. Therefore, to find k such that $P(X > k) = 0.7$, we instead find k such that $P(X < k) = 1 - 0.7 = 0.3$.

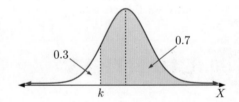

EXERCISE 18D.1

1 Suppose X is normally distributed with mean 20 and standard deviation 3. Illustrate with a sketch and find k such that:

    a $P(X \leqslant k) = 0.3$ **b** $P(X \leqslant k) = 0.9$ **c** $P(X \leqslant k) = 0.5$

    d $P(X > k) = 0.2$ **e** $P(X < k) = 0.62$ **f** $P(X \geqslant k) = 0.13$

2 Suppose Z has the standard normal distribution. Illustrate with a sketch and find k such that:

    a $P(Z \leqslant k) = 0.81$ **b** $P(Z \leqslant k) = 0.58$ **c** $P(Z \leqslant k) = 0.17$

    d $P(Z \geqslant k) = 0.95$ **e** $P(Z \geqslant k) = 0.9$ **f** $P(Z \geqslant k) = 0.41$

3 Suppose X is normally distributed with mean 30 and standard deviation 5, and $P(X \leqslant a) = 0.57$.

    a Using a diagram, determine whether a is greater or less than 30.

    b Use technology to find a.

    c Without using technology, find: **i** $P(X \geqslant a)$ **ii** $P(30 \leqslant X \leqslant a)$

4 Given that X is normally distributed with mean 15 and standard deviation 3, find k such that:

    a $P(X < k) = 0.2$ **b** $P(X > k) = 0.1$

    c $P(15 - k < X < 15 + k) = 0.9$

5 Suppose X is normally distributed with mean 80 and standard deviation 10.

    a Find $P(X \leqslant 72)$.

    b Hence find k such that $P(72 \leqslant X \leqslant k) = 0.1$.

6 Given that $X \sim N(45, 8^2)$, find a such that:

    a $P(35 \leqslant X \leqslant a) = 0.25$ **b** $P(a \leqslant X \leqslant 50) = 0.15$

    c $P(a \leqslant X \leqslant 54) = 0.6$

7 The lengths of a fish species are normally distributed with mean 35 cm and standard deviation 8 cm. The fisheries department has decided that the smallest 10% of the fish are not to be harvested. What is the size of the smallest fish that can be harvested?

8 The lengths of screws produced by a machine are normally distributed with mean 75 mm and standard deviation 0.1 mm. 1% of the screws are rejected because they are too long. What is the length of the smallest screw to be rejected?

9

The volumes of cool drink in bottles filled by a machine are normally distributed with mean 503 mL and standard deviation 0.5 mL. 1% of the bottles are rejected because they are underfilled, and 2% are rejected because they are overfilled. They are otherwise kept for sale. Find, correct to 1 decimal place, the range of volumes in the bottles that are kept.

10 Abbey goes for a morning walk as long as the temperature is not too cold and not too hot. The morning temperatures are normally distributed with mean 20°C and standard deviation 5°C. Given that the lower limit of Abbey's walking temperatures is 11°C, and that she goes for a walk 95% of the time, find the upper limit of Abbey's walking temperatures.

FINDING AN UNKNOWN MEAN OR STANDARD DEVIATION

We always need to convert to z-scores if we are trying to find an unknown mean μ or standard deviation σ.

Example 8
🔊 **Self Tutor**

The weights of an adult scallop population are known to be normally distributed with a standard deviation of 5.9 g. If 15% of scallops weigh less than 58.2 g, find the mean weight of the population.

Let the mean weight of the population be μ g.

If X g denotes the weight of an adult scallop, then
$X \sim N(\mu, 5.9^2)$.

$$P(X \leqslant 58.2) = 0.15$$

$$\therefore \;\; P\left(Z \leqslant \frac{58.2 - \mu}{5.9}\right) = 0.15$$

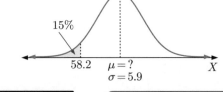

Using the inverse normal function
for $N(0, 1^2)$,

$$\frac{58.2 - \mu}{5.9} \approx -1.0364$$

$$\therefore \;\; 58.2 - \mu \approx -6.1$$

$$\mu \approx 64.3$$

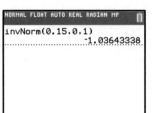

Since we do not know μ we cannot use the inverse normal function directly.

So, the mean weight is approximately 64.3 g.

EXERCISE 18D.2

1 Suppose X is normally distributed with standard deviation 6, and that $P(X < 40) = 0.2$.

   a Would you expect the mean of X to be greater or less than 40? Explain your answer.

   b Find the mean of X.

2 Suppose X is normally distributed with mean 15 and $P(X > 20) = 0.1$. Find the standard deviation of X.

3 The IQs of students at a school are normally distributed with a standard deviation of 15. If 20% of the students have an IQ higher than 125, find the mean IQ of students at the school.

4 The distances an athlete jumps are normally distributed with mean 5.2 m. If 15% of the athlete's jumps are less than 5 m, what is the standard deviation?

5 The weekly income of a bakery is normally distributed with mean £6100. If the weekly income exceeds £6000 85% of the time, what is the standard deviation?

6 The arrival times of buses at a depot are normally distributed with standard deviation 5 minutes. If 10% of the buses arrive before 3:55 pm, find the mean arrival time of buses at the depot.

Example 9

◀ **Self Tutor**

Find the mean and standard deviation of a normally distributed random variable X for which
$P(X \leqslant 20) = 0.1$ and $P(X \geqslant 29) = 0.15$.

$X \sim N(\mu, \sigma^2)$ where we have to find μ and σ.

We start by finding z_1 and z_2 which correspond to
$x_1 = 20$ and $x_2 = 29$.

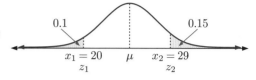

Now $P(X \leqslant x_1) = 0.1$ and $P(X \leqslant x_2) = 0.85$

$\therefore P\left(Z \leqslant \dfrac{20 - \mu}{\sigma}\right) = 0.1$ $\qquad \therefore P\left(Z \leqslant \dfrac{29 - \mu}{\sigma}\right) = 0.85$

$\therefore z_1 = \dfrac{20 - \mu}{\sigma} \approx -1.282$ {technology} $\qquad \therefore z_2 = \dfrac{29 - \mu}{\sigma} \approx 1.036$ {technology}

$\therefore 20 - \mu \approx -1.282\sigma$ .... (1) $\qquad \therefore 29 - \mu \approx 1.036\sigma$ .... (2)

Solving (1) and (2) simultaneously we get $\mu \approx 25.0$ and $\sigma \approx 3.88$.

7 Find the mean and standard deviation of a normally distributed random variable X for which
$P(X \geqslant 35) = 0.32$ and $P(X \leqslant 8) = 0.26$.

8 **a** Find the mean and standard deviation of a normally distributed random variable X for which
$P(X \geqslant 80) = 0.1$ and $P(X \leqslant 30) = 0.15$.

b In a Mathematics examination it was found that 10% of the students scored at least 80, and 15%
scored 30 or less. Assuming the scores are normally distributed, what percentage of students scored
more than 50?

9 The diameters of pistons manufactured by a company are normally distributed. Only those pistons
whose diameters lie between 3.994 cm and 4.006 cm are acceptable.

a Find the mean and the standard deviation of the distribution if 4% of the pistons are rejected as
being too small, and 5% are rejected as being too large.

b Determine the probability that the diameter of a randomly chosen piston measures between 3.997 cm
and 4.003 cm.

10 Circular metal tokens are used to operate a washing
machine in a laundromat. The diameters of the tokens
are normally distributed, and only tokens with diameters
between 1.94 and 2.06 cm will operate the machine. In
the manufacturing process, 2% of the tokens were made
too small, and 3% were made too large.

a Find the mean and standard deviation of the
distribution.

b Find the probability that at most one token out of
a randomly selected sample of 20 will not operate
the machine.

c Jake has 3 tokens. Find the probability that they all have diameter greater than 2 cm.

Game

Click on the icon to play a card game for the normal distribution.

CARD GAME

Investigation 4   The normal approximation to the binomial distribution

You should have seen previously how the **binomial distribution** arises from considering the number of "successes" in a fixed number of independent trials of an experiment.

> Suppose $X \sim B(n, p)$ is the number of successes in n independent trials, each with probability of success p. The probability of getting x successes is:
>
> $$P(X = x) = \binom{n}{x} p^x (1 - p)^{n-x} \quad \text{where} \quad x = 0, 1, 2, ..., n$$

You will have used this formula to calculate probabilities for binomial random variables in cases where the number of trials n is relatively small. As n increases, the probability becomes more difficult to calculate. This is because the binomial coefficient $\binom{n}{x} = \dfrac{n!}{x!(n-x)!}$ becomes very large.

What to do:

1 Click on the icon to access a demonstration which draws the probability distribution of $X \sim B(n, p)$.

DEMO

  a Set $p = 0.5$ and use the sliders to change the value of n. Describe what happens to the distribution of X as n increases.

  b Repeat **a** for p equal to:

  i 0.25   **ii** 0.1   **iii** 0.75   **iv** 0.9

  Comment on your observations.

  c Do you think that it would be *reasonable* to approximate the binomial distribution with a normal distribution? Explain your answer.

2 Click on the icon to access a simulation which generates 1000 random numbers for the binomial random variable $X \sim B(n, p)$.

SIMULATION

  a Set $n = 30$ and $p = 0.5$. Use the simulation to help estimate the mean and standard deviation of X.

  b Repeat **a** for values of n and p of your choosing. Record your results in a table like the one below:

n	p	Mean of X	Standard deviation of X
30	0.5		

  c Calculate np and $\sqrt{np(1-p)}$ for each row of your table in **b**. What do you notice?

  d Suppose we wanted to approximate the distribution of $X \sim B(n, p)$ with a normal distribution. Suggest what values the mean and standard deviation of the normal distribution should take.

3 Consider $X \sim B(50, 0.2)$ and its normal approximation $X_{\text{norm}} \sim N(\mu, \sigma^2)$.

    a Write down expressions for μ and σ.

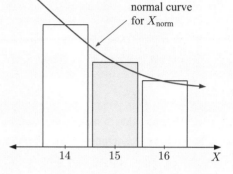

    b Suppose we want to calculate the probability of 15 successes. The diagram alongside shows part of the probability distribution of X with the normal distribution curve of X_{norm} drawn over the top. Explain why $P(X = 15) \approx P(14.5 \leqslant X_{\text{norm}} \leqslant 15.5)$.

    c Describe how you would estimate the following using the normal approximation:

      i $P(X \leqslant 10)$ **ii** $P(X < 25)$

      iii $P(10 \leqslant X < 25)$

4 It is known that 2% of tyres manufactured by a company are unfit for sale. A quality inspector randomly sampled 500 tyres. Use a normal approximation to estimate the probability that at least 10 tyres in the sample will be unfit for sale.

Review set 18A

1 Discuss whether the following variables will be normally distributed:

    a the time students take to read a novel

    b the amount spent on groceries at a supermarket.

2 The normal curve of the $N(3, 2^2)$ distribution is shown below.

PRINTABLE GRAPH

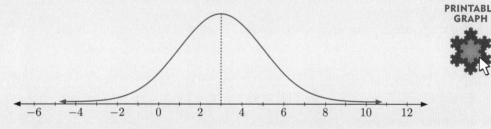

On the same set of axes, sketch the normal curves for:

    a $N(5, 2^2)$ **b** $N(1, 4^2)$ **c** the standard normal distribution

3 The amount of juice Simon can squeeze from his lemons is normally distributed with mean 35 mL and standard deviation 5 mL.

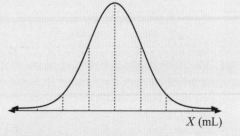

    a Copy and complete this curve.

    b What percentage of the lemons will produce:

      i between 25 mL and 35 mL of juice

      ii at least 45 mL of juice?

4 The average height of 17 year old boys is normally distributed with mean 179 cm and standard deviation 8 cm. Calculate the percentage of 17 year old boys whose heights are:

    a more than 195 cm   **b** between 171 cm and 187 cm   **c** between 163 cm and 195 cm.

5 The weight of the edible part of a batch of Coffin Bay oysters is normally distributed with mean 38.6 grams and standard deviation 6.3 grams.

    a Find the percentage of oysters that weigh between 30 g and 40 g.

    b From a sample of 200 oysters, how many would you expect to weigh more than 50 g?

6 The results of a test are normally distributed. The z-score of Harri's test score is -2.

    a Interpret this z-score with regard to the mean and standard deviation of the test scores.

    b What percentage of students obtained a better score than Harri?

    c The mean test score was 61 and Harri's actual score was 47. Find the standard deviation of the test scores.

7 A random variable X is normally distributed with mean 20.5 and standard deviation 4.3. Find:

    a $P(X \geqslant 22)$ **b** $P(18 \leqslant X \leqslant 22)$ **c** k such that $P(X \leqslant k) = 0.3$.

8 Let X be the weight in grams of bags of sugar filled by a machine. Bags less than 500 grams are considered underweight.

    Suppose $X \sim N(503, 2^2)$.

    a What percentage of bags are underweight?

    b If a quality inspector randomly selects 20 bags, what is the probability that 2 or fewer bags are underweight?

9 The life of a Xenon-brand battery is normally distributed with mean 33.2 weeks and standard deviation 2.8 weeks.

    a Find the probability that a randomly selected battery will last at least 35 weeks.

    b For how many weeks can the manufacturer expect the batteries to last before 8% of them fail?

10 Suppose X is normally distributed with mean 25 and standard deviation 6. Find k such that:

    a $P(X \leqslant k) = 0.7$ **b** $P(X \geqslant k) = 0.4$ **c** $P(20 \leqslant X \leqslant k) = 0.3$

11 The distribution curve shown corresponds to $X \sim N(\mu, \sigma^2)$. Area $A =$ Area $B = 0.2$.

    a Find μ and σ.

    b Calculate:

        i $P(X \leqslant 35)$

        ii $P(23 \leqslant X \leqslant 30)$

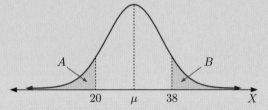

12 Machines A and B both produce nails whose lengths are normally distributed. The lengths of nails from machine A have mean 50.2 mm and standard deviation 1.1 mm. The lengths of nails from machine B have mean 50.6 mm and standard deviation 0.8 mm. Nails which are longer than 52 mm or shorter than 48 mm are rejected.

    a Find the probability of randomly selecting a nail that has to be rejected from:

        i machine A   **ii** machine B.

    b A quality inspector randomly selects a nail from a randomly chosen machine. Find the probability that the nail was made by machine A *given* that it should be rejected.

Review set 18B

1 Sketch the following normal distributions on the same set of axes.

Distribution	Mean (cm)	Standard deviation (cm)
A	22	2
B	20	4
C	25	7

2 The variable X is normally distributed with graph shown.

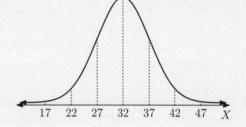

    a State the mean and standard deviation of X.

    b What percentage of values of X are:

       i between 27 and 32

      ii less than 37

     iii greater than 42?

3 The contents of soft drink cans are normally distributed with mean 327 mL and standard deviation 4.2 mL.

    a Find the percentage of cans with contents:

       i less than 318.6 mL   **ii** between 322.8 mL and 339.6 mL.

    b Find the probability that a randomly selected can contains between 327 mL and 331.2 mL.

4 The arm lengths of 18 year old females are normally distributed with mean 64 cm and standard deviation 4 cm.

    a Find the percentage of 18 year old females whose arm lengths are:

       i between 61 cm and 73 cm   **ii** greater than 57 cm.

    b Find the probability that a randomly chosen 18 year old female has an arm length in the range 50 cm to 65 cm.

    c The arm lengths of 70% of the 18 year old females are more than x cm. Find the value of x.

5 The random variable Z has the standard normal distribution $N(0, 1^2)$. Find the value of k for which $P(-k \leqslant Z \leqslant k) = 0.95$.

6 Suppose X is normally distributed with mean 50 and standard deviation 7. Find:

    a $P(46 \leqslant X \leqslant 55)$ **b** $P(X \geqslant 60)$

    c k such that $P(X > k) = 0.23$.

7 In a competition to see who could hold their breath underwater the longest, the times in the preliminary round were normally distributed with mean 150 seconds and standard deviation 12 seconds. If the top 15% of contestants went through to the finals, what time was required to advance?

8 For $X \sim N(12, 2^2)$, find a such that:

    a $P(X < a) = 0.07$ **b** $P(X > a) = 0.2$ **c** $P(a \leqslant X \leqslant 11) = 0.1$

9 X is normally distributed with standard deviation 2.1. Let $Z \sim N(0, 1^2)$.
Given that $P(X > 5.4) = P(Z > -1.7)$, find the mean of X.

10 A normally distributed random variable X has the distribution curve shown. Its mean is 50, and $P(X < 90) \approx 0.975$.

Find the shaded area.

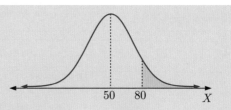

11 The weight of an apple in an apple harvest is normally distributed with mean 300 grams and standard deviation 50 grams. Only apples with weights between 250 grams and 350 grams are considered fit for sale.

a Find the percentage of apples fit for sale.

b In a sample of 100 apples, what is the probability that at least 75 are fit for sale?

12 Giovanni and Beppe are both carrot farmers. The lengths of Giovanni's carrots are normally distributed with mean 22 cm and standard deviation 3.4 cm. The lengths of Beppe's carrots are also normally distributed, with mean 23.5 cm and standard deviation 4.2 cm.

a Find the probability that a carrot is longer than 20 cm, given it comes from:

i Giovanni's farm   **ii** Beppe's farm.

b A buyer randomly selects a carrot from each farmer's crop. Calculate the probability that *neither* carrot is longer than 20 cm.

19

Statistical hypothesis testing

Contents:

Opening problem

Frank is a market gardener who grows tomatoes.

Last year, the average weight of his tomatoes was 106.3 g. The weights were approximately normally distributed with standard deviation 12.41 g.

This year, Frank used a new fertiliser to try to increase the weight of his crop. Frank has collected a random sample of 65 tomatoes and found the mean weight of the sample is 110.1 g.

Things to think about:

a How can Frank use this information to determine whether the new fertiliser was effective?

b How *significant* is Frank's evidence? Is it sufficient to conclude that the fertiliser was effective? Would it be more significant if the sample size was bigger?

A **statistical hypothesis** is a claim about a population parameter.

To assess the reasonableness of a statistical hypothesis, we conduct a **statistical hypothesis test**.

Statistical hypothesis tests have three key components:

- **formulating** statistical hypotheses
- collecting data from a sample and **calculating statistics** to test our hypotheses
- **making decisions** about the population based on what we see in the sample.

In previous years, you should have studied statistical hypotheses about population proportions.

In this Chapter, we will consider hypotheses about **population means** and **correlation coefficients**.

A   THE CENTRAL LIMIT THEOREM

If we want to make inferences about a population mean μ, the most obvious thing to do is to use the **sample mean** $\overline{x}$ to estimate μ.

It is unlikely that two samples from the same population will be exactly the same, so the value of $\overline{x}$ will *vary* between samples. The sample mean is therefore a **random variable**. We use the notation $\overline{X}_n$ to refer to the random variable of sample means for samples of size n.

$\overline{x}$ is the *observed value* of $\overline{X}_n$.

Discussion

Why does the size of the sample affect the distribution of sample means?

Investigation   The distribution of sample means

Before we can define a statistical hypothesis test for μ, we need to determine the distribution of $\overline{X}_n$.

In this Investigation we will explore the distribution of $\overline{X}_n$ for different types of populations.

Click on the icon to run a computer simulation which generates the distribution of sample   **SIMULATION**
means of size n for a given population. The population distribution may be "normal"
(bell-shaped) or be far from normal.

What to do:

1 Set the population distribution to *normal* with $\mu = 0$ and $\sigma = 1$.

    a Use the slider to change the sample size n. Comment on how the *shape* of the distribution of
      $\overline{X}_n$ changes as n increases.

    b Use the information underneath the histogram in the software to copy and complete this table:

Normal distribution, $\mu = 0$, $\sigma = 1$

n	Mean of $\overline{X}_n$	Standard deviation of $\overline{X}_n$
4		
9		
16		
25		
100		

    c What do you notice about the mean of $\overline{X}_n$ as n increases?

    d Calculate $\dfrac{\sigma}{\sqrt{n}}$ for each row of the table. Comment on your results and how they affect the
      shape of the distribution of $\overline{X}_n$.

2 Change the population distribution to *negatively skewed*. Keep $\mu = 0$ and $\sigma = 1$.

    a Copy and complete this table:

Negatively skewed, $\mu = 0$, $\sigma = 1$

n	Mean of $\overline{X}_n$	Standard deviation of $\overline{X}_n$
4		
9		
16		
25		
100		

    b For large values of n, does the distribution of $\overline{X}_n$ resemble that of the population?

    c As n increases, what happens to the mean and standard deviation of $\overline{X}_n$?

    d Comment on the similarities and differences from your observations of the normal distribution
      in **1**.

PRINTABLE
TABLES

3 Repeat **2** for the other non-normal distributions. Try experimenting with different
values of μ and σ, and comment on your findings.

From the **Investigation** you should have found that regardless of the population distribution, as the sample size n increases:

- the distribution of $\overline{X}_n$ becomes symmetric and bell-shaped
- the mean of $\overline{X}_n$ is close to the population mean
- the standard deviation of $\overline{X}_n$ decreases by a factor of $\dfrac{1}{\sqrt{n}}$.

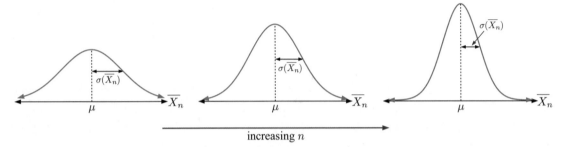

increasing n

This remarkable result is known as the **Central Limit Theorem**.

THE CENTRAL LIMIT THEOREM

> Consider a population with mean μ and standard deviation σ. Let the random variable $\overline{X}_n$ be the mean of a sample of size n drawn from this population.
>
> For sufficiently large n, $\overline{X}_n$ is approximately normally distributed with mean μ and standard deviation $\dfrac{\sigma}{\sqrt{n}}$. The larger the value of n, the better the approximation will be.

We note that:

- A *single* observation from the population is usually denoted with the random variable X. This distinguishes the distribution of the population from that of $\overline{X}_n$.

- It is sensible to ask what is meant by "sufficiently large n". Many texts suggest a "rule of thumb" of $n \geqslant 30$ to indicate n is large enough. However, this value depends on the "normality" of the population, which means how close it is to being bell-shaped.

The distribution of $\overline{X}_n$ becomes more normal as n increases.

- If the population is finite and the n observations are chosen without replacement, then n must also be small compared to the size of the population. This is to ensure that the observations are approximately independent.

- The population may not necessarily be normally distributed. However, if the population *is* normally distributed, then $\overline{X}_n$ will be normally distributed for any value of n.

The Central Limit Theorem allows us to estimate probabilities associated with $\overline{X}_n$ using what we learned in **Chapter 18**.

Example 1

🔊 Self Tutor

A random variable X has mean 60 and standard deviation 4. A sample of size 45 is taken from X. Find the probability that the mean of the sample is less than 59.

Since the sample has size $n = 45$, we apply the Central Limit Theorem.

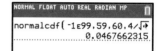

∴ $\overline{X}_{45}$ is approximately normally distributed with mean 60 and

standard deviation $\dfrac{4}{\sqrt{45}}$.

∴ $P(\overline{X}_{45} < 59) \approx 0.0468$

EXERCISE 19A

1 X is a random variable with mean 70 and standard deviation 12. Let $\overline{X}_{64}$ be the sample mean of 64 observations of X.

   a Find the mean and standard deviation of $\overline{X}_{64}$.

   b Is $\overline{X}_{64}$ approximately normally distributed? Explain your answer.

2 The random variable X is normally distributed with mean 50 and standard deviation 9.

   a For the sample means of size 5, $\overline{X}_5$, find the:

      i mean   **ii** standard deviation.

   b Can we say that $\overline{X}_5$ is normally distributed? Explain your answer.

3 Suppose X has mean 80 and standard deviation 9.

   a Find the mean and standard deviation of $\overline{X}_{36}$.

   b Is $\overline{X}_{36}$ approximately normally distributed? Explain your answer.

   c Find:

      i $P(75 \leqslant \overline{X}_{36} \leqslant 81)$ **ii** $P(\overline{X}_{36} > 82)$ **iii** $P(\overline{X}_{36} < 79)$

   d Can we find $P(70 < X < 85)$? Explain your answer.

4 Suppose Y has mean 3.6 and standard deviation 0.7 .

   a Find the mean and standard deviation of $\overline{Y}_{32}$.

   b Is $\overline{Y}_{32}$ approximately normally distributed? Explain your answer.

   c Find:

      i $P(3.5 < \overline{Y}_{32} < 3.7)$ **ii** $P(\overline{Y}_{32} > 3.7)$ **iii** $P(\overline{Y}_{32} < 3.3)$

5 The random variable X is normally distributed with mean 90 and standard deviation 20.

   a Is $\overline{X}_5$ normally distributed? Explain your answer.

   b Would you expect X or $\overline{X}_5$ to have a greater proportion of scores over 100? Explain your answer.

   c Find:

      i $P(X > 100)$ **ii** $P(\overline{X}_5 > 100)$

6 Two histograms are shown below. One is from a distribution X with mean 10 and standard deviation 10. The other is from the distribution of the sample means $\overline{X}_{64}$ selected from the distribution X. Note that the scales are not the same in the two histograms.

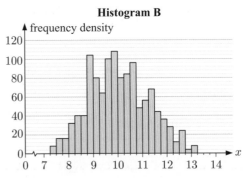

Histogram A

Histogram B

a Which of the two histograms is from $\overline{X}_{64}$? Give reasons for your answer.

b Use the appropriate histogram to estimate $P(\overline{X}_{64} < 9)$.

c Use the appropriate histogram to estimate the probability that $\overline{X}_{64}$ is within one standard deviation of the mean. How does this answer compare with using the normal approximation?

Example 2

◀)) **Self Tutor**

The contents of soft drink cans is normally distributed with mean 328 mL and standard deviation 7.2 mL. Find the probability that:

a an individual can contains less than 325 mL

b a box of 36 cans has average contents less than 325 mL per can.

a The contents X of an individual can is normally distributed with mean 328 mL and standard deviation 7.2 mL.

$\therefore$ $P(X < 325) \approx 0.338$

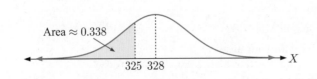

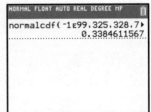

b The average contents $\overline{X}_{36}$ of a box of 36 cans is normally distributed with mean 328 mL per can and standard deviation $\dfrac{7.2}{\sqrt{36}} = 1.2$ mL.

$\therefore$ $P(\overline{X}_{36} < 325) \approx 0.006\,21$

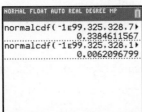

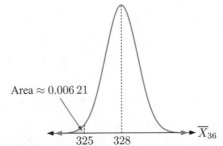

7 Let the random variable X be the IQ of 17 year old girls. Suppose X is normally distributed with mean 105 and standard deviation 15.

   a Find the probability that an individual 17 year old girl has an IQ of more than 110.

   b Find the probability that the mean IQ of a class of twenty 17 year old girls is greater than 110.

8 The energy content of a fruit bar is normally distributed with mean 1067 kJ and standard deviation 61.7 kJ. Find the probability that:

   a an individual bar has more than 1050 kJ of energy

   b the mean energy content of a sample of 30 fruit bars is more than 1050 kJ.

9 Customers at a clothing store are in the shop for a mean time of 18 minutes with standard deviation 5.3 minutes. What is the probability that, in a sample of 37 customers, the mean stay in the shop is between 17 and 20 minutes?

10 The mean sodium content of a box of cheese rings is 1183 mg with standard deviation 88.6 mg. Find the probability that the mean sodium content per box for a sample of 50 boxes lies between 1150 mg and 1200 mg.

11

The weight of oranges is normally distributed with mean 130 g and standard deviation 15 g. The oranges are sold in bags of 9.

   a Find the probability that an individual orange weighs more than 140 g.

   b Find the probability that the average weight of oranges in a bag is more than 140 g.

12 Suppose the duration of human pregnancies can be modelled by a normal distribution with mean 267 days and standard deviation 15 days. If a pregnancy lasts longer than 267 days it is said to be "overdue". If a pregnancy lasts less than 267 days it is said to be "premature".

   a What percentage of pregnancies will be overdue by between 1 and 2 weeks?

   b A certain obstetrician is providing prenatal care for 64 pregnant women.

   i Describe the distribution of $\overline{X}_{64}$.

   ii Find the probability that the mean duration of these 64 patients' pregnancies will be premature by at least one week.

   c Suppose the duration of human pregnancies is not actually normally distributed, but is positively skewed. Does this impact the answers to parts **a** and **b** above?

Discussion

Could the Central Limit Theorem reasonably be referred to as the "Fundamental Theorem of Statistics"?

Historical note

The French mathematician **Abraham de Moivre** is most famous for his formula linking complex numbers and trigonometry, but he was also responsible for the early development of the Central Limit Theorem. In an article from 1733, he used the normal distribution to approximate the distribution of the number of heads resulting from many tosses of a fair coin.

The work of de Moivre was extended in 1812 by country-man **Pierre-Simon Laplace**, but the theorem was not formalised and rigorously proven until the early 20th century work of the Russian mathematicians **Pafnuty Chebyshev**, **Andrey Markov**, and **Aleksandr Lyapunov**.

In 1889, **Sir Francis Galton** wrote about the normal distribution, with particular relevance to the Central Limit Theorem (though that name had not yet been used):

Abraham de Moivre

> *"I know of scarcely anything so apt to impress the imagination as the wonderful form of cosmic order expressed by the law of frequency of error. The law would have been personified by the Greeks if they had known of it. It reigns with serenity and complete self-effacement amidst the wildest confusion. The larger the mob, the greater the apparent anarchy, the more perfect is its sway. It is the supreme law of unreason."*

The name "Central Limit Theorem" was first used in 1920 by the Hungarian mathematician **George Pólya** (1887 - 1985). He used the word "central" because of the importance of the theorem in probability theory.

B   HYPOTHESIS TESTING FOR A POPULATION MEAN (Z-TEST)

Happy Cow Milk produces 1 litre cartons of milk. The company wants to know if the mean volume of milk in each carton μ is actually equal to 1 litre, so they can adjust the machines filling the cartons if necessary. The company therefore wishes to conduct a **statistical hypothesis test** for μ to determine how to proceed.

From previous data, the company knows that the volume of milk in each carton will be normally distributed with standard deviation $\sigma = 1.7$ mL $= 0.0017$ L.

In this Section we describe the steps involved in their hypothesis test.

FORMULATING HYPOTHESES

We consider **two** hypotheses when we conduct a statistical hypothesis test. These hypotheses are called:

- the **null hypothesis**, H_0, which states that the population parameter takes a certain value
- the **alternative hypothesis**, H_1, which states that the value of the population parameter is *different* to the value specified by H_0.

In the Happy Cow Milk example:

- The null hypothesis states that the mean volume of milk is *equal* to 1 litre, so we write H_0: $\mu = 1$.
- The form of the alternative hypothesis depends on what kind of difference the company wants to detect. If the company is concerned about:
  - over-filling, then they would use H_1: $\mu > 1$
  - under-filling, then they would use H_1: $\mu < 1$
  - any change, then they would use H_1: $\mu \neq 1$.

We describe H_1: $\mu > 1$ and H_1: $\mu < 1$ as **1-tailed alternative hypotheses**.

H_1: $\mu \neq 1$ is called a **2-tailed alternative hypothesis**.

THE TEST STATISTIC

When we collect data from a sample, we calculate statistics which can be used to test our hypotheses.

A **test statistic** is a *random variable* that summarises the information in a sample.

The distribution of the test statistic under the assumptions of H_0 is called the **null distribution**.

From **Section A**, we know that for a population that is normally distributed with mean μ and standard deviation σ, the sample mean $\overline{X}_n \sim N\left(\mu, \left(\frac{\sigma}{\sqrt{n}}\right)^2\right)$.

If the population is *not* normally distributed but the sample size is sufficiently large, we can apply the Central Limit Theorem to assume normality.

Under the assumptions of H_0, $\mu = \mu_0$.

$$\therefore \ \overline{X}_n \sim N\left(\mu_0, \left(\frac{\sigma}{\sqrt{n}}\right)^2\right)$$

$\therefore$ the Z-transformation of $\overline{X}_n$, $Z = \dfrac{\overline{X}_n - \mu_0}{\frac{\sigma}{\sqrt{n}}}$,   has the standard normal distribution $N(0, 1^2)$.

Consider a statistical hypothesis test of H_0: $\mu = \mu_0$ for a normally distributed population with standard deviation σ. Given a sample of size n with observed sample mean $\overline{x}$:

- the **test statistic** is $Z = \dfrac{\overline{X}_n - \mu_0}{\frac{\sigma}{\sqrt{n}}}$ which has **observed value** $z = \dfrac{\overline{x} - \mu_0}{\frac{\sigma}{\sqrt{n}}}$

- the **null distribution** is $Z \sim N(0, 1^2)$.

For example, suppose a quality inspector took a sample of 50 Happy Cow Milk milk cartons and found that the sample mean volume of milk $\overline{x} = 1.002$ litres.

The observed value of the test statistic is $z = \dfrac{1.002 - 1}{\frac{0.0017}{\sqrt{50}}} \approx 8.32$.

Since we call the test statistic Z, the hypothesis test about a population mean μ is often called a **Z-test**.

Discussion

Why do you think we use $Z = \dfrac{\overline{X}_n - \mu_0}{\frac{\sigma}{\sqrt{n}}}$ as the test statistic instead of $\overline{X}_n$ directly?

CALCULATING THE p-VALUE

"Extreme" values of the test statistic are unlikely, so observing such a value is evidence against the null hypothesis. However, we need a measure of just *how* extreme a value should be before we conclude that the null hypothesis should be rejected.

> The p-**value** of a test statistic is the probability of that result being observed if H_0 is true.

The meaning of "extreme" depends on whether the alternative hypothesis is 1-tailed or 2-tailed:

- If H_1 is a 1-tailed alternative hypothesis, we use *one* tail of the null distribution to calculate the p-value.

  ▸ If H_1: $\mu < \mu_0$, we use the lower tail.   ▸ If H_1: $\mu > \mu_0$, we use the upper tail.

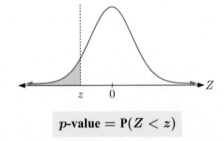

p-value $= P(Z < z)$

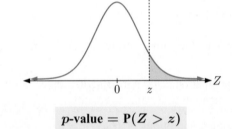

p-value $= P(Z > z)$

- If H_1 is the 2-tailed alternative hypothesis H_1: $\mu \neq \mu_0$, then we must consider *both* tails of the null distribution. We define "extreme" values as those which have probability less than or equal to that of the test statistic.

If $z \geqslant 0$,

$\quad p$-value $= P(Z \geqslant z$ or $Z \leqslant -z)$
$\quad\quad\quad\quad = 2 \times P(Z \geqslant z)$ {symmetry}

If $z < 0$,

$\quad p$-value $= P(Z \geqslant -z$ or $Z \leqslant z)$
$\quad\quad\quad\quad = 2 \times P(Z \geqslant -z)$ {symmetry}

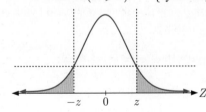

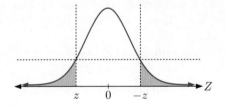

So, for a 2-tailed alternative hypothesis, p-**value** $= 2 \times P(Z \geqslant |z|)$.

MAKING DECISIONS

Although "extreme" values are unlikely to occur, it is still *possible* to observe such values in a sample. We therefore need a rule which defines how much evidence is required to reject the null hypothesis.

> The **significance level** α of a statistical hypothesis test is the largest p-value that would result in rejecting H_0. Any p-value less than or equal to α results in H_0 being rejected.

If a statistical hypothesis test has significance level α, the probability of *incorrectly* rejecting H_0 is α.

For example, suppose Happy Cow Milk was concerned about any change in volume. The quality inspector decided to test the hypothesis H_0: $\mu = 1$ against H_1: $\mu \neq 1$ at a significance level $\alpha = 0.01$.

Using the test statistic $z \approx 8.32$, p-value $\approx 2 \times P(Z > |8.32|)$

$$\approx 8.88 \times 10^{-17}$$

> The smaller the p-value, the more evidence there is against H_0.

Since the p-value is much less than the significance level α, the quality inspector has very strong evidence to reject H_0.

Important: We **must** choose a significance level **before** we test the hypothesis. It is *not* appropriate to calculate a p-value and then select a value of α so that H_0 will be rejected.

SUMMARY OF STEPS FOR A HYPOTHESIS TEST FOR A POPULATION MEAN (Z-TEST)

Step 1:   State the **null hypothesis** H_0: $\mu = \mu_0$ and **alternative hypothesis** H_1.

Step 2:   State the **significance level** α.

Step 3:   Calculate the **observed value** of the **test statistic:** $z = \dfrac{\overline{x} - \mu_0}{\frac{\sigma}{\sqrt{n}}}$.

Step 4:   Calculate the *p*-**value** using $Z \sim N(0, 1^2)$ as follows:

- If H_1: $\mu > \mu_0$, p-value $= P(Z > z)$.
- If H_1: $\mu < \mu_0$, p-value $= P(Z < z)$.
- If H_1: $\mu \neq \mu_0$, p-value $= 2 \times P(Z > |z|)$.

Step 5:   Reject H_0 if p-value $< \alpha$.

Step 6:   Interpret your decision in the context of the problem. Write your conclusion in a sentence.

Example 3   ◀⏺ Self Tutor

The manager of a restaurant chain goes to a seafood wholesaler and inspects a large catch of over 50 000 prawns. It is known that the population is normally distributed with standard deviation 4.2 grams. He will buy the catch if the mean weight exceeds 55 grams per prawn.

A random sample of 60 prawns is taken, and the mean weight is 56.2 grams. At a 5% significance level, should the manager purchase the catch?

Step 1:   Let μ be the population mean weight per prawn.
The hypotheses that should be considered are:
H_0: $\mu = 55$ {the mean weight is 55 grams per prawn}
H_1: $\mu > 55$ {the mean weight exceeds 55 grams per prawn}

Step 2:   The significance level is $\alpha = 0.05$.

Step 3: $\bar{x} = 56.2$ grams, $\sigma = 4.2$ grams, $n = 60$

The observed value of the test statistic is $z = \dfrac{56.2 - 55}{\frac{4.2}{\sqrt{60}}} \approx 2.21$

Step 4:   Since H_1: $\mu > 55$, p-value $= P(Z > z)$ where $Z \sim N(0, 1^2)$
$$\approx P(Z > 2.21)$$
$$\approx 0.0134$$

Step 5:   Since p-value $< 0.05 = \alpha$, we have enough evidence to reject H_0 in favour of H_1 on a 5% significance level.

Step 6:   Since we have accepted H_1, we conclude that the mean weight exceeds 55 grams per prawn. The manager should purchase the catch.

In a real-world problem, it is highly unlikely that the population standard deviation σ will be known, especially if the population mean μ is unknown. We would therefore usually calculate the standard deviation s of our sample, and use it as an estimate of σ.

Example 4   ◀️ Self Tutor

The fat content (in grams) of 30 randomly selected pasties at the local bakery was recorded:

15.1	14.8	13.7	15.6	15.1	16.1	16.6	17.4	16.1	13.9
17.5	15.7	16.2	16.6	15.1	12.9	17.4	16.5	13.2	14.0
17.2	17.3	16.1	16.5	16.7	16.8	17.2	17.6	17.3	14.8

For a mean fat content of pasties made at this bakery μ, conduct a 2-tailed hypothesis test of
H_0: $\mu = 16$ grams on a 10% level of significance.

Step 1: H_0: $\mu = 16$ {the mean fat content is 16 grams}
H_1: $\mu \neq 16$ {the mean fat content is *not* 16 grams}

Step 2:   The significance level is $\alpha = 0.1$.

Step 3:   The sample size $n = 30$.

Using technology, $\bar{x} = 15.9$ and $s \approx 1.36$.

We therefore estimate that $\sigma \approx 1.36$.

The observed value of the test statistic is

$$z \approx \dfrac{15.9 - 16}{\frac{1.36}{\sqrt{30}}} \approx -0.402$$

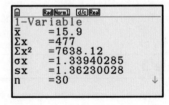

```
          Rad Norm1  d/c Real
1-Variable
x̄     =15.9
Σx    =477
Σx²   =7638.12
σx    =1.33940285
sx    =1.36230028
n     =30                  ↓
```

Step 4:   Assuming that a sample size $n = 30$ is large enough for the Central Limit Theorem to apply, we can assume normality of the test statistic.

Since H_1: $\mu \neq 16$, p-value $= 2 \times P(Z > |z|)$ where $Z \sim N(0, 1^2)$
$$\approx 2 \times P(Z > |-0.402|)$$
$$\approx 0.688$$

Step 5:   Since p-value $> 0.1 = \alpha$, we do not have enough evidence to reject H_0 in favour of H_1 on a 10% significance level. We therefore accept H_0.

Step 6:   Since we have accepted H_0, we cannot conclude that the mean fat content is appreciably different from 16 grams.

EXERCISE 19B

1 A population has known standard deviation $\sigma = 3.97$. A sample of size 36 is taken and the sample mean $\overline{x} = 23.75$. We are required to test the hypothesis H_0: $\mu = 25$ against H_1: $\mu < 25$.

  a Find: **i** the test statistic   **ii** the p-value.

  b What decision should be made at a 5% level?

2 A statistician believes that a population which has a standard deviation of 12.9, has a mean μ that is greater than 80. To test this he takes a random sample of 200 measurements, and finds the sample mean is 83.1. He then performs a hypothesis test with significance level $\alpha = 0.01$.

  a Write down the null and alternative hypotheses.

  b Find the observed value of the test statistic and state the null distribution.

  c Calculate the p-value.

  d Make a decision to reject or not reject H_0.

  e State the conclusion for the test.

3 Bags of salted cashew nuts state their net contents is 100 g. The manufacturer knows that the standard deviation of the population is 1.6 g. A customer claims that the bags have been lighter in recent purchases, so the factory quality control manager decides to investigate. He samples 40 bags and finds that their mean weight is 99.4 g.

Perform a hypothesis test at the 5% level of significance to determine whether the customer's claim is valid.

4 An alpaca breeder wants to produce fleece which is extremely fine. In 2013, his herd had mean fineness 20.3 microns with standard deviation 2.89 microns. The standard deviation remains relatively constant over time. In 2017, a sample of 80 alpacas from the herd was randomly selected, and the mean fineness was 19.2 microns.

Perform a 2-tailed hypothesis test at the 5% level of significance to determine whether the herd fineness has changed.

5 The length of screws produced by a machine is known to be normally distributed with standard deviation $\sigma = 0.08$ cm. The machine is supposed to produce screws with mean length $\mu = 2.00$ cm. A quality controller selects a random sample of 15 screws. She finds that the mean length of the 15 screws is $\overline{x} = 2.04$ cm with sample standard deviation $s = 0.09$ cm. Does this justify the need to adjust the machine on a 2% level of significance?

6

A machine packs sugar into 1 kg bags. It is known that the masses of the bags of sugar are normally distributed with standard deviation 1.5 g. A random sample of eight filled bags was taken and the masses of the bags measured to the nearest gram. Their masses in grams were:

1001, 998, 999, 1002, 1001, 1003, 1002, 1002.

Perform a test at the 1% level, to determine whether the machine under-fills the bags.

7   A market gardener claims that the carrots in his field have a mean weight of more than 50 grams. A prospective buyer will purchase the crop if the market gardener's claim is true. To test this she pulls 20 carrots at random, and finds that their individual weights in grams are:

57.6   34.7   53.9   52.5   61.8   51.5   61.3   49.2   56.8   55.9
57.9   58.8   44.3   58.3   49.3   56.0   59.5   47.0   58.0   47.2

a   Explain why it is reasonable to assume that the carrots' weights are normally distributed.

b   Determine whether the buyer will purchase the crop using a 5% level of significance.

Activity   Multiple testing and statistical fallacy

In many applications of statistical hypothesis testing, it is common to conduct multiple identical or very similar hypothesis tests simultaneously. For example, in genetics an experiment called a **DNA microarray** is used to measure expression levels of thousands of genes, each with their own set of hypotheses to test.

In this Activity, we will investigate the effects of conducting multiple hypothesis tests on the interpretation of individual outcomes.

What to do:

1   A normally distributed population has mean μ and standard deviation $\sigma = 5$.
Consider the following hypotheses for this population: H_0: $\mu = 2$
                                                            H_1: $\mu \neq 2$

Click on the icon to run a computer simulation which generates samples of size 10   **SIMULATION**
from the $N(2, 5^2)$ distribution. The above hypotheses are tested for each sample
at a significance level of α, and a p-value is calculated.

a   Write down the formula used to calculate the observed test statistic given a sample mean $\overline{x}$.

b   Set $\alpha = 0.05$. Copy and complete the following table by generating m samples and counting the number of times H_0 is rejected.

m	Number of times H_0 is rejected	Proportion of samples where H_0 was rejected
20		
50		
100		
500		
1000		
5000		
10 000		

c   Repeat **b** for:

   i $\alpha = 0.1$ **ii** $\alpha = 0.025$ **iii** $\alpha = 0.01$.

Comment on your results.

2   For the simulation in **1**, explain why:

a H_0 is true in every sample that the simulation generates

b   P(incorrectly reject H_0) $= \alpha$ for each sample

c   the *expected number* of samples where H_0 is incorrectly rejected is $m\alpha$.

3 Sabeen is a psychologist. She is writing a journal article about the effect of diet cola on a person's ability to concentrate. To account for possible confounding factors, Sabeen divides her data into 10 different age groups for each gender. For each age group and gender, she conducts a hypothesis test and obtains a p-value.

Sabeen's results are shown in the table below:

Age group	10 - 14	15 - 19	20 - 24	25 - 29	30 - 34
Male	0.296	0.143	0.305	0.378	0.169
Female	0.814	0.022	0.125	0.301	0.432

Age group	35 - 39	40 - 44	45 - 49	50 - 54	55+
Male	0.699	0.221	0.078	0.790	0.423
Female	0.987	0.643	0.448	0.672	0.789

a Which age group and gender do you think Sabeen is most likely to report on in her journal article? Explain your answer.

b Sabeen's colleague Mysha repeats Sabeen's experiment. She samples people exclusively from the group you identified in **a**. Do you think Mysha is likely to replicate Sabeen's results for this group? Explain your answer.

4 Most academic publications report findings which are "statistically significant" as these findings are more likely to yield more interesting results and lead to further research.

a Will all claims in academic publications based on statistical data necessarily be true? Explain your answer.

b Discuss the role that statistical interpretation plays in research ethics.

c The MMR vaccine controversy was caused by a fraudulent paper published in 1998 which claimed a causal relationship between the MMR vaccine and autism in children. The paper has since been retracted after thorough investigation. However, because of the widespread misconceptions that it has caused, it has been cited as "perhaps the most damaging medical hoax of the last 100 years".

    i Research the details of the MMR vaccine controversy. In particular, consider how the authors collected and used the data cited in the original paper.

    ii What other things should a statistician be mindful of when analysing data?

C   CRITICAL REGIONS

In a statistical hypothesis test:

- The **critical region** C is the set of all values of the test statistic which result in H_0 being rejected.
- The **acceptance region** A is the set of all values of the test statistic which result in H_0 being accepted.
- We can make a decision about H_0 using the test statistic directly by comparing it to a **critical value** c which is the value in the critical region which has the largest p-value associated with it.

In a Z-test with significance level α, the critical value is the value of the test statistic for which the p-value $= \alpha$.

1-TAILED ALTERNATIVE HYPOTHESES

For H_1: $\mu < \mu_0$, the critical value c satisfies:

$$P(Z < c) = \alpha$$

Any value in the shaded area has a p-value less than or equal to that of c, so the shaded area corresponds to the critical region.

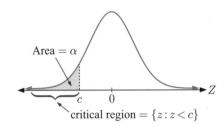

Similarly, for H_1: $\mu > \mu_0$, the critical value c satisfies:

$$P(Z > c) = \alpha$$

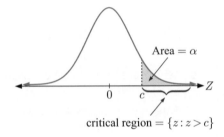

2-TAILED ALTERNATIVE HYPOTHESES

If H_1: $\mu \neq \mu_0$, then the critical value c satisfies:

$$2 \times P(Z > |c|) = \alpha$$

$$\therefore \quad P(Z > |c|) = \frac{\alpha}{2}$$

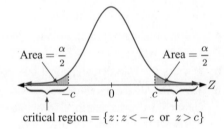

From the diagram it is clear that we need **two** critical values for a 2-tailed alternative. However, because of symmetry, we only actually need to perform one calculation.

Example 5

◀)) **Self Tutor**

Suppose μ is the mean time taken by students to travel to school in minutes. It is known that the population is normally distributed with standard deviation $\sigma = 2.3$ minutes. Miro wants to test the following hypotheses at the 5% level:

$$H_0: \ \mu = 5 \quad \text{\{the mean travel time is 5 minutes\}}$$
$$H_1: \ \mu > 5 \quad \text{\{the mean travel time is \textit{greater} than 5 minutes\}}$$

a For Miro's Z-test, find:

   i the critical value c **ii** the critical region $\mathcal{C}$ **iii** the acceptance region $\mathcal{A}$.

b Miro randomly selects 34 students and finds that the mean travel time for his sample is $\overline{x} = 6.4$ minutes.

   i Calculate the observed value of the test statistic.

   ii Is this sufficient evidence to reject H_0?

a   i We require c such that $P(Z > c) = 0.05$

$$\therefore \quad c \approx 1.64$$

ii $C = \{z : z > c\} \approx \{z : z > 1.64\}$

iii $A = \{z : z \leqslant c\} \approx \{z : z \leqslant 1.64\}$

b   i $\overline{x} = 6.4$ minutes, $\sigma = 2.3$ minutes, $n = 34$

The observed value of the test statistic is $z = \dfrac{6.4 - 5}{\frac{2.3}{\sqrt{34}}}$

$$\approx 3.55$$

ii $3.55 \in C$, so there is sufficient evidence to reject H_0.

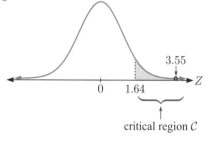

EXERCISE 19C

1 a Explain why the following sets of hypotheses, each tested at the same significance level α, have the same critical value, critical region, and acceptance region.

$$\begin{array}{ll} H_0: \ \mu = 0.1 & H_0: \ \mu = 120 \\ H_1: \ \mu < 0.1 & \text{and} \quad H_1: \ \mu < 120 \end{array}$$

b Could we make the same conclusion in **a** if the alternative hypotheses were both 2-tailed instead? Explain your answer.

2 The tread depth of car tyres is known to be normally distributed with standard deviation $\sigma = 0.5$ mm.

A minimum tread depth of 1.6 mm is required by law. A police officer wants to determine if the mean tyre tread depth of a taxi company's fleet satisfies the legal requirement.

a State the hypotheses that should be tested.

b For this test, and a 2.5% significance level, find:

   i the critical value   **ii** the critical region

   iii the acceptance region.

c The police officer randomly selects 10 taxis and measures the tread depth of each taxi's four tyres. The sample mean is found to be $\overline{x} = 1.51$ mm. Is this sufficient evidence to claim that the mean tread depth of the fleet does not meet the legal requirement?

3 The daily energy intake of students at a school is normally distributed with standard deviation $\sigma = 300$ kJ. Last year the mean energy intake of the students was 9210 kJ. To test whether the mean has changed, 50 students were selected, and their mean energy intake was $\overline{x} = 9100$ kJ.

a For this test, and a 2% significance level, find:

   i the critical region   **ii** the acceptance region.

b Is there sufficient evidence to claim that the mean energy intake has changed?

Example 6

🔊 **Self Tutor**

A population is normally distributed with mean μ and standard deviation $\sigma = 7$. To test the hypothesis H_0: $\mu = 40$ against H_1: $\mu \neq 40$, a random sample of size 60 was taken and found to have mean $\overline{x}$. For what values of $\overline{x}$ will the null hypothesis be rejected at the 5% level?

The observed test statistic is $z = \dfrac{\overline{x} - \mu_0}{\frac{\sigma}{\sqrt{n}}} = \dfrac{\overline{x} - 40}{\frac{7}{\sqrt{60}}}$.

The critical value c satisfies $\quad P(Z < -c \text{ or } Z > c) = 0.05$

$$\therefore \quad P(Z > c) = 0.025$$

$$\therefore \quad c \approx 1.96$$

```
Rad Norm1 ab/c Real
Inverse Normal
 xInv=1.95996398
```

H_0 will be rejected if: $z < -1.96$ $\qquad$ or $\qquad$ $z > 1.96$

$$\therefore \quad \dfrac{\overline{x} - 40}{\frac{7}{\sqrt{60}}} < -1.96 \qquad \text{or} \qquad \dfrac{\overline{x} - 40}{\frac{7}{\sqrt{60}}} > 1.96$$

$$\therefore \quad \overline{x} < 40 - 1.96 \times \tfrac{7}{\sqrt{60}} \quad \text{or} \qquad \overline{x} > 40 + 1.96 \times \tfrac{7}{\sqrt{60}}$$

$$\therefore \quad \overline{x} < 38.2 \qquad \qquad \text{or} \qquad \overline{x} > 41.8$$

So, the null hypothesis will be rejected if $\overline{x} < 38.2$ or $\overline{x} > 41.8$.

---

4 A normally distributed population has mean μ and standard deviation $\sigma = 4$. To test the hypothesis H_0: $\mu = -23$ against H_1: $\mu \neq -23$, a random sample of size 100 was taken and found to have mean $\overline{x}$. For what values of $\overline{x}$ will the null hypothesis be rejected at the 5% level?

5 The volume of orange juice dispensed by a machine is normally distributed with standard deviation 3 mL. A quality controller is required to adjust the machine if the mean volume dispensed is not 504 mL. In a routine test, the quality controller finds the mean volume $\overline{x}$ of 20 randomly selected bottles. For what values of $\overline{x}$ should the quality controller *not* adjust the machine? Use a 2% level of significance.

D   CORRELATION COEFFICIENTS

You should have seen previously how the relationship or **correlation** between two numerical variables can be described using a number called the **product moment correlation coefficient** r.

PROPERTIES OF THE PRODUCT MOMENT CORRELATION COEFFICIENT

- The values of r range from -1 to $+1$.
- The **sign** of r indicates the **direction** of the correlation.
  - If $r > 0$, the variables are **positively correlated**.
    An increase in one variable results in an increase in the other.
  - If $r < 0$, the variables are **negatively correlated**.
    An increase in one variable results in a decrease in the other.
  - If $r = 0$ there is **no correlation** between the variables.
- The **size** of r indicates the **strength** of the correlation.

Positive correlation		
$r = 1$	perfect positive correlation	
$0.95 \leqslant r < 1$	very strong positive correlation	
$0.87 \leqslant r < 0.95$	strong positive correlation	
$0.7 \leqslant r < 0.87$	moderate positive correlation	
$0.5 \leqslant r < 0.7$	weak positive correlation	
$0 < r < 0.5$	very weak positive correlation	

Negative correlation		
$r = -1$	perfect negative correlation	
$-1 < r \leqslant -0.95$	very strong negative correlation	
$-0.95 < r \leqslant -0.87$	strong negative correlation	
$-0.87 < r \leqslant -0.7$	moderate negative correlation	
$-0.7 < r \leqslant -0.5$	weak negative correlation	
$-0.5 < r < 0$	very weak negative correlation	

In previous years, you should have used technology to calculate r for a *sample* of bivariate data. In this Section, we will refer to r as the **sample product moment correlation coefficient**.

When we talk about the correlation between two variables in a *population*, we use the **population product moment correlation coefficent** ρ.

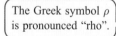
The Greek symbol ρ is pronounced "rho".

HYPOTHESIS TESTING FOR ρ

If no relationship or association exists between two variables, then there is **no correlation** between them, and $\rho = 0$.

In a hypothesis testing context, this is equivalent to saying that the null hypothesis is H_0: $\rho = 0$. The alternative hypothesis would then depend on the *direction* of the correlation that we would like to detect.

If we are interested in:

- **positive** correlation, we use H_1: $\rho > 0$
- **negative** correlation, we use H_1: $\rho < 0$
- **any** correlation, whether positive or negative, we use H_1: $\rho \neq 0$.

A statistical hypothesis test of H_0: $\rho = 0$ against H_1 follows the same general steps as those for testing a population mean μ listed on page **453**. However, in the case of the population product moment correlation coefficient ρ, the definition of the test statistic and p-value are beyond the scope of this course.

Instead, we will use technology to conduct tests about ρ. You should be able to use your calculator to obtain test statistics and p-values for such tests, and interpret these numbers in context.

**GRAPHICS
CALCULATOR
INSTRUCTIONS**

Example 7   ◀) Self Tutor

Consider the bivariate data set:

x	2	2	2	2	3	3	3	4	4	5
y	8	10	12	14	8	10	12	8	10	8

Conduct a hypothesis test at the 5% level of significance to determine whether the variables are negatively correlated.

Step 1: Let ρ be the population product moment correlation coefficient between the variables. We use the hypotheses:

H_0: $\rho = 0$ {there is no correlation between the variables}

H_1: $\rho < 0$ {the variables are negatively correlated}

Step 2: The significance level is $\alpha = 0.05$.

Step 3:

      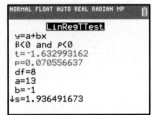

The observed value of the test statistic ≈ -1.63.

Step 4: p-value ≈ 0.0706

Step 5: Since p-value $\geqslant 0.05 = \alpha$, we do not have enough evidence to reject H_0 in favour of H_1 on a 5% significance level. We therefore accept H_0.

Step 6: Since we have accepted H_0, we conclude that the variables are not correlated.

EXERCISE 19D

1 Consider the bivariate data set:

x	1	2	3	4	5	6	7
y	4.1	6.2	4.5	9.8	7.7	5.8	8.8

A statistical hypothesis test is to be conducted to determine whether the variables are positively correlated at the 5% level of significance.

a State the hypotheses to be tested.

b Use technology to calculate the test statistic and p-value.

c Determine the outcome of the hypothesis test.

2 The following table shows the *Mathematics mark* and *Physics mark* for a group of students:

Mathematics mark (x)	13	10	8	14	6	11	10	5	12	13
Physics mark (y)	21	15	14	20	12	16	9	10	17	12

Is there enough evidence to conclude that there is an association between a student's Mathematics mark and Physics mark at the: **a** 5% level   **b** 1% level?

3 The *cholesterol level* in the bloodstream (x) and the *resting heart rate* (y) of 10 people are:

Cholesterol level (x)	5.32	5.54	5.45	5.06	6.13	5.00	4.90	6.00	6.70	4.75
Resting heart rate (y)	55	48	55	53	74	44	49	68	78	51

A researcher claims that there is a positive correlation between *cholesterol level* and *resting heart rate*. Based on this data, is the researcher's claim justified at the 1% level?

4 The table below shows the number of Sudoku puzzles (x) and the number of logic puzzles (y) solved in a three hour period by participants in a psychology experiment.

Number of Sudoku puzzles (x)	5	8	12	15	15	17	20	21	25	27
Number of logic puzzles (y)	3	11	9	6	15	13	25	15	13	20

Carry out a hypothesis test to determine whether the variables are linearly correlated at the 2% level.

5 Consider the data in the table below.

Weight of mother (x kg)	49	46	48	45	46	42	43	40	52	55	68
Birth weight of child (y kg)	3.8	3.1	2.5	3.0	3.2	2.8	3.1	2.9	3.4	2.9	3.9

The critical regions for a test of H_0: $\rho = 0$ against H_1: $\rho \neq 0$ for various significance levels are shown in the table alongside.

α	Critical region
0.1	$(-\infty, -1.83) \cup (1.83, \infty)$
0.05	$(-\infty, -2.26) \cup (2.26, \infty)$
0.02	$(-\infty, -2.82) \cup (2.82, \infty)$
0.01	$(-\infty, -3.25) \cup (3.25, \infty)$

a Use technology to calculate the value of the test statistic for this hypothesis test.

b Use the critical regions to determine the significance level(s) for which H_0 will be rejected.

c Check your answer to **b** using the p-value.

Discussion

For the data set alongside, the sample product moment correlation coefficient $r = 0$.

x	2	4	6	8	10	12	14
y	6	16	22	24	22	16	6

When we construct a scatter diagram of y against x, there clearly *is* a relationship between the variables, but it is quadratic rather than linear.

Discuss the limitations of hypothesis tests for ρ when we want to establish the existence of *any* relationship between the variables.

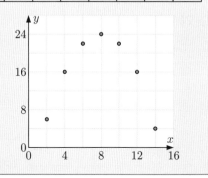

Review set 19A

1 The mature heights of a particular species of plant have mean 21 cm and standard deviation $\sqrt{90}$ cm. A random sample of 40 mature plants is taken and the mean height is calculated. Find the probability that this sample mean lies between 19.5 cm and 24 cm.

2 Suppose X has mean 50 and standard deviation 6.

  a Find the mean and standard deviation of $\overline{X}_{60}$.

  b Is $\overline{X}_{60}$ approximately normally distributed? Explain your answer.

  c Find $P(\overline{X}_{60} > 49)$.

3 Rosario owns an apricot orchard. Last year, the mean weight of his apricots was 90 grams. Rosario fears that severe droughts this year may have reduced the weight of his apricots. To address his concerns, Rosario randomly selected 20 apricots from his current harvest and recorded their weights in grams:

$$88 \quad 72 \quad 93 \quad 71 \quad 86 \quad 94 \quad 70 \quad 99 \quad 86 \quad 80$$
$$92 \quad 93 \quad 88 \quad 78 \quad 83 \quad 72 \quad 79 \quad 75 \quad 78 \quad 84$$

  a Write down the hypotheses that Rosario should test.

  b Are Rosario's concerns justified at the 1% level?

4 Yarni's resting pulse rate was 68 beats per minute for many years. However, with a sensible diet and exercise, she hopes to reduce this rate. After six months, Yarni's mean pulse rate is now 65 beats per minute, with standard deviation 1.732. These statistics were calculated from 42 measurements. Using the p-value, is there sufficient evidence at a 5% level, to conclude that Yarni's pulse rate has decreased?

5 In 2011, the mean weight of a gentoo penguin from a penguin colony was 7.82 kg, with standard deviation 1.83 kg. Exactly one year after these statistics were found, a sample of 48 penguins from the same colony were found to have mean weight 7.55 kg. Is there sufficient evidence, at a 5% level of significance, to suggest that the mean weight in 2012 differs from that in 2011?

6  Quickshave produces disposable razorblades. They claim that the mean number of shaves before a blade has to be thrown away is 13. A researcher wishing to test the claim asks 30 men to supply data on how many shaves they get from one Quickshave blade. The researcher found the sample mean was 12.8.

Given the population standard deviation $\sigma = 1.6$, use a critical region to test the manufacturer's claim at a 2% level.

7 The house prices in a large town are normally distributed. A real estate agent claims that the mean house price is £438 000. To test the real estate agent's claim, 30 recently sold houses were randomly selected and the mean price $\overline{x}$ was calculated. The standard deviation for this sample was £23 500. At a 2% level of significance, what values of $\overline{x}$, to the nearest £500, would support the agent's claim?

8 The bivariate data below compares the height (x) and weight (y) of 11 men.

Height (x cm)	164	167	173	176	177	178	180	180	181	184	192
Weight (y kg)	68	88	72	96	85	89	71	100	83	97	93

At a 5% level of significance, conduct a hypothesis test to determine whether or not the variables are linearly correlated.

Review set 19B

1 The weight X of a packet of rice is normally distributed with mean 500 g and standard deviation 3.5 g.

Let $\overline{X}_{20}$ be the average weight of a sample of 20 packets.

a Find the mean and standard deviation of $\overline{X}_{20}$.

b Is $\overline{X}_{20}$ normally distributed? Explain your answer.

2 The weight W of gourmet sausages produced by Hans is normally distributed with mean 63.4 g and standard deviation 6.33 g.

Suppose $\overline{W}_{10}$ is the mean weight of 10 randomly selected sausages.

a Find the probability that a randomly chosen sausage will weigh 60 grams or less.

b Find the mean and standard deviation of $\overline{W}_{10}$.

c Hence calculate the probability that $\overline{W}_{10} \leqslant 60$ g. Interpret your result.

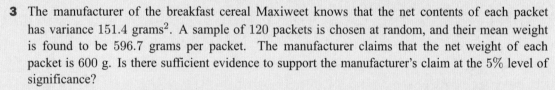

3 The manufacturer of the breakfast cereal Maxiweet knows that the net contents of each packet has variance 151.4 grams^2. A sample of 120 packets is chosen at random, and their mean weight is found to be 596.7 grams per packet. The manufacturer claims that the net weight of each packet is 600 g. Is there sufficient evidence to support the manufacturer's claim at the 5% level of significance?

4 It is claimed that the mean disposable income of households in a country town is £50 per week. To test this claim, 36 households were sampled. It was found that the mean disposable income of the 36 families was £47. If the population standard deviation $\sigma = £12$, test the claim at the 10% level.

5 The distance a golfer can hit a ball is distributed with mean $\mu = 115$ metres, and standard deviation $\sigma = 32$ metres.

After spending time with a professional, the golfer measured the distance of 30 drives. The results for the drives in metres were as follows:

$$
\begin{array}{cccccccccc}
100 & 126 & 93 & 171 & 131 & 94 & 136 & 144 & 138 & 110 \\
168 & 132 & 100 & 49 & 156 & 119 & 119 & 150 & 146 & 139 \\
149 & 145 & 122 & 56 & 140 & 118 & 115 & 73 & 105 & 133
\end{array}
$$

a Assuming that a sample size of 30 is large enough for the Central Limit Theorem to apply, is there enough evidence at the 5% level to claim that the golfer has improved?

b The golfer decided to take another sample of 50 drives. The mean of the 50 drives is the same as in **a**. Does this new sample provide enough evidence to claim that the golfer has improved using the same significance level?

6 Last year a ten-pin bowler scored a mean of 200 each game, with standard deviation 11.36. Her last 35 scores have had mean 196.4, and she is convinced that her form is below that of last year. Assuming her scores are normally distributed, use a critical region to test her claim at a 2% level.

7 A machine is used to fill packets with rice. The contents of each packet weighs X grams. X is normally distributed with mean μ grams and standard deviation 3.71 grams.

The mean weight μ is stated to be 500 g. To check this statement, a sample of 13 packets is selected, and the mean contents $\overline{x}$ is calculated.

The hypotheses for the test are H_0: $\mu = 500$ g and H_1: $\mu \neq 500$ g.

The null hypothesis is to be rejected if $\overline{x} < 498$ or $\overline{x} > 502$.

  a State the distribution of the sample mean $\overline{X}_{13}$.

  b Find the critical value(s) for this test.

  c Show that the significance level for the test is approximately 5.19%.

8 A basketballer takes 20 shots from each of ten different positions marked on the court. The table below shows how far each position is from the goal, and how many shots were successful:

Position	A	B	C	D	E	F	G	H	I	J
Distance from goal (x m)	2	5	3.5	6.2	4.5	1.5	7	4.1	3	5.6
Successful shots (y)	17	6	10	5	8	18	6	8	13	9

Conduct a hypothesis test to determine whether or not the *distance from goal* is negatively correlated with the *number of successful shots* at a 1% level of significance.

ANSWERS

EXERCISE 1A

1 **a** 7, 13, 19, 25, 31, ...   **b** 23, 19, 15, 11, 7, ...
  c 1, 4, 9, 16, 25, ...

2 **a** $t_2 = 3$ **b** $t_5 = 11$ **c** $t_{10} = 29$

3 **a** We start at 4 and increase by 3 each time.
  b $t_1 = 4$, $t_4 = 13$ **c** $t_8 = 25$

4 $t_1 = 7$, $t_2 = 9$, $t_3 = 11$, $t_4 = 13$

5 **a** $t_1 = 1$ **b** $t_5 = 13$ **c** $t_{27} = 79$

6 **a** **B**   **b** $t_{20} = 390$

7 **a** 5, 11, 17, 23   **b** 8, 3, −2, −7   **c** 4, 12, 36, 108
  d 80, 40, 20, 10   **e** 3, 7, 4, −3

8 **a** **B**   **b** **C**   **c** **A**

9 **a** $t_{n+2} = 10 - t_{n+1} = 10 - (10 - t_n) = t_n$
  b $t_{200} = 3$, $t_{375} = 7$

10 **a** $\frac{1}{2}, \frac{2}{3}, \frac{3}{5}, \frac{5}{8}$
  b
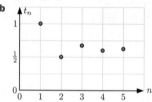

  c **Hint:** As $n \to \infty$, $t_{n+1} \to t$ and $t_n \to t$

11 **a** $\approx 0.707\,11$, $\approx 0.765\,37$, $\approx 0.752\,63$, $\approx 0.755\,36$
  c $\approx 0.754\,88$

12 $\approx 0.682\,33$

EXERCISE 1B

1 **a** arithmetic   **b** not arithmetic   **c** arithmetic
  d not arithmetic

2 **a** $t_1 = 5$, $d = 4$ **b** $t_1 = -4$, $d = 7$
  c $t_1 = 23$, $d = -5$ **d** $t_1 = -6$, $d = -9$

3 **a** **i** $t_1 = 19$, $d = 6$ **ii** $t_n = 6n + 13$
    iii $t_{15} = 103$
  b **i** $t_1 = 101$, $d = -4$ **ii** $t_n = 105 - 4n$
    iii $t_{15} = 45$
  c **i** $t_1 = 8$, $d = 1\frac{1}{2}$ **ii** $t_n = 1\frac{1}{2}n + 6\frac{1}{2}$
    iii $t_{15} = 29$
  d **i** $t_1 = 31$, $d = 5$ **ii** $t_n = 5n + 26$
    iii $t_{15} = 101$
  e **i** $t_1 = 5$, $d = -8$ **ii** $t_n = 13 - 8n$
    iii $t_{15} = -107$
  f **i** $t_1 = a$, $d = d$ **ii** $t_n = a + (n-1)d$
    iii $t_{15} = a + 14d$

4 **a** $t_1 = 6$, $d = 11$ **b** $t_n = 11n - 5$
  c $t_{50} = 545$ **d** yes, $t_{30} = 325$ **e** no

5 **a** $t_1 = 87$, $d = -4$ **b** $t_n = 91 - 4n$
  c $t_{40} = -69$ **d** t_{97}

6 **b** $t_1 = 1$, $d = 3$ **c** $t_{57} = 169$ **d** $t_{150} = 448$

7 **b** $t_1 = 32$, $d = -\frac{7}{2}$ **c** $t_{75} = -227$ **d** $n \geqslant 68$

8 **a** $t_1 = -12$, $d = 7$ **b** $t_{200} = 1381$ **c** no

9 **a** $k = 17\frac{1}{2}$ **b** $k = 4$ **c** $k = 4$ **d** $k = 0$
  e $k = -2$ or 3   **f** $k = -1$ or 3

10 **a** $t_n = 6n - 1$ **b** $t_n = -\frac{3}{2}n + \frac{11}{2}$
  c $t_n = -5n + 36$ **d** $t_n = -\frac{3}{2}n + \frac{1}{2}$

11 5, $6\frac{1}{4}$, $7\frac{1}{2}$, $8\frac{3}{4}$, 10

12 −1, $3\frac{5}{7}$, $8\frac{3}{7}$, $13\frac{1}{7}$, $17\frac{6}{7}$, $22\frac{4}{7}$, $27\frac{2}{7}$, 32

13 **a** 50, $48\frac{1}{2}$, 47, $45\frac{1}{2}$, 44   **b** $t_{35} = -1$

14 **a** Month 1: 5 cars   Month 4: 44 cars
    Month 2: 18 cars   Month 5: 57 cars
    Month 3: 31 cars   Month 6: 70 cars
  b The total number of cars made increases by 13 each month.
    So, the common difference $d = 13$.
  c 148 cars   **d** 20 months

15 **a** $t_1 = 34$, $d = 7$ **b** 111 online friends   **c** 18 weeks

16 **a** Day 1: 97.3 tonnes,   Day 2: 94.6 tonnes,
    Day 3: 91.9 tonnes
  b $d = -2.7$, the cattle eat 2.7 tonnes of hay each day.
  c $t_{25} = 32.5$. After 25 days (that is, July 25th) there will be
    32.5 tonnes of hay left.
  d 16.3 tonnes

EXERCISE 1C

1 **a** $t_1 = 5$, $r = 3$ **b** $t_1 = 72$, $r = \frac{1}{2}$
  c $t_1 = 2$, $r = -4$ **d** $t_1 = 6$, $r = -\frac{1}{3}$

2 **a** $b = 18$, $c = 54$ **b** $b = 2\frac{1}{2}$, $c = 1\frac{1}{4}$
  c $b = 3$, $c = -1\frac{1}{2}$

3 **a** **i** $t_1 = 3$, $r = 2$ **ii** $t_n = 3 \times 2^{n-1}$
    iii $t_9 = 768$
  b **i** $t_1 = 2$, $r = 5$ **ii** $t_n = 2 \times 5^{n-1}$
    iii $t_9 = 781\,250$
  c **i** $t_1 = 512$, $r = \frac{1}{2}$ **ii** $t_n = 512 \times 2^{1-n}$
    iii $t_9 = 2$
  d **i** $t_1 = 1$, $r = 3$ **ii** $t_n = 3^{n-1}$
    iii $t_9 = 6561$
  e **i** $t_1 = 12$, $r = \frac{3}{2}$ **ii** $t_n = 12 \times (\frac{3}{2})^{n-1}$
    iii $t_9 = \dfrac{3^9}{2^6}$
  f **i** $t_1 = \frac{1}{16}$, $r = -2$ **ii** $t_n = \frac{1}{16}(-2)^{n-1}$
    iii $t_9 = 16$

4 **a** $t_1 = 5$, $r = 2$ **b** $t_n = 5 \times 2^{n-1}$, $t_{15} = 81\,920$

5 **a** $t_1 = 12$, $r = -\frac{1}{2}$
  b $t_n = 12 \times (-\frac{1}{2})^{n-1}$, $t_{13} = \frac{3}{1024}$

6 $t_1 = 8$, $r = -\frac{3}{4}$, $t_{10} \approx -0.601$

7 $t_1 = 8$, $r = \frac{1}{\sqrt{2}}$ **Hint:** $t_n = 2^3 \times (2^{-\frac{1}{2}})^{n-1}$

8 **a** $r = 3$ **b** $t_{10} = 2916$ **c** 3 terms

9 **a** $k = \pm 14$ **b** $k = 2$ **c** $k = -2$ or 4

10 **a** $k = -3$ or 4
  b For $k = -3$, next term is $\frac{27}{2}$.
    For $k = 4$, next term is 24.

11 a $t_n = 3 \times 2^{n-1}$ **b** $t_n = 32 \times (-\frac{1}{2})^{n-1}$

c $t_n = 3 \times (\pm\sqrt{2})^{n-1}$ **d** $t_n = 10 \times (\pm\sqrt{2})^{1-n}$

12 a $t_9 = 13\,122$ **b** $t_{14} = 2916\sqrt{3} \approx 5050.66$

c $t_{18} = \frac{3}{32\,768} \approx 0.000\,091\,55$

EXERCISE 1D

1 £9367.58   **2** £3453.07

3 a £20\,977.42   **b** £23\,077.89

4 a £37\,305.85   **b** £7305.85

5 £11\,222.90   **6** £14\,977   **7** £11\,478

8 £22\,054.85   **9** £30\,003.40

10 a   i 1553 ants   **ii** 4823 ants   **b** 13 weeks

11 a ≈ 278 **b** year 2057

12 a   i 73 deer   **ii** 167 deer   **b** ≈ 30.5 years

13 a   i ≈ 2860 **ii** $\approx 184\,000$ **b** ≈ 14.5 years

EXERCISE 1E

1 a $S_3 = 18$ **b** $S_5 = 37$ **c** $S_{12} = 153$ **2** $t_5 = 7$

3 a   i $S_n = \sum_{k=1}^{n}(8k - 5)$ **ii** $S_5 = 95$

b   i $S_n = \sum_{k=1}^{n}(47 - 5k)$ **ii** $S_5 = 160$

c   i $S_n = \sum_{k=1}^{n}12(\frac{1}{2})^{k-1}$ **ii** $S_5 = 23\frac{1}{4}$

d   i $S_n = \sum_{k=1}^{n}2(\frac{3}{2})^{k-1}$ **ii** $S_5 = 26\frac{3}{8}$

e   i $S_n = \sum_{k=1}^{n}\frac{1}{2^{k-1}}$ **ii** $S_5 = 1\frac{15}{16}$

f   i $S_n = \sum_{k=1}^{n}k^3$ **ii** $S_5 = 225$

4 a 24   **b** 27   **c** 10   **d** 25   **e** 168   **f** 310

5 $S_{20} = \sum_{k=1}^{20}(3k - 1) = 610$

7 a $1 + 2 + 3 + \dots + (n-1) + n$
$n + (n-1) + (n-2) + \dots + 2 + 1$

b $S_n = \dfrac{n(n+1)}{2}$ **c** $a = 16,\ b = 3$

8 $S_n = \sum_{k=1}^{n}(2k - 1)$

9 $\sum_{k=1}^{n}(k+1)(k+2) = \dfrac{n(n^2 + 6n + 11)}{3}$,

$\sum_{k=1}^{10}(k+1)(k+2) = 570$

EXERCISE 1F

1 a, b, c 128

2 a 820   **b** $3087\frac{1}{2}$ **c** -1460 **d** -740

3 a 1749   **b** 184   **c** 2115   **d** $1410\frac{1}{2}$

4 a 160   **b** -630 **c** 135   **5** 203   **6** $-115\frac{1}{2}$

7 18 layers

8 a 65 seats   **b** 1914 seats   **c** 47\,850 seats

9 a 14\,025   **b** 71\,071   **c** 3367

10 $t_1 = 56,\ t_2 = 49$ **11** 10, 4, -2 or -2, 4, 10

12 $t_8 = 32$ **13** 2, 5, 8, 11, 14 or 14, 11, 8, 5, 2

14 a $t_1 = 7,\ t_2 = 10$ **b** $t_{20} = 64$ **15** $S_{80} = -80$

16 a

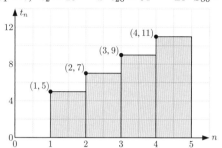

b S_n is the sum of the areas of the first n rectangles.

c   i The left side of each rectangle increases in length by 2 units from the previous rectangle, $t_{n+1} = t_n + 2$.

ii The area of the $(n + 1)$ th rectangle is t_{n+1}.
S_{n+1} is the sum of the areas of the first n rectangles and the $(n + 1)$ th rectangle, $S_{n+1} = S_n + t_{n+1}$.

EXERCISE 1G.1

1 a $\frac{3069}{128} \approx 24.0$ **b** $\approx 189\,134$ **c** $\frac{32\,769}{8192} \approx 4.00$

d ≈ 0.585

2 a $S_n = \dfrac{3 + \sqrt{3}}{2}\left((\sqrt{3})^n - 1\right)$ **b** $S_n = 24\left(1 - (\frac{1}{2})^n\right)$

c $S_n = 1 - (0.1)^n$ **d** $S_n = \frac{40}{3}\left(1 - (-\frac{1}{2})^n\right)$

3 a $t_1 = 3$ **b** $r = \frac{1}{3}$ **c** $t_5 = \frac{1}{27}$ **d** $S_5 = 4\frac{13}{27}$

4 a 3069   **b** $\frac{4095}{1024} \approx 4.00$ **c** $-134\,217\,732$

5 c £26\,361.59

6 a £5790   **b** $S_n = 100\,000((1.05)^n - 1)$

c $\approx$ £40\,710

7 a $S_1 = \frac{1}{2},\ S_2 = \frac{3}{4},\ S_3 = \frac{7}{8},\ S_4 = \frac{15}{16},\ S_5 = \frac{31}{32}$

b $S_n = \dfrac{2^n - 1}{2^n}$ **c** $S_n = 1 - (\frac{1}{2})^n = \dfrac{2^n - 1}{2^n}$

d as $n \to \infty$, $S_n \to 1$

e As $n \to \infty$, the sum of the fractions approaches the area of a 1×1 unit square.

8 $t_4 = \frac{2}{3}$ or 54   **9** $n = 11$

11 a $A_3 = £8000(1.03)^3 - (1.03)^2 R - 1.03R - R$

b $A_8 = £8000(1.03)^8 - (1.03)^7 R - (1.03)^6 R - (1.03)^5 R$
$\qquad - (1.03)^4 R - (1.03)^3 R - (1.03)^2 R - (1.03)R - R$
$\qquad = 0$

c $R = £1139.65$

EXERCISE 1G.2

1 a It is geometric with $t_1 = \frac{3}{10}$ and $r = \frac{1}{10}$, and we are adding all the terms. Therefore it is an infinite geometric series.

b Using **a**, $S = \dfrac{\frac{3}{10}}{1 - \frac{1}{10}} = \frac{3}{9} = \frac{1}{3}$ $\therefore$ $0.\overline{3} = \frac{1}{3}$

2 a $0.\overline{4} = \frac{4}{9}$ **b** $0.\overline{16} = \frac{16}{99}$ **c** $0.\overline{312} = \frac{104}{333}$

4 a 54   **b** 14.175

5 a 1   **b** $4\frac{2}{7}$ **6** $t_1 = 9,\ r = \frac{2}{3}$

7 $t_1 = 8,\ r = \frac{1}{5}$ and $t_1 = 2,\ r = \frac{4}{5}$

8 b $S_n = 19 - 20(0.9)^n$ **c** 19 seconds   **9** 70 cm

10   **a** $0.\overline{9} = \frac{9}{10} + \frac{9}{100} + \frac{9}{1000} + ...$ which is geometric with

     $t_1 = \frac{9}{10}$ and $r = \frac{1}{10}$

     $\therefore \;\; 0.\overline{9} = S = \dfrac{\frac{9}{10}}{1 - \frac{1}{10}} = 1$

  c

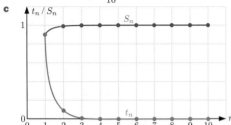

11 $x = \frac{1}{2}$

EXERCISE 1H.1

1   **a** $x^3 - 6x^2 + 12x - 8$

   b $16x^4 + 32x^3 + 24x^2 + 8x + 1$

   c $m^5 + 5m^4n + 10m^3n^2 + 10m^2n^3 + 5mn^4 + n^5$

   d $a^6 - 3a^4 + 3a^2 - 1$ **e** $x^4 - 8x^2 + 24 - \dfrac{32}{x^2} + \dfrac{16}{x^4}$

   f $x^4 + 4x^{\frac{7}{2}} + 6x^3 + 4x^{\frac{5}{2}} + x^2$

2   **a** $\binom{8}{4}(3x)^4(-2)^4$ **b** $\binom{10}{3}x^7(y^2)^3$

   c $\binom{11}{5}(2x)^6\left(-\dfrac{1}{x}\right)^5$ **d** $\binom{16}{6}(\sqrt{x})^{10}\,1^6$

3   **a** $T_{r+1} = \binom{12}{r}\left(\frac{1}{2}x\right)^{12-r}\left(\frac{1}{3}\right)^r$

   b   **i** $\dfrac{55}{39\,366}$ **ii** $\dfrac{11}{972}$ **iii** $\dfrac{55}{3456}$

4   **a**   15   **b**   29 326 171 875

5   **a** $a = \pm\frac{1}{2}$ **b** $a = \pm2$ **c** $a = \frac{1}{\sqrt{2}}$

6   **a** -1400 **b**   24   **c** -200

EXERCISE 1H.2

1   **a** $1 - x + x^2 - x^3 + ...$ **b** $-1 < x < 1$

   c ≈ 0.909 (using technology, $\frac{1}{1.1} = 0.\overline{90}$)

2   **a** $\frac{1}{2} + \frac{1}{16}x + \frac{3}{256}x^2 + \frac{5}{2048}x^3 + ...$ **b** $-4 < x < 4$

   c $\approx 0.527\,031$ (using technology, $\frac{1}{\sqrt{3.6}} \approx 0.527\,046$)

3 $\dfrac{1}{(1-2x)^2} = 1 + 4x + 12x^2 + 32x^3 + ...$

   The expansion is valid for $-\frac{1}{2} < x < \frac{1}{2}$.

   $\dfrac{1}{(0.96)^2} \approx 1.085\,056$

4   **a** $\sqrt{1+x} = 1 + \frac{1}{2}x - \frac{1}{8}x^2 + \frac{1}{16}x^3 - ....$

     The expansion is valid for $-1 < x < 1$.

   b   The series only converges provided $-1 < x < 1$.

   c $\approx 6.082\,763$

5   **a** $\dfrac{1}{(1+2x)^2} = 1 - 4x + 12x^2 - 32x^3 + ...$ **b**   75

REVIEW SET 1A

1   **a** $t_2 = 9$ **b** $t_6 = 19$ **c** $S_4 = 37$

2   **a**   arithmetic, $d = -8$ **b**   geometric, $r = -\frac{1}{2}$ **c**   neither

3 $k = -\frac{11}{2}$ **4**   **b** $t_1 = 6$, $r = \frac{1}{2}$ **c** $t_{16} \approx 0.000\,183$

5 $t_n = \frac{1}{6} \times 2^{n-1}$ or $-\frac{1}{6} \times (-2)^{n-1}$

6   23, 21, 19, 17, 15, 13, 11, 9   **7** $S_9 = 234$

8   **a** $10\frac{4}{5}$ **b** $16 + 8\sqrt{2}$ **9**   **a**   1272   **b** $302\frac{1}{2}$

10   27 metres

11   **a** $t_n = 89 - 3n$ **b** $t_n = \dfrac{2n+1}{n+3}$

   c $t_n = 100(0.9)^{n-1}$

12   **a** $1 + 4 + 9 + 16 + 25 + 36 + 49 = 140$

   b $\frac{4}{3} + \frac{5}{4} + \frac{6}{5} + \frac{7}{6} = \frac{99}{20}$

13   **a** $t_n = 3n + 1$

14 $x = 3$, $y = -1$, $z = \frac{1}{3}$ or $x = \frac{1}{3}$, $y = -1$, $z = 3$

15   **a**   £15 425.20   **b**   £15 453.77

16 $t_1 = 54$, $r = \frac{2}{3}$ and $t_1 = 150$, $r = -\frac{2}{5}$

   $|r| < 1$ in both cases, so the series will converge.

   For $t_1 = 54$, $r = \frac{2}{3}$, $S = 162$

   For $t_1 = 150$, $r = -\frac{2}{5}$, $S = 107\frac{1}{7}$

18   **a** $\binom{9}{3}y^6(-3z)^3$ **b** $\binom{10}{4}(2x)^6\left(-\dfrac{1}{\sqrt{x}}\right)^4$

19   **a** $a = \pm\frac{1}{4}$ **b** $a = \pm2\sqrt{2}$

20   **a** $1 + x + x^2 + x^3$ **b** $-1 < x < 1$

   c $\approx 1.052\,625$ (using technology, $\frac{1}{0.95} \approx 1.052\,632$)

21   **a** $\sqrt{1-x} = 1 - \frac{1}{2}x - \frac{1}{8}x^2 - \frac{1}{16}x^3 - ....$

     The expansion is valid for $-1 < x < 1$.

   b $\approx 3.872\,986$

REVIEW SET 1B

1 $t_1 = 7$, $t_2 = 14$, $t_3 = 28$, $t_4 = 56$

2   **a** $d = -5$ **b** $t_1 = 63$, $d = -5$ **c** $t_{37} = -117$

   d $t_{54} = -202$

3   **a** $t_1 = 3$, $r = 4$ **b** $t_n = 3 \times 4^{n-1}$, $t_9 = 196\,608$

4   **a** $t_n = 73 - 6n$ **b** $t_{34} = -131$

5   **a** $t_{81} = -36$ **b** $t_{35} = -1\frac{1}{2}$ **c** $S_{40} = 375$

6   **a** $S_{12} = 432$ **b** $S_{12} = \dfrac{12\,285}{256} \approx 48.0$

7   **a**   £8337.11   **b**   £8369.33   **c**   £8376.76

8   **a** $k = \pm\frac{2\sqrt{3}}{3}$ **b**   When $k = \frac{2\sqrt{3}}{3}$, $r = \frac{\sqrt{3}}{6}$

                              When $k = -\frac{2\sqrt{3}}{3}$, $r = -\frac{\sqrt{3}}{6}$

9   **a** $\dfrac{1331}{2100} \approx 0.634$ **b** $\dfrac{98}{15} \approx 6.53$

10   **a**   35.5 km   **b**   1183 km

11 $t_{11} = \dfrac{8}{19\,683} \approx 0.000\,406$

12   **a**   70   **b** ≈ 241 **c** $\dfrac{64}{1875}$

14   **a** $0 < x < 1$ (we require $|2x - 1| < 1$)   **b** $35\frac{5}{7}$

15   **a** ≈ 3470 iguanas   **b**   year 2029   **16** $x > -\frac{1}{2}$

17   **a** $T_{r+1} = \binom{9}{r}(3x)^{9-r}4^r$

   b   **i**   5 308 416   **ii**   7 838 208   **iii**   3 919 104

18   **a** $-18\,144$ **b**   0   **19**   17 040

20 $\dfrac{1}{(1+2x)^{\frac{3}{2}}} = 1 - 3x + \frac{15}{2}x^2 - \frac{35}{2}x^3 + ...$

   $\dfrac{1}{(1.1)^{\frac{3}{2}}} \approx 0.866\,563$

21   **a** $\dfrac{1}{(1-3x)^2} = 1 + 6x + 27x^2 + 108x^3 + ...$ **b**   91

EXERCISE 2A

1 **a** $5 - 2x$ **b** $-2x - 2$ **c** 11 **d** -2

2 **a** $-2 - 2x^2$ **b** $1 + 4x^2$ **c** -10 **d** -4

3 **a** $-4x^2 - 16x - 13$ **b** $10 - 2x^2$ **c** 14 **d** $-\frac{73}{16}$

4 **a** $25x - 42$ **b** $\sqrt{8}$ **c** -7 **d** 2

5 **a** $fg(x) = 9 - \sqrt{x^2 + 4}$
   Domain $= \{x : x \in \mathbb{R}\}$,   Range $= \{y : y \leqslant 7\}$
   b 53
   c $ff(x) = 9 - \sqrt{9 - \sqrt{x}}$
   Domain $= \{x : 0 \leqslant x \leqslant 81\}$,   Range $= \{y : 6 \leqslant y \leqslant 9\}$

6 **a** $-6x - 9$ **b** $x = -1$

7 **a** **i** $1 - 9x^2$ **ii** $1 + 6x - 3x^2$ **b** $x = -\frac{1}{9}$

8 **a** $fg(x) = \dfrac{1}{x - 3}$
   Domain $= \{x : x \neq 3\}$,   Range $= \{y : y \neq 0\}$

   b $fg(x) = \sqrt{\sin x}$
   Domain $= \{x : 2k\pi \leqslant x \leqslant (2k + 1)\pi, \ k \in \mathbb{Z}\}$
   Range $= \{y : 0 \leqslant y \leqslant 1\}$

   c $fg(x) = -\dfrac{1}{x^2 + 3x + 2}$
   Domain $= \{x : x \neq -1, \ x \neq -2\}$
   Range $= \{y : y \geqslant 4, \ y < 0\}$

9 **a** $R_g \cap D_f \neq \varnothing$
   b Domain $= \{x : x \in D_g, \ g(x) \in D_f\}$

EXERCISE 2B

1 **a** **i**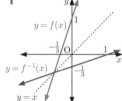

   b **i**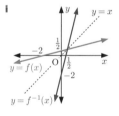

   ii, iii $f^{-1}(x) = \dfrac{x - 1}{3}$ **ii, iii** $f^{-1}(x) = 4x - 2$

2 **a** **i** $f^{-1}(x) = \dfrac{x - 5}{2}$ **b** **i** $f^{-1}(x) = -2x + \frac{3}{2}$

   ii   **ii**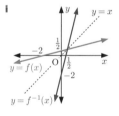

   c **i** $f^{-1}(x) = x - 3$ **ii**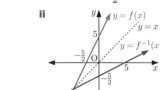

3 **a**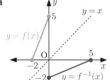

f:
Domain $= \{x : -2 \leqslant x \leqslant 0\}$
Range $= \{y : 0 \leqslant y \leqslant 5\}$
f^{-1}:
Domain $= \{x : 0 \leqslant x \leqslant 5\}$
Range $= \{y : -2 \leqslant y \leqslant 0\}$

   b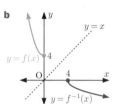

f:
Domain $= \{x : x \leqslant 0\}$
Range $= \{y : y \geqslant 4\}$
f^{-1}:
Domain $= \{x : x \geqslant 4\}$
Range $= \{y : y \leqslant 0\}$

   c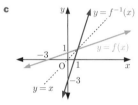

f:
Domain $= \{x : x \in \mathbb{R}\}$
Range $= \{y : y \in \mathbb{R}\}$
f^{-1}:
Domain $= \{x : x \in \mathbb{R}\}$
Range $= \{y : y \in \mathbb{R}\}$

   d

f:
Domain $= \{x : x \in \mathbb{R}\}$
Range $= \{y : y > 0\}$
f^{-1}:
Domain $= \{x : x > 0\}$
Range $= \{y : y \in \mathbb{R}\}$

   e

f:
Domain $= \{x : x \in \mathbb{R}\}$
Range $= \{y : y \in \mathbb{R}\}$
f^{-1}:
Domain $= \{x : x \in \mathbb{R}\}$
Range $= \{y : y \in \mathbb{R}\}$

   f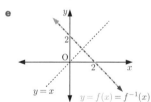

f:
Domain $= \{x : x \in \mathbb{R}\}$
Range $= \{y : y \in \mathbb{R}\}$
f^{-1}:
Domain $= \{x : x \in \mathbb{R}\}$
Range $= \{y : y \in \mathbb{R}\}$

4 $(f^{-1})^{-1}(x) = 2x - 5 = f(x)$

5 $f(x) = x$ and $f(x) = -x + c, \ c \in \mathbb{R}$

6 **a** $\{(2, 1), (4, 2), (5, 3)\}$ **b** inverse does not exist
   c $\{(1, 2), (0, -1), (2, 0), (3, 1)\}$
   d $\{(-1, -1), (0, 0), (1, 1)\}$

7 **a**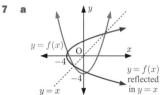

b no   **c** yes, it is $f^{-1}: x \mapsto \sqrt{x+4}$

8

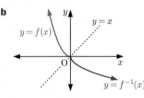

9 $f^{-1}: x \mapsto \dfrac{1}{x}$ $\therefore$ $f = f^{-1}$

10 a If a horizontal line cuts f more than once, a vertical line will cut its reflection in $y=x$ more than once, and so the reflection of f in $y=x$ will not be a function.

  b i is the only one.

  c ii Domain $= \{x : x \geqslant 1\}$ (or $\{x : x \leqslant 1\}$)

   iii Domain $= \{x : x \geqslant 1\}$ (or $\{x : x \leqslant -2\}$)

11 a $f^{-1}(x) = -\sqrt{x}$ **b**

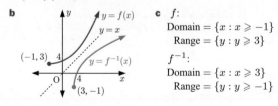

12 a

Every vertical line cuts the graph once. So, it is a function.
A horizontal line above the vertex cuts the graph **twice**. So, it does not have an inverse.

  b For $x \geqslant 2$, all horizontal lines cut the graph at most once.
  $\therefore$ $g(x)$ has an inverse.
  Hint: Inverse is $x = y^2 - 4y + 3$ for $y \geqslant 2$

  c g: Domain $= \{x : x \geqslant 2\}$, Range $= \{y : y \geqslant -1\}$
   g^{-1}: Domain $= \{x : x \geqslant -1\}$, Range $= \{y : y \geqslant 2\}$

  d Hint: Find $gg^{-1}(x)$ and $g^{-1}g(x)$ and show that they both equal x.

13 a $f^{-1}(x) = \sqrt{x-3} - 1$, $x \geqslant 3$

  b   **c** f:
Domain $= \{x : x \geqslant -1\}$
Range $= \{y : y \geqslant 3\}$
f^{-1}:
Domain $= \{x : x \geqslant 3\}$
Range $= \{y : y \geqslant -1\}$

14 a 10 **c** $x = 3$ **15 a i** 25 **ii** 16 **b** $x = 1$

16 $f^{-1}g^{-1}(x) = \dfrac{x+3}{8}$ and $(gf)^{-1}(x) = \dfrac{x+3}{8}$

17 b, c, d, and **e** are self-inverse functions.

18 a B is $(f(x), x)$

20 a Hint: Find $fg(x)$ and $gf(x)$.

  b i $f^{-1}(x) = \dfrac{\ln x - 1}{2}$ **ii** $f^{-1}(x) = 1 - e^x$

21 a

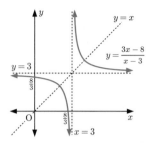

$y = \dfrac{3x - 8}{x - 3}$ is symmetrical about $y = x$

$\therefore$ f is a self-inverse function.

  b $f^{-1}(x) = \dfrac{3x - 8}{x - 3}$ and $f(x) = \dfrac{3x - 8}{x - 3}$

  $\therefore$ $f = f^{-1}$ $\therefore$ f is a self-inverse function

EXERCISE 2C

2 a $\dfrac{1}{x+1} - \dfrac{1}{x+2}$ **b** $\dfrac{-2}{x-2} + \dfrac{3}{x-3}$

  c $\dfrac{1}{x+2} + \dfrac{2}{x-4}$ **d** $\dfrac{1}{x+2} + \dfrac{2}{x-2}$

  e $\dfrac{3}{4(x+1)} + \dfrac{5}{4(x-3)}$ **f** $\dfrac{-1}{x-2} - \dfrac{4}{x+6}$

REVIEW SET 2A

1 a $2x^2 + 1$ **b** $4x^2 - 12x + 11$ **c** -1

2 a $6x - 3$ **b** $x = 1$

3 a $1 - 2\sqrt{x}$ **b** $\sqrt{1 - 2x}$ **c** 3

4 a $f^{-1}(x) = \dfrac{x-2}{4}$ **b** $f^{-1}(x) = \dfrac{3 - 4x}{5}$

5 $f^{-1}h^{-1}(x) = (hf)^{-1}(x) = x - 2$

6 a $h^{-1}(x) = 4 + \sqrt{x-3}$, $x \geqslant 3$

7 a, d

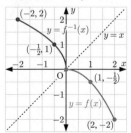

  b Any horizontal line cuts the graph at most once.
  c $g^{-1}(x) = -3 - \sqrt{x+2}$, $x \geqslant -2$
  e Range of $g = \{y : y \geqslant -2\}$
  f Domain of $g^{-1} = \{x : x \geqslant -2\}$,
   Range of $g^{-1} = \{y : y \leqslant -3\}$

8 a

  b Range $= \{y : 0 \leqslant y \leqslant 2\}$

  c i $x = \sqrt{3}$ **ii** $x = -\frac{1}{2}$

9 a

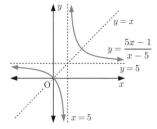

$y = \dfrac{5x-1}{x-5}$ is symmetrical about $y = x$

$\therefore$ f is a self-inverse function.

b $f^{-1}(x) = \dfrac{5x-1}{x-5}$ and $f(x) = \dfrac{5x-1}{x-5}$

$\therefore$ $f = f^{-1}$ $\therefore$ f is a self-inverse function.

10 a $\dfrac{2}{3(x+5)} + \dfrac{1}{3(x-1)}$ **b** $\dfrac{-2}{2x+1} + \dfrac{2}{x+3}$

REVIEW SET 2B

1 a $-4x^2 + 4x + 2$ **b** $5 - 2x^2$ **c** 2

2 a $fg(x) = \dfrac{1}{(x^2 - 4x + 3)^2}$

Domain $= \{x : x \neq 3, \ x \neq 1\}$,   Range $= \{y : y > 0\}$

b $fg(x) = \dfrac{1}{\cos x}$

Domain $= \{x : x \neq \frac{\pi}{2} + k\pi, \ k \in \mathbb{Z}\}$

Range $= \{y : y \leqslant -1, \ y \geqslant 1\}$

3 a i $6x^2 - 3x + 5$ **ii** $18x^2 + 57x + 45$

b $x = -\frac{5}{11}$

4 a $fg(x) = \ln \sqrt{x}$

Domain $= \{x : x > 0\}$,   Range $= \{y : y \in \mathbb{R}\}$

b $gf(x) = \sqrt{\ln x}$

Domain $= \{x : x \geqslant 1\}$,   Range $= \{y : y \geqslant 0\}$

5 a ⬛   **b** ⬛

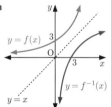

6 a $f^{-1}(x) = \dfrac{7-x}{4}$ **b** $f^{-1}(x) = \dfrac{5x-3}{2}$

7 a   **b** $f^{-1}(x) = \dfrac{x+7}{2}$

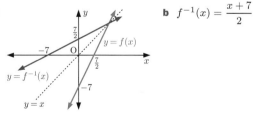

8 $f^{-1}h^{-1}(x) = (hf)^{-1}(x) = \dfrac{4x+6}{15}$

9 a $gf(x) = \dfrac{2}{3x+1}$ **b** $x = -\frac{1}{2}$

c i vertical asymptote $x = -\frac{1}{3}$,

horizontal asymptote $y = 0$

ii

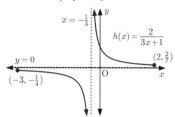

iii Range $= \{y : y \leqslant -\frac{1}{4} \ \text{ or } \ y \geqslant \frac{2}{7}\}$

10 a $\dfrac{1}{x-1} - \dfrac{3}{x-3}$ **b** $\dfrac{-10}{x+3} + \dfrac{11}{x+4}$

EXERCISE 3A

1 a $\frac{\pi}{2}$ **b** $\frac{\pi}{3}$ **c** $\frac{\pi}{6}$ **d** $\frac{\pi}{10}$ **e** $\frac{\pi}{20}$

f $\frac{3\pi}{4}$ **g** $\frac{5\pi}{4}$ **h** $\frac{3\pi}{2}$ **i** 2π **j** 4π

k $\frac{7\pi}{4}$ **l** 3π **m** $\frac{\pi}{5}$ **n** $\frac{4\pi}{9}$ **o** $\frac{23\pi}{18}$

2 a $\approx 0.641^c$ **b** $\approx 2.39^c$ **c** $\approx 5.55^c$ **d** $\approx 3.83^c$

e $\approx 6.92^c$

3 a $36°$ **b** $108°$ **c** $135°$ **d** $10°$ **e** $20°$

f $140°$ **g** $18°$ **h** $27°$ **i** $210°$ **j** $22.5°$

4 a $\approx 114.59°$ **b** $\approx 87.66°$ **c** $\approx 49.68°$

d $\approx 182.14°$ **e** $\approx 301.78°$

5 a F   **b** B   **c** D   **d** A   **e** E   **f** C

6 a

Degrees	0	45	90	135	180	225	270	315	360
Radians	0	$\frac{\pi}{4}$	$\frac{\pi}{2}$	$\frac{3\pi}{4}$	π	$\frac{5\pi}{4}$	$\frac{3\pi}{2}$	$\frac{7\pi}{4}$	2π

b

Deg.	0	30	60	90	120	150	180	210	240	270	300	330	360
Rad.	0	$\frac{\pi}{6}$	$\frac{\pi}{3}$	$\frac{\pi}{2}$	$\frac{2\pi}{3}$	$\frac{5\pi}{6}$	π	$\frac{7\pi}{6}$	$\frac{4\pi}{3}$	$\frac{3\pi}{2}$	$\frac{5\pi}{3}$	$\frac{11\pi}{6}$	2π

EXERCISE 3B

1 a 7 cm   **b** 12 cm   **c** ≈ 13.0 m

2 a 6 cm^2 **b** 48 cm^2 **c** ≈ 8.21 cm^2

3 a arc length ≈ 49.5 cm,   area ≈ 223 cm^2

b arc length ≈ 23.0 cm,   area ≈ 56.8 cm^2

4 a $\approx 0.686^c$ **b** 0.6^c

5 a $\theta = 0.75^c$, area $= 24$ cm^2 **b** $\theta = 1.68^c$, area $= 21$ cm^2

c $\theta \approx 2.32^c$, area $= 126.8$ cm^2

6 a ≈ 3.14 m   **b** ≈ 9.30 m^2

7 a ≈ 5.91 cm   **b** ≈ 18.9 cm

8 a $\alpha \approx 18.43$ **b** $\theta \approx 143.1$ **c** ≈ 387.3 m^2

9 a ≈ 11.7 cm   **b** $r \approx 11.7$ **c** ≈ 37.7 cm   **d** $\approx 3.23^c$

10 ≈ 25.9 cm   **11 b** ≈ 1 h 41 min   **12** ≈ 227 m^2

EXERCISE 3C

1

θ (radians)	0	$\frac{\pi}{2}$	π	$\frac{3\pi}{2}$	2π	$\frac{5\pi}{2}$
sine	0	1	0	-1	0	1
cosine	1	0	-1	0	1	0
tangent	0	undef.	0	undef.	0	undef.

2 a i $\frac{1}{\sqrt{2}} \approx 0.707$ **ii** $\frac{\sqrt{3}}{2} \approx 0.866$

b

θ (radians)	$\frac{\pi}{6}$	$\frac{\pi}{4}$	$\frac{\pi}{3}$	$\frac{3\pi}{4}$	$\frac{5\pi}{6}$	$\frac{7\pi}{6}$	$\frac{4\pi}{3}$	$\frac{7\pi}{4}$
θ (degrees)	30°	45°	60°	135°	150°	210°	240°	315°
sine	$\frac{1}{2}$	$\frac{1}{\sqrt{2}}$	$\frac{\sqrt{3}}{2}$	$\frac{1}{\sqrt{2}}$	$\frac{1}{2}$	$-\frac{1}{2}$	$-\frac{\sqrt{3}}{2}$	$-\frac{1}{\sqrt{2}}$
cosine	$\frac{\sqrt{3}}{2}$	$\frac{1}{\sqrt{2}}$	$\frac{1}{2}$	$-\frac{1}{\sqrt{2}}$	$-\frac{\sqrt{3}}{2}$	$-\frac{\sqrt{3}}{2}$	$-\frac{1}{2}$	$\frac{1}{\sqrt{2}}$
tangent	$\frac{1}{\sqrt{3}}$	1	$\sqrt{3}$	-1	$-\frac{1}{\sqrt{3}}$	$\frac{1}{\sqrt{3}}$	$\sqrt{3}$	-1

3 a

Quadrant	Degree measure	Radian measure	$\cos\theta$	$\sin\theta$	$\tan\theta$
1	$0° < \theta < 90°$	$0 < \theta < \frac{\pi}{2}$	+ve	+ve	+ve
2	$90° < \theta < 180°$	$\frac{\pi}{2} < \theta < \pi$	-ve	+ve	-ve
3	$180° < \theta < 270°$	$\pi < \theta < \frac{3\pi}{2}$	-ve	-ve	+ve
4	$270° < \theta < 360°$	$\frac{3\pi}{2} < \theta < 2\pi$	+ve	-ve	-ve

    b **i** 1 and 4   **ii** 2 and 3   **iii** 3   **iv** 2

4 a $\cos\frac{25\pi}{9} = \cos\left(\frac{7\pi}{9} + 2\pi\right) = \cos\frac{7\pi}{9}$

    b $\sin\left(-\frac{2\pi}{7}\right) = \sin\left(\frac{12\pi}{7} - 2\pi\right) = \sin\frac{12\pi}{7}$

    c $\tan\frac{15\pi}{11} = \tan\left(\frac{4\pi}{11} + \pi\right) = \tan\frac{4\pi}{11}$

5 a The points have the same y-coordinate.

    b The x-coordinates of the points have the same magnitude but are opposite in sign.

6 Q has coordinates $(\cos(-\theta), \sin(-\theta))$.

    a The y-coordinates of P and Q have the same magnitude but are opposite in sign.
    $\therefore\ \sin(-\theta) = -\sin\theta$

    b P and Q have the same x-coordinate.
    $\therefore\ \cos(-\theta) = \cos\theta$

7 a **i** The angle between [OP] and the positive x-axis is $\frac{\pi}{2} - \theta$.

    $\therefore$ P is $\left(\cos\left(\frac{\pi}{2} - \theta\right),\ \sin\left(\frac{\pi}{2} - \theta\right)\right)$

    ii In $\triangle$OXP, $\sin\theta = \dfrac{XP}{OP} = \dfrac{XP}{1}$

    $\therefore$ XP $= \sin\theta$

    iii In $\triangle$OXP, $\cos\theta = \dfrac{OX}{OP} = \dfrac{OX}{1}$

    $\therefore$ OX $= \cos\theta$

    b **i** $\cos\left(\frac{\pi}{2} - \theta\right) = $ XP $= \sin\theta$
    ii $\sin\left(\frac{\pi}{2} - \theta\right) = $ OX $= \cos\theta$

    c **i** $\cos\frac{\pi}{5} = \sin\frac{3\pi}{10} \approx 0.809$
    ii $\sin\frac{2\pi}{5} = \cos\frac{\pi}{10} \approx 0.951$

EXERCISE 3D

1

	a	b	c	d	e
$\sin\theta$	$\frac{1}{\sqrt{2}}$	$\frac{1}{\sqrt{2}}$	$-\frac{1}{\sqrt{2}}$	0	$-\frac{1}{\sqrt{2}}$
$\cos\theta$	$\frac{1}{\sqrt{2}}$	$-\frac{1}{\sqrt{2}}$	$\frac{1}{\sqrt{2}}$	-1	$-\frac{1}{\sqrt{2}}$
$\tan\theta$	1	-1	-1	0	1

2

	a	b	c	d	e
$\sin\beta$	$\frac{1}{2}$	$\frac{\sqrt{3}}{2}$	$-\frac{1}{2}$	$-\frac{\sqrt{3}}{2}$	$-\frac{1}{2}$
$\cos\beta$	$\frac{\sqrt{3}}{2}$	$-\frac{1}{2}$	$-\frac{\sqrt{3}}{2}$	$\frac{1}{2}$	$\frac{\sqrt{3}}{2}$
$\tan\beta$	$\frac{1}{\sqrt{3}}$	$-\sqrt{3}$	$\frac{1}{\sqrt{3}}$	$-\sqrt{3}$	$-\frac{1}{\sqrt{3}}$

3 a $\cos\frac{2\pi}{3} = -\frac{1}{2}$, $\sin\frac{2\pi}{3} = \frac{\sqrt{3}}{2}$, $\tan\frac{2\pi}{3} = -\sqrt{3}$

    b $\cos\left(-\frac{\pi}{4}\right) = \frac{1}{\sqrt{2}}$, $\sin\left(-\frac{\pi}{4}\right) = -\frac{1}{\sqrt{2}}$, $\tan\left(-\frac{\pi}{4}\right) = -1$

4 a $\cos\frac{\pi}{2} = 0$, $\sin\frac{\pi}{2} = 1$ **b** $\tan\frac{\pi}{2}$ is undefined

5 a $\frac{3}{4}$ **b** $\frac{1}{4}$ **c** 3   **d** $\frac{1}{4}$ **e** $-\frac{1}{4}$ **f** 1
    g $\sqrt{2}$ **h** $\frac{1}{2}$ **i** $\frac{1}{2}$ **j** 2   **k** -1 **l** $-\sqrt{3}$

6 a $\frac{\pi}{6}, \frac{5\pi}{6}$ **b** $\frac{\pi}{3}, \frac{2\pi}{3}$ **c** $\frac{\pi}{4}, \frac{7\pi}{4}$
    d $\frac{2\pi}{3}, \frac{4\pi}{3}$ **e** $\frac{3\pi}{4}, \frac{5\pi}{4}$ **f** $\frac{4\pi}{3}, \frac{5\pi}{3}$

7 a $\frac{\pi}{4}, \frac{5\pi}{4}$ **b** $\frac{3\pi}{4}, \frac{7\pi}{4}$ **c** $\frac{\pi}{3}, \frac{4\pi}{3}$
    d $0, \pi, 2\pi$ **e** $\frac{\pi}{6}, \frac{7\pi}{6}$ **f** $\frac{2\pi}{3}, \frac{5\pi}{3}$

8 a $\frac{\pi}{6}, \frac{11\pi}{6}, \frac{13\pi}{6}, \frac{23\pi}{6}$ **b** $\frac{7\pi}{6}, \frac{11\pi}{6}, \frac{19\pi}{6}, \frac{23\pi}{6}$ **c** $\frac{3\pi}{2}, \frac{7\pi}{2}$

9 a $\theta = \frac{\pi}{3}, \frac{5\pi}{3}$ **b** $\theta = \frac{\pi}{3}, \frac{2\pi}{3}$ **c** $\theta = \pi$
    d $\theta = \frac{\pi}{2}$ **e** $\theta = \frac{3\pi}{4}, \frac{5\pi}{4}$ **f** $\theta = \frac{\pi}{2}, \frac{3\pi}{2}$
    g $\theta = 0, \pi, 2\pi$ **h** $\theta = \frac{\pi}{4}, \frac{3\pi}{4}, \frac{5\pi}{4}, \frac{7\pi}{4}$
    i $\theta = \frac{5\pi}{6}, \frac{11\pi}{6}$ **j** $\theta = \frac{\pi}{3}, \frac{2\pi}{3}, \frac{4\pi}{3}, \frac{5\pi}{3}$

10 a $\theta = k\pi,\ k \in \mathbb{Z}$ **b** $\theta = \frac{\pi}{2} + k\pi,\ k \in \mathbb{Z}$

EXERCISE 3E

1 a $\cos\theta = \pm\frac{2\sqrt{2}}{3}$ **b** $\cos\theta = \pm\frac{\sqrt{5}}{3}$ **c** $\cos\theta = 0$
    d $\cos\theta = \pm1$

2 a $\sin\theta = \pm\frac{\sqrt{15}}{4}$ **b** $\sin\theta = \pm\frac{2\sqrt{6}}{5}$ **c** $\sin\theta = \pm\frac{4}{5}$
    d $\sin\theta = 0$

3 a $\sin\theta = \frac{3}{5}$ **b** $\cos\theta = -\frac{3}{5}$ **c** $\tan\theta = -\frac{1}{\sqrt{15}}$
    d $\sin\theta = \frac{2\sqrt{10}}{7}$ **e** $\tan\theta = -\frac{\sqrt{21}}{2}$ **f** $\cos\theta = -\frac{\sqrt{13}}{7}$

4 a $\cos\theta = \frac{1}{\sqrt{5}}$, $\sin\theta = \frac{2}{\sqrt{5}}$
    b $\cos\theta = -\frac{3}{\sqrt{10}}$, $\sin\theta = \frac{1}{\sqrt{10}}$
    c $\cos\theta = -\frac{4}{\sqrt{17}}$, $\sin\theta = -\frac{1}{\sqrt{17}}$
    d $\cos\theta = \sqrt{\frac{5}{14}}$, $\sin\theta = -\frac{3}{\sqrt{14}}$

EXERCISE 3F

1 a $\theta \approx 0.322$ or 3.46 **b** $\theta \approx 1.13$ or 5.16
    c $\theta \approx 0.656$ or 2.49 **d** $\theta \approx 1.82$ or 4.46
    e $\theta \approx 3.03$ or 6.17 **f** $\theta \approx 3.31$ or 6.12

2 a $\theta \approx -1.67$ or 1.67 **b** $\theta \approx 0.927$ or 2.21
    c $\theta \approx -2.16$ or 0.983 **d** $\theta \approx -0.644$ or 0.644
    e $\theta \approx -0.695$ or 2.45 **f** $\theta \approx -2.45$ or -0.690

EXERCISE 3G

1 a $\approx 2\theta + \theta^2$ **b** $\approx \frac{3}{2}\theta^2 - 1$ **c** $\approx \frac{3}{2}\theta^2 + 2\theta - 3$
    d $\approx 2 - 2\theta^2$ **e** $\approx 5 - 4\theta^2$ **f** $\approx \frac{1}{4} + \frac{2}{3}\theta - \frac{1}{2}\theta^2$

3 $\approx -1 + 4\theta - \frac{9}{2}\theta^2$ **5** ≈ 9990 m

REVIEW SET 3A

1 a $\frac{2\pi}{3}$ **b** $\frac{5\pi}{4}$ **c** $\frac{5\pi}{6}$ **d** 3π

2 **a** $72°$ **b** $225°$ **c** $140°$ **d** $330°$

3

4 12 cm

5 **a**

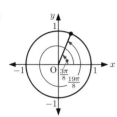

b

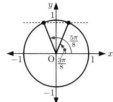

c

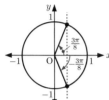

d

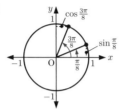

6

	a	b	c
$\sin \theta$	$\frac{1}{2}$	$-\frac{\sqrt{3}}{2}$	$-\frac{1}{\sqrt{2}}$
$\cos \theta$	$-\frac{\sqrt{3}}{2}$	$-\frac{1}{2}$	$-\frac{1}{\sqrt{2}}$
$\tan \theta$	$-\frac{1}{\sqrt{3}}$	$\sqrt{3}$	1

7 **a** **i** $60°$ **ii** $\frac{\pi}{3}$ **b** $\frac{\pi}{3}$ units **c** $\frac{\pi}{6}$ units^2

8 $\frac{1}{\sqrt{15}}$ **9** **a** $\frac{\sqrt{3}}{2}$ **b** 0 **c** $\frac{1}{2}$

10 **a** $\cos x = -\frac{2}{\sqrt{7}}$ **b** $\sin x = -\sqrt{\frac{3}{7}}$ **11** $\tan \theta = \sqrt{\frac{2}{5}}$

12 **a** $\theta \approx 1.91$ or 4.37 **b** $\theta \approx 0.848$ or 2.29
  c $\theta \approx 2.03$ or 5.18

13 **a** $\approx 1 + \frac{3}{2}\theta^2$ **b** $\approx 4 - 4\theta^2$

REVIEW SET 3B

1 **a** $\approx 1.239^c$ **b** $\approx 2.175^c$ **c** $\approx -2.478^c$

2 **a** $\approx 171.89°$ **b** $\approx 83.65°$ **c** $\approx 24.92°$
  d $\approx -302.01°$

3 ≈ 111 cm^2

4 **a** $\cos \frac{3\pi}{2} = 0$, $\sin \frac{3\pi}{2} = -1$
  b $\cos\left(-\frac{\pi}{2}\right) = 0$, $\sin\left(-\frac{\pi}{2}\right) = -1$

5 **a** $\frac{5\pi}{6}, \frac{7\pi}{6}$ **b** $\frac{\pi}{4}, \frac{3\pi}{4}$ **c** $\frac{5\pi}{6}, \frac{11\pi}{6}$

6 **a** $\theta = \pi$ **b** $\theta = \frac{\pi}{3}, \frac{2\pi}{3}, \frac{4\pi}{3}, \frac{5\pi}{3}$

7 perimeter ≈ 28.9 cm, area ≈ 38.0 cm^2

9 **a** $\frac{\sqrt{7}}{4}$ **b** $-\frac{\sqrt{7}}{3}$ **c** $\frac{3}{4}$

10 **a** $2\frac{1}{2}$ **b** $1\frac{1}{2}$ **c** $-\frac{1}{2}$

12 $\approx -1 + \frac{23}{8}\theta^2$

EXERCISE 4A

1 **a** vertical translation 1 unit downwards
  b horizontal translation $\frac{\pi}{4}$ to the right
  c vertical dilation, scale factor 2
  d horizontal dilation, scale factor $\frac{1}{4}$
  e vertical dilation, scale factor $\frac{1}{2}$
  f horizontal dilation, scale factor 4
  g reflection in the x-axis
  h translation $\frac{\pi}{3}$ units right, 2 units upwards

2 **a** $\frac{2\pi}{5}$ **b** $\frac{10\pi}{3}$ **c** 2 **d** $\frac{2\pi}{3}$ **e** 6π **f** 100

3 **a** $b = \frac{2}{5}$ **b** $b = 3$ **c** $b = \frac{1}{6}$ **d** $b = \frac{\pi}{2}$ **e** $b = \frac{\pi}{50}$

4 **a** maximum 4, minimum -4
  b maximum 8, minimum 2
  c maximum -2, minimum -6

5 $|a| =$ amplitude, $b = \dfrac{2\pi}{\text{period}}$, $c =$ horizontal translation, $d =$ vertical translation

6 **a**

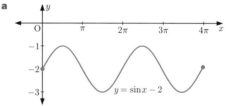

  b

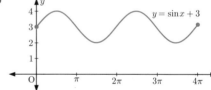

  c

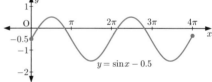

  d

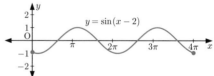

  e

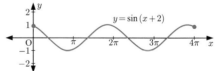

  f

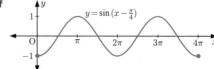

g

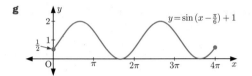

$y = \sin\left(x - \frac{\pi}{6}\right) + 1$

h

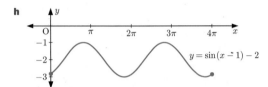

$y = \sin(x - 1) - 2$

i

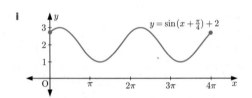

$y = \sin\left(x + \frac{\pi}{4}\right) + 2$

j

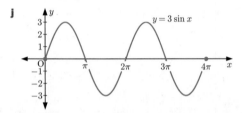

$y = 3\sin x$

k

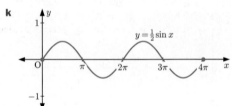

$y = \frac{1}{2}\sin x$

l

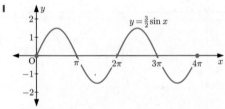

$y = \frac{3}{2}\sin x$

m

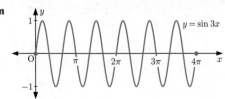

$y = \sin 3x$

n

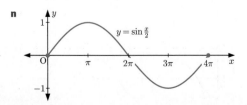

$y = \sin\frac{x}{2}$

o

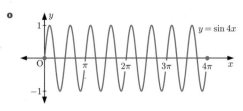

$y = \sin 4x$

7 a

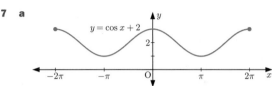

$y = \cos x + 2$

b

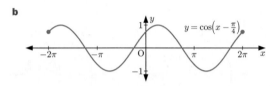

$y = \cos\left(x - \frac{\pi}{4}\right)$

c

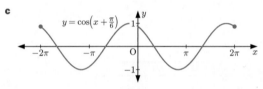

$y = \cos\left(x + \frac{\pi}{6}\right)$

d

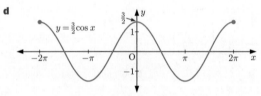

$y = \frac{3}{2}\cos x$

e

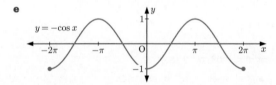

$y = -\cos x$

f

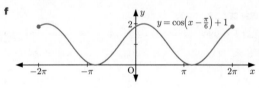

$y = \cos\left(x - \frac{\pi}{6}\right) + 1$

g

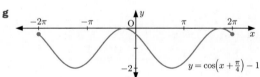

$y = \cos\left(x + \frac{\pi}{4}\right) - 1$

h

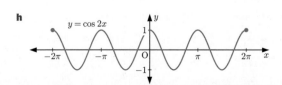

$y = \cos 2x$

i

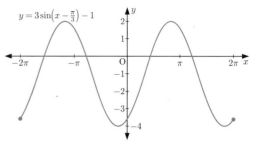

$y = \cos \frac{x}{2}$

8 a

$y = 4\sin x$

$\left(\frac{5\pi}{6}, 2\right)$

$\left(\frac{7\pi}{4}, -2\sqrt{2}\right)$

b i $y = 2$ **ii** $y = -2\sqrt{2} \approx -2.83$

9 a $d > 3$ **b** $d < -3$ **c** $-3 < d < 3$

10 a A horizontal dilation with scale factor $\frac{1}{3}$, then a vertical dilation with scale factor 2.

b A vertical dilation with scale factor 2, then a reflection in the x-axis.

c A vertical dilation with scale factor 3, then a translation 5 units downwards.

d A horizontal dilation with scale factor $\frac{1}{2}$, then a translation $\frac{\pi}{6}$ units left.

11 a

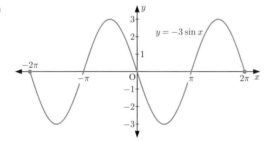

$y = -3\sin x$

b

$y = \cos 2x + 1$

c

$y = \frac{1}{2}\sin\left(x + \frac{\pi}{6}\right) - \frac{1}{3}$

d

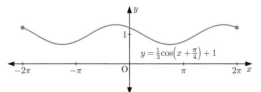

$y = \frac{1}{3}\cos\left(x + \frac{\pi}{4}\right) + 1$

e

$y = 3\sin\left(x - \frac{\pi}{3}\right) - 1$

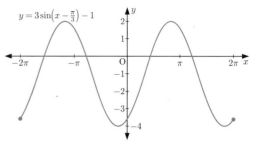

f

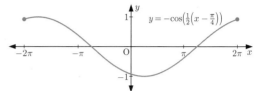

$y = -\cos\left(\frac{1}{2}\left(x - \frac{\pi}{4}\right)\right)$

12 a b, c, d **b** c, d **c** a, d

13 a $y = \sin x - 2$ **b** $y = \sin 3x$

c $y = \sin\left(x + \frac{\pi}{2}\right)$ **d** $y = 2\sin x + 1$

e $y = 4\sin\frac{x}{2} - 1$ **f** $y = 6\sin\left(\frac{2\pi x}{5}\right)$

14 a $y = 2\cos 2x$ **b** $y = \cos\frac{x}{2} + 2$ **c** $y = -5\cos\left(\frac{\pi x}{3}\right)$

EXERCISE 4B

1 a A translation $\frac{\pi}{4}$ units right.

b A horizontal dilation with scale factor $\frac{1}{2}$, then a translation 1 unit downwards.

c A vertical dilation with scale factor $\frac{1}{2}$, then a reflection in the x-axis.

d No transformation is needed.

2 a $\frac{\pi}{3}$ **b** 1   **c** $\frac{3\pi}{2}$ **d** $\frac{\pi}{n}$

3 a i $\frac{k\pi}{2}$, $k \in \mathbb{Z}$ **ii** $x = \frac{\pi}{4} + \frac{k\pi}{2}$, $k \in \mathbb{Z}$

b i $\frac{2\pi}{3} + k\pi$, $k \in \mathbb{Z}$ **ii** $x = \frac{\pi}{6} + k\pi$, $k \in \mathbb{Z}$

c i $\frac{\pi}{6} + 2k\pi$, $k \in \mathbb{Z}$ **ii** $x = \frac{7\pi}{6} + 2k\pi$, $k \in \mathbb{Z}$

4 a

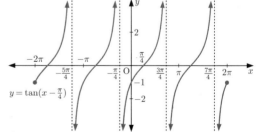

$y = \tan\left(x - \frac{\pi}{4}\right)$

b

$y = \frac{1}{2}\tan\frac{x}{4}$

c

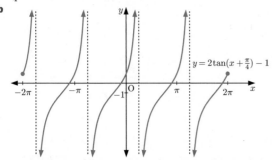

$y = 3\tan(x - \frac{\pi}{9})$

5 $p = \frac{1}{2}, \quad q = 1$ **6** $a = \frac{3}{2}, \quad b = -\frac{2\pi}{15} + 2k\pi, \quad k \in \mathbb{Z}$

7 **a** A vertical dilation with scale factor 2, then a translation $\frac{\pi}{4}$ units left and 1 unit downwards.

b

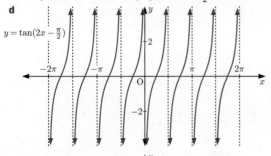

$y = 2\tan(x + \frac{\pi}{4}) - 1$

8 **a** **i** $\tan\left(2x - \frac{\pi}{2}\right)$ **ii** $2\tan x - \frac{\pi}{2}$

   b **i** $\frac{1}{\sqrt{3}}$ **ii** $-\frac{\pi}{2}$

   c **i** period $\frac{\pi}{2}$,   vertical asymptotes $x = \dfrac{k\pi}{2}, \quad k \in \mathbb{Z}$

     ii period π,   vertical asymptotes $x = \dfrac{\pi}{2} + k\pi, \quad k \in \mathbb{Z}$

   d

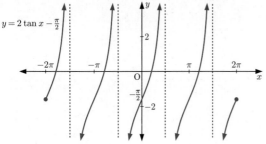

$y = \tan(2x - \frac{\pi}{2})$

$y = 2\tan x - \frac{\pi}{2}$

EXERCISE 4C

1 **a** $\frac{2}{\sqrt{3}}$ **b** $-\frac{1}{\sqrt{3}}$ **c** $-\frac{2}{\sqrt{3}}$ **d** undefined

2 **a** $\csc x = \frac{5}{3}, \quad \sec x = \frac{5}{4}, \quad \cot x = \frac{4}{3}$

   b $\csc x = -\frac{3}{\sqrt{5}}, \quad \sec x = \frac{3}{2}, \quad \cot x = -\frac{2}{\sqrt{5}}$

3 **a** $\sin x = -\frac{\sqrt{7}}{4}, \quad \tan x = -\frac{\sqrt{7}}{3}, \quad \csc x = -\frac{4}{\sqrt{7}},$

     $\sec x = \frac{4}{3}, \quad \cot x = -\frac{3}{\sqrt{7}}$

   b $\cos x = -\frac{\sqrt{5}}{3}, \quad \tan x = \frac{2}{\sqrt{5}}, \quad \csc x = -\frac{3}{2},$

     $\sec x = -\frac{3}{\sqrt{5}}, \quad \cot x = \frac{\sqrt{5}}{2}$

   c $\sin x = \frac{\sqrt{21}}{5}, \quad \cos x = \frac{2}{5}, \quad \tan x = \frac{\sqrt{21}}{2},$

     $\csc x = \frac{5}{\sqrt{21}}, \quad \cot x = \frac{2}{\sqrt{21}}$

   d $\sin x = \frac{1}{2}, \quad \cos x = -\frac{\sqrt{3}}{2}, \quad \tan x = -\frac{1}{\sqrt{3}},$

     $\sec x = -\frac{2}{\sqrt{3}}, \quad \cot x = -\sqrt{3}$

   e $\sin \beta = -\frac{1}{\sqrt{5}}, \quad \cos \beta = -\frac{2}{\sqrt{5}}, \quad \csc \beta = -\sqrt{5},$

     $\sec \beta = -\frac{\sqrt{5}}{2}, \quad \cot \beta = 2$

   f $\sin \theta = -\frac{3}{5}, \quad \cos \theta = -\frac{4}{5}, \quad \tan \theta = \frac{3}{4},$

     $\csc \theta = -\frac{5}{3}, \quad \sec \theta = -\frac{5}{4}$

4

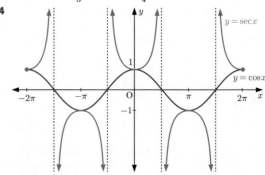

$y = \sec x$

$y = \cos x$

5

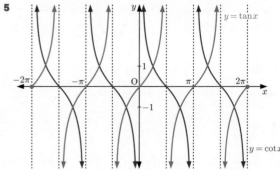

$y = \tan x$

$y = \cot x$

6

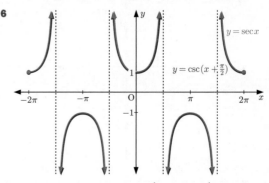

$y = \sec x$

$y = \csc(x + \frac{\pi}{2})$

$\sin(x + \frac{\pi}{2}) = \cos x \quad \therefore \quad \dfrac{1}{\sin(x + \frac{\pi}{2})} = \dfrac{1}{\cos x}$

$\therefore \quad \csc(x + \frac{\pi}{2}) = \sec x$

EXERCISE 4D

1 **a** 0 **b** $-\frac{\pi}{2}$ **c** $\frac{\pi}{4}$ **d** $-\frac{\pi}{4}$ **e** $\frac{\pi}{6}$ **f** $\frac{5\pi}{6}$

    g $\frac{\pi}{3}$ **h** $\frac{3\pi}{4}$ **i** $-\frac{\pi}{6}$ **j** ≈ -0.874

    k ≈ 1.24 **l** ≈ -1.55

2 **a** $(0,\,0)$ **b** $(0,\,0)$ **c** $(0.739,\,0.739)$

3 **a** horizontal asymptotes $y = -\frac{\pi}{2},\ y = \frac{\pi}{2}$

    b No, $y = \sin x$ and $y = \cos x$ do not have horizontal asymptotes.

4 **a** $\arcsin(\sin\frac{\pi}{3}) = \frac{\pi}{3}$ **b** $\arccos(\cos(-\frac{\pi}{6})) = \frac{\pi}{6}$

    c $\tan(\arctan(0.3)) = 0.3$ **d** $\cos(\arccos(-\frac{1}{2})) = -\frac{1}{2}$

    e $\arctan(\tan\pi) = 0$ **f** $\arcsin(\sin\frac{4\pi}{3}) = -\frac{\pi}{3}$

EXERCISE 4E

1 **a** $x = \frac{\pi}{3}$ or $\frac{5\pi}{3}$ **b** $x = \frac{5\pi}{4}$ or $\frac{7\pi}{4}$ **c** $x = \frac{\pi}{6}$ or $\frac{7\pi}{6}$

    d $x = \frac{3\pi}{2}$ **e** $x = \frac{\pi}{2}$ or $\frac{3\pi}{2}$ **f** $x = 0,\ \pi,$ or 2π

2 **a** $x = \frac{\pi}{3}$ or $\frac{2\pi}{3}$ **b** $x = \pi$ **c** $x = \frac{\pi}{4}$ or $\frac{5\pi}{4}$

3 **a** $x = \frac{2\pi}{3}, \frac{4\pi}{3}, \frac{8\pi}{3},$ or $\frac{10\pi}{3}$ **b** $x = \frac{\pi}{4}, \frac{3\pi}{4}, \frac{9\pi}{4},$ or $\frac{11\pi}{4}$

    c $x = \frac{\pi}{4}, \frac{5\pi}{4}, \frac{9\pi}{4},$ or $\frac{13\pi}{4}$

4 **a** $x = -\frac{2\pi}{3}, -\frac{\pi}{3}, \frac{4\pi}{3},$ or $\frac{5\pi}{3}$

    b $x = -\frac{5\pi}{4}, -\frac{3\pi}{4}, \frac{3\pi}{4},$ or $\frac{5\pi}{4}$

    c $x = -\frac{5\pi}{4}, -\frac{\pi}{4}, \frac{3\pi}{4},$ or $\frac{7\pi}{4}$

5 **a** $x = \frac{\pi}{6}, \frac{5\pi}{6}, \frac{7\pi}{6},$ or $\frac{11\pi}{6}$ **b** $x = \frac{\pi}{2}$ or $\frac{3\pi}{2}$

    c $x = \frac{\pi}{3}, \frac{2\pi}{3}, \frac{4\pi}{3},$ or $\frac{5\pi}{3}$

6 **a** $0 \leqslant 2x \leqslant 4\pi$ **b** $0 \leqslant \frac{x}{4} \leqslant \frac{\pi}{2}$

    c $\frac{\pi}{2} \leqslant x + \frac{\pi}{2} \leqslant \frac{5\pi}{2}$ **d** $-\frac{\pi}{6} \leqslant x - \frac{\pi}{6} \leqslant \frac{11\pi}{6}$

    e $-\frac{\pi}{2} \leqslant 2\left(x - \frac{\pi}{4}\right) \leqslant \frac{7\pi}{2}$ **f** $-2\pi \leqslant -x \leqslant 0$

7 **a** $-3\pi \leqslant 3x \leqslant 3\pi$ **b** $-\frac{\pi}{4} \leqslant \frac{x}{4} \leqslant \frac{\pi}{4}$

    c $-\frac{3\pi}{2} \leqslant x - \frac{\pi}{2} \leqslant \frac{\pi}{2}$ **d** $-\frac{3\pi}{2} \leqslant 2x + \frac{\pi}{2} \leqslant \frac{5\pi}{2}$

    e $-2\pi \leqslant -2x \leqslant 2\pi$ **f** $0 \leqslant \pi - x \leqslant 2\pi$

8 **a** $x = \frac{\pi}{3}, \frac{5\pi}{3},$ or $\frac{7\pi}{3}$

    b $x = \frac{\pi}{6}, \frac{5\pi}{6}, \frac{7\pi}{6}, \frac{11\pi}{6}, \frac{13\pi}{6},$ or $\frac{17\pi}{6}$

    c $x = 0, \frac{4\pi}{3},$ or 2π

9 **a** $x = \frac{7\pi}{12}, \frac{11\pi}{12}, \frac{19\pi}{12},$ or $\frac{23\pi}{12}$

    b $x = \frac{\pi}{18}, \frac{11\pi}{18}, \frac{13\pi}{18}, \frac{23\pi}{18}, \frac{25\pi}{18},$ or $\frac{35\pi}{18}$

    c $x = \frac{\pi}{6}, \frac{2\pi}{3}, \frac{7\pi}{6},$ or $\frac{5\pi}{3}$ **d** $x = \frac{\pi}{2}$ or $\frac{3\pi}{2}$

    e $x = \frac{4\pi}{3}$ **f** $x = \frac{3\pi}{4}$

10 **a** $x = \frac{\pi}{9}, \frac{2\pi}{9}, \frac{4\pi}{9}, \frac{5\pi}{9}, \frac{7\pi}{9}, \frac{8\pi}{9}, \frac{10\pi}{9}, \frac{11\pi}{9}, \frac{13\pi}{9}, \frac{14\pi}{9},$ $\frac{16\pi}{9},$ or $\frac{17\pi}{9}$

    b $x = \frac{\pi}{4}, \frac{3\pi}{4}, \frac{5\pi}{4},$ or $\frac{7\pi}{4}$ **c** $x = \frac{\pi}{3}$ or $\frac{5\pi}{3}$

11 **a** $x = \frac{3\pi}{4}$ or $\frac{7\pi}{4}$ **b** $x = \frac{\pi}{12}, \frac{5\pi}{12}, \frac{3\pi}{4}, \frac{13\pi}{12}, \frac{17\pi}{12},$ or $\frac{7\pi}{4}$

    c $x = \frac{\pi}{6}, \frac{2\pi}{3}, \frac{7\pi}{6},$ or $\frac{5\pi}{3}$

12 **a** $x = -\frac{5\pi}{3}, -\pi, \frac{\pi}{3},$ or π **b** $x = 0, \frac{3\pi}{2},$ or 2π

    c $x = 0, \frac{\pi}{4}, \frac{\pi}{2}, \frac{3\pi}{4},$ or π **d** $x = 0, \frac{\pi}{6}, \pi, \frac{7\pi}{6},$ or 2π

13 $x = \frac{\pi}{3}$ or $\frac{4\pi}{3}$

    a $x = \frac{\pi}{2}$ or $\frac{3\pi}{2}$

    b $x = \frac{\pi}{12}, \frac{\pi}{3}, \frac{7\pi}{12}, \frac{5\pi}{6}, \frac{13\pi}{12}, \frac{4\pi}{3}, \frac{19\pi}{12},$ or $\frac{11\pi}{6}$

    c $x = \frac{\pi}{3}, \frac{2\pi}{3}, \frac{4\pi}{3},$ or $\frac{5\pi}{3}$

14 **a** $x = \frac{\pi}{3}$ or $\frac{5\pi}{3}$ **b** $x = \frac{5\pi}{4}$ or $\frac{7\pi}{4}$

    c $x = \frac{5\pi}{12}, \frac{7\pi}{12}, \frac{17\pi}{12},$ or $\frac{19\pi}{12}$ **d** $x = \frac{13\pi}{12}$ or $\frac{19\pi}{12}$

    e $x = \frac{3\pi}{4}$ or $\frac{7\pi}{4}$ **f** $x = \frac{5\pi}{24}, \frac{17\pi}{24}, \frac{29\pi}{24},$ or $\frac{41\pi}{24}$

15 **a** $x = 0, \frac{\pi}{6}, \frac{5\pi}{6}, \pi,$ or 2π **b** $x = \frac{\pi}{3}, \frac{\pi}{2}, \frac{3\pi}{2},$ or $\frac{5\pi}{3}$

    c $x = 0, \frac{2\pi}{3}, \frac{4\pi}{3},$ or 2π **d** $x = \frac{7\pi}{6}, \frac{3\pi}{2},$ or $\frac{11\pi}{6}$

    e $x = \frac{\pi}{6}, \frac{\pi}{2},$ or $\frac{5\pi}{6}$ **f** $x = \frac{\pi}{6}, \frac{5\pi}{6}, \frac{7\pi}{6},$ or $\frac{11\pi}{6}$

    g $x = \frac{\pi}{3}, \frac{2\pi}{3}, \frac{4\pi}{3},$ or $\frac{5\pi}{3}$

16 **a** $x = 1$ **b** $x = -\frac{\sqrt{3}}{2}$ **c** $x = -\frac{1}{\sqrt{2}}$

    d $x = -\frac{1}{2}$ **e** no solution exists   **f** $x = 0$

EXERCISE 4F

1 **a** **i** 7500 grasshoppers   **ii** 10 300 grasshoppers

    b 10 500 grasshoppers, when $t = 4$ weeks

    c **i** at $t = 1\frac{1}{3}$ weeks and $6\frac{2}{3}$ weeks

       ii at $t = 9\frac{1}{3}$ weeks

    d $2.51 \leqslant t \leqslant 5.49$

2 **a** 1 m above ground   **b** at $t = 1\frac{1}{2}$ min   **c** 3 min

    d

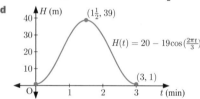

    e $\approx 0.570 \leqslant t \leqslant \approx 2.43$

3 **a** 400 water buffalo

    b **i** 577 water buffalo   **ii** 400 water buffalo

    c 650, which is the maximum population.

    d 150, after 3 years   **e** $t \approx 0.262$ years

4 **a** **i** true   **ii** true   **b** 116.8 pence L^{-1}

    c on the 5th, 11th, 19th, and 25th days

    d 98.6 pence L^{-1} on the 1st and 15th days

5 **a** $T \approx 6.5\sin\left(\frac{\pi}{6}(t - 4.5)\right) + 20.5$

    b The model fits reasonably well but not perfectly.

6 **a** $T \approx 4.5\cos\left(\frac{\pi}{6}(t - 2)\right) + 11.5$

7 **a** $H \approx 7\sin(0.507(t - 3.1))$

    b

      (graph of $H \approx 7\sin(0.507(t - 3.1))$, with values 3.1, 9.3, 12.4 on the t-axis and 7, -7 on the H-axis)

8 **a** $T \approx 9.5\sin\left(\frac{\pi}{6}(t - 10.5)\right) - 9.5$

    b A reasonable fit but not perfect.

9 **a** $H(t) = 3\cos\left(\frac{\pi}{2}t\right) + 4$ **b** $t \approx 1.46$ s

REVIEW SET 4A

1 **a** minimum $= 0$, maximum $= 2$

    b minimum $= -2$, maximum $= 2$

2 **a** 10π **b** $\frac{\pi}{2}$ **c** 4π **d** $\frac{\pi}{3}$

3

Function	Period	Amplitude	Range
$y = -3\sin\frac{x}{4} + 1$	8π	3	$-2 \leqslant y \leqslant 4$
$y = 3\cos\pi x$	2	3	$-3 \leqslant y \leqslant 3$

4 **a**

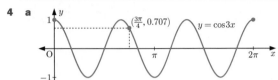

b $y = \frac{1}{\sqrt{2}} \approx 0.707$

5 **a**

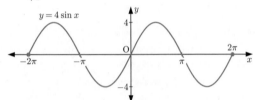

b

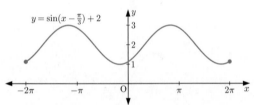

c

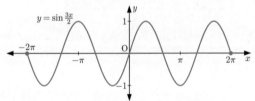

d

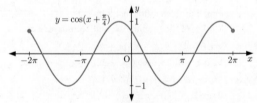

e

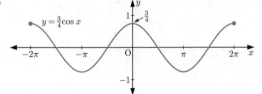

f

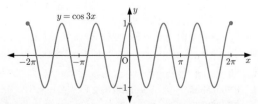

6 **a** A vertical dilation with scale factor 3, then a horizontal dilation with scale factor $\frac{1}{2}$.

b A translation $\frac{\pi}{3}$ units right and 1 unit downwards.

c A reflection in the x-axis, then a horizontal dilation with scale factor $\frac{1}{2}$.

d A vertical dilation with scale factor 2, then a horizontal dilation with scale factor 2, then a translation $\frac{\pi}{2}$ units right and $\frac{1}{2}$ unit upwards.

7 **a** $y = -4\cos 2x$ **b** $y = \cos\frac{\pi x}{4} + 2$

8 **a**

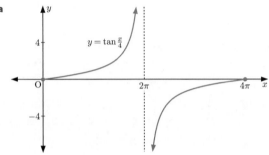

b

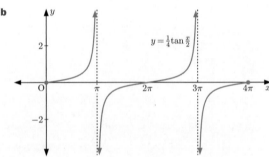

9 **a** A horizontal dilation with scale factor $\frac{1}{3}$, then a vertical translation 2 units upwards.

b $\frac{\pi}{3}$

c

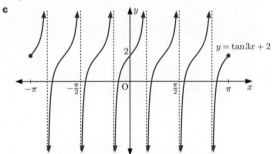

10 **a** $\sin x = \frac{2\sqrt{2}}{3}$, $\tan x = 2\sqrt{2}$, $\csc x = \frac{3}{2\sqrt{2}}$, $\sec x = 3$, $\cot x = \frac{1}{2\sqrt{2}}$

b $\sin x = -\frac{4}{\sqrt{41}}$, $\cos x = -\frac{5}{\sqrt{41}}$, $\csc x = -\frac{\sqrt{41}}{4}$, $\sec x = -\frac{\sqrt{41}}{5}$, $\cot x = \frac{5}{4}$

11 **a** $x = \frac{7\pi}{6}$ or $\frac{11\pi}{6}$ **b** $x = \frac{\pi}{4}$ or $\frac{7\pi}{4}$

     c $x = \frac{\pi}{3}, \frac{2\pi}{3}, \frac{4\pi}{3}$, or $\frac{5\pi}{3}$ **d** $x = \frac{\pi}{4}$ or $\frac{7\pi}{4}$

12 **a** $x = \frac{\pi}{8}, \frac{3\pi}{8}, \frac{5\pi}{8}, \frac{7\pi}{8}, \frac{9\pi}{8}, \frac{11\pi}{8}, \frac{13\pi}{8},$ or $\frac{15\pi}{8}$

b $x = \frac{3\pi}{2}$ **c** $x = \frac{\pi}{6}, \frac{5\pi}{6}, \frac{7\pi}{6},$ or $\frac{11\pi}{6}$

13 **a** $x = 0, \frac{3\pi}{2}, 2\pi, \frac{7\pi}{2},$ or 4π **b** $x = \frac{\pi}{6}, \frac{2\pi}{3}, \frac{7\pi}{6},$ or $\frac{5\pi}{3}$

14 **a** 5000   **b** 3000, 7000

c $0.5 < t < 2.5$ and $6.5 < t \leqslant 8$

15 **a** maximum: $-5°C$,   minimum: $-79°C$

b $T \approx 37\sin(0.008\,98n) - 42$ **c** ≈ 700 Mars days

REVIEW SET 4B

1 **a** A translation $\frac{\pi}{3}$ units right and 1 unit upwards.

b A horizontal dilation with scale factor $\frac{1}{3}$.

2 **a** 6π **b** $\frac{\pi}{4}$

3 **a** $b = \frac{1}{3}$ **b** $b = 24$ **c** $b = \frac{2\pi}{9}$

4 **a** minimum $= -8$,   maximum $= 2$

b minimum $= \frac{2}{3}$,   maximum $= 1\frac{1}{3}$

5 **a**

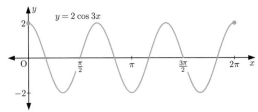

$y = 2\cos 3x$

b

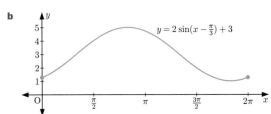

$y = 2\sin\left(x - \frac{\pi}{3}\right) + 3$

c

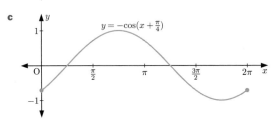

$y = -\cos\left(x + \frac{\pi}{4}\right)$

d

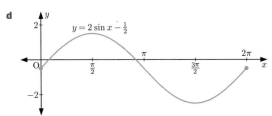

$y = 2\sin x - \frac{1}{2}$

e

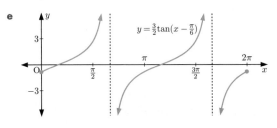

$y = \frac{3}{2}\tan\left(x - \frac{\pi}{6}\right)$

f

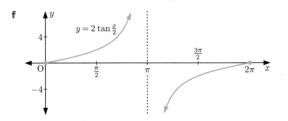

$y = 2\tan\frac{x}{2}$

6 **a** $y = 4\sin x + 6$ **b** $y = 4\cos\left(x - \frac{\pi}{2}\right) + 6$

7 $a = \frac{3}{2}$, $b = -\frac{1}{2}$

8 **a**

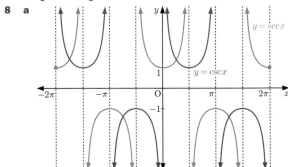

$y = \sec x$ $y = \csc x$

b translation $\frac{\pi}{2}$ units right

9 **a** $\frac{\pi}{4}$ **b** $\frac{\pi}{6}$ **c** $-\frac{\pi}{6}$

10 **a** $x = \frac{4\pi}{9}, \frac{5\pi}{9}, \frac{10\pi}{9}, \frac{11\pi}{9}, \frac{16\pi}{9},$ or $\frac{17\pi}{9}$ **b** $x = \frac{5\pi}{3}$

c $x = \frac{\pi}{12}, \frac{7\pi}{12}, \frac{13\pi}{12},$ or $\frac{19\pi}{12}$

11 **a** $x = -\frac{\pi}{2}$ or $\frac{\pi}{2}$ **b** $x = -\frac{2\pi}{3}, -\frac{\pi}{6}, \frac{\pi}{3},$ or $\frac{5\pi}{6}$

c $x = -\frac{2\pi}{3}, -\frac{\pi}{3}, \frac{\pi}{3},$ or $\frac{2\pi}{3}$ **d** $x = -\frac{5\pi}{6}$ or $\frac{\pi}{6}$

12 **a** $\frac{4\pi}{9}, \frac{5\pi}{9}, \frac{10\pi}{9}, \frac{11\pi}{9}, \frac{16\pi}{9},$ or $\frac{17\pi}{9}$ **b** $\frac{3\pi}{4}, \frac{7\pi}{4},$ or $\frac{11\pi}{4}$

13 **a** $x = \frac{\sqrt{3}}{2}$ **b** $x = 2 + \frac{1}{\sqrt{3}}$

14 **a** 28 milligrams per m^3 **b** 8:00 am Monday

15 **a** $a \approx 7.05$, $b \approx \frac{\pi}{6}$, $c \approx 4.5$, $d \approx 14.75$

b The model fits reasonably well but not perfectly.

EXERCISE 5A.1

1 **a** $2\sin\theta$ **b** $3\cos\theta$ **c** $2\sin\theta$ **d** $\sin\theta$

e $-2\tan\theta$ **f** $-3\cos^2\theta$

2 **a** $2\tan x$ **b** $\tan^2 x$ **c** $\sin x$ **d** $\cos x$

e $5\sin x$ **f** $2\sec x$

3 **a** 1   **b** 1   **c** $\dfrac{\cos x}{\sin^2 x}$ **d** $\cos x$ **e** $\cos x$

4 **a** $2\cos\theta$ **b** 0   **c** $-\tan\theta$ **d** $\cot\theta$

e $2\cos\theta$ **f** $\tan\theta$ **g** $\tan\theta$ **h** $2\tan\theta$

EXERCISE 5A.2

1 **a** 3   **b** -2 **c** -1 **d** $3\cos^2\theta$

e $4\sin^2\theta$ **f** $\cos\theta$ **g** $-\sin^2\theta$ **h** $-\cos^2\theta$

i $-2\sin^2\theta$ **j** 1   **k** $\sin\theta$ **l** $\sin\theta$

3 **a** $1 + 2\sin\theta + \sin^2\theta$ **b** $\sin^2\alpha - 4\sin\alpha + 4$

c $\sec^2\alpha - 2\tan\alpha$ **d** $1 + 2\sin\alpha\cos\alpha$

e $1 - 2\sin\beta\cos\beta$ **f** $-4 + 4\cos\alpha - \cos^2\alpha$

4 **a** $-\tan^2\beta$ **b** 1   **c** $\sin^2\alpha$

d $\sin^2 x - \tan^2 x$ **e** 13   **f** $\cos^2\theta$ **g** 0

EXERCISE 5A.3

1 a $(1 + \sin\theta)(1 - \sin\theta)$ **b** $(\sin\alpha + \cos\alpha)(\sin\alpha - \cos\alpha)$
c $(\tan\alpha + 1)(\tan\alpha - 1)$ **d** $\sin\beta(2\sin\beta - 1)$
e $\cos\phi(2 + 3\cos\phi)$ **f** $3\sin\theta(\sin\theta - 2)$
g $(\tan\theta + 3)(\tan\theta + 2)$ **h** $(2\cos\theta + 1)(\cos\theta + 3)$
i $(3\cos\alpha + 1)(2\cos\alpha - 1)$ **j** $\tan\alpha(3\tan\alpha - 2)$
k $(\sec\beta + \csc\beta)(\sec\beta - \csc\beta)$
l $(2\cot x - 1)(\cot x - 1)$
m $(2\sin x + \cos x)(\sin x + 3\cos x)$

2 a $1 + \sin\alpha$ **b** $\tan\beta - 1$ **c** $\cos\phi - \sin\phi$
d $\cos\phi + \sin\phi$ **e** $\dfrac{1}{\sin\alpha - \cos\alpha}$ **f** $\dfrac{\cos\theta}{2}$
g $\sin\theta$ **h** $\cos\theta$ **i** $\sec\theta + 1$

4 a $x = \frac{\pi}{6}, \frac{5\pi}{6}$, or $\frac{3\pi}{2}$ **b** no real solutions
c $x = \frac{\pi}{6}$ or $\frac{5\pi}{6}$ **d** $x \approx 0.730,\ 2.41,\ 3.87$, or 5.55

EXERCISE 5B

2 a $\frac{24}{25}$ **b** $-\frac{7}{25}$ **c** $-\frac{24}{7}$ **3 a** $-\frac{7}{9}$ **b** $\frac{1}{9}$
4 a $\cos\alpha = -\frac{\sqrt{5}}{3}$ **b** $\sin 2\alpha = \frac{4\sqrt{5}}{9}$
5 a $\sin\beta = -\frac{\sqrt{21}}{5}$ **b** $\sin 2\beta = -\frac{4\sqrt{21}}{25}$
6 a $\frac{1}{3}$ **b** $\frac{2\sqrt{2}}{3}$ **7 a** $\tan A = -\frac{7}{3}$ **b** $\tan A = \frac{3}{2}$
8 $\tan\frac{\pi}{8} = \sqrt{2} - 1$ **9** $\frac{3}{2}$
10 a $\sin 2\alpha$ **b** $2\sin 2\alpha$ **c** $\frac{1}{2}\sin 2\alpha$ **d** $\cos 2\beta$
e $-\cos 2\phi$ **f** $\cos 2N$ **g** $-\cos 2M$ **h** $\cos 2\alpha$
i $-\cos 2\alpha$ **j** $\sin 4A$ **k** $\sin 6\alpha$ **l** $\cos 8\theta$
m $-\cos 6\beta$ **n** $\cos 10\alpha$ **o** $-\cos 6D$ **p** $\cos 4A$
q $\cos\alpha$ **r** $-2\cos 6P$
12 a $x = 0, \frac{2\pi}{3}, \pi, \frac{4\pi}{3}$, or 2π **b** $x = \frac{\pi}{2}$ or $\frac{3\pi}{2}$
c $x = 0, \pi$, or 2π
14 a $\cos A = \frac{7}{10}$ **b** $\cos A = \frac{3}{4}$
16 a $x = 0, \frac{2\pi}{3}, \frac{4\pi}{3}$, or 2π **b** $x = \frac{\pi}{3}$ or $\frac{5\pi}{3}$
c $x = \frac{\pi}{2}, \frac{7\pi}{6}$, or $\frac{11\pi}{6}$
d $x = 0, \frac{\pi}{6}, \frac{\pi}{2}, \frac{5\pi}{6}, \pi, \frac{7\pi}{6}, \frac{3\pi}{2}, \frac{11\pi}{6}$, or 2π **e** $x = \frac{\pi}{2}$
f $x = \frac{\pi}{4}$ **g** $x = 0, \frac{7\pi}{6}, \frac{11\pi}{6}$, or 2π

EXERCISE 5C

2 a $\cos\theta$ **b** $-\sin\theta$ **c** $\sin\theta$
d $-\cos\alpha$ **e** $-\sin A$ **f** $-\sin\theta$
g $\dfrac{1 + \tan\theta}{1 - \tan\theta}$ **h** $\dfrac{1 + \tan\theta}{1 - \tan\theta}$ **i** $\tan\theta$
3 a $\frac{1}{2}\sin\theta + \frac{\sqrt{3}}{2}\cos\theta$ **b** $\frac{\sqrt{3}}{2}\sin\theta - \frac{1}{2}\cos\theta$
c $-\frac{1}{\sqrt{2}}\sin\theta + \frac{1}{\sqrt{2}}\cos\theta$ **d** $-\frac{\sqrt{3}}{2}\sin\theta + \frac{1}{2}\cos\theta$
4 a $\cos\theta$ **b** $\sin 3A$ **c** $\sin(B - A)$
d $\cos(\alpha - \beta)$ **e** $-\cos(\phi + \theta)$ **f** $2\sin(\alpha - \beta)$
g $\tan\theta$ **h** $\tan 3A$
7 a $2 + \sqrt{3}$ **b** $-2 - \sqrt{3}$ **8** $\frac{7}{17}$ **9** 7
10 $\dfrac{9 + 5\sqrt{2}}{2}$ **11** $\sqrt{3}$ **12** $\tan A = -\frac{1}{21}$
13 $\tan\alpha = \frac{25}{62}$ **14** $\frac{1}{8}$

15 a $\cos 2\alpha$ **b** $-\sin 3\phi$ **c** $\cos\beta$ **d** -1
e $\tan 2A$
16 a $\frac{54 - 25\sqrt{5}}{22}$ **b** $-4\sqrt{5}$ **17** $\tan A = \pm 1$
18 $\tan(A + B + C)$
$= \dfrac{\tan A + \tan B + \tan C - \tan A \tan B \tan C}{1 - \tan A \tan B - \tan A \tan C - \tan B \tan C}$
21 b $\theta = -\frac{8\pi}{9}, -\frac{4\pi}{9}, -\frac{2\pi}{9}, \frac{2\pi}{9}, \frac{4\pi}{9}$, or $\frac{8\pi}{9}$
22 a $\sin 3\theta = -4\sin^3\theta + 3\sin\theta$
b $\theta = 0, \frac{\pi}{4}, \frac{3\pi}{4}, \pi, \frac{5\pi}{4}, \frac{7\pi}{4}, 2\pi, \frac{9\pi}{4}, \frac{11\pi}{4}$, or 3π
23 $k = 2, \quad a = \frac{\pi}{6}$
24 a $2\cos x + 2\sin x = 2\sqrt{2}\cos\left(x + \frac{7\pi}{4}\right)$
b $x = \frac{7\pi}{12}$ or $\frac{23\pi}{12}$
25 a A vertical dilation with scale factor $\sqrt{10}$, then a translation of ≈ 0.322 units left.
b greatest value $= 12$, least value $= 2$
27 a $2\cos x - 5\sin x \approx \sqrt{29}\cos(x + 1.19)$
b $x \approx 0.761$ or π
d $x \approx 0.761$ (the solution $x = \pi$ has been lost)
28 c i $\frac{1}{2}\sin 4\theta + \frac{1}{2}\sin 2\theta$ **ii** $\frac{1}{2}\sin 7\alpha + \frac{1}{2}\sin 5\alpha$
iii $\sin 6\beta + \sin 4\beta$ **iv** $2\sin 5\theta + 2\sin 3\theta$
v $3\sin 7\alpha - 3\sin\alpha$ **vi** $\frac{1}{6}\sin 8A - \frac{1}{6}\sin 2A$
29 c i $\frac{1}{2}\cos 5\theta + \frac{1}{2}\cos 3\theta$ **ii** $\frac{1}{2}\cos 8\alpha + \frac{1}{2}\cos 6\alpha$
iii $\cos 4\beta + \cos 2\beta$ **iv** $3\cos 8x + 3\cos 6x$
v $\frac{3}{2}\cos 5P + \frac{3}{2}\cos 3P$ **vi** $\frac{1}{8}\cos 6x + \frac{1}{8}\cos 2x$
30 c i $\frac{1}{2}\cos 2\theta - \frac{1}{2}\cos 4\theta$ **ii** $\frac{1}{2}\cos 5\alpha - \frac{1}{2}\cos 7\alpha$
iii $\cos 4\beta - \cos 6\beta$ **iv** $2\cos 3\theta - 2\cos 5\theta$
v $5\cos 6A - 5\cos 10A$ **vi** $\frac{1}{10}\cos 4M - \frac{1}{10}\cos 10M$
31 a $\sin A\cos A = \frac{1}{2}\sin 2A$, $\cos^2 A = \frac{1}{2}\cos 2A + \frac{1}{2}$,
$\sin^2 A = \frac{1}{2} - \frac{1}{2}\cos 2A$
c $\cos S + \cos D = 2\cos\left(\frac{S+D}{2}\right)\cos\left(\frac{S-D}{2}\right)$
d $\cos D - \cos S = 2\sin\left(\frac{S+D}{2}\right)\sin\left(\frac{S-D}{2}\right)$
32 a $2\sin 3x\cos 2x$ **b** $2\cos 5A\cos 3A$ **c** $-2\sin 2\alpha\sin\alpha$
d $2\cos 4\theta\sin\theta$ **e** $-2\sin 4\alpha\sin 3\alpha$ **f** $2\sin 5\alpha\cos 2\alpha$
g $2\sin 3B\sin B$ **h** $2\cos\left(x + \frac{h}{2}\right)\sin\frac{h}{2}$
i $-2\sin\left(x + \frac{h}{2}\right)\sin\frac{h}{2}$

REVIEW SET 5A

1 a $\cos\theta$ **b** $-\sin\theta$ **c** $\cos\theta$
d $1 - \cos\theta$ **e** $\dfrac{1}{\sin\alpha + \cos\alpha}$ **f** $\dfrac{-\cos\alpha}{2}$
2 a $-\frac{\sqrt{7}}{4}$ **b** $\frac{3\sqrt{7}}{8}$ **c** $-\frac{1}{8}$
d $-3\sqrt{7}$ **e** $-\sqrt{\frac{4 - \sqrt{7}}{8}}$ **f** $\sqrt{\frac{4 + \sqrt{7}}{8}}$
5 a $\frac{-\sqrt{2} - \sqrt{6}}{4}$ **b** $2 - \sqrt{3}$ **6** 60 m
7 $x = 0, \frac{\pi}{4}, \pi, \frac{5\pi}{4}$, or 2π **9 c** $x = \frac{16}{3}$
11 $3\sin x - 5\cos x \approx \sqrt{34}\cos(x + 3.68)$
12 a $2\sin x + \sqrt{3}\cos x \approx \sqrt{7}\sin(x + 0.714)$
b i $A = \sqrt{7}$ **ii** $b \approx 2.43$

REVIEW SET 5B

1 **a** $\sec x$ **b** $\sin x$ **c** $\cos x$

2 **a** $\cos\theta$ **b** $-\sin\theta$ **c** $5\cos^2\theta$ **d** $-\cos\theta$
   e $\csc\theta$ **f** $\sin 2\theta$

3 **a** $\frac{120}{169}$ **b** $\frac{119}{169}$ **c** $\frac{120}{119}$

6 $x = -\frac{2\pi}{3}, -\frac{\pi}{2}, -\frac{\pi}{3}$, or $\frac{\pi}{2}$ **7** $\sin\left(\theta + \frac{\pi}{6}\right) = \frac{3\sqrt{3}-\sqrt{7}}{8}$

8 $\tan\theta = \frac{9}{19}$ **9** $3\sin x + 4\cos x \approx 5\sin(x + 0.927)$

10 1.5 m

11 **b** $y = 2\sec 2x$ has range $(-\infty, -2] \cup [2, \infty)$

$$\therefore \quad \frac{1}{1 + \sqrt{2}\sin x} + \frac{1}{1 - \sqrt{2}\sin x} = 1 \quad \text{has no solutions.}$$

EXERCISE 6A

1 **a** If you do not get wet then it is not raining.
   b If it is not Thursday then penguins cannot fly.
   c If you are not aged 18 or over then you do not drink alcohol.
   d If k is prime then k is not divisible by 5.
   e If a and b are not even then $a^2 + b^2 \neq c^2$.

3 **Hint:** The contrapositive is "if a and b are rational then the product ab is rational".

4 **Hint:** The contrapositive is "if n is a perfect square then $n \equiv 0$ or $1 \bmod 4$".

EXERCISE 6B

1 **Hint:** Suppose $\log_2 5 = \frac{p}{q}$, where $p, q \in \mathbb{Z}$, $q \neq 0$.

2 **Hint:** Suppose $\log_3 5 = \frac{p}{q}$, where $p, q \in \mathbb{Z}$, $q \neq 0$.

3 **Hint:** If p^2 is a multiple of 3, then p must be a multiple of 3.

4 **Hint:** If p^2 is a multiple of 4, then p is not necessarily a multiple of 4.

6 **Hint:** Use **Exercise 6A** question **4**.

7 No, we could find another solution by reversing the 2s and 3s. So, the solution is not unique, and the sudoku would not be well-posed.

8 No (**Hint:** Consider the first player employing the same strategy S.)

9 **Hint:** Suppose there is a largest prime p, then consider $N = 4(3 \times 7 \times 11 \times \ldots \times p) - 1$.

REVIEW SET 6A

1 **Hint:** The contrapositive is "if n is odd then n^3 is odd".

3 **c** **Hint:** Use your answer to part **b**.

4 **a** definitely a THOG   **b** definitely not a THOG
   c definitely not a THOG   **d** definitely a THOG

REVIEW SET 6B

1 **Hint:** The contrapositive is "if either a or b are divisible by n then ab is divisible by n".

4 **b** **Hint:** $4n^2 - m^2 = (2n + m)(2n - m)$

EXERCISE 7A

1 **a** $f'(x) = 2x$
   b $f'(3) = 6$. The gradient of the tangent to $f(x) = x^2$ at the point where $x = 3$, is 6.

2 **a** $f'(x) = 1$ **b** $f'(x) = 0$ **c** $f'(x) = -2x$
   d $f'(x) = 6x^2$

3 **a** $f'(x) = -\frac{3}{x^2}$ **b** $f'(x) = -\frac{2}{x^3}$ **c** $f'(x) = \frac{6}{x^4}$

4 **a** $f'(x) = -\frac{4}{x^3}$ **b** **i** 4   **ii** $-\frac{1}{2}$

c
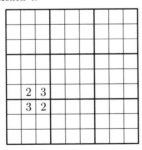

gradient $= 4$ gradient $= -\frac{1}{2}$

5 **a** $f'(x) = \frac{1}{\sqrt{x}}$ **b** $f'(x) = \frac{3\sqrt{x}}{2}$ **c** $f'(x) = \frac{5x\sqrt{x}}{2}$

6 **a** $\frac{dy}{dx} = -\frac{1}{2x\sqrt{x}}$ **b** $-\frac{1}{16}$

7 **a**

$f(x)$	$f'(x)$
$1 = x^0$	0
$x = x^1$	1
x^2	$2x$
x^3	$3x^2$
$\frac{1}{x} = x^{-1}$	$-\frac{1}{x^2} = -x^{-2}$
$\frac{1}{x^2} = x^{-2}$	$-\frac{2}{x^3} = -2x^{-3}$
$\sqrt{x} = x^{\frac{1}{2}}$	$\frac{1}{2\sqrt{x}} = \frac{1}{2}x^{-\frac{1}{2}}$
$x\sqrt{x} = x^{\frac{3}{2}}$	$\frac{3\sqrt{x}}{2} = \frac{3}{2}x^{\frac{1}{2}}$

b If $f(x) = x^n$, then $f'(x) = nx^{n-1}$.

8 **a** $\frac{dy}{dx} = 2x - 3$ **b** $\frac{dy}{dx} = 2 + \frac{1}{x^2}$ **c** $\frac{dy}{dx} = 3 + \frac{1}{\sqrt{x}}$

EXERCISE 7B

1 **a** $f'(x) = 5x^4$ **b** $f'(x) = 9x^8$
   c $f'(x) = 18x^5$ **d** $f'(x) = -14x^6$
   e $f'(x) = 2x^3$ **f** $f'(x) = 3x^2 + 4$
   g $f'(x) = 2x - 6$ **h** $f'(x) = 8x + 3$
   i $f'(x) = 16x$ **j** $f'(x) = -6x$
   k $f'(x) = 3x^2 - 2$ **l** $f'(x) = x^2 - 10x + 4$
   m $f'(x) = -5 - 4x$ **n** $f'(x) = 6x^5 + \frac{4}{3}x^3 - 7$
   o $f'(x) = 3x^2 - 8x$ **p** $f'(x) = 2x + 2$

2 **a** $f'(t) = 6t^5$ **b** $f'(t) = 9t^2$
   c $f'(t) = 2t - 4$ **d** $f'(t) = 2t^3 - \frac{1}{3}$
   e $f'(t) = -1 - \frac{5}{2}t^4$ **f** $f'(t) = 2t - 4$

3 a $\dfrac{dy}{dx} = -\dfrac{3}{x^4}$ **b** $\dfrac{dy}{dx} = -\dfrac{4}{x^5}$ **c** $\dfrac{dy}{dx} = -\dfrac{7}{x^8}$

d $\dfrac{dy}{dx} = -\dfrac{4}{x^2}$ **e** $\dfrac{dy}{dx} = \dfrac{10}{x^6}$ **f** $\dfrac{dy}{dx} = -\dfrac{10}{x^3}$

g $\dfrac{dy}{dx} = 3 - \dfrac{2}{x^2}$ **h** $\dfrac{dy}{dx} = -\dfrac{1}{x^2} + \dfrac{8}{x^3}$

i $\dfrac{dy}{dx} = -\dfrac{6}{x^3} + \dfrac{28}{x^5}$ **j** $\dfrac{dy}{dx} = -\dfrac{1}{2x^2}$ **k** $\dfrac{dy}{dx} = \dfrac{4}{3x^3}$

l $\dfrac{dy}{dx} = 14x + \dfrac{5}{4x^6}$ **m** $\dfrac{dy}{dx} = 1 - \dfrac{6}{x^2}$

n $\dfrac{dy}{dx} = -\dfrac{2}{x^3} + \dfrac{15}{x^4}$ **o** $\dfrac{dy}{dx} = \dfrac{4}{x^2} - \dfrac{6}{x^3}$

4 a $11\frac{3}{4}$ **b** $4\frac{1}{2}$ **c** $-\frac{1}{2}$ **d** 4

5 a $f'(x) = \frac{1}{2}x^{-\frac{1}{2}} = \dfrac{1}{2\sqrt{x}}$ **b** $f'(x) = 2x^{-\frac{1}{2}} = \dfrac{2}{\sqrt{x}}$

c $f'(x) = \frac{1}{5}x^{-\frac{4}{5}} = \dfrac{1}{5\sqrt[5]{x^4}}$ **d** $f'(x) = -\frac{3}{2}x^{\frac{1}{2}} = -\dfrac{3\sqrt{x}}{2}$

e $f'(x) = -x^{-\frac{3}{2}} = -\dfrac{1}{x\sqrt{x}}$

f $f'(x) = 2 + 4x^{-\frac{1}{2}} = 2 + \dfrac{4}{\sqrt{x}}$

g $f'(x) = 15x^{\frac{3}{2}} = 15x\sqrt{x}$

h $f'(x) = -7x^{-2} + 2x^{-\frac{3}{2}} = -\dfrac{7}{x^2} + \dfrac{2}{x\sqrt{x}}$

i $f'(x) = \frac{3}{2}x^{-\frac{1}{2}} + 4x^{-\frac{3}{2}} = \dfrac{3}{2\sqrt{x}} + \dfrac{4}{x\sqrt{x}}$

j $f'(x) = -2x^{-\frac{4}{3}} = -\dfrac{2}{\sqrt[3]{x^4}}$

k $f'(x) = \frac{15}{2}x^{-\frac{7}{2}} = \dfrac{15}{2x^3\sqrt{x}}$

l $f'(x) = -\frac{15}{2}x^{-\frac{5}{2}} + 2x = -\dfrac{15}{2x^2\sqrt{x}} + 2x$

6 a $\dfrac{dy}{dx} = 4x^{-\frac{1}{2}} = \dfrac{4}{\sqrt{x}}$ **b** $\dfrac{dy}{dt} = 3t^{-\frac{3}{2}} = \dfrac{3}{t\sqrt{t}}$

c $\dfrac{dT}{dx} = \frac{1}{3}x^{-\frac{2}{3}} + 4x^{-3} = \dfrac{1}{3\sqrt[3]{x^2}} + \dfrac{4}{x^3}$

d $\dfrac{dP}{du} = -5u^{-2} - 15u^{\frac{1}{2}} = -\dfrac{5}{u^2} - 15\sqrt{u}$

7 a $\dfrac{dy}{dx} = 2 - 2x^{-\frac{1}{2}} = 2 - \dfrac{2}{\sqrt{x}}$ **b** $\frac{4}{3}$

8 a $f'(x) = 2x - 7 - 8x^{-\frac{3}{2}} = 2x - 7 - \dfrac{8}{x\sqrt{x}}$

b $f'(4) = 0$

∴ gradient of tangent to $f(x) = x^2 - 7x + \dfrac{16}{\sqrt{x}}$ at

$x = 4$ is zero, so the tangent is horizontal.

c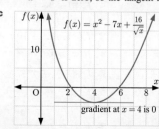
gradient at $x = 4$ is 0

---

1 a $g(f(x)) = (2x + 7)^2$ **b** $g(f(x)) = 2x^2 + 7$

c $g(f(x)) = \sqrt{3 - 4x}$ **d** $g(f(x)) = 3 - 4\sqrt{x}$

e $g(f(x)) = \dfrac{2}{x^2 + 3}$ **f** $g(f(x)) = \dfrac{4}{x^2} + 3$

2 Note: There may be other answers.

a $g(x) = x^3$, $f(x) = 3x + 10$

b $g(x) = x^5$, $f(x) = 7 - 2x$

c $g(x) = \dfrac{1}{x}$, $f(x) = 2x + 4$

d $g(x) = \sqrt{x}$, $f(x) = x^2 - 3x$

e $g(x) = \dfrac{1}{x^4}$, $f(x) = 5x - 1$

f $g(x) = \dfrac{10}{x^3}$, $f(x) = 3x - x^2$

1 a u^{-2}, $u = 2x - 1$ **b** $u^{\frac{1}{2}}$, $u = x^2 - 3x$

c $2u^{-\frac{1}{2}}$, $u = 2 - x^2$ **d** $u^{\frac{1}{3}}$, $u = x^3 - x^2$

e $4u^{-3}$, $u = 3 - x$ **f** $10u^{-1}$, $u = x^2 - 3$

2 a $\dfrac{dy}{dx} = 4(2x + 3) = 8x + 12$ **b** $\dfrac{dy}{dx} = 8x + 12$

3 a $\dfrac{dy}{dx} = 8(4x - 5)$ **b** $\dfrac{dy}{dx} = 2(5 - 2x)^{-2}$

c $\dfrac{dy}{dx} = \frac{1}{2}(3x - x^2)^{-\frac{1}{2}}(3 - 2x)$

d $\dfrac{dy}{dx} = -12(1 - 3x)^3$ **e** $\dfrac{dy}{dx} = -18(5 - x)^2$

f $\dfrac{dy}{dx} = \frac{1}{3}(2x^3 - x^2)^{-\frac{2}{3}}(6x^2 - 2x)$

g $\dfrac{dy}{dx} = -60(5x - 4)^{-3}$

h $\dfrac{dy}{dx} = 5(x^2 - 5x + 8)^4(2x - 5)$

i $\dfrac{dy}{dx} = 6\left(x^2 - \dfrac{2}{x}\right)^2\left(2x + \dfrac{2}{x^2}\right)$

4 a $-\frac{1}{\sqrt{3}}$ **b** -18 **c** -8 **d** -4 **e** $-\frac{3}{32}$ **f** 0

5 $a = 3,\ b = 1$ **6** $a = 2,\ b = 1$ **7** $a = 2,\ b = 3$

8 a $\dfrac{dy}{dx} = 3x^2$, $\dfrac{dx}{dy} = \frac{1}{3}y^{-\frac{2}{3}}$ **Hint:** Substitute $y = x^3$

b $\dfrac{dy}{dx} \times \dfrac{dx}{dy} = \dfrac{dy}{dy}$ {chain rule} $= 1$

9 a $\dfrac{dx}{dy} = 3y^2 + 1$ **b** $\frac{1}{13}$

1 a $f'(x) = 2x - 1$ **b** $f'(x) = 4x + 2$

c $f'(x) = 2x(x + 1)^{\frac{1}{2}} + \frac{1}{2}x^2(x + 1)^{-\frac{1}{2}}$

d $f'(x) = 2x + 2$

e $f'(x) = (x^2 - 1)^{\frac{1}{2}} + x^2(x^2 - 1)^{-\frac{1}{2}}$

f $f'(x) = (x + 1)^2 + 2x(x + 1)$

2 a $\dfrac{dy}{dx} = 2x(2x - 1) + 2x^2$

b $\dfrac{dy}{dx} = 4(2x + 1)^3 + 24x(2x + 1)^2$

c $\dfrac{dy}{dx} = 2x(3 - x)^{\frac{1}{2}} - \frac{1}{2}x^2(3 - x)^{-\frac{1}{2}}$

d $\frac{dy}{dx} = \frac{1}{2}x^{-\frac{1}{2}}(x-3)^2 + 2\sqrt{x}(x-3)$

e $\frac{dy}{dx} = 10x(3x^2-1)^2 + 60x^3(3x^2-1)$

f $\frac{dy}{dx} = \frac{1}{2}x^{-\frac{1}{2}}(x-x^2)^3 + 3\sqrt{x}(x-x^2)^2(1-2x)$

3 a -48 **b** $406\frac{1}{4}$ **c** $\frac{13}{3}$ **d** $\frac{11}{2}$

4 b $x=3$ or $\frac{3}{5}$ **c** $x \leqslant 0$ **5** $x=-1$ and $x=-\frac{5}{3}$

6 a $x=\frac{2}{3}$ **b** $x=0$

EXERCISE 7E

1 a $\frac{dy}{dx} = \frac{7}{(2-x)^2}$ **b** $\frac{dy}{dx} = \frac{2x^2+2x}{(2x+1)^2}$

c $\frac{dy}{dx} = \frac{-x^2-3}{(x^2-3)^2}$ **d** $\frac{dy}{dx} = \frac{2x+1}{2\sqrt{x}(1-2x)^2}$

e $\frac{dy}{dx} = \frac{3x^2-6x+9}{(3x-x^2)^2}$ **f** $\frac{dy}{dx} = \frac{2-3x}{2(1-3x)^{\frac{3}{2}}}$

2 a 1 **b** 1 **c** $-\frac{7}{324}$ **d** $-\frac{28}{27}$

3 $f'(x) = \frac{x-2}{2(x-1)^{\frac{3}{2}}}$

4 a $\frac{dy}{dx} = -\frac{1}{(x+1)^2}$

b At $x=-2$, $\frac{dy}{dx} = -1$. At $x=0$, $\frac{dy}{dx} = -1$.
So, the gradient of each tangent is -1
∴ the tangents are parallel.

5 b i never $\{\frac{dy}{dx}$ is undefined at $x=-1\}$
ii $x \leqslant 0$ and $x=1$

6 b i $x=-3$ and $x=2$ **ii** $x=-\frac{1}{2}$

7 b i $x=-2\pm\sqrt{11}$ **ii** $x=-2$

8 a $\frac{dx}{dy} = \frac{y^2-10y-1}{(y-5)^2}$ **b** $\frac{18}{5}$

9 $Q(x) = u(x)[v(x)]^{-1}$
$Q'(x) = u'(x)[v(x)]^{-1} + u(x)(-1)[v(x)]^{-2}v'(x)$
$\hspace{4cm}$ {product rule}
$\hspace{1cm} = \frac{u'(x)}{v(x)} - \frac{u(x)v'(x)}{[v(x)]^2}$
$\hspace{1cm} = \frac{u'(x)v(x) - u(x)v'(x)}{[v(x)]^2}$

EXERCISE 7F

1 a $f'(x) = 4e^{4x}$ **b** $f'(x) = e^x$

c $f'(x) = -2e^{-2x}$ **d** $f'(x) = \frac{1}{2}e^{\frac{x}{2}}$

e $f'(x) = -e^{-\frac{x}{2}}$ **f** $f'(x) = 2e^{-x}$

g $f'(x) = 2e^{\frac{x}{2}} + 3e^{-x}$ **h** $f'(x) = \frac{e^x - e^{-x}}{2}$

i $f'(x) = -2xe^{-x^2}$ **j** $f'(x) = -\frac{e^{\frac{1}{x}}}{x^2}$

k $f'(x) = 20e^{2x}$ **l** $f'(x) = 40e^{-2x}$

m $f'(x) = 2e^{2x+1}$ **n** $f'(x) = \frac{1}{4}e^{\frac{x}{4}}$

o $f'(x) = -4xe^{1-2x^2}$ **p** $f'(x) = -0.02e^{-0.02x}$

2 a $e^x + xe^x$ **b** $3x^2e^{-x} - x^3e^{-x}$ **c** $\frac{xe^x - e^x}{x^2}$

d $\frac{1-x}{e^x}$ **e** $2xe^{3x} + 3x^2e^{3x}$ **f** $\frac{xe^x - \frac{1}{2}e^x}{x\sqrt{x}}$

g $20e^{-0.5x} - 10xe^{-0.5x}$ **h** $\frac{e^x + 2 + 2e^{-x}}{(e^{-x}+1)^2}$

3 a $\frac{dy}{dx} = 4e^x(2+e^x)^3$ **b** $\frac{dy}{dx} = \frac{e^x}{2\sqrt{e^x-1}}$

c $\frac{dy}{dx} = \frac{3}{2}(e^x + e^{-x})^{\frac{1}{2}}(e^x - e^{-x})$

d $\frac{dy}{dx} = -e^{2x}(e^{2x}+2)^{-\frac{3}{2}}$

4 a 108 **b** -1 **c** $\frac{9}{\sqrt{19}}$ **d** $-\frac{4}{e^3}$

5 a $\{x: x \leqslant \ln 6\}$ **b i** $(\ln 2, 2)$ **ii** $-\frac{1}{2}$

6 $k=-9$

7 a $\frac{dy}{dx} = 2^x \ln 2$

c i $\frac{dy}{dx} = 5^x \ln 5$ **ii** $\frac{dy}{dx} = 8 \times 10^x \ln 10$

8 $P(0, 0)$ or $P(2, \frac{4}{e^2})$

EXERCISE 7G

1 a $\frac{dy}{dx} = \frac{1}{x}$ **b** $\frac{dy}{dx} = \frac{2}{2x+1}$ **c** $\frac{dy}{dx} = \frac{1-2x}{x-x^2}$

d $\frac{dy}{dx} = -\frac{2}{x}$ **e** $\frac{dy}{dx} = 2x\ln x + x$ **f** $\frac{dy}{dx} = \frac{1-\ln x}{2x^2}$

g $\frac{dy}{dx} = e^x \ln x + \frac{e^x}{x}$ **h** $\frac{dy}{dx} = \frac{2\ln x}{x}$

i $\frac{dy}{dx} = \frac{1}{2x\sqrt{\ln x}}$ **j** $\frac{dy}{dx} = \frac{e^{-x}}{x} - e^{-x}\ln x$

k $\frac{dy}{dx} = \frac{\ln(2x)}{2\sqrt{x}} + \frac{1}{\sqrt{x}}$ **l** $\frac{dy}{dx} = \frac{\ln x - 2}{\sqrt{x}(\ln x)^2}$

m $\frac{dy}{dx} = \frac{4}{1-x}$ **n** $\frac{dy}{dx} = \ln(x^2+1) + \frac{2x^2}{x^2+1}$

o $\frac{dy}{dx} = \frac{1-2\ln x}{x^3}$

2 a $\frac{dy}{dx} = \ln 5$ **b** $\frac{dy}{dx} = \frac{3}{x}$ **c** $\frac{dy}{dx} = \frac{4x^3+1}{x^4+x}$

d $\frac{dy}{dx} = \frac{1}{x-2}$ **e** $\frac{dy}{dx} = \frac{6}{2x+1}[\ln(2x+1)]^2$

f $\frac{dy}{dx} = \frac{1-\ln(4x)}{x^2}$ **g** $\frac{dy}{dx} = -\frac{1}{x}$

h $\frac{dy}{dx} = \frac{1}{x\ln x}$ **i** $\frac{dy}{dx} = \frac{-1}{x(\ln x)^2}$

3 a $\frac{dy}{dx} = \frac{1}{2x-1}$ **b** $\frac{dy}{dx} = \frac{-2}{2x+3}$

c $\frac{dy}{dx} = 1 + \frac{1}{2x}$ **d** $\frac{dy}{dx} = \frac{1}{x} - \frac{1}{2(2-x)}$

e $\frac{dy}{dx} = \frac{1}{x+3} - \frac{1}{x-1}$ **f** $\frac{dy}{dx} = \frac{2}{x} + \frac{1}{3-x}$

4 a $f'(x) = \frac{9}{3x-4}$ **b** $f'(x) = \frac{1}{x} + \frac{2x}{x^2+1}$

c $f'(x) = \dfrac{2x+2}{x^2+2x} - \dfrac{1}{x-5}$

d $f'(x) = \dfrac{3}{x} - \dfrac{1}{x+4} - \dfrac{1}{x-1}$

5 a 2   **b** $-\frac{5}{3}$ **6 a** $\dfrac{dx}{dy} = \ln y + \dfrac{2}{y} + 1$ **b** $\frac{1}{3}$

7 $a = 3$, $b = \dfrac{1}{e}$ **8** $a = 4$, $b = e^3$

EXERCISE 7H

2 a $\dfrac{dy}{dx} = 2\cos 2x$ **b** $\dfrac{dy}{dx} = \cos x - \sin x$

c $\dfrac{dy}{dx} = -3\sin 3x - \cos x$ **d** $\dfrac{dy}{dx} = \cos(x+1)$

e $\dfrac{dy}{dx} = 2\sin(3-2x)$ **f** $\dfrac{dy}{dx} = 5\sec^2 5x$

g $\dfrac{dy}{dx} = \frac{1}{2}\cos\frac{x}{2} + 3\sin x$ **h** $\dfrac{dy}{dx} = 3\pi\sec^2 \pi x$

i $\dfrac{dy}{dx} = 4\cos x + 2\sin 2x$

3 a $2x - \sin x$ **b** $\sec^2 x - 3\cos x$

c $e^x \cos x - e^x \sin x$ **d** $-e^{-x}\sin x + e^{-x}\cos x$

e $\dfrac{\cos x}{\sin x}$ **f** $2e^{2x}\tan x + e^{2x}\sec^2 x$

g $3\cos 3x - 8\sin 2x$ **h** $-\frac{1}{2}\sin\frac{x}{2}$

i $6\sec^2 2x$ **j** $\cos x - x\sin x$

k $\dfrac{x\cos x - \sin x}{x^2}$ **l** $\tan x + x\sec^2 x$

4 a $\dfrac{\sin x}{\cos^2 x}$ **b** $-\dfrac{\cos x}{\sin^2 x}$ **c** $-\dfrac{1}{\sin^2 x}$

5 a $2x\cos(x^2)$ **b** $-\dfrac{1}{2\sqrt{x}}\sin(\sqrt{x})$ **c** $-\dfrac{\sin x}{2\sqrt{\cos x}}$

d $2\sin x\cos x$ **e** $-3\sin x\cos^2 x$

f $-\sin x\sin 2x + 2\cos x\cos 2x$ **g** $\sin x\sin(\cos x)$

h $-12\sin 4x\cos^2 4x$ **i** $-\dfrac{3\cos 3x}{\sin^2 3x}$

j $\dfrac{2\sin 2x}{\cos^2 2x}$ **k** $-\dfrac{8\cos 2x}{\sin^3 2x}$ **l** $\dfrac{-12}{\cos^2\frac{x}{2}\tan^4\frac{x}{2}}$

6 a $-\frac{9}{8}$ **b** 0

7 a $f'(x) = 0$ **b** $f(x) = 2(\cos^2 x + \sin^2 x) + 1$
$= 2(1) + 1$
$= 3$, a constant

8 a tangent **B**

b $\dfrac{dy}{dx} = -\sin x + 4\cos 2x$

When $x = \frac{\pi}{6}$, $\dfrac{dy}{dx} = \frac{3}{2}$

When $x = \frac{7\pi}{6}$, $\dfrac{dy}{dx} = \frac{5}{2}$ which is $> \frac{3}{2}$ ✓

EXERCISE 7I

1 a $f''(x) = 8$ **b** $f''(x) = -6x + 4$

c $f''(x) = \frac{3}{4}x^{-\frac{1}{2}} - \frac{1}{4}x^{-\frac{3}{2}}$ **d** $f''(x) = 108 - 162x$

e $f''(x) = -12x - 34$ **f** $f''(x) = \dfrac{20}{(2x-1)^3}$

2 a $\dfrac{d^2y}{dx^2} = \dfrac{5}{x^3}$ **b** $\dfrac{d^2y}{dx^2} = -30x^4 + 108x^2 - 54$

c $\dfrac{d^2y}{dx^2} = 2 + \dfrac{2}{(1-x)^3}$ **d** $\dfrac{d^2y}{dx^2} = 3e^x$

e $\dfrac{d^2y}{dx^2} = \dfrac{-x^2 e^{-x} - 2xe^{-x} + 2 - 2e^{-x}}{x^3}$

f $\dfrac{d^2y}{dx^2} = -\dfrac{x^3 - 3x^2 - 6x - 6}{x^3 e^x}$

3 a $f(1) = 3e - 2$ **b** $f'(1) = 3e - 2$ **c** $f''(1) = 3e$

4 a $x = 1$ **b** $x = 0$ or $x \approx -0.924$

5 a $\dfrac{d^2y}{dx^2} = 2\cos x - x\sin x$

b $\dfrac{d^2y}{dx^2} = \dfrac{6\cos^2 x}{x^4} + \dfrac{4\sin 2x - 2}{x^3} - \dfrac{2\cos 2x}{x^2}$

c $\dfrac{d^2y}{dx^2} = -2e^{-x}\cos x$

7 b $f''(x) = 3\sin x\cos 2x + 6\cos x\sin 2x$

8 a $\dfrac{d^2y}{dx^2} = \dfrac{1}{x^2}$ **b** $\dfrac{d^2y}{dx^2} = \dfrac{1}{x}$ **c** $\dfrac{d^2y}{dx^2} = \dfrac{2}{x^2}(1 - \ln x)$

EXERCISE 7J

1 a $2\dfrac{dy}{dx}$ **b** $-3\dfrac{dy}{dx}$ **c** $6y^2\dfrac{dy}{dx}$

d $\dfrac{4}{y^2}\dfrac{dy}{dx}$ **e** $4y^3\dfrac{dy}{dx}$ **f** $\dfrac{1}{2\sqrt{y}}\dfrac{dy}{dx}$

g $-\dfrac{2}{y^3}\dfrac{dy}{dx}$ **h** $y + x\dfrac{dy}{dx}$ **i** $2xy + x^2\dfrac{dy}{dx}$

j $y^3 + 3xy^2\dfrac{dy}{dx}$

2 a $\dfrac{dy}{dx} = -\dfrac{x}{y}$ **b** $\dfrac{dy}{dx} = -\dfrac{x}{3y}$

c $\dfrac{dy}{dx} = \dfrac{x}{y}$ **d** $\dfrac{dy}{dx} = \dfrac{2x}{3y^2}$

e $\dfrac{dy}{dx} = \dfrac{-2x - y}{x}$ **f** $\dfrac{dy}{dx} = \dfrac{3x^2 - 2y}{2x}$

g $\dfrac{dy}{dx} = \dfrac{2}{x^3} - \dfrac{y}{x}$ **h** $\dfrac{dy}{dx} = \dfrac{-1 - 2y}{2x + 2y}$

i $\dfrac{dy}{dx} = \dfrac{1 - 3x^2 - y^2}{2xy - 4y}$ **j** $\dfrac{dy}{dx} = \dfrac{y}{x - y^2}$

k $\dfrac{dy}{dx} = \dfrac{1}{\sin y}$ **l** $\dfrac{dy}{dx} = \dfrac{-e^y}{\cos y + xe^y - 2}$

4 a $\dfrac{dy}{dx} = \dfrac{3 - y^2}{2xy - 1}$ **b i** -3 **ii** $-\frac{1}{7}$

5 a $-\frac{1}{4}$ **b** -2 **c** 1   **d** $-\frac{1}{9}$ **e** $\frac{9}{8}$ **f** $\frac{1}{2}$

6 a $\dfrac{dy}{dx} = \dfrac{4x^3 y + y^3}{2x^4 + xy^2}$

b i $P(1, 1)$, $Q(1, -2)$
ii At P, gradient $= \frac{5}{3}$.   At Q, gradient $= -\frac{8}{3}$.

REVIEW SET 7A

1 $f'(x) = -\dfrac{2}{x^2}$

2 a $f'(x) = 6x^5 - 5$ **b** $f'(x) = 12x^2 - \dfrac{2}{x^3}$

c $f'(x) = 3x^{\frac{1}{2}} = 3\sqrt{x}$

3 a $\dfrac{dy}{dx} = 6x - 4x^3$ **b** $\dfrac{dy}{dx} = 1 + \dfrac{1}{x^2}$

c $\dfrac{dy}{dx} = 2x(x-2)^{\frac{1}{2}} + \dfrac{1}{2}x^2(x-2)^{-\frac{1}{2}}$

4 a $f'(x) = (x^2+1)^{-\frac{3}{2}}$ **b** $(0, 0)$

5 a $\dfrac{dy}{dx} = 3x^2 e^{x^3+2}$ **b** $\dfrac{dy}{dx} = \dfrac{1}{x+3} - \dfrac{2}{x}$

c $\dfrac{dy}{dx} = 3x^2 e^{2x} + 2x^3 e^{2x}$

7 a $5 + 3x^{-2}$ **b** $4(3x^2 + x^{\frac{1}{2}})^3 (6x + \frac{1}{2}x^{-\frac{1}{2}})$

c $2x(1-x^2)^3 - 6x(x^2+1)(1-x^2)^2$

8 $(-2, 19)$ and $(1, -2)$ **9 a** -15 **b** 3

10 a $(5\cos 5x)\ln x + \dfrac{\sin 5x}{x}$

b $\cos x \cos 2x - 2\sin x \sin 2x$

c $-2e^{-2x}\tan x + e^{-2x}\sec^2 x$

11 $\dfrac{\sqrt{3}}{2}$

12 a $f'(x) = \dfrac{-x^2 + 6x - 3}{e^x}$ **b** $\dfrac{2}{e}$ **c** $x = 3 \pm \sqrt{6}$

13 a $f'(x) = 8x(x^2+3)^3$

b $g'(x) = \dfrac{\frac{1}{2}x(x+5)^{-\frac{1}{2}} - 2(x+5)^{\frac{1}{2}}}{x^3}$

c $h'(x) = \dfrac{6e^{4x} - 8xe^{4x}}{(1-2x)^2}$

14 a $f(\frac{\pi}{2}) = 1$ **b** $f'(\frac{\pi}{2}) = 0$ **c** $f''(\frac{\pi}{2}) = 2$

15 a $\dfrac{dy}{dx} = \cos^2 x - \sin^2 x$

b **Hint:** $\cos 2x = \cos^2 x - \sin^2 x$ **c** $-\frac{1}{2}$

16 a $\dfrac{d^2y}{dx^2} = \dfrac{3}{2}x^2 + x - \dfrac{1}{2}$ **b** $\dfrac{d^2y}{dx^2} = -2e^{-x} + xe^{-x}$

17 $\dfrac{dy}{dx} = \sec^2 x$

18 a $\dfrac{dy}{dx} = -\dfrac{9x^2}{2y}$ **b** $\dfrac{dy}{dx} = \dfrac{1+y^2}{2-2xy}$ **c** $\dfrac{dy}{dx} = \dfrac{3-y}{x-2y}$

19 a $\dfrac{dy}{dx} = \dfrac{x}{y}$

b at $(-4, 1)$, gradient $= -4$;   at $(8, 7)$, gradient $= \frac{8}{7}$

REVIEW SET 7B

1 a $f'(x) = 3$ **b** $f'(x) = 6\sqrt{x}$

2 a $\dfrac{dy}{dx} = 6x^2 - 12x + 7$ **b** $\dfrac{dy}{dx} = -\dfrac{3}{x^2} + \dfrac{15}{x^4}$

c $\dfrac{dy}{dx} = -\dfrac{5}{x^{\frac{4}{3}}}$

3 a $\dfrac{dy}{dx} = 3x^2(1-x^2)^{\frac{1}{2}} - x^4(1-x^2)^{-\frac{1}{2}}$

b $\dfrac{dy}{dx} = \dfrac{(2x-3)(x+1)^{\frac{1}{2}} - \frac{1}{2}(x^2-3x)(x+1)^{-\frac{1}{2}}}{x+1}$

4 a $\dfrac{dy}{dx} = e^x + xe^x$ **b** $(1, e)$

5 a $f'(x) = \dfrac{e^x}{e^x + 3}$ **b** $f'(x) = \dfrac{3}{x+2} - \dfrac{1}{x}$

6 a $f'(x) = 1 - \dfrac{2}{x^2}$ **b i** -1 **ii** $\frac{1}{2}$

c

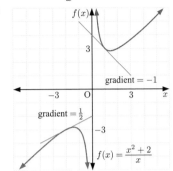

gradient $= -1$

gradient $= \frac{1}{2}$

$f(x) = \dfrac{x^2+2}{x}$

7 when $x = 1$, $\dfrac{dy}{dx} = 0$

8 a $\dfrac{dy}{dx} = \dfrac{3x^2 - 3}{x^3 - 3x}$ **b** $\dfrac{dy}{dx} = \dfrac{e^x(x-2)}{x^3}$

c $\dfrac{dy}{dx} = 2e^{2x}\sin x + e^{2x}\cos x$

9 a $f''(x) = 24x^2 - 24x - 18$ **b** $x = -\frac{1}{2}, \frac{3}{2}$

10 a $10 - 10\cos 10x$ **b** $\tan x$

c $(5\cos 5x)\ln(2x) + \dfrac{\sin 5x}{x}$

11 a $\dfrac{28}{9}$ **b** 8 **12** $a = 4$, $b = e^2$

13 a $-\frac{1}{4}$ **b** **Hint:** Show that $\dfrac{dy}{dx} = -\dfrac{2\sin x + 1}{(\sin x + 2)^2}$.

14 b i $x = \frac{1}{2}$ **ii** $x \leqslant 0$

15 a $\dfrac{d^2y}{dx^2} = -\dfrac{10}{(1-2x)^3}$ **b** $\dfrac{d^2y}{dx^2} = 6x + \dfrac{3}{4}x^{-\frac{5}{2}}$

17 a $\dfrac{dy}{dx} = -\dfrac{4x^3}{15y^2}$ **b** $\dfrac{dy}{dx} = \dfrac{2x - e^x}{2e^{2y}}$

c $\dfrac{dy}{dx} = \dfrac{\cos x - x\sin x}{\cos y}$

18 a $\dfrac{dy}{dx} = -\dfrac{x}{2y}$

b i P has y-coordinate 4, Q has y-coordinate $-\sqrt{\dfrac{39}{2}}$

ii at P, gradient $= -\frac{1}{2}$;   at Q, gradient $= \dfrac{3}{\sqrt{78}}$

EXERCISE 8A

1 a $f'(x) = 2x - 4$ **b** $y = -2x - 1$

2 a $y = 4x + 6$ **b** $y = 3x - 4$ **c** $y = 10x - 11$

d $y = \dfrac{19}{18}x - \dfrac{3}{2}$ **e** $y = \dfrac{1}{3}x + \dfrac{5}{3}$ **f** $y = \dfrac{1}{9}x + \dfrac{4}{9}$

g $y = -\dfrac{1}{2}x + \dfrac{3}{4}$ **h** $y = \dfrac{1}{27}x + \dfrac{4}{9}$

3 a $y = -\dfrac{1}{5}x + \dfrac{17}{5}$ **b** $y = x + 1$ **c** $y = 4 - x$

d $y = -\dfrac{4}{11}x + \dfrac{78}{11}$ **e** $y = -16x - \dfrac{65}{4}$ **f** $x = 2$

g $y = -\dfrac{4}{57}x + \dfrac{1042}{57}$ **h** $y = \dfrac{1}{2}x + \dfrac{1}{2}$

4 a $y = 4 - 2x$ **b** $y = -\dfrac{9}{62}x + \dfrac{1259}{186}$

5 a $k = 1$ **b** $y = 8x - 7$ **c** $\frac{7}{8}$ **6** $a = 4$, $b = 3$

7 a $\dfrac{dy}{dx} = -\dfrac{3x^2}{2y}$ **b** $y = 4x - 22$

8 a $y = -e^{-2}x + 3e^{-2}$ **b** $y = -\frac{1}{3}x - \frac{1}{3} + \ln 3$

    c $y = 4ex - e$ **d** $y = \frac{e^2}{2}x - \frac{3}{2}$ **e** $y = 3ex - 5e$

9 a i -1 **ii** -2 **iii** -3 **b** y-intercept is $-a$

10 x-intercept is $\frac{2}{3}$, y-intercept is $-2e$

11 a $\dfrac{dx}{dy} = e^y(y + 2)$ **b** $x - 4e^2y = -5e^2$

12 a $y = x$ **b** $y = x$ **c** $y = -\frac{1}{2}x + \frac{\pi}{12} + \frac{\sqrt{3}}{2}$
    d $y = 1$ **e** $y = 2$

13 Hint: Show that there are no tangents which have gradient $= 0$.

14 $\frac{1}{2}(\frac{\pi}{2} + 1)^2$ units^2

15 a $\sqrt{2}x - y = \sqrt{2}\left(\frac{\pi}{4} - 1\right)$ **b** $2x + y = \frac{2\pi}{3} + \sqrt{3}$

16 a $x - 2\sqrt{3}y = \frac{\pi}{6} - 4\sqrt{3}$ **b** $\sqrt{6}x + y = \pi\sqrt{6} + \sqrt{2}$

17 $(-1, 2)$ **18** $(3, 50)$ **19** $(-1, -2)$

20 $(-1, -2)$ and $(2, 1)$

21 a $P(1) = 1 - 3 - 1 + 3 = 0$
    The zeros of $P(x)$ are $x = 1, -1,$ and 3.

    b

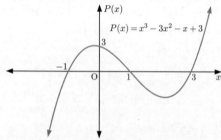

    c $y = -x - 1$ **d** $(-1, 0)$
    e Hint: Let $P(x) = \alpha(x - a)(x - b)(x - c)$.

22 a $y = (2a + 4)x - a^2$
    b $y = 12x - 16$ with contact point $(4, 32)$ and
    $y = -4$ with contact point $(-2, -4)$

23 $y = 5x - 15$ or $y = -7x - 3$

24 a $y = e^ax + e^a(1 - a)$ **b** $y = ex$

25 $x = 0$ **26** $a = \frac{1}{4}$, point of intersection $(\frac{1}{2}, \frac{\sqrt{3}}{2})$

27 $b = 3$ **28** $\approx 63.43°$

29 a Hint: They must have the same y-coordinate at $x = b$ and
    the same gradient.

    c $a = \frac{1}{2e}$ **d** $y = e^{-\frac{1}{2}}x - \frac{1}{2}$

30 a $x + \sqrt{3}y = 2$, $x - \sqrt{3}y = 2$ **b** $(2, 0)$

31 a $y = 2asx - as^2$, $y = 2atx - at^2$

EXERCISE 8B

1 a $-4 < x \leqslant 2$ **b** $-1 \leqslant x \leqslant 4$ **c** $-4 < x < -1$
    d $2 < x < 4$

2 a $f'(x) = 3x^2 - 12x$
    $= 3x(x - 4)$

    b increasing for $x \leqslant 0$ and $x \geqslant 4$
    decreasing for $0 \leqslant x \leqslant 4$

3 a increasing for $x \geqslant -3$, decreasing for $x \leqslant -3$
    b increasing for $-1 \leqslant x \leqslant 1$,
    decreasing for $x \leqslant -1$ and $x \geqslant 1$

    c increasing for $x > 0$, never decreasing
    d increasing for $x \geqslant -\frac{1}{6}$, decreasing for $x \leqslant -\frac{1}{6}$
    e never increasing, decreasing for $x > 0$
    f increasing for $x \leqslant 0$ and $x \geqslant 2$,
    decreasing for $0 \leqslant x \leqslant 2$

4 In this case, $f(x)$ is only defined when $x > 0$.
    $\therefore$ $f'(x)$ is only defined when $x > 0$.

5 a increasing for all $x \in \mathbb{R}$, never decreasing
    b increasing for $x > -2$, never decreasing
    c never increasing, decreasing for all $x \in \mathbb{R}$
    d increasing for $x \geqslant -1$, decreasing for $x \leqslant -1$
    e increasing for $x \geqslant 1$, decreasing for $0 < x \leqslant 1$
    f increasing for $x \geqslant e^{-\frac{1}{3}}$, decreasing for $0 < x \leqslant e^{-\frac{1}{3}}$

6 a

    $\begin{array}{ccccc} - & | & + & | & - \\ & -1 & & 1 & \end{array}$ $f'(x)$, x

    b increasing for $-1 \leqslant x \leqslant 1$,
    decreasing for $x \leqslant -1$ and $x \geqslant 1$

7 a

    $\begin{array}{ccccc} - & | & + & \vdots & - \\ & -1 & & 1 & \end{array}$ $f'(x)$, x

    b increasing for $-1 \leqslant x < 1$,
    decreasing for $x \leqslant -1$ and $x > 1$

8 a

    $\begin{array}{ccccccc} - & | & + & \vdots & + & | & - \\ & -1 & & 1 & & 3 & \end{array}$ $f'(x)$, x

    b increasing for $-1 \leqslant x < 1$ and $1 < x \leqslant 3$,
    decreasing for $x \leqslant -1$ and $x \geqslant 3$

9 a increasing for $x \geqslant \sqrt{3}$ and $x \leqslant -\sqrt{3}$,
    decreasing for $-\sqrt{3} \leqslant x < -1$ and $-1 < x < 1$ and
    $1 < x \leqslant \sqrt{3}$
    b increasing for $x \leqslant 0$, decreasing for $x \geqslant 0$
    c increasing for $x \geqslant 0$, decreasing for $x \leqslant 0$
    d increasing for $x \geqslant 2$,
    decreasing for $x < 1$ and $1 < x \leqslant 2$
    e increasing for $x \geqslant 0$, decreasing for $x \leqslant 0$
    f increasing for $x \leqslant -1$,
    decreasing for $-1 \leqslant x < 0$ and $x > 0$
    g increasing for $x \geqslant e^{-2}$, decreasing for $0 < x \leqslant e^{-2}$
    h increasing for $-\frac{\pi}{2} + 2k\pi \leqslant x \leqslant \frac{\pi}{2} + 2k\pi$, $k \in \mathbb{Z}$,
    decreasing for $\frac{\pi}{2} + 2k\pi \leqslant x \leqslant \frac{3\pi}{2} + 2k\pi$, $k \in \mathbb{Z}$
    i increasing for all $x \in \mathbb{R}$, never decreasing

10 $-1 \leqslant k \leqslant 0$

EXERCISE 8C

1 a A is a local minimum, B is a local maximum
    b $f'(x) = -3x^2 + 6x + 9$
    $= -3(x - 3)(x + 1)$

    $\begin{array}{ccccc} - & | & + & | & - \\ & -1 & & 3 & \end{array}$ $f'(x)$, x

    c A$(-1, -12)$ and B$(3, 20)$

2 b

    $\begin{array}{ccccc} - & | & - & | & + \\ & -1 & & 2 & \end{array}$ $f'(x)$, x

    c increasing for $x \geqslant 2$, decreasing for $x \leqslant 2$

d $(-1, 13)$ is a stationary inflection

$(2, -14)$ is a local minimum

e

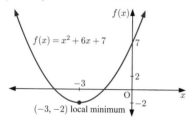

$f(x) = x^4 - 6x^2 - 8x + 10$

$(-1, 13)$ stationary inflection

$(2, -14)$ local minimum

3 a $(-3, -2)$ is a local minimum

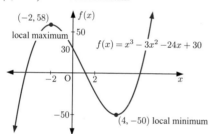

$f(x) = x^2 + 6x + 7$

$(-3, -2)$ local minimum

b $(-2, 58)$ is a local maximum,

$(4, -50)$ is a local minimum

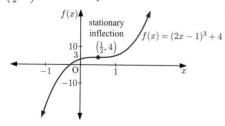

$(-2, 58)$ local maximum

$f(x) = x^3 - 3x^2 - 24x + 30$

$(4, -50)$ local minimum

c $\left(\frac{1}{2}, 4\right)$ is a stationary inflection

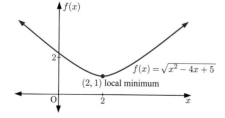

stationary inflection

$f(x) = (2x - 1)^3 + 4$

$\left(\frac{1}{2}, 4\right)$

d $(2, 1)$ is a local minimum

$f(x) = \sqrt{x^2 - 4x + 5}$

$(2, 1)$ local minimum

e $(0, 0)$ is a local maximum,

$(2, 4)$ is a local minimum

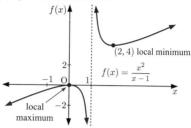

$(2, 4)$ local minimum

$f(x) = \dfrac{x^2}{x - 1}$

local maximum

f $\left(-1, -\frac{1}{2}\right)$ is a local minimum,

$\left(1, \frac{1}{2}\right)$ is a local maximum

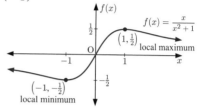

$f(x) = \dfrac{x}{x^2 + 1}$

$\left(1, \frac{1}{2}\right)$ local maximum

$\left(-1, -\frac{1}{2}\right)$ local minimum

4 a $\left(1, \dfrac{1}{e}\right)$ is a local maximum

b $\left(-2, \dfrac{4}{e^2}\right)$ is a local maximum, $(0, 0)$ is a local minimum

c $(1, e)$ is a local minimum

d $(-1, e)$ is a local maximum

5 a $a = 3$, $b = 6$ **b** local minimum   **6 a** $x > 0$

7 a $\left(\dfrac{\pi}{2}, 1\right)$ is a local maximum,

$\left(\dfrac{3\pi}{2}, -1\right)$ is a local minimum

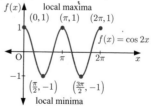

$\left(\dfrac{\pi}{2}, 1\right)$ local maximum

$f(x) = \sin x$

$\left(\dfrac{3\pi}{2}, -1\right)$ local minimum

b $(0, 1)$, $(\pi, 1)$, and $(2\pi, 1)$ are local maxima,

$\left(\dfrac{\pi}{2}, -1\right)$ and $\left(\dfrac{3\pi}{2}, -1\right)$ are local minima

local maxima

$(0, 1)$ $(\pi, 1)$ $(2\pi, 1)$

$f(x) = \cos 2x$

$\left(\dfrac{\pi}{2}, -1\right)$ $\left(\dfrac{3\pi}{2}, -1\right)$

local minima

c $(0, 0)$, $(\pi, 0)$, and $(2\pi, 0)$ are local minima,

$\left(\dfrac{\pi}{2}, 1\right)$ and $\left(\dfrac{3\pi}{2}, 1\right)$ are local maxima

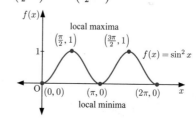

local maxima

$\left(\dfrac{\pi}{2}, 1\right)$ $\left(\dfrac{3\pi}{2}, 1\right)$

$f(x) = \sin^2 x$

$(0, 0)$ $(\pi, 0)$ $(2\pi, 0)$

local minima

d $\left(\frac{\pi}{2}, e\right)$ is a local maximum,

$\left(\frac{3\pi}{2}, \frac{1}{e}\right)$ is a local minimum

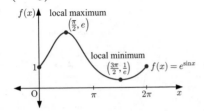

e $\left(\frac{3\pi}{4}, -\sqrt{2}\right)$ is a local minimum,

$\left(\frac{7\pi}{4}, \sqrt{2}\right)$ is a local maximum

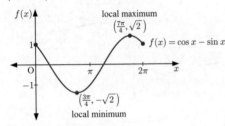

f $\left(\frac{\pi}{6}, \frac{3\sqrt{3}}{2}\right)$ is a local maximum,

$\left(\frac{5\pi}{6}, -\frac{3\sqrt{3}}{2}\right)$ is a local minimum,

$\left(\frac{3\pi}{2}, 0\right)$ is a stationary inflection

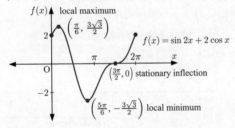

9 $a = \frac{\sqrt{e}}{2}, \ b = -\frac{1}{8}$

10 Hint: Show that $\dfrac{\ln x}{x}$ has only one stationary point, which is a local maximum.

11 Hint: Show that $f(x) \geqslant 1$ for all $x > 0$.

12 a $x = \frac{\pi}{2}, \frac{3\pi}{2}$

b $(0, 1)$ and $(2\pi, 1)$ are local minima, $(\pi, -1)$ is a local maximum

c $f(x)$ has period 2π

d

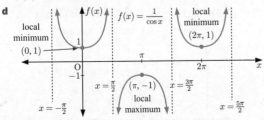

13 a $\dfrac{dy}{dx} = \dfrac{e^x(x+1)}{1-2y}$ **b** $y = x+1$

c $\left(-1, \dfrac{1+\sqrt{1+\frac{4}{e}}}{2}\right)$ and $\left(-1, \dfrac{1-\sqrt{1+\frac{4}{e}}}{2}\right)$

EXERCISE 8D

1 a concave up for all $x \in \mathbb{R}$

b concave up for $x \leqslant 0$, concave down for $x \geqslant 0$

c concave down for all $x \neq 0$

d concave up for all $x \in \mathbb{R}$

e concave up for $0 < x \leqslant 2$, concave down for $x \geqslant 2$

f concave up for $x > -2$, concave down for $x < -2$

g concave up for $x < -1$, concave down for $x > -1$

h concave up for $0 < x \leqslant 0.753$ and $x > 2$, concave down for $0.753 \leqslant x < 2$

2 a increasing for $-1 \leqslant x \leqslant 1$

b decreasing for $x \leqslant -1$ and $x \geqslant 1$

c concave up for $-\sqrt{3} \leqslant x \leqslant 0$ and $x \geqslant \sqrt{3}$

d concave down for $x \leqslant -\sqrt{3}$ and $0 \leqslant x \leqslant \sqrt{3}$

EXERCISE 8E

1 a

Point	$f(x)$	$f'(x)$	$f''(x)$
A	0	$-$	$+$
B	$-$	0	$+$
C	0	$+$	0
D	$+$	0	0

b B is a local minimum.

c C is a non-stationary inflection point, D is a stationary inflection point

2 a no points of inflection **b** stationary inflection at $(0, 2)$

c non-stationary inflection at $(2, 3)$

d stationary inflection at $(0, 2)$

non-stationary inflection at $\left(-1\frac{1}{3}, 11\frac{13}{27}\right)$

e no points of inflection

f stationary inflection at $(-2, -3)$

g non-stationary inflection at $(1, 9)$

h non-stationary inflections at $(-1, 5)$ and $(1, 5)$

i non-stationary inflections at $(-\sqrt{3}, 4)$ and $(\sqrt{3}, 4)$

j non-stationary inflection at $\left(\frac{1}{e^2}, -\frac{1}{2}\right)$

3 a i local minimum at $\left(\frac{5}{2}, -\frac{9}{4}\right)$

ii no points of inflection

iii increasing for $x \geqslant \frac{5}{2}$, decreasing for $x \leqslant \frac{5}{2}$

iv concave up for all $x \in \mathbb{R}$

v

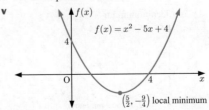

b i local maximum at $\left(-\frac{8}{3}, \frac{256}{27}\right)$, local minimum at $(0, 0)$

ii non-stationary inflection at $\left(-\frac{4}{3}, \frac{128}{27}\right)$

iii increasing for $x \leqslant -\frac{8}{3}$ and $x \geqslant 0$,
decreasing for $-\frac{8}{3} \leqslant x \leqslant 0$

iv concave up for $x \geqslant -\frac{4}{3}$, concave down for $x \leqslant -\frac{4}{3}$

v

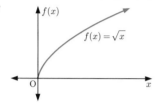

c i no turning points
ii no points of inflection
iii increasing for $x > 0$,   never decreasing
iv concave down for $x > 0$,   never concave up

v

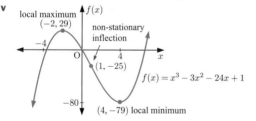

d i local maximum at $(-2, 29)$
local minimum at $(4, -79)$
ii non-stationary inflection at $(1, -25)$
iii increasing for $x \leqslant -2$ and $x \geqslant 4$
decreasing for $-2 \leqslant x \leqslant 4$
iv concave down for $x \leqslant 1$,   concave up for $x \geqslant 1$

v

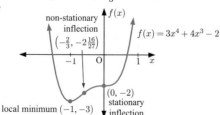

e i local minimum at $(-1, -3)$
ii non-stationary inflection at $\left(-\frac{2}{3}, -2\frac{16}{27}\right)$
stationary inflection at $(0, -2)$
iii increasing for $x \geqslant -1$,   decreasing for $x \leqslant -1$
iv concave down for $-\frac{2}{3} \leqslant x \leqslant 0$
concave up for $x \leqslant -\frac{2}{3}$ and $x \geqslant 0$

v

local minimum $(-1, -3)$

f i local minimum at $(1, 0)$
ii no points of inflection
iii increasing for $x \geqslant 1$,   decreasing for $x \leqslant 1$

iv concave up for all $x \in \mathbb{R}$

v

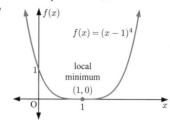

g i local minima at $(-\sqrt{2}, -1)$ and $(\sqrt{2}, -1)$,
local maximum at $(0, 3)$
ii non-stationary inflections at $\left(\sqrt{\frac{2}{3}}, \frac{7}{9}\right)$ and
$\left(-\sqrt{\frac{2}{3}}, \frac{7}{9}\right)$
iii increasing for $-\sqrt{2} \leqslant x \leqslant 0$ and $x \geqslant \sqrt{2}$
decreasing for $x \leqslant -\sqrt{2}$ and $0 \leqslant x \leqslant \sqrt{2}$
iv concave down for $-\sqrt{\frac{2}{3}} \leqslant x \leqslant \sqrt{\frac{2}{3}}$
concave up for $x \leqslant -\sqrt{\frac{2}{3}}$ and $x \geqslant \sqrt{\frac{2}{3}}$

v

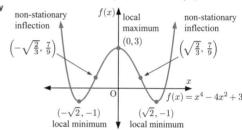

h i no turning points
ii no points of inflection
iii increasing for $x > 0$,   never decreasing
iv concave down for $x > 0$,   never concave up

v

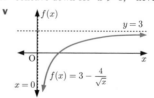

4 a x-intercept $\ln\sqrt{3}$, y-intercept -2
b $f'(x) = 2e^{2x} > 0$ for all $x \in \mathbb{R}$
c $f''(x) = 4e^{2x} > 0$ for all $x \in \mathbb{R}$
d as $x \to -\infty$, $e^{2x} \to 0$ $\therefore$ $e^{2x} - 3 \to -3^{+}$
e

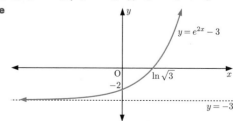

5 a $f(x)$: x-intercept $\ln 3$, y-intercept -2
$g(x)$: x-intercept $\ln\left(\frac{5}{3}\right)$, y-intercept -2

b $f(x)$: as $x \to \infty$, $f(x) \to \infty$
 as $x \to -\infty$, $f(x) \to -3^+$
 $g(x)$: as $x \to \infty$, $g(x) \to 3^-$
 as $x \to -\infty$, $g(x) \to -\infty$

c

$f(x)$ is increasing and concave up for all $x \in \mathbb{R}$.

$g(x)$ is increasing and concave down for all $x \in \mathbb{R}$.

d $(0, -2)$ and $(\ln 5, 2)$

e

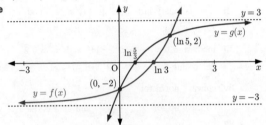

6 **a** x-intercept $\ln\sqrt{3}$, y-intercept -2

b $f'(x) = e^x + 3e^{-x} > 0$ for all $x \in \mathbb{R}$

c y is concave down below the x-axis and concave up above the x-axis.

d

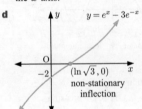

7 **a** x-intercept is $\dfrac{e^3 + 1}{2} \approx 10.5$

b no, $\therefore$ there is no y-intercept

c Domain $= \{x : x > \tfrac{1}{2}\}$ **d** gradient $= 2$

e $f''(x) = \dfrac{-4}{(2x-1)^2} < 0$ for all $x > \tfrac{1}{2}$, so $f(x)$ is concave down.

f

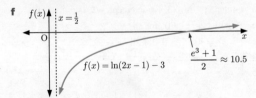

8 **a** $x \neq 0$

b $f'(x) = \dfrac{2}{x}$

$f(x)$ is increasing for $x > 0$ and decreasing for $x < 0$.

$f''(x) = -\dfrac{2}{x^2}$

$f(x)$ is concave down for all $x \neq 0$.

c

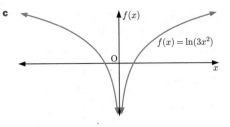

9 **a** $f(x)$ does not have any x or y-intercepts.

b as $x \to \infty$, $f(x) \to \infty$, as $x \to -\infty$, $f(x) \to 0^-$

c local minimum at $(1, e)$

d **i** $x > 0$ **ii** $x < 0$

e

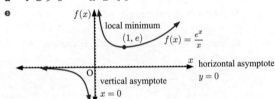

f $2x + ey = -3$

10 **a** local maximum at $\left(0, \dfrac{1}{\sqrt{2\pi}}\right)$

increasing for $x \leqslant 0$, decreasing for $x \geqslant 0$

b non-stationary inflections at $\left(-1, \dfrac{1}{\sqrt{2e\pi}}\right)$ and $\left(1, \dfrac{1}{\sqrt{2e\pi}}\right)$

c as $x \to \infty$, $f(x) \to 0^+$
as $x \to -\infty$, $f(x) \to 0^+$

d

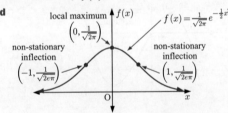

11 **a** The inflection points coincide with the x-intercepts.

b non-stationary inflection points at $\left(\dfrac{\pi}{2}, 0\right)$ and $\left(\dfrac{3\pi}{2}, 0\right)$

c **i** $\pi \leqslant x \leqslant 2\pi$ **ii** $0 \leqslant x \leqslant \pi$ **iii** $\dfrac{\pi}{2} \leqslant x \leqslant \dfrac{3\pi}{2}$

 iv $0 \leqslant x \leqslant \dfrac{\pi}{2}$ and $\dfrac{3\pi}{2} \leqslant x \leqslant 2\pi$

d

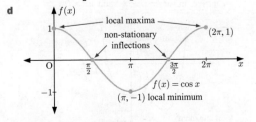

12 **a** **i** **Hint:** Show that $f'(t) = Ae^{-bt}(1 - bt)$

 ii **Hint:** Show that $f''(t) = Abe^{-bt}(bt - 2)$

b

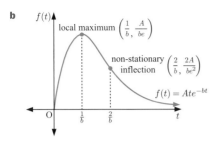

local maximum $\left(\frac{1}{b}, \frac{A}{be}\right)$

non-stationary $\left(\frac{2}{b}, \frac{2A}{be^2}\right)$
inflection

$f(t) = Ate^{-bt}$

13 a y-intercept is $\dfrac{C}{1+A}$

c

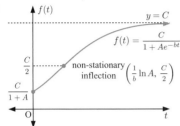

$y = C$

$f(t) = \dfrac{C}{1+Ae^{-bt}}$

$\dfrac{C}{2}$

non-stationary $\left(\frac{1}{b}\ln A, \frac{C}{2}\right)$
inflection

$\dfrac{C}{1+A}$

EXERCISE 8F

1 a i $Q(0) = 100$ **ii** $Q(25) = 50$ **iii** $Q(100) = 0$

   b i decreasing by 1 unit per year

   ii decreasing by $\frac{1}{\sqrt{2}}$ units per year

   c $Q'(t) = -\dfrac{5}{\sqrt{t}} < 0$ for all $t > 0$

2 a 0.5 m

   b i ≈ 15.8 m   **ii** ≈ 21.7 m   **iii** ≈ 24.9 m

   c $t = 0$: 6.9 m per year, $t = 5$: 1.725 m per year,
   $t = 10$: 0.767 m per year

   d $\dfrac{dH}{dt} = \dfrac{172.5}{(t+5)^2} > 0$ for all $t \geqslant 0$

   The tree will continue to grow forever.

3 a i 4500 euros   **ii** 4000 euros

   b i decreasing at ≈ 210.22 euros per km h^{-1}

   ii increasing at ≈ 11.31 euros per km h^{-1}

   c ≈ 79.4 km h^{-1}

4 a $\dfrac{dV}{dt} = -1250\left(1 - \dfrac{t}{80}\right)$ L min^{-1}

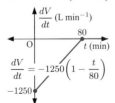

$\dfrac{dV}{dt} = -1250\left(1 - \dfrac{t}{80}\right)$

   b $t = 0$ when the tap was first opened

   c $\dfrac{d^2V}{dt^2} = \dfrac{125}{8} > 0$

   This shows that the rate of change of V is constantly
   increasing, so the outflow is increasing at a constant rate.

5 a The near part of the lake is 2 km from the sea, the furthest
   part is 3 km.

b $\dfrac{dy}{dx} = \dfrac{3}{10}x^2 - x + \dfrac{3}{5}$

   When $x = \frac{1}{2}$, $\dfrac{dy}{dx} = 0.175$, height of hill is increasing as
   gradient is positive.

   When $x = 1\frac{1}{2}$, $\dfrac{dy}{dx} = -0.225$, height of hill is decreasing
   as gradient is negative.

   c ≈ 2.55 km from the sea, ≈ 63.1 m deep

6 a $k = \frac{1}{50}\ln 2 \approx 0.0139$

   b i 20 grams   **ii** ≈ 14.3 grams   **iii** ≈ 1.95 grams

   c ≈ 216 hours or ≈ 9 days

   d i ≈ -0.0693 g h^{-1} **ii** $\approx -2.64 \times 10^{-7}$ g h^{-1}

   e Hint: You should find $\dfrac{dW}{dt} = -\frac{1}{50}\ln 2 \times 20e^{-\frac{t}{50}\ln 2}$

7 a $k = \frac{1}{15}\ln\left(\frac{19}{3}\right) \approx 0.123$ **b** $100°$C

   c $c = -k \approx -0.123$

   d i decreasing at $\approx 11.7°$C min^{-1}

   ii decreasing at $\approx 3.42°$C min^{-1}

   iii decreasing at $\approx 0.998°$C min^{-1}

8 a ≈ 43.9 cm   **b** ≈ 10.4 years

   c i growing at ≈ 5.45 cm per year

   ii growing at ≈ 1.88 cm per year

9 a When $t = 0$, $A = 0$ litres

   b i $k = \dfrac{\ln 2}{3}$ (≈ 0.231)

   ii ≈ 0.728 litres of alcohol produced per hour

10 a ≈ 5.15 m   **b** rising

11 a i 0 volts   **ii** 340 volts

   b i $-34\,000\pi$ volts per second   **ii** 0 volts per second

12 a 2000 bees   **b** $\approx 37.8\%$ **c** yes, 3000 bees

   d $B'(t) = \dfrac{2595}{e^{1.73t}(1 + 0.5e^{-1.73t})^2} > 0$ for all $t \geqslant 0$

   $\therefore$ $B(t)$ is increasing over time.

   e ≈ 0.0806 bees per month

   f

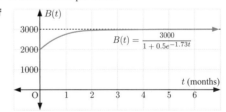

$B(t) = \dfrac{3000}{1 + 0.5e^{-1.73t}}$

EXERCISE 8G

1 b 15 m $\times$ 30 m

2 a volume $= x^2 y = 2000$ cm^3

   c ≈ 15.9 cm $\times \approx 15.9$ cm $\times \approx 7.94$ cm

   d ≈ 756 cm^2

3 a area $= 4x\sqrt{25 - x^2}$ cm^2 **b** $5\sqrt{2}$ cm $\times 5\sqrt{2}$ cm

4 20 kettles   **5** C$\left(\frac{1}{\sqrt{2}}, e^{-\frac{1}{2}}\right)$

6 b $\theta \approx 1.91$, $A \approx 237$ cm^2

7 a $E'(t) = 750e^{-1.5t}(1 - 1.5t)$

   b at approximately 40 minutes after the injection

8 c $\theta = \frac{\pi}{6}$, area ≈ 130 cm^2 **9** $3\frac{1}{3}$ km

10 **c** $\theta = \frac{\pi}{6}$ **d**

e **i** Row from P to Q at an angle of $\frac{\pi}{6}$ to the diameter of the lake, then walk from Q to R.
ii Walk from P to R.

11 **a** X is between A and C.
c $x \approx 2.67$ This is the distance in km from A to X which minimises the time taken to get from B to C.

12 **d** $\approx 63.7\%$

13 **c** $L \approx 7.02$ m.   This is the length of metal tube which can be carried around the corner from one corridor to the other without bending.

EXERCISE 8H

1 **a** $b^3 \dfrac{da}{dt} + 3ab^2 \dfrac{db}{dt} = 0$

b a is decreasing at 7.5 units per second.

2 **a** $A = x^2$ **b** $\dfrac{dA}{dt} = 2x \dfrac{dx}{dt}$ **c** 24 cm^2 per second

3 **a** $xy = 100$ **b** $y \dfrac{dx}{dt} + x \dfrac{dy}{dt} = 0$

c **i** 0.25 cm per minute   **ii** 1 cm per minute

4 **a** 4π m^2 per second   **b** 8π m^2 per second

5 increasing at 0.375 m per minute

6 decreasing at 0.16 m^3 per minute   **7** $\frac{20}{3}$ cm per minute

8 decreasing at $\frac{250}{13} \approx 19.2$ m s^{-1} **9** ≈ 1.35 cm per minute

10 **a** 0.2 m s^{-1} **b** $\frac{4}{45}$ m s^{-1}

11 **a** increase   **b** increasing at $\frac{2\pi}{3\sqrt{3}} \approx 1.21$ cm per minute

12 $\frac{\sqrt{2}}{100}$ radians per second

13 **a** decreasing at $\frac{3}{100}$ radians per second

b decreasing at $\frac{1}{100}$ radians per second

14 increasing at 0.128 radians per second

15 increasing at 0.12 radians per minute

16 increasing at ≈ 24.3 m s^{-1}

17 **a** $\frac{\sqrt{3}}{2}\pi$ cm s^{-1} **b** 0 cm s^{-1}

18 **a** $\frac{200}{\sqrt{13}}\pi$ radians per second   **b** 100π radians per second

19 **b** $\frac{\sqrt{3}}{120}$ m per minute

REVIEW SET 8A

1 **a** $y = 7x - 14$ **b** $y = -1$ **c** $y = \dfrac{3}{e}x + \dfrac{1}{e}$

d $y = \dfrac{2}{e}x$

2 **a** $y = -\frac{8}{3}x + \frac{44}{3}$ **b** $y = -\dfrac{1}{6e^2}x + \dfrac{18e^4 + 1}{6e^2}$

4 **a** $\dfrac{dy}{dx} = \dfrac{2x - y}{x + 3y^2 - 2}$ **b** $y = \frac{3}{2}x - \frac{5}{2}$

5 **a** $x \ne -3$ **b** x-intercept $\frac{2}{3}$, y-intercept $-\frac{2}{3}$

c $f'(x) = \dfrac{11}{(x+3)^2}$

d no stationary points

6 $(-2, -25)$

7 **a** $y = 4\sqrt{3}x - \dfrac{4\pi + 2\sqrt{3}}{\sqrt{3}}$ **b** $y = -\dfrac{1}{\sqrt{2}}x + \dfrac{\pi + 4}{2\sqrt{2}}$

8 **a** $y = -\dfrac{1}{2e^{2a}}x + \dfrac{a}{2e^{2a}} + e^{2a}$ **b** $y \approx -1.116x$

9 $k = 3$, point of intersection $(3, 12)$, or
$k = 147$, point of intersection $(-5, 84)$

10 **a** increasing for $x \le -\sqrt{2}$ and $x \ge \sqrt{2}$,
decreasing for $-\sqrt{2} \le x \le \sqrt{2}$
b increasing for $x \ge 1$, decreasing for $x \le 1$
c increasing for all $x \in \mathbb{R}$

11 **a** local maximum at $(1, 3)$, local minimum at $(\frac{1}{3}, \frac{77}{27})$
b local maximum at $(-6, -12)$, local minimum at $(0, 0)$

12 **a** $(-\pi, -1)$ is a local minimum, $(\pi, 1)$ is a local maximum

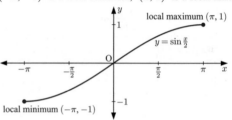

b $(-\pi, 1)$, $(0, 1)$, and $(\pi, 1)$ are local maxima,
$(-\frac{\pi}{2}, 0)$ and $(\frac{\pi}{2}, 0)$ are local minima

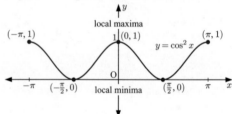

c $\left(-\frac{5\pi}{6}, \frac{3}{2}\right)$ and $\left(-\frac{\pi}{6}, \frac{3}{2}\right)$ are local maxima,
$\left(-\frac{\pi}{2}, 1\right)$ and $\left(\frac{\pi}{2}, -3\right)$ are local minima

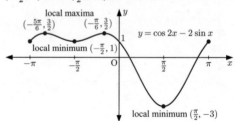

13 **a** concave up for $x \ge \frac{4}{3}$, concave down for $x \le \frac{4}{3}$
b concave up for $x \le -3$,
concave down for $-3 \le x < 0$ and $x > 0$
c concave up for $-4 < x \le 2$ and $x > 0$,
concave down for $x < -4$ and $-2 \le x < 0$

14 **a** $x > 0$
b $f'(x) = 1 + \dfrac{1}{x}$ $f''(x) = -\dfrac{1}{x^2}$

$f(x)$ is increasing and concave downwards for all $x > 0$.

c

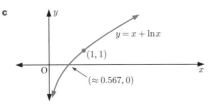

15 **a** $f'(x) = e^{x\sqrt{3}}(\cos x + \sqrt{3}\sin x)$

 b $x = \frac{5\pi}{6}$ or $\frac{11\pi}{6}$

 c

 d **i** $0 \leqslant x \leqslant \frac{5\pi}{6}$ and $\frac{11\pi}{6} \leqslant x \leqslant 2\pi$

  ii $\frac{5\pi}{6} \leqslant x \leqslant \frac{11\pi}{6}$

16 **a** $(0, \ln 5)$ is a local minimum

 b $(-\sqrt{5}, \ln 10)$ and $(\sqrt{5}, \ln 10)$ are non-stationary inflections

 c increasing for $x \geqslant 0$, decreasing for $x \leqslant 0$

 d concave up for $-\sqrt{5} \leqslant x \leqslant \sqrt{5}$

  concave down for $x \leqslant -\sqrt{5}$ and $x \geqslant \sqrt{5}$

 e

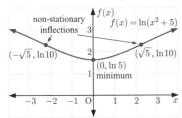

17 **a** 60 cm **b** **i** ≈ 4.24 years **ii** ≈ 201 years

 c **i** 16 cm per year **ii** ≈ 1.95 cm per year

18 **a** £20 000 **b** £146.53 per year

19 x-coordinate of P is $\ln a$.

20 **a** **i** 5 km **ii** $2\sqrt{10}$ km

21 **a** $A = \frac{\sqrt{3}}{4}x^2$ **b** $\frac{45\sqrt{3}}{2}$ cm^2 per minute

22 **a** $V(r) = \frac{8}{9}\pi r^3$ m^3

 b $\frac{dr}{dt} = -\frac{8}{375\pi} \approx -0.006\,79$ m min^{-1}

23 **a** $p(x) = a(x-r)^2 + (2ar+b)(x-r) + ar^2 + br + c$

 c The tangent to $p(x)$ at $x = r$ is

  $p(x) = p(r) + p'(r)(x - r)$. We can use the form in **a** to find $p(r)$, $p'(r)$, and hence a formula for the tangent to $p(x)$ at $x = r$.

REVIEW SET 8B

1 **a** $y = 31x - 43$ **b** $y = -\frac{1}{128}x + \frac{41}{128}$

 c $y = 3x + \frac{3\sqrt{3} - \pi}{2}$ **d** $y = \frac{3}{4}x + \frac{1}{2}$

2 **a** $x = 0$ **b** $x + 2e^e y = e + 4e^{2e}$

3 **a** $\frac{dy}{dx} = \frac{7\pi \cos \frac{\pi x}{2} + 8x}{4y}$ **b** $y = \frac{3}{2}x - \frac{9}{2}$

4 **a** local minimum at $(0, 1)$ **b** as $x \to \infty$, $f(x)$

c $f''(x) = e^x$

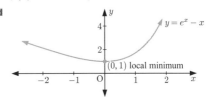

 $f(x)$ is concave up for all $x \in \mathbb{R}$.

d

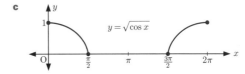

5 $\frac{3267}{152} \approx 21.5$ units^2 **6** $(\frac{1}{2}, \frac{1}{4})$

7 **a** $2x + 3y = \frac{2\pi}{3} + 2\sqrt{3}$ **b** $x + 2\sqrt{2}y = \frac{\pi}{2} + 2$

8 $y = \frac{3}{4}x + \frac{5}{4}$ **9** $a = \frac{5}{2}$, $b = -\frac{3}{2}$

10 **a**

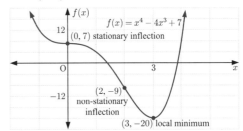

11 $(1 - a, e^{a-1})$ is a local maximum **12** $a = 2$, $b = 3$

13 **a** stationary inflection at $(0, 9)$,

  non-stationary inflection at $(\frac{3}{2}, \frac{63}{16})$

 b non-stationary inflections at $(-1, 8)$ and $(\frac{3}{2}, \frac{313}{16})$

14 **a** $0 \leqslant x \leqslant \frac{\pi}{2}$ and $\frac{3\pi}{2} \leqslant x \leqslant 2\pi$

 b $f'(x) = -\frac{\sin x}{2\sqrt{\cos x}}$

  increasing for $\frac{3\pi}{2} \leqslant x \leqslant 2\pi$, decreasing for $0 \leqslant x \leqslant \frac{\pi}{2}$

 c

15 **a** $f'(x) = 4x^3 - 12x^2$

  $f''(x) = 12x^2 - 24x$

 b $(3, -20)$ is a local minimum

 c $(0, 7)$ is a stationary inflection

  $(2, -9)$ is a non-stationary inflection

 d **i** $x \geqslant 3$ **ii** $x \leqslant 3$ **iii** $x \leqslant 0$ and $x \geqslant 2$

  iv $0 \leqslant x \leqslant 2$

 e

16 **a** **i** local maximum at $(0, 16)$, local minima at $(-2, 0)$ and $(2, 0)$

  stationary inflections at $(-\frac{2}{\sqrt{3}}, \frac{64}{9})$ and

iii

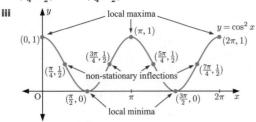

$y = (x+2)^2(x-2)^2$

local maximum $(0, 16)$

$\left(-\frac{2}{\sqrt{3}}, \frac{64}{9}\right)$ non-stationary inflections $\left(\frac{2}{\sqrt{3}}, \frac{64}{9}\right)$

local minima

$(-2, 0)$ O $(2, 0)$

b **i** local maxima at $(0, 1)$, $(\pi, 1)$, and $(2\pi, 1)$,
local minima at $\left(\frac{\pi}{2}, 0\right)$ and $\left(\frac{3\pi}{2}, 0\right)$

ii non-stationary inflections at $\left(\frac{\pi}{4}, \frac{1}{2}\right)$, $\left(\frac{3\pi}{4}, \frac{1}{2}\right)$,
$\left(\frac{5\pi}{4}, \frac{1}{2}\right)$, and $\left(\frac{7\pi}{4}, \frac{1}{2}\right)$

iii

local maxima

$(0, 1)$ $(\pi, 1)$ $y = \cos^2 x$
$(2\pi, 1)$

$\left(\frac{3\pi}{4}, \frac{1}{2}\right)$ $\left(\frac{5\pi}{4}, \frac{1}{2}\right)$

$\left(\frac{\pi}{4}, \frac{1}{2}\right)$ non-stationary inflections $\left(\frac{7\pi}{4}, \frac{1}{2}\right)$

O $\left(\frac{\pi}{2}, 0\right)$ π $\left(\frac{3\pi}{2}, 0\right)$ 2π

local minima

c **i** local maximum at $(0, 4)$

ii non-stationary inflections at $\left(-\frac{1}{\sqrt{2}}, 4\sqrt{\frac{2}{3}}\right)$ and
$\left(\frac{1}{\sqrt{2}}, 4\sqrt{\frac{2}{3}}\right)$

iii

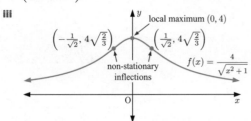

local maximum $(0, 4)$

$\left(-\frac{1}{\sqrt{2}}, 4\sqrt{\frac{2}{3}}\right)$ $\left(\frac{1}{\sqrt{2}}, 4\sqrt{\frac{2}{3}}\right)$

$f(x) = \dfrac{4}{\sqrt{x^2+1}}$

non-stationary inflections

O

17 **a** $v(0) = 0 \text{ cm s}^{-1}$, $v\left(\frac{1}{2}\right) = -\pi \text{ cm s}^{-1}$, $v(1) = 0 \text{ cm s}^{-1}$,
$v(1\frac{1}{2}) = \pi \text{ cm s}^{-1}$, $v(2) = 0 \text{ cm s}^{-1}$

b $0 \leqslant t \leqslant 1$, $2 \leqslant t \leqslant 3$, $4 \leqslant t \leqslant 5$, and so on
So, for $2n \leqslant t \leqslant 2n+1$, $n \in \{0, 1, 2, 3, ...\}$

18 **a** ≈ 50 wasps   **b** ≈ 367 wasps

c $P'(t) = \dfrac{25\,000\,000}{e^{0.5t}(1 + 1000e^{-0.5t})^2}$

d $P'(t) > 0$ for all $t \geqslant 0$
So, the population is always increasing.

e **i** ≈ 67.6 wasps/week   **ii** ≈ 2810 wasps/week

19 $x \approx 2.11$ **20** $\approx 3.60 \text{ m s}^{-1}$ **21** **b** $\frac{1}{\sqrt{2}}$ metres

22 $\frac{20\sqrt{10}}{3} \approx 21.1$ m per minute

23 **b** **i** $y = -\dfrac{2b}{a}x + b$ **ii** when $x = 0$, $y = b$

iii **Hint:** Let P'' be the point on the line $y = -b$ where
the distance to P is shortest.
Show that $FP = P''P$.

c **i** **Hint:** Show that $\triangle FPP'$ is congruent to $\triangle P''P$

ii **Hint:** Show that the tangents meet at $\left(\dfrac{a+c}{2}, \right.$

---

1 **a** $\dfrac{d}{dx}\left(x^{\frac{3}{2}}\right) = \frac{3}{2}x^{\frac{1}{2}} = \frac{3}{2}\sqrt{x}$

$\displaystyle\int \sqrt{x}\,dx = \frac{2}{3}x^{\frac{3}{2}} + c$

b $\dfrac{d}{dx}\left(x^{-\frac{1}{3}}\right) = -\frac{1}{3}x^{-\frac{4}{3}}$

$\displaystyle\int x^{-\frac{4}{3}}\,dx = -3x^{-\frac{1}{3}} + c$

c $\dfrac{d}{dx}\left(x^{n+1}\right) = (n+1)x^n$

$\displaystyle\int x^n\,dx = \frac{x^{n+1}}{n+1} + c, \quad n \neq -1$

2 **a** $\dfrac{d}{dx}\left(e^{3x}\right) = 3e^{3x}$ **b** $\dfrac{d}{dx}\left(e^{\frac{x}{2}}\right) = \frac{1}{2}e^{\frac{x}{2}}$

$\displaystyle\int e^{3x}\,dx = \frac{1}{3}e^{3x} + c$ $\displaystyle\int e^{\frac{x}{2}}\,dx = 2e^{\frac{x}{2}} + c$

c $\dfrac{d}{dx}\left(e^{kx}\right) = ke^{kx}$

$\displaystyle\int e^{kx}\,dx = \frac{1}{k}e^{kx} + c, \quad k \neq 0$

3 **a** $\dfrac{d}{dx}(\sin x) = \cos x$ **b** $\dfrac{d}{dx}(\cos x) = -\sin x$

$\displaystyle\int \cos x\,dx = \sin x + c$ $\displaystyle\int \sin x\,dx = -\cos x + c$

4 **a** $\dfrac{1}{x}$ **b** $-\dfrac{1}{x}$ **c** $\ln|x| + c$

5 **a** $\dfrac{d}{dx}(\sin 3x) = 3\cos 3x$

$\displaystyle\int \cos 3x\,dx = \frac{1}{3}\sin 3x + c$

b $\dfrac{d}{dx}\left(\cos\left(\frac{\pi}{3} - x\right)\right) = \sin\left(\frac{\pi}{3} - x\right)$

$\displaystyle\int \sin\left(\frac{\pi}{3} - x\right)dx = \cos\left(\frac{\pi}{3} - x\right) + c$

c $\dfrac{d}{dx}\left(e^{3x+1}\right) = 3e^{3x+1}$

$\displaystyle\int e^{3x+1}\,dx = \frac{1}{3}e^{3x+1} + c$

d $\dfrac{d}{dx}\left(6e^{-2x}\right) = -12e^{-2x}$

$\displaystyle\int e^{-2x}\,dx = -\frac{1}{2}e^{-2x} + c$

e $\dfrac{d}{dx}\left(\sqrt{5x-1}\right) = \frac{5}{2}(5x-1)^{-\frac{1}{2}} = \dfrac{5}{2\sqrt{5x-1}}$

$\displaystyle\int \frac{1}{\sqrt{5x-1}}\,dx = \frac{2}{5}\sqrt{5x-1} + c$

f $\dfrac{d}{dx}\left(x^4 - x^2\right) = 4x^3 - 2x$

$\displaystyle\int (4x^3 - 2x)\,dx = x^4 - x^2 + c$

g $\dfrac{d}{dx}\left((2x+1)^4\right) = 8(2x+1)^3$

$\displaystyle\int (2x+1)^3\,dx = \tfrac{1}{8}(2x+1)^4 + c$

6 a $k\,f(x)$ **7 a** $f(x) + g(x)$

EXERCISE 9B

1 a $\tfrac{2}{3}x^3 - \tfrac{3}{2}x^2 + x + c$ **b** $-\tfrac{1}{4}x^4 + \tfrac{4}{3}x^3 - 3x + c$

c $\tfrac{1}{4}x^2 + \tfrac{1}{3}x^3 + \tfrac{1}{4}x^4 + c$ **d** $\tfrac{4}{3}x^3 + \ln|x| + c$

e $\tfrac{4}{3}x^{\frac{3}{2}} - 6x^{\frac{1}{2}} + c$ **f** $\tfrac{1}{3}\ln|x| + \dfrac{2}{x} + c$

g $\tfrac{2}{5}x^{\frac{5}{2}} - 9x + c$ **h** $-6x^{-\frac{1}{2}} + \tfrac{4}{5}x^{\frac{5}{4}} + c$

2 a $2e^x - \tfrac{3}{2}x^2 + c$ **b** $4\ln|x| + \tfrac{1}{3}x^3 - e^x + c$

c $5e^x + \tfrac{1}{6}x^3 + c$

3 a $-3\cos x - 2x + c$ **b** $2x^2 - 2\sin x + c$

c $-\cos x - 2\sin x + e^x + c$ **d** $\tfrac{2}{7}x^{\frac{7}{2}} + 10\cos x + c$

e $\tfrac{1}{9}x^3 - \tfrac{1}{6}x^2 + \sin x + c$ **f** $\cos x + \tfrac{4}{3}x^{\frac{3}{2}} + c$

4 a $3x^3 - 6x^2 + 4x + c$ **b** $\tfrac{1}{2}x^2 - 2x + \ln|x| + c$

c $3\ln|x| - \tfrac{1}{2}x^2 + c$ **d** $\ln|x| - \tfrac{1}{2}x^2 + c$

e $-\dfrac{4}{x} + 4\ln|x| + x + c$

f $\tfrac{1}{4}x^4 + \tfrac{3}{2}x^2 + 3\ln|x| - \dfrac{1}{2x^2} + c$

g $\tfrac{2}{3}x^{\frac{3}{2}} + 8x^{-\frac{1}{2}} + c$ **h** $x - 4\ln|x| - \dfrac{10}{x} + c$

i $\tfrac{1}{5}x^5 + x^4 + 2x^3 + 2x^2 + x + c$

5 a $\tfrac{2}{3}x^{\frac{3}{2}} + \tfrac{1}{2}\sin x + c$ **b** $2e^x + 4\cos x + c$

c $3\sin x + \cos x + c$

6 a $y = 2x - \ln|x| + c$ **b** $y = -\cos x + 2\sin x + c$

c $y = 2e^x - 5x + \tfrac{1}{2}x^2 + c$

7 a $\tfrac{3}{2}x^2 - 2x - \dfrac{5}{x} + c$ **b** $\tfrac{1}{5}x^5 - x^2 - \dfrac{1}{x} + c$

c $\tfrac{2}{5}x^{\frac{5}{2}} - \tfrac{8}{3}x^{\frac{3}{2}} + 4x^{\frac{1}{2}} + c$ **d** $\tfrac{18}{7}x^{\frac{7}{2}} - \tfrac{12}{5}x^{\frac{5}{2}} + \tfrac{2}{3}x^{\frac{3}{2}} + c$

8 $\dfrac{d}{dx}(e^x\sin x) = e^x\sin x + e^x\cos x$

$\displaystyle\int e^x(\sin x + \cos x)\,dx = e^x\sin x + c$

9 $\dfrac{d}{dx}(e^{-x}\sin x) = -e^{-x}\sin x + e^{-x}\cos x$

$\displaystyle\int \dfrac{\cos x - \sin x}{e^x}\,dx = e^{-x}\sin x + c$

10 $\dfrac{d}{dx}(x\cos x) = \cos x - x\sin x$

$\displaystyle\int x\sin x\,dx = \sin x - x\cos x + c$

11 $\dfrac{d}{dx}\left(\ln(3x^2+1)\right) = \dfrac{6x}{3x^2+1}$

$\displaystyle\int \dfrac{2x}{3x^2+1}\,dx = \tfrac{1}{3}\ln(3x^2+1) + c$

EXERCISE 9C

1 a $f(x) = 3x - \tfrac{1}{2}x^2 + 2$ **b** $f(x) = \tfrac{2}{3}x^3 - 9x + \tfrac{19}{3}$

c $f(x) = \tfrac{2}{3}x^{\frac{3}{2}} - 2x + \tfrac{8}{3}$ **d** $f(x) = \ln|x| + 1$

2 a $f(x) = \tfrac{1}{3}x^3 - 4\sin x + 3$

b $f(x) = 2\sin x + 3\cos x - 2\sqrt{2}$

c $f(x) = \tfrac{2}{3}x^{\frac{3}{2}} + 2\cos x - 4$

d $f(x) = e^x + 3\sin x - e^\pi$

3 $y = x - e^x + 2e^3 - 3$ **4** $f(x) = 3x^3 + 1$

5 $f(x) = -\cos x + \tfrac{1}{2}x^2 - \left(\tfrac{15}{4\pi} + \tfrac{\pi}{6}\right)x + 2$

EXERCISE 9D

1 a $\tfrac{1}{8}(2x+5)^4 + c$ **b** $\dfrac{1}{2(3-2x)} + c$

c $\dfrac{-2}{3(2x-1)^3} + c$ **d** $\tfrac{1}{32}(4x-3)^8 + c$

e $\tfrac{2}{9}(3x-4)^{\frac{3}{2}} + c$ **f** $-4\sqrt{1-5x} + c$

g $-\tfrac{3}{5}(1-x)^5 + c$ **h** $-2\sqrt{3-4x} + c$

i $-\dfrac{5}{6(3x-2)^2} + c$

2 $y = \tfrac{1}{3}(2x-7)^{\frac{3}{2}} + 2$ **3** $(-8, -19)$

4 a $\tfrac{1}{2}(2x-1)^3 + c$ **b** $\tfrac{1}{12}(4x-5)^3 + c$

c $-\tfrac{1}{12}(1-3x)^4 + c$ **d** $-\tfrac{1}{15}(2-5x)^3 + c$

e $-\tfrac{8}{3}(5-x)^{\frac{3}{2}} + c$ **f** $\tfrac{1}{35}(7x+1)^5 + c$

5 $y = \tfrac{1}{2}x^2 - \dfrac{5}{1-x} - 7$

6 a $-\tfrac{1}{3}\cos 3x + c$ **b** $-\tfrac{1}{2}\sin(-4x) + x + c$ **c** $6\sin\tfrac{x}{2} + c$

d $-\tfrac{3}{2}\cos 2x + e^{-x} + c$ **e** $-\cos\left(2x + \tfrac{\pi}{6}\right) + c$

f $3\sin\left(\tfrac{\pi}{4} - x\right) + c$ **g** $\tfrac{1}{2}\sin 2x - \tfrac{1}{2}\cos 2x + c$

h $-\tfrac{2}{3}\cos 3x + \tfrac{5}{4}\sin 4x + c$ **i** $\tfrac{1}{16}\sin 8x + 3\cos x + c$

7 a $2e^x + \tfrac{5}{2}e^{2x} + c$ **b** $\tfrac{3}{5}e^{5x-2} + c$

c $-\tfrac{1}{3}e^{7-3x} + c$ **d** $\tfrac{1}{2}e^{2x} + 2x - \tfrac{1}{2}e^{-2x} + c$

e $-\tfrac{1}{2}e^{-2x} - 4e^{-x} + 4x + c$

f $\tfrac{1}{3}e^{3x} - 10e^x - 25e^{-x} + c$

8 $y = x - 2e^x + \tfrac{1}{2}e^{2x} + \tfrac{11}{2}$

9 $p = -\tfrac{1}{4}$, $f(x) = \tfrac{1}{2}\cos\tfrac{x}{2} + \tfrac{1}{2}$ **11** $f(x) = -e^{-2x} + 4$

12 $y = \tfrac{2}{3}x^{\frac{3}{2}} - \tfrac{1}{8}e^{-4x} + \tfrac{1}{8}e^{-4} - \tfrac{2}{3}$

13 $\displaystyle\int (\sin x + \cos x)^2\,dx = x - \tfrac{1}{2}\cos 2x + c$

14 a $\sin^2 x = \tfrac{1}{2} - \tfrac{1}{2}\cos 2x$, $\cos^2 x = \tfrac{1}{2} + \tfrac{1}{2}\cos 2x$

b i $\tfrac{1}{2}x - \tfrac{1}{4}\sin 2x + c$ **ii** $\tfrac{1}{2}x + \tfrac{1}{4}\sin 2x + c$

15 a $6\ln|x+4| + c$ **b** $x - x^2 + 4\ln|x-3| + c$

16 a $\tfrac{1}{2}\ln|2x-1| + c$ **b** $-\tfrac{5}{3}\ln|1-3x| + c$

c $4x + \tfrac{1}{5}\ln|5x-2| + c$ **d** $-e^{-x} - 2\ln|2x+1| + c$

e $\ln|x+2| + 2\ln|x-3| + c$

f $5\ln|x-6| - \tfrac{2}{3}\ln|3x-1| + c$

17 Both are correct. Recall that:

$$\dfrac{d}{dx}\left(\ln(Ax)\right) = \dfrac{d}{dx}\left(\ln A + \ln x\right) = \dfrac{1}{x},\quad A,\,x > 0$$

18 $\int \dfrac{4x+1}{x-1}\,dx = 4x + 5\ln|x-1| + c$

19 **a** $\frac{5}{2}x + \frac{1}{4}\sin 2x + c$ **b** $\frac{1}{2}x - \frac{1}{4}\sin 2x + 2x^2 + c$

   c $\frac{3}{2}x + \frac{1}{8}\sin 4x + c$ **d** $\frac{5}{2}x + \frac{1}{12}\sin 6x + c$

   e $\frac{1}{4}x + \frac{1}{32}\sin 8x + c$ **f** $\frac{3}{2}x + 2\sin x + \frac{1}{4}\sin 2x + c$

   g $x - \frac{1}{2}\sin 2x + \cos x + c$

   h $\frac{51}{2}x + 10\cos x - \frac{1}{4}\sin 2x + c$

   i $\sin 2x + 4\sin x + 3x + c$

   j $\frac{11}{2}x + 6\cos x - \frac{9}{4}\sin 2x + c$

20 **c**   **i** $\tan x - x + c$ **ii** $\tan x - 10x + c$

   iii $\frac{1}{2}\tan x - \frac{11}{2}x + c$ **iv** $\tan x - \frac{1}{2}x - \frac{1}{4}\sin 2x + c$

EXERCISE 9E

2 **a** 18 **b** 18 **c** $-\frac{51}{4}$ **d** $\frac{2}{15}$ **e** $21 + \ln 4$ **f** $\frac{101}{6}$

3 **a** $\frac{341}{5}$ **b** 10 **c** $\frac{124}{9}$ **d** $\frac{14}{3}$ **e** $\frac{1}{5}$ **f** 2

4 $\ln 4$

5 **a** $-\ln 3$ **b** $\ln 3$ **c** $\frac{4}{3}\ln 2$ **d** $2\ln\left(\frac{13}{5}\right)$

6 **a** $e - 1$ **b** $2e^3 - 11$ **c** $14 - \frac{1}{2}e^4 + \frac{1}{2}e$

   d $\frac{1}{3}(e^6 - 1)$ **e** $e - 1$ **f** $\frac{3}{2}$

7 **a** $\frac{5}{6}$ **b** $-\dfrac{1}{2e^4} - \dfrac{3}{2e^2} + \dfrac{2}{e} + 1$ **c** $\frac{6}{35}$ **d** $\frac{7}{2} + \ln 2$

8 **a** 2 **b** $\frac{1}{2}$ **c** $\frac{1}{2}$ **d** $\frac{1}{3}$ **e** 1 **f** $\frac{1}{\sqrt{2}}$

9 **a** $\frac{\pi}{8} + \frac{1}{4}$ **b** $\frac{\pi}{4}$ **c** $-\frac{1}{6}$ **d** $\frac{5}{3} + \frac{3\sqrt{3}}{8} + \frac{\pi}{3}$

10 $m = -5$

REVIEW SET 9A

1 $\dfrac{d}{dx}\left(\sin\left(\frac{\pi}{3} - 2x\right)\right) = -2\cos\left(\frac{\pi}{3} - 2x\right)$

$\displaystyle\int \cos\left(\frac{\pi}{3} - 2x\right)\,dx = -\frac{1}{2}\sin\left(\frac{\pi}{3} - 2x\right) + c$

2 $\dfrac{d}{dx}\left(\sqrt{x^2 - 4}\right) = \dfrac{x}{\sqrt{x^2 - 4}}$

$\displaystyle\int \dfrac{x}{\sqrt{x^2 - 4}}\,dx = \sqrt{x^2 - 4} + c$

3 **a** $-\frac{3}{5}x^5 + 2x^3 + c$ **b** $\frac{3}{2}x^2 - x + \dfrac{1}{x} + c$

   c $\frac{4}{3}x^3 - \frac{8}{5}x^{\frac{5}{2}} + \frac{1}{2}x^2 + c$ **d** $4e^x - 3\ln|x| + c$

   e $-\frac{1}{4}\cos(4x - 5) + c$ **f** $-\frac{1}{3}e^{4-3x} + c$

4 **a** $y = -3e^{-x} - 2\cos\left(\frac{\pi}{2} - x\right) + c$

   b $y = \frac{1}{4}\sin 4x - \frac{1}{6}x^3 + c$

5 $f(x) = \frac{3}{2}e^{2x} + \frac{1}{2}$

6 $\dfrac{d}{dx}\left(x^2 \sin x\right) = 2x\sin x + x^2 \cos x$

$\dfrac{d}{dx}\left(x\cos x\right) = \cos x - x\sin x$

$\displaystyle\int x^2 \cos x\,dx = (x^2 - 2)\sin x + 2x\cos x + c$

7 **a** $\frac{1}{2}x^2 - 7\ln|x| + c$ **b** $\frac{1}{2}e^{2x-3} - \frac{2}{3}\ln|3x - 1| + c$

   c $-\frac{1}{12}(4 - 3x)^4 + \frac{1}{2}\cos(-2x) + c$

8 $a = 6\sqrt{2}$, $f(x) = 2\sqrt{2}\sin 3x - 1$ **9** $m \approx 1.86$

10 **a** $\frac{112}{9}$ **b** $\sqrt{2}$ **c** $2\ln 3$ **11** $a = \frac{1}{2}\ln 2$

12 $\dfrac{d}{dx}\left(e^{-2x}\sin x\right) = e^{-2x}(\cos x - 2\sin x)$

$\displaystyle\int_0^{\frac{\pi}{2}} \left[e^{-2x}(\cos x - 2\sin x)\right]\,dx = e^{-\pi}$

REVIEW SET 9B

1 $\dfrac{d}{dx}\left(\ln(2x + 1)\right) = \dfrac{2}{2x + 1}$

$\displaystyle\int \dfrac{1}{2x + 1}\,dx = \frac{1}{2}\ln|2x + 1| + c$

2 $\displaystyle\int \left(3x^2 + x\right)^2 (6x + 1)\,dx = \frac{1}{3}(3x^2 + x)^3 + c$

3 **a** $\frac{1}{2}x^4 - \frac{5}{2}x^2 + 7x + c$ **b** $\frac{3}{2}x^2 - \ln|x| + c$

   c $x - x^3 + \frac{3}{5}x^5 - \frac{1}{7}x^7 + c$ **d** $-2e^{-x} + 3x + c$

   e $2\sin 2x + c$ **f** $9x + 3e^{2x-1} + \frac{1}{4}e^{4x-2} + c$

4 **a** $y = \frac{1}{3}x^3 - x^2 + 3x - 2\ln|x| - \dfrac{1}{x} + c$

   b $y = \frac{1}{3}\sin(3x - \pi) - 2\cos x + c$

5 $f(x) = 2\ln|x| - x + e + 2 - \ln 4$

6 $y = -\frac{3}{68}x^3 + \frac{21}{68}(x - 1)^{\frac{3}{2}} + \frac{275}{68}$

7 $f(x) = -2\sqrt{4 - 3x} + 8$

8 $\displaystyle\int (\sin x - \cos x)^2\,dx = x + \frac{1}{2}\cos 2x + c$

9 **a** $-\frac{1}{2}\ln|3 - 2x| + c$ **b** $\frac{4}{5}\ln|5x + 1| + c$

   c $\frac{3}{2}x + 2\cos x - \frac{1}{4}\sin 2x + c$

10 **a** $\frac{2}{3}\left(\sqrt{5} - \sqrt{2}\right)$ **b** $4\ln\left(\frac{4}{3}\right)$ **c** $\frac{3\pi}{4} + 1 + 2\sqrt{2}$

11 $b = \frac{\pi}{4}$ or $\frac{3\pi}{4}$ **12** **b** $\frac{1}{32}\sin 4x + \frac{1}{4}\sin 2x + \frac{3}{8}x + c$

EXERCISE 10A

1 **b** $3\ln|x + 2| - \ln|x - 2| + c$

2 **b** $\frac{1}{2}\ln|2x - 1| - \frac{1}{2}\ln|2x + 1| + c$

3 **a** $\dfrac{8}{2x - 3} - \dfrac{4}{x + 1}$ **b** $4\ln|2x - 3| - 4\ln|x + 1| + c$

4 **a** $\dfrac{5}{2(x + 1)} - \dfrac{3}{2(x - 3)}$

   b $\frac{5}{2}\ln|x + 1| - \frac{3}{2}\ln|x - 3| + c$

5 **a** $\ln|x - 2| - \ln|x + 1| + c$

   b $3\ln|x + 3| + \ln|x - 1| + c$

   c $\frac{1}{2}\ln|x| - \frac{3}{2}\ln|x + 2| + c$

   d $\frac{7}{3}\ln|x + 1| - \frac{1}{3}\ln|x - 2| + c$

   e $-\frac{5}{14}\ln|2x + 1| - \frac{1}{7}\ln|x - 3| + c$

   f $\frac{10}{57}\ln|3x - 5| + \frac{3}{19}\ln|2x + 3| + c$

6 **a** $13\ln 2 - 7\ln 3$ **b** $2\ln 3 - 2\ln 2$

   c $\frac{5}{3}\ln 10 - \ln 2$ **d** $-\frac{3}{20}\ln 3 - \frac{1}{8}\ln 5$

EXERCISE 10B.1

1 **a** $\frac{1}{5}(x^3 + 1)^5 + c$ **b** $\frac{1}{3}e^{x^3 + 1} + c$

   c $\frac{1}{5}\sin^5 x + c$ **d** $\sin(x^2 - 3) + c$

2 **a** $\frac{1}{4}(2+x^4)^4+c$ **b** $2\sqrt{x^2+3}+c$

c $-\dfrac{1}{3(2x^3-1)^3}+c$ **d** $\frac{2}{3}(x^3+x)^{\frac{3}{2}}+c$

e $\frac{1}{5}(x^3+2x+1)^5+c$ **f** $\dfrac{1}{8(1-x^2)^4}+c$

g $-\dfrac{1}{27(3x^3-1)^3}+c$ **h** $-\dfrac{1}{2(x^2+4x-3)}+c$

i $\frac{1}{5}(x^2+x)^5+c$

3 **a** $e^{1-2x}+c$ **b** $e^{x^2}+c$ **c** $2e^{\sqrt{x}}+c$
d $-e^{x-x^2}+c$

4 **a** $\ln\left|x^2+1\right|+c$ **b** $-\frac{1}{2}\ln\left|2-x^2\right|+c$

c $\ln\left|x^2-3x\right|+c$ **d** $2\ln\left|x^3-x\right|+c$

e $-2\ln\left|5x-x^2\right|+c$ **f** $-\frac{1}{3}\ln\left|x^3-3x\right|+c$

5 **a** $-\frac{1}{9}(3-x^3)^3+c$ **b** $4\ln|\ln x|+c$

c $-\frac{1}{3}(1-x^2)^{\frac{3}{2}}+c$ **d** $-\frac{1}{2}e^{1-x^2}+c$

e $-\ln\left|x^3-x\right|+c$ **f** $\frac{1}{4}(\ln x)^4+c$

6 **a** $\frac{1}{5}\sin^5 x+c$ **b** $-\frac{1}{6}\cos^6 x+c$

c $-2(\cos x)^{\frac{1}{2}}+c$ **d** $-\ln|\cos x|+c$

e $\frac{2}{3}(\sin x)^{\frac{3}{2}}+c$ **f** $-(2+\sin x)^{-1}+c$

g $\frac{1}{2}\sec^2 x+c$ **h** $\ln|1-\cos x|+c$

i $\frac{1}{2}\ln|\sin 2x-3|+c$

7 **a** $-\frac{1}{3}\sin^3 x+\sin x+c$

b $-\cos x+\frac{2}{3}\cos^3 x-\frac{1}{5}\cos^5 x+c$

c $\frac{1}{5}\sin^5 x-\frac{1}{7}\sin^7 x+c$ **d** $\frac{1}{8}\sin^4 2x+c$

8 **a** $-e^{\cos x}+c$ **b** $-\frac{1}{2}\cos(x^2)+c$
c $\ln|\sin x-\cos x|+c$

9 **a** $\frac{1}{3}\ln|\sin 3x|+c$ **b** $\sec x+c$ **c** $-\csc x+c$
d $\frac{1}{3}\sec 3x+c$ **e** $-2\csc\frac{x}{2}+c$ **f** $-\frac{1}{3}\csc^3 x+c$

10 **a** $\frac{1}{3}(2x+1)^{\frac{3}{2}}+c$ **b** $\frac{1}{4}\left[\ln(x^2+7)\right]^2+c$

c $\frac{2}{5}(x-3)^{\frac{5}{2}}+2(x-3)^{\frac{3}{2}}+c$

d $\frac{2}{7}(x-16)^{\frac{7}{2}}+\frac{64}{5}(x-16)^{\frac{5}{2}}+\frac{512}{3}(x-16)^{\frac{3}{2}}+c$

e $-(3-x^2)^{\frac{3}{2}}+\frac{1}{5}(3-x^2)^{\frac{5}{2}}+c$

f $2\ln(1+|\ln x|^2)+c$

11 $2x^{\frac{1}{2}}-3x^{\frac{1}{3}}+6x^{\frac{1}{6}}-6\ln(x^{\frac{1}{6}}+1)+c$

EXERCISE 10B.2

1 **a** $20\frac{1}{4}$ **b** $\frac{1}{12}$ **c** $\frac{1}{3}(e^2-e)\approx 1.56$
d $20\frac{1}{3}$ **e** $\frac{1}{4}(e^{-2}-e^{-8})\approx 0.0337$
f $\frac{1}{2}\ln(\frac{2}{7})\approx -0.626$ **g** $\frac{1}{2}(\ln 2)^2\approx 0.240$
h 0 **i** $2\ln 7\approx 3.89$

2 **a** $2-\sqrt{2}$ **b** $\frac{1}{24}$ **c** $\frac{1}{2}\ln 2$ **d** $\ln 2$ **e** $\ln 2$ **f** $\frac{1}{4}$

3 $\dfrac{3^{n+1}}{2n+2}$, $\ n\neq -1$, undefined for $n=-1$

4 **a** $\frac{28\sqrt{3}}{5}-\frac{44\sqrt{2}}{15}$ **b** $\frac{48\sqrt{6}-54}{5}$ **c** $\frac{1054\sqrt{3}}{35}$

EXERCISE 10C

1 **a** xe^x-e^x+c **b** $-x\cos x+\sin x+c$
c $\frac{1}{3}x^3\ln x-\frac{1}{9}x^3+c$ **d** $-\frac{1}{3}x\cos 3x+\frac{1}{9}\sin 3x+c$
e $\frac{1}{2}x\sin 2x+\frac{1}{4}\cos 2x+c$

2 **a** $x\ln x-x+c$ **b** $x(\ln x)^2-2x\ln x+2x+c$

3 **a** $-x^2e^{-x}-2xe^{-x}-2e^{-x}+c$
b $\frac{1}{2}e^x(\sin x+\cos x)+c$ **c** $-\frac{1}{2}e^{-x}(\cos x+\sin x)+c$
d $-x^2\cos x+2x\sin x+2\cos x+c$

4 **a** $u^2e^u-2ue^u+2e^u+c$
b $x(\ln x)^2-2x\ln x+2x+c$

5 **a** $-\frac{1}{4}\cos 2x+c$ **b, c** $\frac{1}{2}\sin^2 x+c$

6 **a** $\frac{1}{2}x^2(\ln(x^2)-1)+c$ **b** $\frac{1}{2}e^2+\frac{1}{2}$

REVIEW SET 10A

1 **b** $3\ln 5-2\ln 2$

2 **a** $\frac{1}{3}\ln|x+2|+\frac{2}{3}\ln|x-4|+c$
b $\frac{1}{2}\ln|2x-5|+\ln|x+1|+c$

3 **a** $\frac{1}{4}(x^2+1)^4+c$ **b** **i** $\frac{15}{4}$ **ii** $-\frac{609}{8}$

4 **a** $\frac{1}{8}\sin^8 x+c$ **b** $-\frac{1}{2}\ln|\cos 2x|+c$
c $e^{\sin x}+c$ **d** $\frac{2}{9}(3x+2)^{\frac{3}{2}}+c$

5 **a** $\frac{1}{3}x^3(\ln(x^3)-1)+c$ **b** $\frac{1}{5}(1+e^{2\pi})$

6 **a** e^2-2e **b** $\frac{5}{18}$

7 **a** $\frac{1}{2}\ln\left|x^2-9\right|+c$ **b** $\frac{1}{2}\ln|x+3|+\frac{1}{2}\ln|x-3|+c$
c $\ln\left|\cos\left(\arcsin\left(\frac{x}{3}\right)\right)\right|+c$

REVIEW SET 10B

1 **a** $\dfrac{2}{x-6}-\dfrac{3}{2x+1}$ **b** $2\ln|x-6|-\frac{3}{2}\ln|2x+1|+c$

2 $-15\ln 3$

3 **a** $2\sqrt{x^2-5}+c$ **b** $\frac{1}{3}\sec^3 x+c$ **c** $-2e^{-x^2}+c$

4 **a** $\frac{5}{54}$ **b** $\frac{19}{384}$ **c** $\frac{14\sqrt{7}}{3}$

5 **a** $\frac{1}{2}e^{-x}(\sin x-\cos x)+c$ **b** $e^x(x^2-2x+2)+c$

6 $\dfrac{\cos^{1-\frac{n}{2}}x}{\frac{n}{2}-1}+c$, for $n\neq 2$, $-\ln|\cos x|+c$, for $n=2$

7 The argument has not accounted for the constant of integration c.

EXERCISE 11A

1 **a** 6 units^2 **b** 6 units^2

2 **a** 30 units^2 **b** $4\frac{1}{2}$ units^2 **c** $13\frac{1}{2}$ units^2

3 **a** $5\frac{1}{3}$ units^2 **b** $12\frac{2}{3}$ units^2

4 **a** $\frac{1}{3}$ units^2 **b** $63\frac{3}{4}$ units^2 **c** $2\frac{1}{6}$ units^2

5 **a** A$(-2, 0)$, B$(3, 0)$ **b** $20\frac{5}{6}$ units^2 **6** 9 units^2

8 1 unit^2 **9** $\frac{2}{3}$ units^2

10 **a** region A
b A: $\frac{1}{2}$ units^2, B: $1-\frac{1}{\sqrt{2}}\approx 0.293$ units^2
$\therefore$ region A is larger.

11 **a** $\frac{1}{2}$ units^2 **b** $(e-1)$ units^2 **c** $4\frac{1}{2}$ units
d $\left(2e-\dfrac{2}{e}\right)$ units^2 **e** 18 units^2

12 b $\left(\dfrac{\sin 2}{1 + \cos 2}\right)$ units^2 $(\approx 1.56$ units$^2)$

13 b $(8\ln 2 - 3)$ units^2 $(\approx 2.55$ units$^2)$

14 a $b = \left(\dfrac{3}{2}\right)^{\frac{2}{3}} \approx 1.3104$ **b** $a = \sqrt{3}$

  c $k = \dfrac{3e^{0.4} - 1}{2} \approx 1.7377$ **d** $k = 2\sqrt{5}$

EXERCISE 11B

1 a $4\frac{1}{2}$ units^2 **b** $(1 + e^{-2})$ units^2
  c 2 units^2 **d** $2\frac{1}{4}$ units^2

2 a $40\frac{1}{2}$ units^2 **b** 8 units^2 **c** 8 units^2

3 a $\displaystyle\int_3^5 f(x)\,dx = -$ (area between $x = 3$ and $x = 5$)

  b $\displaystyle\int_1^3 f(x)\,dx - \int_3^5 f(x)\,dx + \int_5^7 f(x)\,dx$

4 Region B is larger.

  $\displaystyle\int_{-2}^4 f(x)\,dx =$ area of region $A + (-$ area of region $B)$
  $\qquad\qquad\qquad\qquad$ {region B is below the x-axis}
  $\qquad\qquad = $ area of region $A -$ area of region B
  $\qquad\qquad = -6$

  $\therefore$ area of region $B > $ area of region A

5 $k = \frac{3}{2}$

6 a
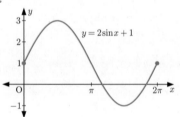

  b $\left(2\sqrt{3} - \frac{2\pi}{3}\right)$ units^2 $(\approx 1.37$ units$^2)$

7 a $33\frac{1}{3}$ cm^2 **b** ≈ 66.7 L

8 a $\frac{1}{3}\sin(x^3) + c$

EXERCISE 11C

1 a
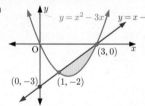
     b $(1, -2)$ and $(3, 0)$

     c $1\frac{1}{3}$ units^2

2 a $10\frac{2}{3}$ units^2 **b** $\frac{1}{3}$ units^2 **3** $\frac{1}{2}$ units^2

4
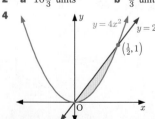

     enclosed area $= \frac{1}{12}$ units^2

5 a $A(\frac{1}{2}, \frac{5}{2})$, $B(2, 1)$ **b** $\left(\frac{21}{8} - \frac{5}{2}\ln\left(\frac{5}{2}\right)\right)$ units^2

6
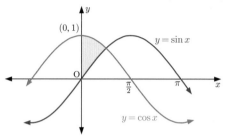

  enclosed area $= (\sqrt{2} - 1)$ units^2

7 a C_1 is $y = 4\sin x$, C_2 is $y = \sin x$ **b** 6 units^2

8 a $A = \displaystyle\int_{-2}^{-1}(x^3 - 7x - 6)\,dx + \int_{-1}^3(-x^3 + 7x + 6)\,dx$

  b $32\frac{3}{4}$ units^2

9 a
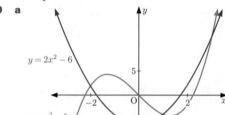

  b $x = -2$, 1, and 3 **c** $21\frac{1}{12}$ units^2

10 a 8 units^2 **b** $101\frac{3}{4}$ units^2 **11** $k \approx 2.3489$

12 a
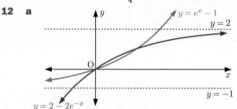

  b $(0, 0)$ and $(\ln 2, 1)$ **c** $(3\ln 2 - 2)$ units^2

13 a $\frac{\pi}{4}$ **b** $\ln\sqrt{2}$ units^2

14 a C_1 is $y = \sin 2x$, C_2 is $y = \sin x$
  b $A(\frac{\pi}{3}, \frac{\sqrt{3}}{2})$ **c** $2\frac{1}{2}$ units^2

15 a C_1 is $y = \cos^2 x$, C_2 is $y = \cos 2x$
  b $A(0, 1)$, $B(\frac{\pi}{4}, 0)$, $C(\frac{\pi}{2}, 0)$, $D(\frac{3\pi}{4}, 0)$, $E(\pi, 1)$

16 a
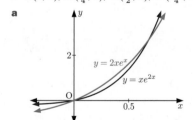

  b 0 and $\ln 2$ **c** $\left(2\ln 2 - \frac{5}{4}\right)$ units^2

EXERCISE 11D

1 a 20 cars per minute   **b** $\approx 8{:}05$ am

c

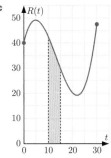

$\int_{10}^{15} R(t)\, dt$ represents the total number of cars going past

the pedestrian crossing from 8:10 am to 8:15 am.

d 1031 cars

2 a i ≈ 1.65 L per minute   **ii** ≈ 1.09 L per minute

b The rate of water leaking into the kayak is greater than the rate of water being bailed from the kayak after 2 minutes. So, the amount of water in the kayak is increasing after 2 minutes.

c i $\int_{0}^{3} R_1(t)\, dt \approx 3.72$

About 3.72 litres of water have leaked into the kayak in the first 3 minutes.

ii $\int_{2}^{5} R_2(t)\, dt \approx 5.27$

About 5.27 litres of water have been bailed out of the kayak from $t = 2$ minutes to $t = 5$ minutes.

iii $\int_{0}^{8} [R_1(t) - R_2(t)]\, dt \approx 5.09$

There are about 5.09 litres of water in the kayak 8 minutes after striking the rock.

d ≈ 6.31 litres

3 a C_1 is $y = 3 \sin \frac{\pi t}{10}$, C_2 is $y = \sin \frac{\pi t}{10}$ **b** $\frac{40}{\pi}$ units^2

c The area in **b** represents the total amount of energy that enters the greenhouse in the first 10 hours.

4 a

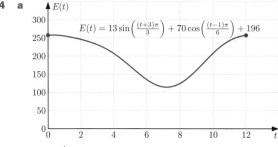

$E(t) = 13 \sin\left(\frac{(t+3)\pi}{3}\right) + 70 \cos\left(\frac{(t-1)\pi}{6}\right) + 196$

b i $\int_{3}^{4} E(t)\, dt \approx 220.12$

The power consumption of the United Kingdom in April is about 220.12 TWh.

ii $\int_{5}^{8} E(t)\, dt \approx 392.96$

The power consumption of the United Kingdom from June 1st to September 1st is about 392.96 TWh.

iii $\int_{0}^{12} E(t)\, dt = 2352$

The yearly power consumption of the United Kingdom is 2352 TWh.

REVIEW SET 11A

1 a 9 units^2 **b** $(3\pi + 1)$ units^2 **c** 36 units^2

d $(3 \ln 3 - 2)$ units^2

2 a 39 units^2 **b** $\left(\frac{16}{3} - 2\sqrt{3}\right)$ units^2 **c** 3 units^2

3 a

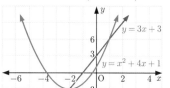

b $(-2, -3)$ and $(1, 6)$

c $4\frac{1}{2}$ units^2

4 No, total area shaded $= \int_{-1}^{1} f(x)\, dx - \int_{1}^{3} f(x)\, dx$.

5 $k = \sqrt[3]{16}$ **6** $(\sin\sqrt{5} - \sin\sqrt{17})$ units^2

7 a $a = \ln 3$ **b** $b = \ln 5$ **8** $40\frac{1}{2}$ units^2

9 a

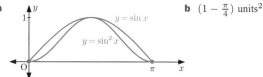

b $(1 - \frac{\pi}{4})$ units^2

10 a i $\int_{0}^{\frac{1}{2}} R_2(t)\, dt \approx 0.66$

About 660 millilitres of water leaks from the watering can in the first 30 seconds.

ii $\int_{0}^{1} [R_1(t) - R_2(t)]\, dt \approx 5.03$

There are about 5.03 litres of water in the watering can after 1 minute.

b ≈ 199 seconds

REVIEW SET 11B

1 a 22 units^2 **b** $\ln 3$ units^2 **c** $\dfrac{e^4 - 1}{2}$ units^2

d $\frac{4\pi}{3}$ units^2

2 $4\frac{1}{2}$ units^2 **3** $(3 - \ln 4)$ units^2

4 a

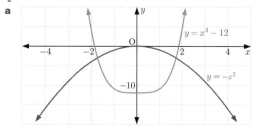

b $(-\sqrt{3}, -3)$ and $(\sqrt{3}, -3)$ **c** $\frac{92\sqrt{3}}{5}$ units^2

5 $k = \frac{4}{3}$ **6** $m = \frac{\pi}{3}$ **7** $21\frac{1}{12}$ units^2

8 a $a = -3$ **b** A has x-coordinate $\sqrt[3]{4}$

9 **a** $-\frac{1}{2}\cos(x^2) + c$

10 **a** $\displaystyle\int_5^{12} E(t)\,dt \approx 10.66$

The solar energy transferred into Callum's solar panels in the morning is about 10.66 kWh.

b $\displaystyle\int_{12}^{20} E(t)\,dt \approx 12.88$

The solar energy transferred into Callum's solar panels in the afternoon is about 12.88 kWh.

c $\displaystyle\int_5^{20} E(t)\,dt \approx 23.54$

The solar energy transferred into Callum's solar panels over a whole day is about 23.54 kWh.

EXERCISE 12A

2 **a** B   **b** C   **c** A

4 **b**

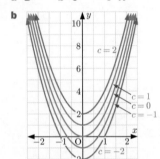

c $y = 2x^2 - \frac{3}{2}$

d $y = 4x - \frac{7}{2}$

5 **b**

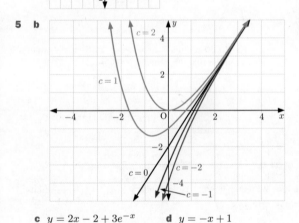

c $y = 2x - 2 + 3e^{-x}$ **d** $y = -x + 1$

EXERCISE 12B

1 **a** $y = x^4 + c$ **b** $y = \frac{1}{3}x^3 + 3x^2 + c$

c $y = \frac{1}{3}e^{3x} + 4x + c$ **d** $y = \sin x - \frac{1}{2}\cos 2x + c$

e $x = \frac{1}{2}t + \frac{1}{4}\sin 2t + c$

f $y = \ln|x + 4| - \frac{2}{3}\ln|3x - 5| + c$

g $M = \ln\left|t^3 - 4\right| + c$ **h** $y = -\sqrt{25 - x^2} + c$

i $f(t) = -\frac{1}{2}e^{-t^2 + 1} + 2t + c$ **j** $S = \frac{2}{3}(\ln t)^{\frac{3}{2}} + c$

k $y = x - \frac{1}{3}\sin 3x + c$ **l** $y = \frac{2}{3}x^{\frac{3}{2}} - 2\ln|x| + c$

2 **a** $y = \frac{3}{2}x^2 - 2x + 5$ **b** $y = \frac{1}{3}e^{3x} + x - \frac{1}{3}$

c $y = \ln|x| + \ln 6$

3 **a** $y = e^{2t} + e^{-t} - 2$ **b** $M = \frac{1}{2}\sin 2\alpha + 3\cos\alpha + 2$

c $P = 2x\sin x + 2\cos x - \frac{\pi}{6}$

4 $f(-2) = 2$

5 **a** $y = e^x + e^{-x} - 1$ **b**

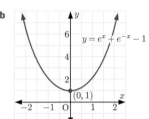

c $3x - 2y = 3\ln 2 - 3$

6 $f(x) = \frac{1}{2}x^2 + \frac{1}{x} - \frac{3}{2}$ **7** $y = \ln(x^2 + 2) + \ln 3$

8 £4250

9 **a** $P(x) = 15x - 0.015x^2 - 650$ pounds

b maximum profit is £3100, when 500 plates are made

c $46 \leqslant x \leqslant 954$ (you cannot produce part of a plate)

10 $\approx 225°C$

11 **a** $\approx 400°C$ **b** ≈ 0.387 m   **c** ≈ 0.187 m

12 **a** $y = \left(\frac{0.01}{3}x^3 - \frac{0.005}{12}x^4 - \frac{0.08}{3}x\right)$ metres

b ≈ 3.33 cm;   yes, it seems reasonable that the maximum sag occurs when $x = 2$.

c 2.375 cm   **d** $\approx 1.05°$

EXERCISE 12C

1 **a** $y = \sqrt[3]{\frac{3}{2}x^2 + c}$ **b** $y = \ln(x^2 + c)$ **c** $y = Ae^{\frac{3}{2}x^2}$

d $y = \left(\frac{x^2}{2} + c\right)^2$ **e** $y = Ae^{-\cos x}$

f $y = \left(-\frac{1}{4}x^2 + c\right)^2 - 1$ **g** $y = Ax$

h $y = -\ln(c - x^3)$ **i** $y = A(x - 1) - 2$

2 **a** $y = Ae^x$ **b** $y = \pm\sqrt{2x + c}$

c $y = Ae^t + 4$ **d** $P = \left(\frac{3}{2}t + c\right)^2$

e $Q = Ae^t - \frac{3}{2}$ **f** $t = Q^2 + 3Q + c$

3 **a** $y = A\sqrt[3]{3x + 1}$ **b** $y = Ae^{2x} + 2$

c $y = \sqrt[3]{3\ln(x^2 + 5) + c}$ **d** $y = Ae^{2\sqrt{4 - x}} + 1$

e $y = \dfrac{1}{-\frac{1}{2}x^2 + 2x + c}$ **f** $y = \left(\frac{9}{2}\ln(x^2 + 5) + c\right)^{\frac{2}{3}}$

4 **a** $y = \sqrt[3]{\frac{9}{2}x^2 + 1}$ **b** $y = \frac{1}{36}(x - 26)^2$

c $y = e^{x + \frac{1}{3}x^3}$ **d** $y = \arcsin\left(\frac{3}{2}x^2 - \frac{3}{2}\right)$

e $y = \left(\frac{9}{2}\sin 2x + 3\sqrt{3}\right)^{\frac{2}{3}}$

f $y = \ln\left[\sqrt[4]{2x^2 + 4x + 1}\left(e^2 + 3\right) - 3\right]$

5 **b** $y = \dfrac{1 - x}{(x + 1)^2}$

6 **a** $\dfrac{5}{x - 4} + \dfrac{3}{x + 3}$

b $y = \frac{1}{4}\left(\ln\left|(x - 4)^5(x + 3)^3\right| + c\right)^2$

7 $y = -\dfrac{1}{\ln\left|\frac{(x+2)^2(x-1)^3}{4}\right| + 2}$ **8** $y = Ae^x\left(\dfrac{x - 1}{x + 1}\right)$

9 a $P = 40e^{\frac{1}{2}t}$ **b** ≈ 803 rabbits

10 a $I = 350e^{-0.4t}$ **b** ≈ 47.4 milliamps

     c ≈ 7.16 milliseconds

11 ≈ 5.19 minutes   **12** 0.8%

13 a $(V_0 - V)$ is the amount of water remaining

     so $\dfrac{dV}{dt} \propto (V_0 - V)$

     $\therefore \quad \dfrac{dV}{dt} = k(V_0 - V)$ for some constant k

     b $\approx 17.7\%$

14 a ≈ 12.5 minutes   **b** 12 midnight

15 a $I(t) = Ae^{-\frac{100}{3}t} + 2$ **b** $I(t) = 2 - 2e^{-\frac{100}{3}t}$

     c 2 amps   **d** ≈ 0.138 seconds

16 4 hours

17 a $\dfrac{dh}{dt} = \dfrac{r^2}{\pi h^2 - 2\pi r h}$

     b   **i** ≈ 14.4 hours   **ii** ≈ 20.9 hours

EXERCISE 12D

1 a $P = \dfrac{200}{1 + 9e^{-0.2t}}$ **b** $P \approx 90.2$

     c as $t \to \infty$, $P \to 200$

     d

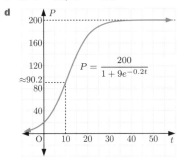

2 a The population P grows logistically over time t.

     b The graph would increase exponentially at first, but then level off to approach the maximum value P.

     c $P = \dfrac{3000}{1 + 5e^{-0.1t}}$

     d   **i** ≈ 924 rodents   **ii** ≈ 23 years   **iii** 3000 rodents

3 a $A = \dfrac{500}{1 + 24e^{-0.1t}}$ **b** ≈ 118 cm^2 **c** 500 cm^2

4 a $N = \dfrac{600}{1 + 299e^{-0.8t}}$ **b** ≈ 21 people

     c 600 people   **d** $\approx 5{:}08$ pm

5 a $N \approx \dfrac{10^{30}}{1 + (5 \times 10^{27})e^{-kt}}$ **b** $k \approx 1.12 \times 10^6$

     c after $\approx 6.09 \times 10^{-5}$ seconds

6 a The population of foxes increased quickly at first, but later levelled off to approach a maximum.

     b   **i** $A = 95\,000$ **ii** $F \approx \dfrac{95\,000}{1 + \frac{94\,986}{14}e^{-0.146t}}$

     c $\approx 85\,100$ foxes   **d**   **i** ≈ 1894 **ii** ≈ 1911

e

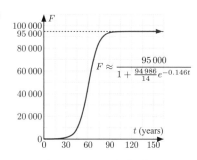

f In 1905, it appears as an inflection on the graph.

REVIEW SET 12A

2 a $y = \frac{3}{4}\sin 2x - \frac{1}{2}x + c$ **b** $y = 3x + \frac{1}{2}e^{-2x} + c$

     c $y = \frac{1}{2}\ln|2x + 1| + 2$ **d** $y = \frac{1}{2}(t^2 + e^{t^2} + 3e - 1)$

3 a $P(x) = 20x - \frac{1}{8}x^2 - 250$ pounds

     b £550 per day when 80 vases are made.

     c between 14 and 146 vases per day inclusive

4 a   **i** 160 spectators per minute

       ii ≈ 249 spectators per minute

     b $\dfrac{d}{dt}\left(\dfrac{1}{1 + 4e^{-0.05t}}\right) = \dfrac{0.2e^{-0.05t}}{(1 + 4e^{-0.05t})^2}$

       $\displaystyle\int S'(t)\,dt = \dfrac{20\,000}{1 + 4e^{-0.05t}} + c$

     c $\displaystyle\int_0^{60} S'(t)\,dt \approx 12\,700$

       About 12 700 spectators in total entered the stadium between noon and 1 pm.

     d $\approx 19\,500$ spectators

5 a $y = Ae^{\frac{5}{3}x^3}$ **b** $y = \dfrac{-1}{x^2 - x + c}$

     c $x = \left(-\frac{9}{16}\cos 2t + c\right)^{\frac{2}{3}}$

6 $y = \ln(e - 3x)$ **7 a** $\dfrac{dM}{dt} = kM$ **b** ≈ 93.2 days

8 a $t = -100\ln(10C) + \dfrac{0.1 - C}{2.66 \times 10^{-4}}$

     b ≈ 509 seconds ≈ 8 minutes 29 seconds

9 a $N = \dfrac{694}{1 + 693e^{-t}}$ **b** ≈ 51 people   **c** ≈ 7.49 weeks

REVIEW SET 12B

2 a $\dfrac{dy}{dx} = ce^x = (ce^x + 1) - 1 = y - 1$ ✓

     b

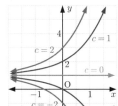

     c $y = 3e^x + 1$

     d $y = 3x + 4$

3 a $y = \ln|e^x - 2| + c$ **b** $y = -\frac{1}{4}\sin\left(\frac{\pi}{3} - 2x\right) - \frac{\sqrt{3}}{8}$

5 a $y = \dfrac{1}{\sqrt[3]{c - 6x}}$ **b** $P = A\sqrt{t^2 + 1}$

6 a $y = \left(\dfrac{1}{2}x + 2\right)^2$ **b** $y = e^{\sin x - 3}$

7 $y = \dfrac{(x + 2)^2}{25(x - 2)}$

8 a $T = 65e^{-0.1t} + 20$ **b** $\approx 63.6°C$

 c as $t \to \infty$, $T \to 20$

 d

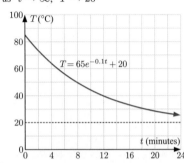

 e i ≈ 3.68 minutes **ii** ≈ 8.11 minutes

9 a $P(t) \approx \dfrac{200\,000}{1 + 7e^{-0.129t}}$ **b** $\approx 42\,800$ ostriches

 c after ≈ 15.1 years

10 a $\dfrac{dV}{dt} = -k\sqrt{h}$ **b** $V = 2 \times 2 \times h$, $\dfrac{dh}{dt} = -\dfrac{k}{4}\sqrt{h}$

 c 20 minutes

EXERCISE 13A.1

1 $\sqrt{2} \approx 1.414$ **2** $\pi \approx 3.1416$ **3** $x \approx 0.161$

4 a $f(-5) = 218$, $f(5) = -22$

 b $a_1 = -5$, $b_1 = 5$

x	$f(x)$
-5	218
-4	122
-3	58
-2	20
-1	2
0	-2
1	2
2	8
3	10
4	2
5	-22

 c A change in sign occurs three times, so there are three solutions to $f(x) = 0$ on the domain $-5 \leqslant x \leqslant 5$.

 d $x \approx -0.76,\ 0.64,\ 4.12$

EXERCISE 13A.2

1 $\sqrt{3} \approx 1.732$ **2** $e \approx 2.718\,28$

3 At least 20 iterations are needed. The interval width must be $< 10^{-6}$.

4 a $x \approx 2.0252$ **b** $x \approx 1.1304$

5 a $x \approx 3.6201$

 b

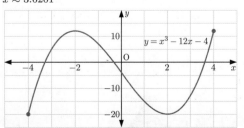

 c No, the method in **a** did not find the solutions $x \approx -3.2836$ and $x \approx -0.3365$.

6 b Hint: Consider the cases where there are 0, 1, and 2 distinct stationary points. In each case, consider how many zeros the cubic may have.

EXERCISE 13A.3

1 $\sqrt{7} \approx 2.645\,75$ **2** $\ln 2 \approx 0.693\,15$

3 a $x \approx 0.25$ **b** $x \approx 1.172\,60$ **4** $x \approx 0.160\,71$

5 a $x \approx 3.6201$

 b Yes. Using interval bisection, the first $x^* = 0$. Using linear interpolation, the first $x^* = 1$. Now $f(0)$ and $f(1)$ are both negative while $f(4)$ is positive. Hence in both cases we are drawn to the only zero on $[0, 4]$.

 c i $x \approx -3.2836$ **ii** no

  iii No, it is not always more efficient. The efficiency of linear interpolation depends on the shape of the graph over the interval.

EXERCISE 13B

1 a

n	x_n
0	0
1	0.707\,106\,78
2	0.568\,527\,31
3	0.598\,220\,84
4	0.591\,983\,78
5	0.593\,299\,30
6	0.593\,022\,07
7	0.593\,080\,50
8	0.593\,068\,19
9	0.593\,070\,78
10	0.593\,070\,24

 b $x \approx 0.593\,070\,24$

 The analytic solution is
 $x = \dfrac{-1 + \sqrt{33}}{8} \approx 0.593\,070\,33$.

2 a $x = 1 + \arctan x$ **b** $x \approx 2.1323$

3 b $x \approx 3.8823$

 c

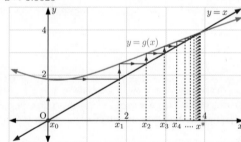

4 a $|g_1'(\sqrt{5})| \approx 5.47$, so $|g_1'(x)| > 1$ in the interval around the solution.

 c $\sqrt{5} \approx 2.236$

EXERCISE 13C

1 a $x \approx 1.732$ **b** $x \approx -1.000$ **c** $x \approx 2.750$

 d $x \approx -1.693$

2 a

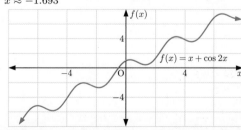

b $x \approx -0.5149$

c $x_0 = 1$ is further from the solution, and at first we move away from the solution. In this case the algorithm takes much longer to find the solution.

3 a i $x \approx -0.247$ **ii** $x \approx 1.445$ **iii** $x \approx 2.802$

b

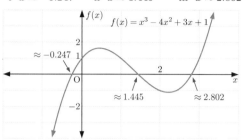

4 a local minimum at $(0, -10)$, local maximum at $(3\frac{1}{3}, 8\frac{14}{27})$

b $x_0 = -1$, $x_0 = 1$, and $x_0 = 4$

c ≈ -1.264, ≈ 1.756, ≈ 4.508

d

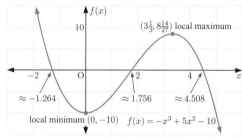

EXERCISE 13D.1

1 a ≈ 2.28125 **b** ≈ 3.03125 **c** ≈ 2.671875

$\int_0^1 (-x^2+4x+1)\,dx = 2\frac{2}{3}$, the approximation using midpoints is the most accurate.

2 a ≈ -5.28125 **b** ≈ -6.03125 **c** ≈ -5.671875

$\int_{-4}^{-3} (x^2 + 4x - 4)\,dx = -5\frac{2}{3}$, the approximation using midpoints is the most accurate.

3 a ≈ 0.9609 **b** ≈ 0.9589 **c** ≈ 0.9576

$\int_1^2 \sin x\,dx = \cos 1 - \cos 2 \approx 0.9564$, the approximations become more accurate as the number of intervals increases.

4 a ≈ 0.9103 **b** ≈ 0.9100 **c** ≈ 0.9097

$\int_2^3 \ln x\,dx \approx 0.9095$, the approximations become more accurate as the number of intervals increases.

5 ≈ 4.5768

$\int_0^4 5xe^{-x}\,dx = 5 - \frac{25}{e^4} \approx 4.5421$

EXERCISE 13D.2

1 a ≈ 2.3479

$\int_2^4 \frac{2}{\sqrt{x}}\,dx = 8 - 4\sqrt{2} \approx 2.3431$

b ≈ 7.25

$\int_1^3 (-x^2 + 6x - 4)\,dx = 7\frac{1}{3}$

2 a ≈ 5.9855 **b** ≈ 0.1346

3 a i ≈ 1.0144 **ii** ≈ 0.992

b $\int_{-0.6}^1 (x^3 - 2x^2 + 1)\,dx = \frac{1888}{1875} \approx 1.0069$

The midordinate values numerical approximation was more accurate.

4 a ≈ -0.7848, ≈ 2.1852 **b** ≈ 6.94 units^2

REVIEW SET 13A

1 $\sqrt{6} \approx 2.449$ **2** $x \approx 1.260$

3 a $f(-1) = 5$, $f(1) = -1$, $f(3) = 9$

b $x \approx 1.21$

c The function is not continuous on $[-1, 1]$, since $f(x)$ is undefined when $x = 0$.

4 a

n	x_n
0	0
1	0.745 355 992
2	0.687 559 614
3	0.692 213 871
4	0.691 840 230
5	0.691 870 233
6	0.691 867 824
7	0.691 868 017
8	0.691 868 001
9	0.691 868 003
10	0.691 868 003

b $x \approx 0.691\,868\,003$

The analytic solution is $x = \frac{-1+\sqrt{181}}{18} \approx 0.691\,868\,003$

5 b $x_1 = \ln 6 \approx 1.792$, $x_2 = \ln(7 - \ln 6) \approx 1.650$

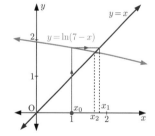

6 $x \approx -1.325$

7 a $f(0) > 0$, $f(\pi) < 0$, and $f(x)$ is continuous on $[0, \pi]$.
∴ there is a value of x between 0 and π such that $f(x) = 0$.

b $x \approx 1.93$

c The tangent at $x = 0$ is horizontal, and so it does not cut the x-axis.

8 a ≈ 2.59375 **b** ≈ 3.09375 **c** ≈ 2.828125

$\int_2^3 (x^2 - 3x + 4)\,dx = \frac{17}{6} \approx 2.833$, the approximation using midpoints is the most accurate.

9 ≈ 6.7955

10 a ≈ 0.4740

b $\displaystyle\int_0^{\frac{\pi}{2}} \sin x \cos x \, dx = \frac{1}{2}$

Our approximation was reasonably accurate, but would improve if more intervals were used.

REVIEW SET 13B

1 $\frac{\pi}{2} \approx 1.571$

2 **a** $f(0) < 0$, $f(10) > 0$, and $f(x)$ is continuous on $[0, 10]$.
   $\therefore$ there is a value of x between 0 and 10 such that $f(x) = 0$.

   b $x \approx 8.29$

   c **i** $f(0.3) = 0.017$
      $\therefore$ there are 2 more solutions to $f(x) = 0$ in $[0, 1]$.

      ii $x \approx 0.28$, $x \approx 0.43$, and $x \approx 8.29$

3 **b** **i** $x_1 \approx 5.874$, $x_2 \approx 5.759$ **ii** $x \approx 5.773$

4 **a**

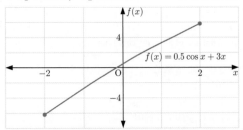

b **i, ii** $x \approx -0.164$

c Linear interpolation obtained the solution more quickly. Linear interpolation is particularly efficient if the function is approximately linear on the interval, as was the case here.

5 **a** $x = 1 + \cos \frac{x}{5}$ **b** $x \approx 1.927$

6 **b** $x \approx 1.534$

7 **a**

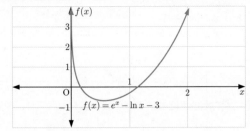

b $x \approx 1.1419$

c If $x_0 = 0.5$, $x_1 \approx -1.37$, and $f(x_1)$ is not defined.

d $x \approx 0.1612$

8 **a** ≈ 9.3079 **b** ≈ 9.2646

$\displaystyle\int_2^8 \ln x \, dx = 8 \ln 8 - 2 \ln 2 - 6 \approx 9.2492$, the approximation

becomes more accurate as the number of intervals increases.

9 **a** ≈ 3.0620

   b This is a reasonable approximation of π.

   $\displaystyle\int_0^2 \sqrt{4 - x^2} \, dx$ is the

   area of a quarter of a circle with radius 2.

   Area $= \dfrac{\pi \times 2^2}{4} = \pi$

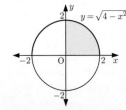

10 **a** ≈ 6.2365

   b over estimate, joining the interval endpoints with straight line segments would give a larger area than the shaded area.

   c $\displaystyle\int_{-2}^{-1} x^4 \, dx = \frac{31}{5} = 6.2$

EXERCISE 14A

1 **a**

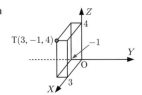

b $\overrightarrow{OT} = \begin{pmatrix} 3 \\ -1 \\ 4 \end{pmatrix}$

c $OT = \sqrt{26}$ units

2 **a**

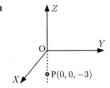

$OP = 3$ units

b $P(0, -1, 2)$

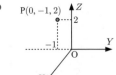

$OP = \sqrt{5}$ units

c

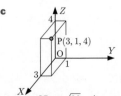

$OP = \sqrt{26}$ units

d $P(-1, -2, 3)$

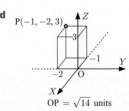

$OP = \sqrt{14}$ units

3 **a** $a = 5$, $b = 6$, $c = -6$ **b** $a = 4$, $b = 2$, $c = 1$

4 **a** $k = \pm\frac{\sqrt{11}}{4}$ **b** $k = \pm\frac{2}{3}$

5 **a** $m = \pm 3$ **b** $m = \pm\sqrt{11}$

EXERCISE 14B

1 **a** $\begin{pmatrix} 3 \\ 1 \\ -2 \end{pmatrix}$ **b** $\begin{pmatrix} 1 \\ -3 \\ 4 \end{pmatrix}$ **c** $\begin{pmatrix} 1 \\ 4 \\ -9 \end{pmatrix}$

   d $\begin{pmatrix} -1 \\ \frac{3}{2} \\ -\frac{7}{2} \end{pmatrix}$ **e** $\begin{pmatrix} 1 \\ -4 \\ 7 \end{pmatrix}$ **f** $\begin{pmatrix} 4 \\ 2 \\ -2 \end{pmatrix}$

2 **a** $\sqrt{10}$ **b** $\sqrt{6}$ **c** $2\sqrt{10}$ **d** $2\sqrt{10}$

   e $-3\sqrt{6}$ **f** $3\sqrt{6}$ **g** $3\sqrt{2}$ **h** $\sqrt{14}$

3 **a** $\begin{pmatrix} 0 \\ 0 \\ 2 \end{pmatrix}$ **b** $\begin{pmatrix} 4 \\ -10 \\ 3 \end{pmatrix}$ **c** $\sqrt{38}$ units

   d $\sqrt{3}$ units   **e** $\begin{pmatrix} \sqrt{11} \\ -3\sqrt{11} \\ 2\sqrt{11} \end{pmatrix}$ **f** $\begin{pmatrix} -\frac{1}{\sqrt{11}} \\ \frac{1}{\sqrt{11}} \\ \frac{3}{\sqrt{11}} \end{pmatrix}$

4 **a** $a = \frac{1}{3}$, $b = 2$, $c = 1$

   b $a = 1$, $b = -1$, $c = 2$

   c $a = 4$, $b = -1$, $c = -19$

   d $a = 0$, $b = -1$, $c = -1$ or $a = -\frac{1}{2}$, $b = 0$, $c = -1$

EXERCISE 14C

1 **a** $x = \frac{1}{2}q$ **b** $x = 2n$ **c** $x = -\frac{1}{3}p$

  d $x = \frac{1}{2}(r - q)$ **e** $x = \frac{1}{5}(4s - t)$ **f** $x = 12m - 3n$

2 **a** $x = \begin{pmatrix} 1 \\ -\frac{5}{2} \\ 2 \end{pmatrix}$ **b** $x = \begin{pmatrix} 6 \\ -15 \\ 12 \end{pmatrix}$ **c** $x = \begin{pmatrix} -\frac{1}{2} \\ \frac{5}{4} \\ -1 \end{pmatrix}$

3 **a** $x = \begin{pmatrix} 4 \\ -6 \\ -5 \end{pmatrix}$ **b** $x = \begin{pmatrix} 1 \\ -\frac{2}{3} \\ \frac{5}{3} \end{pmatrix}$ **c** $x = \begin{pmatrix} \frac{3}{2} \\ -1 \\ \frac{5}{2} \end{pmatrix}$

EXERCISE 14D

1 $\overrightarrow{OA} = \begin{pmatrix} 3 \\ 1 \\ 0 \end{pmatrix}$, $\overrightarrow{OB} = \begin{pmatrix} -1 \\ 1 \\ 2 \end{pmatrix}$, $\overrightarrow{AB} = \begin{pmatrix} -4 \\ 0 \\ 2 \end{pmatrix}$

2 $\overrightarrow{AB} = \begin{pmatrix} 3 \\ 4 \\ -2 \end{pmatrix}$, $AB = \sqrt{29}$ units

3 **a** $\overrightarrow{AB} = \begin{pmatrix} 4 \\ -1 \\ -3 \end{pmatrix}$, $\overrightarrow{BA} = \begin{pmatrix} -4 \\ 1 \\ 3 \end{pmatrix}$

  b $|\overrightarrow{AB}| = \sqrt{26}$ units, $|\overrightarrow{BA}| = \sqrt{26}$ units

4 **a** $\overrightarrow{NM} = \begin{pmatrix} 5 \\ -4 \\ -1 \end{pmatrix}$ **b** $\overrightarrow{MN} = \begin{pmatrix} -5 \\ 4 \\ 1 \end{pmatrix}$

  c $MN = \sqrt{42}$ units

5 **a** $\overrightarrow{AB} = 4i - 5j + 3k$ **b** $\sqrt{50}$ units

6 **a** $\sqrt{13}$ units   **b** $\sqrt{14}$ units   **c** 3 units

7 $\overrightarrow{AC} = -i - 2k$

8 **a** $A(0, y, 0)$ **b** $(0, 2, 0)$ and $(0, -4, 0)$

10 $C(5, 1, -8)$, $D(8, -1, -13)$, $E(11, -3, -18)$

11 $B(0, 3, 5)$,   radius $= \sqrt{3}$ units

12 **a** $\overrightarrow{AB} = \begin{pmatrix} 3 \\ -2 \\ -2 \end{pmatrix}$, $AB = \sqrt{17}$ units

  b $\overrightarrow{AC} = \begin{pmatrix} -2 \\ -1 \\ -5 \end{pmatrix}$, $AC = \sqrt{30}$ units

  c $\overrightarrow{CB} = \begin{pmatrix} 5 \\ -1 \\ 3 \end{pmatrix}$, $CB = \sqrt{35}$ units

  d ABC is scalene, and not right angled.

14 **a** right angled   **b** straight line (not a triangle)

15 **a** $|\overrightarrow{AB}| = \sqrt{158}$ units, $|\overrightarrow{BC}| = \sqrt{129}$ units,
     $|\overrightarrow{AC}| = \sqrt{29}$ units,   and $29 + 129 = 158$
  b area ≈ 30.6 units^2

16 **a** $r = 2,\ s = 4,\ t = -7$ **b** $r = -4,\ s = 0,\ t = 3$

17 **a** $\overrightarrow{AB} = \begin{pmatrix} 2 \\ -5 \\ -1 \end{pmatrix}$, $\overrightarrow{DC} = \begin{pmatrix} 2 \\ -5 \\ -1 \end{pmatrix}$

  b ABCD is a parallelogram.

18 **a** $S(-2, 8, -3)$ **b** midpoints are at $(-\frac{1}{2}, 3, 1)$

19 **a** parallelogram   **b** not a parallelogram

20 **a** $R(3, 1, 6)$ **b** $X(2, -1, 0)$

21 **a** $\overrightarrow{BD} = \frac{1}{2}a$ **b** $\overrightarrow{AB} = b - a$ **c** $\overrightarrow{BA} = -b + a$

  d $\overrightarrow{OD} = b + \frac{1}{2}a$ **e** $\overrightarrow{AD} = b - \frac{1}{2}a$ **f** $\overrightarrow{DA} = \frac{1}{2}a - b$

22 **a** $\begin{pmatrix} -1 \\ 5 \\ -1 \end{pmatrix}$ **b** $\begin{pmatrix} -3 \\ 4 \\ -2 \end{pmatrix}$ **c** $\begin{pmatrix} -3 \\ 6 \\ -5 \end{pmatrix}$

EXERCISE 14E

1 **a** $r = 3,\ s = -9$ **b** $r = -6,\ s = -4$

2 **a** $\overrightarrow{PQ} \parallel \overrightarrow{RS}$, $PQ = 4RS$

  b $\overrightarrow{AB} \parallel \overrightarrow{CD}$, $AB = \frac{2}{3}CD$,   opposite direction

  c L, M, and N are collinear, and $LN = 3LM$

3 $\overrightarrow{QS} = 2\overrightarrow{PR}$, $PR : QS = 1 : 2$

4 **a** $\frac{1}{\sqrt{13}}(2i + 3k)$ **b** $\frac{1}{\sqrt{6}}(-i + 2j - k)$

  c $\frac{1}{3}(2i - 2j + k)$

5 **a** $\pm\frac{1}{3}\begin{pmatrix} 2 \\ -1 \\ -2 \end{pmatrix}$ **b** $\pm\frac{2}{3}\begin{pmatrix} -2 \\ -1 \\ 2 \end{pmatrix}$

6 **a** $\sqrt{2}\begin{pmatrix} -1 \\ 4 \\ 1 \end{pmatrix}$ **b** $\frac{5}{3}\begin{pmatrix} 1 \\ 2 \\ 2 \end{pmatrix}$

7 **a** $\left(-\frac{13}{5}, 3, \frac{14}{5}\right)$ **b** $\left(2 + \frac{4}{\sqrt{6}}, -1 - \frac{4}{\sqrt{6}}, 4 + \frac{8}{\sqrt{6}}\right)$

8 **c** $a = 7,\ b = -1$ **d** $a = -\frac{7}{2},\ b = -\frac{21}{2}$

EXERCISE 14F

1 **Hint:**
Consider the parallelogram.
Find $\overrightarrow{AB}$ and $\overrightarrow{OC}$, etc.

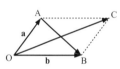

4 **a** $\overrightarrow{OX} = \frac{1}{2}(a + b)$, $\overrightarrow{BZ} = \frac{1}{2}a - b$ **b** $\overrightarrow{OP} = \frac{1}{3}a + \frac{1}{3}b$

  c $\overrightarrow{YP} = \frac{1}{3}a - \frac{1}{6}b$, $\overrightarrow{PA} = \frac{2}{3}a - \frac{1}{3}b$

EXERCISE 14G

1 **a** **i** $\begin{pmatrix} x \\ y \end{pmatrix} = \begin{pmatrix} 1 \\ 4 \end{pmatrix} + \lambda\begin{pmatrix} -1 \\ 2 \end{pmatrix}$, $\lambda \in \mathbb{R}$

   ii $x = 1 - \lambda,\ y = 4 + 2\lambda,\ \lambda \in \mathbb{R}$ **iii** $2x + y = 6$

  b **i** $\begin{pmatrix} x \\ y \end{pmatrix} = \begin{pmatrix} 5 \\ 2 \end{pmatrix} + \lambda\begin{pmatrix} -2 \\ 5 \end{pmatrix}$, $\lambda \in \mathbb{R}$

   ii $x = 5 - 2\lambda,\ y = 2 + 5\lambda,\ \lambda \in \mathbb{R}$ **iii** $5x + 2y = 29$

  c **i** $\begin{pmatrix} x \\ y \end{pmatrix} = \begin{pmatrix} -6 \\ 0 \end{pmatrix} + \lambda\begin{pmatrix} 3 \\ 7 \end{pmatrix}$, $\lambda \in \mathbb{R}$

   ii $x = -6 + 3\lambda,\ y = 7\lambda,\ \lambda \in \mathbb{R}$ **iii** $7x - 3y = -42$

  d **i** $\begin{pmatrix} x \\ y \end{pmatrix} = \begin{pmatrix} 0 \\ 2 \end{pmatrix} + \lambda\begin{pmatrix} 1 \\ 3 \end{pmatrix}$, $\lambda \in \mathbb{R}$

   ii $x = \lambda,\ y = 2 + 3\lambda,\ \lambda \in \mathbb{R}$ **iii** $y = 2 + 3x$

  e **i** $\begin{pmatrix} x \\ y \end{pmatrix} = \begin{pmatrix} 3 \\ 0 \end{pmatrix} + \lambda\begin{pmatrix} 4 \\ 2 \end{pmatrix}$, $\lambda \in \mathbb{R}$

   ii $x = 3 + 4\lambda,\ y = 2\lambda,\ \lambda \in \mathbb{R}$ **iii** $x - 2y = 3$

  f **i** $\begin{pmatrix} x \\ y \end{pmatrix} = \begin{pmatrix} -2 \\ 5 \end{pmatrix} + \lambda\begin{pmatrix} 6 \\ -11 \end{pmatrix}$, $\lambda \in \mathbb{R}$

   ii $x = -2 + 6\lambda,\ y = 5 - 11\lambda,\ \lambda \in \mathbb{R}$

   iii $11x + 6y = 8$

2 **a** $x = 4 - \lambda$, $y = -3 + 2\lambda$, $\lambda \in \mathbb{R}$

  b Points are: $(4, -3), (3, -1), (2, 1), (5, -5), (7, -9)$

3 **a** $(0, 8)$

  b It is parallel to $\begin{pmatrix} -1 \\ 3 \end{pmatrix}$ and in the opposite direction.

  c $\begin{pmatrix} x \\ y \end{pmatrix} = \begin{pmatrix} 0 \\ 8 \end{pmatrix} + \mu \begin{pmatrix} 1 \\ -3 \end{pmatrix}$, $\mu \in \mathbb{R}$

4 **a** **i** $\begin{pmatrix} x \\ y \\ z \end{pmatrix} = \begin{pmatrix} 1 \\ 3 \\ -7 \end{pmatrix} + \lambda \begin{pmatrix} 2 \\ 1 \\ 3 \end{pmatrix}$, $\lambda \in \mathbb{R}$

   ii $x = 1 + 2\lambda$, $y = 3 + \lambda$, $z = -7 + 3\lambda$, $\lambda \in \mathbb{R}$

   iii $\dfrac{x-1}{2} = y - 3 = \dfrac{z+7}{3}$

  b **i** $\begin{pmatrix} x \\ y \\ z \end{pmatrix} = \begin{pmatrix} 0 \\ 1 \\ 2 \end{pmatrix} + \lambda \begin{pmatrix} 1 \\ 1 \\ -2 \end{pmatrix}$, $\lambda \in \mathbb{R}$

   ii $x = \lambda$, $y = 1 + \lambda$, $z = 2 - 2\lambda$, $\lambda \in \mathbb{R}$

   iii $x = y - 1 = \dfrac{-z+2}{2}$

  c **i** $\begin{pmatrix} x \\ y \\ z \end{pmatrix} = \begin{pmatrix} -2 \\ 2 \\ 1 \end{pmatrix} + \lambda \begin{pmatrix} 1 \\ 0 \\ 0 \end{pmatrix}$, $\lambda \in \mathbb{R}$

   ii $x = -2 + \lambda$, $y = 2$, $z = 1$, $\lambda \in \mathbb{R}$

   iii $y = 2$, $z = 1$

  d **i** $\begin{pmatrix} x \\ y \\ z \end{pmatrix} = \begin{pmatrix} 0 \\ 2 \\ -1 \end{pmatrix} + \lambda \begin{pmatrix} 2 \\ -1 \\ 3 \end{pmatrix}$, $\lambda \in \mathbb{R}$

   ii $x = 2\lambda$, $y = 2 - \lambda$, $z = -1 + 3\lambda$, $\lambda \in \mathbb{R}$

   iii $\dfrac{x}{2} = -y + 2 = \dfrac{z+1}{3}$

  e **i** $\begin{pmatrix} x \\ y \\ z \end{pmatrix} = \begin{pmatrix} 3 \\ 2 \\ -1 \end{pmatrix} + \lambda \begin{pmatrix} 0 \\ 0 \\ 1 \end{pmatrix}$, $\lambda \in \mathbb{R}$

   ii $x = 3$, $y = 2$, $z = -1 + \lambda$, $\lambda \in \mathbb{R}$

   iii $x = 3$, $y = 2$

5 **a** $\begin{pmatrix} x \\ y \\ z \end{pmatrix} = \begin{pmatrix} 1 \\ 2 \\ 1 \end{pmatrix} + \lambda \begin{pmatrix} -2 \\ 1 \\ 1 \end{pmatrix}$, $\lambda \in \mathbb{R}$

  b $\begin{pmatrix} x \\ y \\ z \end{pmatrix} = \begin{pmatrix} 0 \\ 1 \\ 3 \end{pmatrix} + \lambda \begin{pmatrix} 3 \\ 0 \\ -4 \end{pmatrix}$, $\lambda \in \mathbb{R}$

  c $\begin{pmatrix} x \\ y \\ z \end{pmatrix} = \begin{pmatrix} 1 \\ 2 \\ 5 \end{pmatrix} + \lambda \begin{pmatrix} 0 \\ -3 \\ 0 \end{pmatrix}$, $\lambda \in \mathbb{R}$

  d $\begin{pmatrix} x \\ y \\ z \end{pmatrix} = \begin{pmatrix} 0 \\ 1 \\ -1 \end{pmatrix} + \lambda \begin{pmatrix} 5 \\ -2 \\ 4 \end{pmatrix}$, $\lambda \in \mathbb{R}$

6 **a** $\begin{pmatrix} -2 \\ 0 \\ 3 \end{pmatrix}$ **b** $\begin{pmatrix} -1 \\ 1 \\ -3 \end{pmatrix}$ **c** $\begin{pmatrix} 3 \\ 2 \\ 1 \end{pmatrix}$ **d** $\begin{pmatrix} -2 \\ 4 \\ 3 \end{pmatrix}$

7 **a** $\left(-\frac{1}{2}, \frac{9}{2}, 0\right)$ **b** $(0, 4, 1)$ **c** $(4, 0, 9)$

8 **a** (x_0, y_0, z_0) **b** $\begin{pmatrix} l \\ m \\ n \end{pmatrix}$

  c $\dfrac{x - x_0}{l} = \dfrac{y - y_0}{m} = \dfrac{z - z_0}{n}$, $l, m, n \neq 0$

9 $(0, 7, 3)$ and $\left(\frac{20}{3}, -\frac{19}{3}, -\frac{11}{3}\right)$

10 **a** $(1, 2, 3)$ **b** $\left(\frac{7}{3}, \frac{2}{3}, \frac{8}{3}\right)$

EXERCISE 14H

1 **a** **i** $(-4, 3, 0)$ **ii** $\begin{pmatrix} 12 \\ 5 \\ 6 \end{pmatrix}$ **iii** $\approx 14.3 \text{ m s}^{-1}$

  b **i** $(3, 0, 4)$ **ii** $\begin{pmatrix} 2 \\ -1 \\ -2 \end{pmatrix}$ **iii** 3 m s^{-1}

2 **a** $\begin{pmatrix} 2\sqrt{66} \\ \frac{7\sqrt{66}}{2} \\ \frac{\sqrt{66}}{2} \end{pmatrix}$ **b** $\begin{pmatrix} -12 \\ 30 \\ -84 \end{pmatrix}$

3 **a** $\begin{pmatrix} -3 \\ 1 \\ -0.5 \end{pmatrix}$ **b** $\approx 19.2 \text{ km h}^{-1}$

  c $\begin{pmatrix} x \\ y \\ z \end{pmatrix} = \begin{pmatrix} 6 \\ 9 \\ 3 \end{pmatrix} + t \begin{pmatrix} -3 \\ 1 \\ -0.5 \end{pmatrix}$, $t \in \mathbb{R}$ **d** 1 hour

EXERCISE 14I.1

1 **a** $y = 1 - 5x$ **b** $y = \frac{7}{2} - \frac{1}{2}x$ **c** $y^2 = x^3$

  d $y^2 = 16x$ **e** $y = 2 - \dfrac{6}{x-1}$ **f** $y = (x+2)^{\frac{2}{3}} + 3$

2 **a**

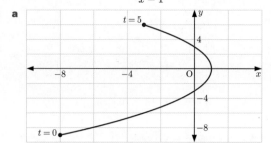

  b **i** $(1, 0)$ **ii** $(0, 3)$ and $(0, -3)$
  c $y^2 = 9(1 - x)$

3 **a** $x^2 + y^2 = 9$, a circle centred at $(0, 0)$ with radius 3 units.

  b $(x+1)^2 + (y-2)^2 = 4$, a circle centred at $(-1, 2)$ with radius 2 units.

  c $x^2 + y^2 = 1$, the unit circle

  d $x = -\sqrt{1 - y^2}$, the left half of the unit circle

4 **a** **i**

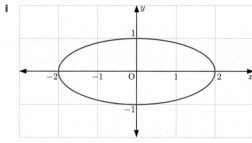

   ii An ellipse centred at $(0, 0)$ with width 4 units and height 2 units.

   iii $\dfrac{x^2}{4} + y^2 = 1$

b i

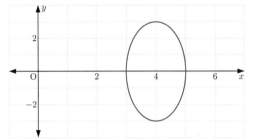

ii An ellipse centred at $(4, 0)$ with width 2 units and height 6 units.

iii $(x-4)^2 + \dfrac{y^2}{9} = 1$

c i

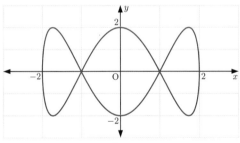

ii A section of a parabola.   **iii** $y = x^2 - 1$

5 a

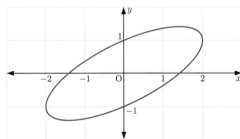

$y^2 = -(x^2-4)(x^2-1)^2$

b

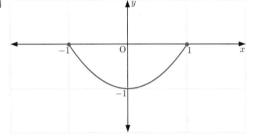

$x^2 - 2xy + 2y^2 - 2 = 0$

c

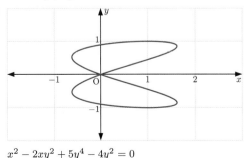

$x^2 - 2xy^2 + 5y^4 - 4y^2 = 0$

d

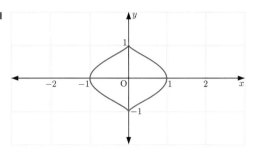

$x^{\frac{2}{3}} + y^2 = 1$

6 A helix curve in three-dimensional space.

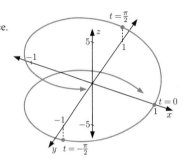

EXERCISE 14I.2

1 a $\dfrac{dy}{dx} = \dfrac{2t+1}{3t^2}$ **b i** P$(-2, 2)$ **ii** $y = x + 4$

c $y = \dfrac{5}{12}x + \dfrac{47}{12}$

2 a $\dfrac{dy}{dx} = -2\cot\theta$ **b** $y = -2\sqrt{3}x + 4$ **c** $\frac{4}{3}\sqrt{3}$ units^2

3 a P$(0, -\frac{3}{4})$, Q$(0, 8)$ **b** $(-6\frac{1}{4}, \frac{1}{2})$

4 a $\dfrac{dy}{dx} = -\dfrac{2\cos 2\theta}{\sin\theta}$

b ii $\theta = \frac{\pi}{3}$ or $\frac{5\pi}{3}$

iii A$\left(\frac{1}{2}, \frac{\sqrt{3}}{2}\right)$, B$\left(\frac{1}{2}, -\frac{\sqrt{3}}{2}\right)$

5 a $\dfrac{dy}{dx} = \dfrac{(\cos^2 t - \sin^2 t)(1+\sin^2 t) - 2\sin^2 t\cos^2 t}{-\sin t(1+\sin^2 t) + 2\sin t\cos^2 t}$

b i $t = \frac{\pi}{2}, \frac{3\pi}{2}$

ii At $t = \frac{\pi}{2}$, tangent is $y = x$, at $t = \frac{3\pi}{2}$, tangent is $y = -x$

c i $\sin^2 t = \frac{1}{3}$ **ii** $\frac{1}{\sqrt{2}}$

EXERCISE 14J

1 a $x(t) = -3t^2 + 9t - 3$, $y(t) = -9t^2 + 12t - 1$, $0 \leqslant t \leqslant 1$

c Q$(\frac{3}{4}, 2\frac{3}{4})$ **d** R$(1\frac{2}{3}, 3)$

2 a $x(t) = -4t^3 + 12t^2 - 9t + 4$, $y(t) = 6t^3 - 15t^2 + 6t + 1$, $0 \leqslant t \leqslant 1$

b $x(t) = 4t^3 + 6t^2 - 9t + 2$, $y(t) = -16t^3 + 33t^2 - 15t + 4$, $0 \leqslant t \leqslant 1$

c $x(t) = -8t^3 + 3t^2 + 3t - 1$, $y(t) = -3t^3 + 15t^2 - 6t + 2$, $0 \leqslant t \leqslant 1$

3 a S$(-5, -1)$, E$(5, -13)$

b

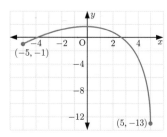

c lowest point is $(5, -13)$, highest point is $\left(-\frac{5}{27}, 1\frac{2}{3}\right)$

d left-most is $(-5, -1)$, right-most is $(5, -13)$

4 a $S(1, 4)$, $E\left(1\frac{1}{6}, 4\frac{1}{2}\right)$

b left-most is $\left(\frac{23}{24}, 4\frac{3}{8}\right)$, right-most is $\left(1\frac{1}{6}, 4\frac{1}{2}\right)$

c highest is $\left(1\frac{1}{6}, 4\frac{1}{2}\right)$, lowest is $(1, 4)$

5 a $x(t) = 9t^3 - 9t^2 - 3t + 1$, $y(t) = 20t^3 - 24t^2 + 6t + 2$, $0 \leqslant t \leqslant 1$

b **i** highest point is $(-2, 4)$,
   lowest point is $\approx (-2.26, 1.25)$
   ii left-most is $\approx (-2.55, 1.71)$, right-most is $(1, 2)$

6 a $F(2.2, 2.8)$, $G(4.8, 4.6)$, $H(6.6, 2.6)$, $I(3.76, 3.88)$, $J(5.88, 3.4)$, $P(5.032, 3.592)$

b $x(t) = -3t^3 + 3t^2 + 6t + 1$, $y(t) = -3t^3 - 6t^2 + 9t + 1$, $0 \leqslant t \leqslant 1$

7 a $x(t) = 9t^3 - 9t^2 + 6t - 2$, $y(t) = 11t^3 - 21t^2 + 15t - 3$, $z(t) = -4t^3 + 9t^2 - 3t + 1$, $0 \leqslant t \leqslant 1$

c $\left(-\frac{1}{8}, \frac{5}{8}, 1\frac{1}{4}\right)$

REVIEW SET 14A

1 a

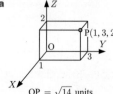

$OP = \sqrt{14}$ units

b

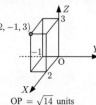

$OP = \sqrt{14}$ units

2 a $k = \pm\frac{\sqrt{3}}{4}$ **b** $k = \pm\sqrt{\frac{11}{128}}$

3 $a = 1$, $b = 1$, $c = -5$ **4** $\mathbf{x} = \begin{pmatrix} 1 \\ -\frac{5}{3} \\ -\frac{2}{3} \end{pmatrix}$

5 a $\begin{pmatrix} -6 \\ 1 \\ 3 \end{pmatrix}$ **b** $\sqrt{46}$ units **c** $\left(-1, 3\frac{1}{2}, \frac{1}{2}\right)$

6 $AB = AC = \sqrt{53}$ units and $BC = \sqrt{46}$ units
$\therefore \triangle ABC$ is isosceles.

7 $c = \frac{50}{3}$ **8 a** 5 units **b** $2\mathbf{i} - 5\mathbf{j} + 4\mathbf{k}$

9 $m = 5$, $n = -\frac{1}{2}$ **11** $-\frac{5}{\sqrt{14}}\begin{pmatrix} 3 \\ 2 \\ -1 \end{pmatrix}$

13 a $\begin{pmatrix} x \\ y \end{pmatrix} = \begin{pmatrix} -6 \\ 3 \end{pmatrix} + t\begin{pmatrix} 4 \\ -3 \end{pmatrix}$, $t \in \mathbb{R}$

b $x = -6 + 4t$, $y = 3 - 3t$, $t \in \mathbb{R}$ **c** $3x + 4y = -6$

14 $m = 10$

15 a $\begin{pmatrix} x \\ y \\ z \end{pmatrix} = \begin{pmatrix} 2 \\ -1 \\ 3 \end{pmatrix} + \lambda\begin{pmatrix} -2 \\ 2 \\ -4 \end{pmatrix}$, $\lambda \in \mathbb{R}$

b $\left(2 - \frac{2}{\sqrt{6}}, -1 + \frac{2}{\sqrt{6}}, 3 - \frac{4}{\sqrt{6}}\right)$ and
$\left(2 + \frac{2}{\sqrt{6}}, -1 - \frac{2}{\sqrt{6}}, 3 + \frac{4}{\sqrt{6}}\right)$

16 a $x = 4 - 2\lambda$, $y = 2 - \lambda$, $z = -1 + 6\lambda$, $\lambda \in \mathbb{R}$
b $x = 4 + 5\lambda$, $y = 2 + 2\lambda$, $z = -1 + 2\lambda$, $\lambda \in \mathbb{R}$

17 $72\begin{pmatrix} \frac{3}{\sqrt{29}} \\ -\frac{4}{\sqrt{29}} \\ \frac{2}{\sqrt{29}} \end{pmatrix}$

18 $4x^4 - 36x^2 + 81y^2 = 0$

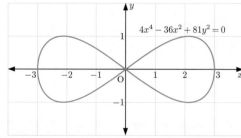

19 a $P(1, 0)$, $Q(2, 6)$ **b** 4 **c** $y = 8x - 10$
d $y = 2x^2 - 2$

20 a $S(1, 3)$, $E(1, 1)$ **b** $(0.625, 2.625)$

REVIEW SET 14B

1 $a = -1$, $b = \frac{-1 \pm \sqrt{17}}{2}$, $c = 10$

2 a $\begin{pmatrix} 3 \\ -3 \\ 11 \end{pmatrix}$ **b** $\begin{pmatrix} 7 \\ -3 \\ -26 \end{pmatrix}$ **c** $\sqrt{74}$ units

3 a $\mathbf{x} = \begin{pmatrix} -1 \\ \frac{1}{3} \\ \frac{2}{3} \end{pmatrix}$ **b** $\mathbf{x} = \begin{pmatrix} 1 \\ -10 \\ 2 \end{pmatrix}$ **4** $\begin{pmatrix} 8 \\ -8 \\ 7 \end{pmatrix}$

5 a $\overrightarrow{PQ} = \begin{pmatrix} -3 \\ 12 \\ 3 \end{pmatrix}$ **b** $\sqrt{162}$ units **c** $\sqrt{61}$ units

6 $k = \pm\frac{1}{2}$ **7** $(0, 0, 1)$ and $(0, 0, 9)$

8 $|\overrightarrow{KL}| = \sqrt{59}$ units, $|\overrightarrow{LM}| = \sqrt{50}$ units, $|\overrightarrow{KM}| = 3$ units, and $50 + 9 = 59$

9 $a = -2$, $b = 0$ **10** $\pm\frac{4}{\sqrt{14}}(3\mathbf{i} - 2\mathbf{j} + \mathbf{k})$

11 a $\overrightarrow{CD} = (k - 1)\mathbf{b} - \mathbf{a}$

12 a $AB = AC = \sqrt{94}$ units and $BC = \sqrt{86}$ units
$\therefore \triangle ABC$ is isosceles.
b $D(-8, 1, -6)$

13 a $(5, 2)$ **b** $\begin{pmatrix} 4 \\ 10 \end{pmatrix}$ is a non-zero scalar multiple of $\begin{pmatrix} 2 \\ 5 \end{pmatrix}$
c $\begin{pmatrix} x \\ y \end{pmatrix} = \begin{pmatrix} 5 \\ 2 \end{pmatrix} + s\begin{pmatrix} 4 \\ 10 \end{pmatrix}$, $s \in \mathbb{R}$

14 $\begin{pmatrix} x \\ y \end{pmatrix} = \begin{pmatrix} 0 \\ 8 \end{pmatrix} + \lambda\begin{pmatrix} 5 \\ 4 \end{pmatrix}$, $\lambda \in \mathbb{R}$

15 a i $\begin{pmatrix} x \\ y \\ z \end{pmatrix} = \begin{pmatrix} 2 \\ -3 \\ 1 \end{pmatrix} + \lambda \begin{pmatrix} 4 \\ 2 \\ -1 \end{pmatrix}, \ \lambda \in \mathbb{R}$

ii $x = 2 + 4\lambda, \ y = -3 + 2\lambda, \ z = 1 - \lambda, \ \lambda \in \mathbb{R}$

b i $\begin{pmatrix} x \\ y \\ z \end{pmatrix} = \begin{pmatrix} -1 \\ 6 \\ 3 \end{pmatrix} + \lambda \begin{pmatrix} 6 \\ -8 \\ -3 \end{pmatrix}, \ \lambda \in \mathbb{R}$

ii $x = -1 + 6\lambda, \ y = 6 - 8\lambda, \ z = 3 - 3\lambda, \ \lambda \in \mathbb{R}$

16 a $(-10, \ 5, \ 12)$ **b** $\begin{pmatrix} 6 \\ 14 \\ -0.4 \end{pmatrix}$ **c** ≈ 15.2 m s^{-1}

d 30 seconds **e** $(170, \ 425, \ 0)$

17 a i

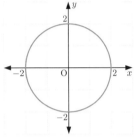

ii $x^2 + y^2 = 4$

b i

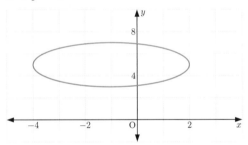

ii $\left(\dfrac{x+1}{3}\right)^2 + \left(\dfrac{y-5}{2}\right)^2 = 1$

c i

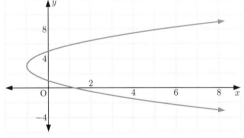

ii $y^2 - 6y - 4x + 5 = 0$

18 a $P\left(-1, \ \dfrac{\sqrt{3}}{2}\right)$ **b** $y = \dfrac{1}{2\sqrt{3}}x + \dfrac{2}{\sqrt{3}}$

c $\approx (-1.79, \ 0.45), \ \approx (1.79, \ -0.45)$

19 a $S(1, \ 4), \quad E(0, \ 11)$ **b** $\left(-\dfrac{1}{8}, \ 7\dfrac{3}{8}\right)$

20 a $x(t) = -7t^3 - 3t^2 + 3t + 3, \quad y(t) = 12t^2 - 9t - 1,$
$0 \leqslant t \leqslant 1$

b highest point is $(-4, \ 2)$, lowest point is $\approx (3.33, \ -2.69)$

c left-most point is $(-4, \ 2)$,
right-most point is $\approx (3.45, \ -2.53)$

EXERCISE 15A.1

1 a $v(t) = 2t - 6$ m s^{-1}

$a(t) = 2$ m s^{-2}

b $s(0) = 7$ m, $v(0) = -6$ m s^{-1}, $a(0) = 2$ m s^{-2}
Initially, the object is 7 m to the right of O, moving to the left at 6 m s^{-1}, with acceleration 2 m s^{-2}.

c 2 m to the left of O

d

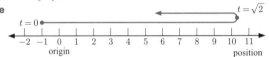

e $0 \leqslant t \leqslant 3$

2 a $v(t) = 12 - 6t^2$ cm s^{-1}, $a(t) = -12t$ cm s^{-2}

b $s(0) = -1$ cm, $v(0) = 12$ cm s^{-1}, $a(0) = 0$ cm s^{-2}
The particle started 1 cm to the left of the origin and was travelling to the right at a constant speed of 12 cm s^{-1}.

c $t = \sqrt{2}$ s, $s(\sqrt{2}) = (8\sqrt{2} - 1)$ cm

d i $t \geqslant \sqrt{2}$ s **ii** never

e

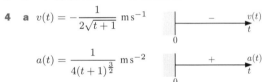

3 b i 69.58 m s^{-1} **ii** ≈ 247 m

4 a $v(t) = -\dfrac{1}{2\sqrt{t+1}}$ m s^{-1}

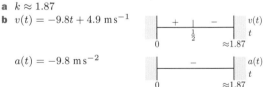

$a(t) = \dfrac{1}{4(t+1)^{\frac{3}{2}}}$ m s^{-2}

b $s(0) = 3$ m, $v(0) = -\frac{1}{2}$ m s^{-1}, $a(0) = \frac{1}{4}$ m s^{-2}
Initially, the particle is 3 m to the right of O, moving to the left at $\frac{1}{2}$ m s^{-1} with acceleration $\frac{1}{4}$ m s^{-2}.

c After 3 seconds, the particle is 2 m to the right of O, moving to the left at $\frac{1}{4}$ m s^{-1}, with acceleration $\frac{1}{32}$ m s^{-2}.

d The particle's speed is continuously decreasing.

5 a $k \approx 1.87$

b $v(t) = -9.8t + 4.9$ m s^{-1}

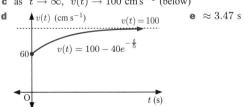

$a(t) = -9.8$ m s^{-2}

c i decreasing **ii** increasing

6 a $v(t) = 100 - 40e^{-\frac{t}{5}}$ cm s^{-1}, $a(t) = 8e^{-\frac{t}{5}}$ cm s^{-2}

b $s(0) = 200$ cm to the right of origin
$v(0) = 60$ cm s^{-1}, $a(0) = 8$ cm s^{-2}

c as $t \to \infty$, $v(t) \to 100$ cm s^{-1} (below)

d

e ≈ 3.47 s

7 **b** $v(t) = 1 - \dfrac{2}{2t+1}$ cm s^{-1}

   c **i** $t \geqslant \frac{1}{2}$ **ii** $0 \leqslant t \leqslant \frac{1}{2}$ **e** $1 + \ln\left(\frac{4}{5}\right)$ cm

8 **a** $x(0) = -1$ cm, $v(0) = 0$ cm s^{-1}, $a(0) = 2$ cm s^{-2}

   b At $t = \frac{\pi}{4}$ seconds, the particle $(\sqrt{2} - 1)$ cm to the left of O, is moving to the right at $\sqrt{2}$ cm s^{-1}, with acceleration $\sqrt{2}$ cm s^{-2}.

   c changes direction when $t = \pi$, $x(\pi) = 3$ cm

   d increasing for $0 \leqslant t \leqslant \frac{\pi}{2}$ and $\pi \leqslant t \leqslant \frac{3\pi}{2}$

9 **a** **i** right   **ii** left   **b** $v(t) = 4\cos\frac{t}{2}$ m s^{-1}

   c **i** left   **ii** right

10 **a**

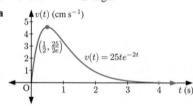

   c $0 \leqslant t \leqslant \frac{1}{2}$ **d** $t \geqslant \frac{1}{2}$

EXERCISE 15A.2

1 **a** $10\frac{1}{6}$ m   **b** $4\frac{1}{2}$ m to the left

2 **a** $s(t) = \frac{1}{3}t^3 + \frac{3}{2}t^2 + 2t$ cm   **b** $53\frac{1}{3}$ cm

   c $53\frac{1}{3}$ cm to the right

3 **a** $s(t) = 32.4t - 4.9t^2$ m   **b** ≈ 53.6 m

4 **a** 1 m   **b** $\frac{\sqrt{3}+2}{4}$ m

5 **a**

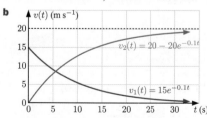

   b **i** ≈ 15.2 miles per hour   **ii** ≈ 19.7 miles per hour

   c ≈ 40.3 miles

6 **a** lion: ≈ 13.6 m s^{-1},   zebra: ≈ 1.90 m s^{-1}

   b

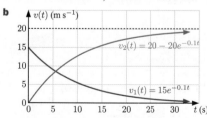

The lion's speed $v_1(t)$ decreases over time whereas the zebra's speed $v_2(t)$ increases over time.

   c $\displaystyle\int_0^3 v_1(t)\,dt = 150 - 150e^{-0.3} \approx 38.9$

The lion has travelled about 38.9 m in the first 3 seconds.

d $\displaystyle\int_0^3 [v_1(t) - v_2(t)]\,dt = 290 - 350e^{-0.3}$
≈ 30.7

In the first 3 seconds, the lion has gained about 30.7 m on the zebra.

e At the time when $v_1(t) = v_2(t)$, the lion and the zebra will be moving at the same speed. Since the lion's speed decreases over time and the zebra's speed increases over time, the zebra will be faster than the lion after that time. So, they will be closest at the point when their speeds are equal.

f $-10\ln\left(\frac{4}{7}\right) \approx 5.60$ s

g No, the lion was about 1.92 m from the zebra at their closest point.

7 **a** **i** 1 m s^{-1} **ii** $\frac{1}{4}$ m s^{-1}

   b $v(t) = \cos^2 t \geqslant 0$ for all t. The object can only change direction when there is a change in the sign of $v(t)$.

   c $s(t) = \frac{1}{2}t + \frac{1}{4}\sin 2t$ m

   d **i** $\frac{\pi}{4}$ m right of the origin

   ii $\left(\frac{\sqrt{3}}{8} + \frac{7\pi}{12}\right)$ m right of the origin

   e $a(t) = -\sin 2t$ m s^{-2}

   f **i** $\frac{\pi}{2} \leqslant t \leqslant \pi$, $\frac{3\pi}{2} \leqslant t \leqslant 2\pi$, and $\frac{5\pi}{2} \leqslant t \leqslant 3\pi$

   ii $0 \leqslant t \leqslant \frac{\pi}{2}$, $\pi \leqslant t \leqslant \frac{3\pi}{2}$, and $2\pi \leqslant t \leqslant \frac{5\pi}{2}$

   g $\frac{\pi - \sqrt{3}}{4}$ m

8 **a** $s(t) = -4t + \frac{2}{3}t^{\frac{3}{2}}$ m   **b** $t = 16$

   c ≈ 10.5 m left of the origin   **d** ≈ 32.2 m

9 **a** 40 m s^{-1} **b** ≈ 47.8 m s^{-1} **c** $2\ln 2 \approx 1.39$ seconds

   d as $t \to \infty$, $v(t) \to 50$ from below

   e $a(t) = 5e^{-0.5t}$ and as $e^x > 0$ for all x, $a(t) > 0$ for all t.

   f

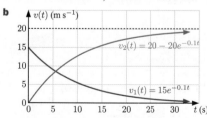

   g ≈ 134 m

10 **a** $v(t) = -\dfrac{1}{(t+1)^2} + 1$ m s^{-1}

   b $s(t) = \dfrac{1}{t+1} + t - 1$ m

   c The particle is $\frac{4}{3}$ m to the right of the origin, moving to the right at $\frac{8}{9}$ m s^{-1}, and accelerating at $\frac{2}{27}$ m s^{-2}.

11 **a** Show that $v(t) = 100 - 80e^{-\frac{1}{20}t}$ m s^{-1} and as $t \to \infty$, $v(t) \to 100$ m s^{-1}.

   b ≈ 370 m

12 **a** $v(t) = 10\sqrt{t+5} - 5\sqrt{5}$ cm s^{-1}

   b $\displaystyle\int_0^4 v(t)\,dt = 180 - \frac{160\sqrt{5}}{3} \approx 60.7$

The droplet travels ≈ 60.7 cm in the first 4 seconds.

13 **a** $\approx 5.28 \text{ cm s}^{-1}$ **b** $\approx 5.07 \text{ cm s}^{-2}$

c $s(t) = -\frac{5}{29}e^{-2t}(9\sin 5t + 8\cos 5t) + \frac{330}{29}$ cm

d ≈ 0.221 cm

EXERCISE 15B.1

1 **a** ≈ 60.0 m   **b** 34.3 m s^{-1}

2 **a** **i** ≈ 4.52 s   **ii** ≈ 90.4 m

b **i** ≈ 1.01 s   **ii** ≈ 0.707 m

c **i** 10 s   **ii** 25 m

3 ≈ 5.85 m   **4** **a** ≈ 1.39 s   **b** $\approx 18.8 \text{ m s}^{-1}$

5 **a** $\approx 1.96 \text{ m s}^{-1}$ **b** $\approx 6.26 \text{ m s}^{-1}$ **c** $\approx 0.128 \text{ m s}^{-1}$

6 $\approx 21.7 \text{ m s}^{-1}$

7 **a** The stones are the same height above the river and they have the same vertical velocity component.

b The vertical component of each stone's motion has constant acceleration $-g$, and each stone is initially at rest.

c No, the stone which is thrown has a larger horizontal velocity component. The horizontal velocity of the dropped stone is zero.

EXERCISE 15B.3

1 **a** initial speed $\approx 62.6 \text{ m s}^{-1}$,
initial angle of trajectory $\approx 16.7°$

b ≈ 16.5 m

c No, the golf ball would be slowed by air resistance and would have a lower maximum height.

2 **a** $\begin{pmatrix} 60\sqrt{3} \\ 60 \end{pmatrix} \text{ m s}^{-1}$ **b** ≈ 6.12 s   **c** ≈ 184 m

3 3.025 m

4 **a** 2 s   **b** $14\sqrt{3}$ m   **c** $\approx 17.8 \text{ m s}^{-1}$ **5** 15 m

6 **a** $t = \dfrac{u_y}{g}$ **b** maximum height $= \dfrac{u_y^2}{2g}$

c $t = \dfrac{2u_y}{g}$ **d** horizontal distance travelled $= \dfrac{2u_x u_y}{g}$

7 **a** horizontal distance travelled $= \dfrac{s^2 \sin 2\theta}{g}$

b $\theta = 45°$

8 ≈ 74.4 m

9 **a** The projectile moves in a parabolic arc.

b **i** $45°$ **ii** $\approx 48.7 \text{ m s}^{-1}$ **iii** 60.5 m
iv ≈ 7.03 s

10 **a** $\approx 485 \text{ m s}^{-1} \approx 1080$ miles per hour   **b** 6 km

EXERCISE 15B.4

3 **b** $\sqrt{u_x^2 + u_y^2}$

REVIEW SET 15A

1 **a** $v(t) = 15 + \dfrac{120}{(t+1)^3} \text{ cm s}^{-1}$, $a(t) = -\dfrac{360}{(t+1)^4} \text{ cm s}^{-2}$

b At $t = 3$, the particle is 41.25 cm to the right of O, moving to the right at $\approx 16.9 \text{ cm s}^{-1}$, and decelerating at $\approx 1.41 \text{ cm s}^{-2}$.

c speed is never increasing

2 **a** $x(0) = 3$ cm, $x'(0) = 2 \text{ cm s}^{-1}$, $x''(0) = 0 \text{ cm s}^{-2}$

b $t = \frac{\pi}{4}$ s and $\frac{3\pi}{4}$ s   **c** 4 cm

3 **a** 1 m   **b** 6 m   **c** 4 m

4
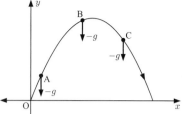

5 **a** $v(0) = 25 \text{ m s}^{-1}$, $v(3) = 4 \text{ m s}^{-1}$

b as $t \to \infty$, $v(t) \to 0$ from above

c
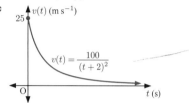

d $\displaystyle\int_0^2 v(t)\, dt = 25$

The boat travels a total distance of 25 m in the first 2 seconds after its engine is turned off.

e 3 seconds

6 **a** $\approx 4.37 \text{ m s}^{-1}$ **b** 9 m s^{-1} **7** $\approx 3.46 \text{ m s}^{-1}$

8 **a** $\begin{pmatrix} \frac{27}{2} \\ \frac{27\sqrt{3}}{2} \end{pmatrix} \text{ m s}^{-1}$ **b** ≈ 4.77 s

c ≈ 64.4 m   **d** ≈ 27.9 m

9 ≈ 168 m   **10** $\theta \approx 63.4°$

REVIEW SET 15B

1 **a** $v(t) = -8e^{-\frac{t}{10}} - 40 \text{ m s}^{-1}$
$a(t) = \frac{4}{5}e^{-\frac{t}{10}} \text{ m s}^{-2}$ $\{t \geqslant 0\}$

b $s(0) = 80$ m, $v(0) = -48 \text{ m s}^{-1}$, $a(0) = 0.8 \text{ m s}^{-2}$

c
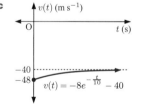

d $t = 10 \ln 2$ seconds

2 **a** $v(0) = 0 \text{ cm s}^{-1}$, $v(\frac{1}{2}) = -\pi \text{ cm s}^{-1}$, $v(1) = 0 \text{ cm s}^{-1}$,
$v(1\frac{1}{2}) = \pi \text{ cm s}^{-1}$, $v(2) = 0 \text{ cm s}^{-1}$

b $0 \leqslant t \leqslant 1$, $2 \leqslant t \leqslant 3$, $4 \leqslant t \leqslant 5$, and so on
So, for $2n \leqslant t \leqslant 2n + 1$, $n \in \{0, 1, 2, 3, ...\}$

3 **a** $s(t) = -\frac{1}{96}t^4 - \frac{1}{24}t^2 + 2$ m   **b** ≈ 3.46 s

4 **a** Tyson

b $\displaystyle\int_0^5 v_1(t)\, dt \approx 42.0$

Tyson has travelled about 42.0 m in the first 5 seconds of the race.

c $s_1(t) = 10t + 8e^{-1.25t} - 8$ m

$s_2(t) = 10.5t + 10.5e^{-t} - 10.5$ m

d Tyson   **e** **Hint:** Find t such that $s_1(t) = 100$.

f Maurice

5 **a** **i** ≈ 4.12 s   **ii** ≈ 6.17 m

b **i** ≈ 6.50 s   **ii** ≈ 5.39 m

6 **a** $\begin{pmatrix} \frac{9}{2} \\ \frac{9\sqrt{3}}{2} \end{pmatrix}$ m s^{-1} **b** ≈ 0.795 s   **c** ≈ 3.10 m

7 ≈ 9140 m   **8** **a** ≈ 5.47 s   **b** ≈ 123 m

EXERCISE 16A

1 **a** $\mathbf{u} = (3\mathbf{i} + 4\mathbf{j})$ N

$|\mathbf{u}| = 5$ N, direction $\approx 53.1°$ to the horizontal.

b $\mathbf{u} = (-5\mathbf{i} - 12\mathbf{j})$ N

$|\mathbf{u}| = 13$ N, direction $\approx 67.4°$ to the left of $\mathbf{a}$.

c $\mathbf{u} = (-5\mathbf{i} + 5\mathbf{j})$ N

$|\mathbf{u}| = 5\sqrt{2}$ N, direction $= 45°$ to the right of $\mathbf{b}$.

d $\mathbf{u} = (8.6\mathbf{i} - 10.3\mathbf{j})$ N

$|\mathbf{u}| \approx 13.4$ N, direction $\approx 39.9°$ to the left of $\mathbf{a}$.

e $|\mathbf{u}| = \sqrt{13}$ N, direction $\approx 56.3°$ to the right of $\mathbf{a}$.

f $|\mathbf{u}| = 3$ N, direction $\approx 53.1°$ to the right of $\mathbf{a}$.

2 **a** $(\mathbf{i} + \mathbf{j})$ N   **b** $(\mathbf{i} + 4\mathbf{j})$ N   **c** $\mathbf{i}$ N

d $(3\mathbf{i} + 5\mathbf{j})$ N   **e** $-\mathbf{i}$ N

3 **a** ≈ 2.65 N, $\approx 40.9°$ to the right of $\mathbf{b}$.

b ≈ 7.45 N, $\approx 12.0°$ to the right of $\mathbf{a}$.

c ≈ 6.48 N, $\approx 19.1°$ to the left of $\mathbf{a}$.

d ≈ 5.12 N, $\approx 32.8°$ to the left of $\mathbf{a}$.

4 $90°$

5 **a** ≈ 11.0 N, $\approx 14.0°$ to the left of $\mathbf{a}$.

b ≈ 6.11 N, $\approx 24.3°$ to the left of $\mathbf{b}$.

EXERCISE 16B

1 $\mathbf{a} + \mathbf{b} + \mathbf{c} = \mathbf{0}$ **2** **a** $\mathbf{b} = -\mathbf{a}$ **b** $\mathbf{c} = -\mathbf{a} - \mathbf{b}$

3 $k = 3$ or -2

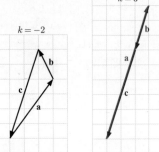

4 $x = -5$, $y = -3$, $z = -2$ **5** $b = 2$, $c = -3$

6 **a** $\mathbf{c} = \begin{pmatrix} -5 \\ 1 \end{pmatrix}$ **b** $\mathbf{c} = \begin{pmatrix} -5 \\ 1 \end{pmatrix}$

7 $x = 2$, $y = -4$ **8** $-4\mathbf{j}$

9 **Hint:** If the particle is in equilibrium then $\mathbf{a}$, $\mathbf{b}$, and $\mathbf{c}$ form a closed triangle.

10 **a** $|\mathbf{b}| = |\mathbf{c}| = 4$ N   **b** $|\mathbf{b}| \approx 2.20$ N, $|\mathbf{c}| \approx 2.69$ N

11 650 N, bearing $\approx 323°$

EXERCISE 16C.1

1 horizontal component $= |\mathbf{T}| \cos\theta$,

vertical component $= |\mathbf{T}| \sin\theta$

2 **a** $\mathbf{v} \approx 9.19\mathbf{i} + 7.71\mathbf{j}$ **b** $\mathbf{v} \approx -1.04\mathbf{i} + 5.91\mathbf{j}$

c $\mathbf{v} \approx 2.96\mathbf{i} - 6.34\mathbf{j}$

3 **a** $\dfrac{mg}{\tan\theta}$ N

b **i** 0 N   **ii** as $\theta \to 0°$, force needed $\to \infty$

EXERCISE 16C.2

1 **a** parallel component ≈ 50.7 N

perpendicular component ≈ 189.3 N

b parallel component ≈ 73.7 N

perpendicular component ≈ 26.8 N

2 $|\mathbf{F}| \approx 250$ N

3 **a** ≈ 56.8 N perpendicular to the surface

b ≈ 56.8 N up the slope

4 ≈ 33.5 N at $20°$ to the horizontal   **5** $T \approx 12\,800$ N

6 **a** $\approx 19.5°$ **b** ≈ 311 N   **c** ≈ 878 N

7 **a** $T = \dfrac{35g}{\cos\frac{\theta}{2}}$ **b** $\theta \geqslant 172°$

c When the anchors are $172°$ apart, there is as much tension in the ropes as a single rope holding a 500 kg mass.

8 ≈ 5.62 m s^{-2} down the slope   **9** $\theta \approx 5.86°$

EXERCISE 16D

1 **a**

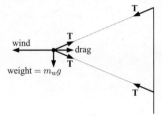

b The forces on the windsock are:
- weight due to gravity vertically downwards
- the breeze horizontally left
- drag horizontally right
- tension in the strings along the line of the strings.

The windsock is "held up" by the upward vertical component of the tension in the string. This must be equal and opposite to the weight and the downward vertical component of the tension.

2 $|\mathbf{F}| \approx 83.2$ N, contact force ≈ 34.4 N vertically upwards, tension $= 117.6$ N

3 **a**

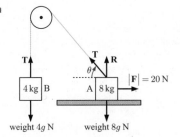

b ≈ 14.4 N horizontally left   **c** $\theta \approx 59.3°$

4 $m \approx 4.71$

EXERCISE 16E

1 **a** 6.86 N   **b** 274.4 N   **2** 220.5 N

3 3.04 m s^{-2} **4** ≈ 5.48 N   **5** ≈ 4.46 N

6 **b** ≈ 0.755 m s^{-2} **c** ≈ 21.6 N

7 **a** $T = \dfrac{\mu mg}{\cos\theta + \mu\sin\theta}$ **b** ≈ 364 N

8 No, the block will accelerate at ≈ 4.33 m s^{-2}.

9 $\mu \approx 0.577$ **10** $\theta \approx 26.6°$

11 **a** ≈ 4.02 m s^{-2}

   b **i** ≈ 20.1 N   **ii** ≈ 36.1 N

   c **i** ≈ 4.77 m s^{-2} **ii** ≈ 21.5 m

12 minimum $|\mathbf{P}| = \dfrac{mg(\sin\theta + \mu\cos\theta)}{\cos\theta - \mu\sin\theta}$

EXERCISE 16F

1 **a** 0 N m   **b** 20 N m   **c** 60 N m

2 **a** 0 N m   **b** −16 N m   **c** −64 N m

3 0.9 N m   **4** 8.2 N m   **5** −9 N m

6 **a** ≈ −4.65 N m

   b Every point on the base has the same perpendicular distance from the line of the force.

EXERCISE 16G

1 The moment about P, B, and C is also 0 N m

2 **a** $|\mathbf{F}| = 9$ N, $x = \frac{40}{9} \approx 4.44$

   b $|\mathbf{F}| = 7$ N, $x = \frac{24}{7} \approx 3.43$

3 **a** $|\mathbf{F}| = 5$ N, $x = 4.96$ **b** $|\mathbf{F}| = 2$ N, $x = 1.62$

   c $|\mathbf{F}| = 7$ N, $x = \frac{34}{7} \approx 4.86$

   d $|\mathbf{F}| = 9.5$ N, $x = \frac{49}{38} \approx 1.29$

4 **a** $|\mathbf{F}| = 10$ N, $x = 5$

   b $|\mathbf{F}| = 1.5$ N, $x = \frac{45}{26} \approx 1.73$

   c $|\mathbf{F}| = 3.5$ N, $x = 4$

   d $|\mathbf{F}| = 14$ N, $x = \frac{8}{3} \approx 2.67$

5 **a** possible

   b impossible (taking moments from the rightmost point, all forces act clockwise)

   c possible

   d impossible (taking moments from the centre, all forces act clockwise)

   e possible

6 $|\mathbf{a}| = 65g$ N, $|\mathbf{b}| = 13g$ N

7 **a** $-4g$ N m   **b** $m = 2$ **c** $10g$ N

EXERCISE 16H

1 **a** $\mathbf{R}_A$ is the reaction force at A.
    $\mathbf{R}_B$ is the reaction force at B.
    $\mathbf{F}_A$ is the frictional force at A.
    $\mathbf{F}_B$ is the frictional force at B.

   b **i** No, the force of the person is mg N downwards, regardless of the position of the person.

      ii Yes, for example if the person moves closer to A, the moment about B decreases.

   c **i** $R_A = F_B$ **ii** $F_A + R_B = mg$

      iii $-\frac{3}{2}mg + 3R_B - 4F_B = 0$

      iv $\frac{3}{2}mg - 4R_A - 3F_A = 0$

   d **Hint:** Show that substituting the results from **i** and **ii** into **iii** gives the result in **iv**.

   e **i** $R_A = \frac{3}{8}mg$, $R_B = mg$, $F_B = \frac{3}{8}mg$

   f No

2 **a**

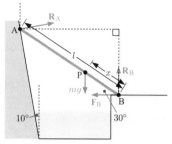

   $\mathbf{R}_A$ is the reaction force at A.
   $\mathbf{R}_B$ is the reaction force at B.
   $\mathbf{F}_B$ is the frictional force at B.
   mg is the weight of the person.

   b $F_B = R_A \cos 10°$
       $R_B + R_A \sin 10° = mg$
       $mg(l - x)\cos 30° + F_B(l\sin 30°) = R_B(l\cos 30°)$
       $mg(x\cos 30°) = R_A(l\sin 40°)$

   c $\mu = \dfrac{x\cos 30° \cos 10°}{l\sin 40° - x\cos 30° \sin 10°}$

   d Yes, as x increases, the minimum value of μ required to maintain equilibrium also increases.

3 **b** $R_A = \dfrac{mgl}{2a}$

4 **a** **Hint:** Resolve the vertical forces.

   d No, we have 3 equations with 4 unknowns.

REVIEW SET 16A

1 **a** $|\mathbf{u}| \approx 1.92$ N, direction ≈ 27.9° to the left of **a**.

   b $|\mathbf{u}| \approx 9.22$ N, direction ≈ 49.4° to the right of **a**.

2 $-\frac{9}{2}\mathbf{i} + \left(\frac{3\sqrt{3}}{2} - 2\right)\mathbf{j}$ **3** $\mathbf{a} + \mathbf{b} + \mathbf{c} = 0$

4 $\mathbf{F} = \begin{pmatrix} -2 \\ -2 \end{pmatrix}$ **5** $|\mathbf{a}| = 2\sqrt{3}$ N, $|\mathbf{b}| = 2$ N

6 **a** **i** ≈ 8.31 N   **ii** ≈ 13.3 N

   b ≈ 13.3 N perpendicular to the surface

7 ≈ 3.19 m s^{-2} **8** ≈ 0.817 m s^{-2}

9 **a**

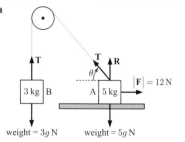

   weight = $3g$ N   weight = $5g$ N

   b ≈ 14.1 N horizontally left   **c** $\theta \approx 65.9°$

10 ≈ 20.1 N   **11** **a** −28 N m   **b** −16 N m   **c** 0 N m

12 8 N m   **13** $|\mathbf{F}| = 5$ N, applied upwards, 4 m from A

14 **a** $\mathbf{R}_A$ is the reaction force at A.
    $\mathbf{R}_B$ is the reaction force at B.
    $\mathbf{F}_B$ is the frictional force at B.

b $R_A = \frac{2}{3}mg$, $R_B = mg$, $F_B = \frac{2}{3}mg$ **c** $\frac{2}{3}$

d No, in reality the wall would not be smooth.

15 $|\mathbf{F}| = 6$ N, $x = \frac{11}{6} \approx 1.83$

REVIEW SET 16B

1 a $(-\mathbf{i} + 6\mathbf{j})$ N   **b** $(-\mathbf{i} - \mathbf{j})$ N

2 a ≈ 7.21 N, $\approx 73.9°$ to the right of $\mathbf{a}$.

 b ≈ 6.48 N, $\approx 18.4°$ to the left of $\mathbf{a}$.

3 $a = 5$, $b = -13$

4 a $\mathbf{v} \approx -5.71\mathbf{i} - 1.85\mathbf{j}$ **b** $\mathbf{v} \approx -2.03\mathbf{i} + 4.35\mathbf{j}$

5 $|\mathbf{F}| \approx 303$ N   **6** $T = 2$ **7** $\theta \approx 17.8$

8 $m \approx 2.89$ **9 a** 5.096 N   **b** 40.768 N

10 minimum $\mu \approx 0.325$

11 Hint: Since the objects are connected, and each move in line with the string, they will move with equal acceleration.

12 -7.2 Nm

13 $|\mathbf{F}| = 14$ N, applied downwards, $\frac{15}{7} \approx 2.14$ cm from A.

14 $|\mathbf{F}| = 4$ N, $x = \frac{16}{3} \approx 5.33$

15 left leg: $82.45g$ N upwards

  right leg: $60.55g$ N upwards

EXERCISE 17A

1 a 7510

 b i ≈ 0.325 **ii** ≈ 0.653 **iii** ≈ 0.243

2 a $\frac{232}{447} \approx 0.519$ **b** $\frac{197}{447} \approx 0.441$ **c** $\frac{25}{447} \approx 0.0559$

 d $\frac{250}{447} \approx 0.559$

3 a

	Junior	Middle	Senior	Total
Sport	131	164	141	436
No sport	28	81	176	285
Total	159	245	317	721

 b i $\frac{436}{721} \approx 0.605$ **ii** $\frac{131}{721} \approx 0.182$ **iii** $\frac{257}{721} \approx 0.356$

4 a $\frac{743}{1235} \approx 0.602$ **b** $\frac{148}{1235} \approx 0.120$ **c** $\frac{1085}{1235} \approx 0.879$

 d $\frac{795}{1235} \approx 0.644$

EXERCISE 17B

1 a $\frac{1}{4}$ **b** $\frac{1}{2}$ **c** 0

2 a $P(A \cap B) = P(A)\,P(B)$ **b** $P(A \mid B) = P(A)$

 c If A and B are independent events, the probability of A does not depend on B.

3 $\frac{1}{2}$

4 a

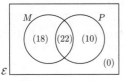

22 study both

 b i $\frac{9}{25}$ **ii** $\frac{11}{20}$

5 a

	Brown	Other	Total
Dark	15	8	23
Fair	3	14	17
Total	18	22	40

 b i $\frac{3}{8}$

  ii $\frac{15}{23}$

6 a

	Sunburnt	Not sunburnt	Total
Bitten	5	17	22
Not bitten	18	10	28
Total	23	27	50

 b i $\frac{14}{25}$ **ii** $\frac{4}{5}$ **iii** $\frac{5}{23}$ **iv** $\frac{9}{14}$

7 $\frac{7}{8}$

8 a $\frac{13}{20}$ **b** $\frac{7}{20}$ **c** $\frac{11}{50}$ **d** $\frac{7}{25}$ **e** $\frac{4}{7}$ **f** $\frac{1}{4}$

EXERCISE 17C

1 a $\frac{7}{15}$ **b** $\frac{7}{30}$ **2 a** $\frac{14}{55}$ **b** $\frac{1}{55}$

3 a $\frac{3}{100}$ **b** $\frac{3}{100} \times \frac{2}{99} \approx 0.000\,606$

 c $\frac{3}{100} \times \frac{2}{99} \times \frac{1}{98} \approx 0.000\,006\,18$

 d $\frac{97}{100} \times \frac{96}{99} \times \frac{95}{98} \approx 0.912$

4 a $\frac{4}{7}$ **b** $\frac{2}{7}$ **5 a** $\frac{10}{21}$ **b** $\frac{1}{21}$

EXERCISE 17D

1 a $A =$ the marble is taken from urn A

  $B =$ the marble is taken from urn B

  $R =$ a red marble is selected

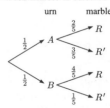

 b $\frac{3}{5}$ **c** $\frac{2}{3}$

2 a $\frac{23}{50}$ **b** $\frac{14}{23}$ **3** $\frac{70}{163}$

4 a 0.15

 b $B =$ the boy eats his lunch

  $G =$ the girl eats her lunch

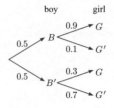

 c 0.65

5 a 0.0484   **b** ≈ 0.393 **6** $\frac{2}{3}$

7 $\frac{187}{460} \approx 0.407$ **8 a** $\frac{325}{833} \approx 0.390$ **b** $\frac{787}{833} \approx 0.945$

REVIEW SET 17A

1 a

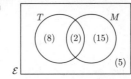

 b i $\frac{1}{15}$

  ii $\frac{2}{17}$

2 a

	Female	Male	Total
Smoker	20	40	60
Non-smoker	70	70	140
Total	90	110	200

b **i** $\frac{7}{20}$ **ii** $\frac{1}{2}$ **c** **i** ≈ 0.121 **ii** ≈ 0.422

3 **a** $\frac{5}{8}$ **b** $\frac{1}{4}$ **4** **a** $\frac{1}{6}$ **b** $\frac{2}{3}$

5 **a** **i**

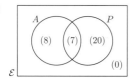

ii

	Acting classes	No acting classes	Total
Piano lessons	7	20	27
No piano lessons	8	0	8
Total	15	20	35

b **i** $\frac{7}{15}$ **ii** $\frac{20}{27}$

6 $\frac{5}{9}$ **7** $\frac{5}{8}$

REVIEW SET 17B

1 **a**

	Men	Women	Total
Prefer coffee	35	26	61
Prefer tea	15	24	39
Total	50	50	100

b

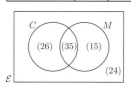

c $\frac{35}{61}$

2 **a** 0.93   **b** 0.8   **c** 0.2   **d** 0.65

3 **a** $\frac{4}{500} \times \frac{3}{499} \times \frac{2}{498} \approx 0.000\,000\,193$

b $1 - \frac{496}{500} \times \frac{495}{499} \times \frac{494}{498} \approx 0.0239$

4 **a** 0.2588   **b** ≈ 0.703

5 **a**

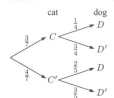

b **i** $\frac{3}{28}$

ii $\frac{23}{35}$

6 **a** $\frac{31}{70}$ **b** $\frac{21}{31}$ **7** $\frac{1}{2}$

EXERCISE 18A.1

1 **B**, **D**, and **F**

2 **a** The diameters may be affected by:
   - the type of lathe used
   - the steadiness of the woodworker's hand
   - the operating speed of the lathe.

b The scores may be affected by:
   - the time spent studying
   - natural ability (for example, memory, learning ability)
   - general knowledge.

c The times may be affected by:
   - the distance that the students live from their school
   - walking speed
   - physical fitness
   - the terrain.

3 **a** The variable is not likely to be normally distributed as it is more likely that there would be more people younger than the mean age than there are older. The distribution may be positively skewed.

b The variable is likely to be normally distributed as the long jumper is likely to jump the same distance consistently, but it will vary due to factors such as the speed at which the long jumper runs before the jump, and the positioning of their body before hitting the sand.

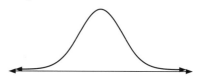

c The variable is not likely to be normally distributed as each number has the same chance of being drawn. The distribution should be uniform.

d The variable is likely to be normally distributed as the lengths of the carrots will be generally centred around the mean, but will vary due to factors such as soil quality, different weather conditions, harvest times, and so on.

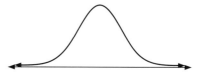

e The variable is not likely to be normally distributed. People are most likely to be served quite quickly. The distribution is likely to be negatively skewed.

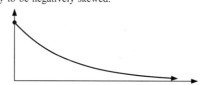

f The variable is not likely to be normal as it is a discrete variable. Each egg has the same probability of being brown, so the distribution is binomial.

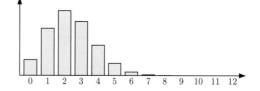

g The variable is not likely to be normally distributed as it is a discrete variable. Most families will have 0 - 2 children, and there will be much fewer families with more than 2 children. The distribution will be positively skewed.

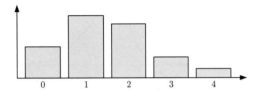

h The variable is not likely to be normally distributed as there will tend to be many more shorter buildings than tall buildings in a city. The distribution will be positively skewed.

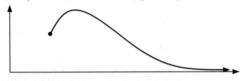

EXERCISE 18A.2

1 **a** B   **b** D   **c** A   **d** C

2

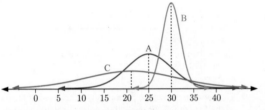

3 **a** $f(x_1) = \dfrac{1}{3\sqrt{2\pi}}\, e^{-\frac{1}{2}\left(\frac{x_1-4}{3}\right)^2}$

b $f(x_2) = \dfrac{1}{3\sqrt{2\pi}}\, e^{-\frac{1}{2}\left(\frac{x_2-6}{3}\right)^2}$

c normally distributed with mean 6 and standard deviation 3

d

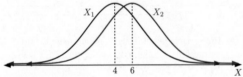

EXERCISE 18B.1

1 **a** **i** 40   **ii** 25

b **i** 1 standard deviation above the mean
   ii 2 standard deviations below the mean
   iii 3 standard deviations above the mean

c

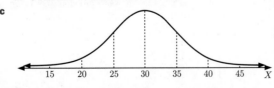

d $\approx 34.13\%$ **e** ≈ 0.1359

2 **a** mean $= 20$,   standard deviation $= 4$
   b **i** $\approx 34.13\%$ **ii** $\approx 13.59\%$ **iii** $\approx 2.28\%$

3 **a**

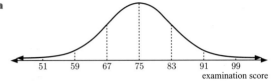

b **i** $\approx 15.87\%$ **ii** $\approx 2.28\%$ **iii** $\approx 81.85\%$

4 **a** ≈ 0.6826 **b** ≈ 0.0228

5 **a** **i** $\approx 34.13\%$ **ii** $\approx 47.72\%$
   b **i** ≈ 0.0228 **ii** ≈ 0.8413
   c ≈ 68 students   **d** $k \approx 178$

6 **a** ≈ 459 babies   **b** ≈ 446 babies

7 **a** ≈ 41 days   **b** ≈ 254 days   **c** ≈ 213 days

8 **a** ≈ 5 competitors   **b** ≈ 32 competitors
   c ≈ 137 competitors

9 **a** $\mu = 176$ g, $\sigma = 24$ g   **b** $\approx 81.85\%$

10 **a** **i** $\approx 84.13\%$ **ii** $\approx 2.28\%$
   b **i** ≈ 0.0215 **ii** ≈ 0.9544 **c** ≈ 0.0223

EXERCISE 18B.2

1 **a**

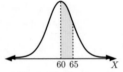

b

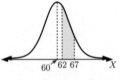

$P(60 \leqslant X \leqslant 65) \approx 0.341$ $P(62 \leqslant X \leqslant 67) \approx 0.264$

c

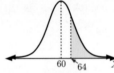

d

$P(X \geqslant 64) \approx 0.212$ $P(X \leqslant 68) \approx 0.945$

e

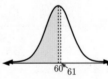

f

$P(X \leqslant 61) \approx 0.579$ $P(57.5 \leqslant X \leqslant 62.5)$
≈ 0.383

2 **a** ≈ 0.334 **b** ≈ 0.166 **3** ≈ 0.378

4 **a** ≈ 0.303 **b** ≈ 0.968 **c** ≈ 0.309

5 **a** ≈ 0.0509 **b** $\approx 52.1\%$ **c** ≈ 47 eels

6 **a** **i** $\approx 90.4\%$ **ii** $\approx 4.78\%$ **b** £4160

7 **a** **i** $\approx 12.7\%$ **ii** $\approx 52.0\%$
   b **i** 21.6 kL   **ii** ≈ 76 customers

8 **a** **i** $\approx 21.5\%$ **ii** $\approx 95.2\%$
   b **i** Enrique   **ii** Damien

9 **a** $\approx 10.3\%$ **b** ≈ 0.456

10 **a** $\approx 84.1\%$ **b** ≈ 0.880

EXERCISE 18C.1

1 a Emma's z-scores: Geography ≈ 1.61
    English ≈ 1.82 Biology $= 0.9$
    Mandarin ≈ 2.33 Mathematics ≈ 2.27

b Mandarin, Mathematics, English, Geography, Biology

c The scores in each of Emma's classes are normally distributed.

2 a Sergio's z-scores:
Physics ≈ -0.463,   Chemistry ≈ 0.431,
Mathematics ≈ 0.198,   German ≈ 0.521,
Biology ≈ -0.769

b German, Chemistry, Mathematics, Physics, Biology

3 a Frederick's z-scores: 50 m freestyle ≈ 1.95
                     100 m backstroke ≈ -1.07
                     200 m breaststroke ≈ -0.578
                     100 m butterfly ≈ 0.345

b Lower times are better as they indicate that the person swims faster.

c 100 m backstroke, 200 m breaststroke, 100 m butterfly, 50 m freestyle

EXERCISE 18C.2

1 a

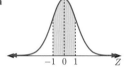

$P(-1 < Z < 1) \approx 0.683$

b

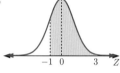

$P(-1 \leqslant Z \leqslant 3) \approx 0.840$

c

$P(-1 < Z < 0) \approx 0.341$

d

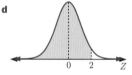

$P(Z < 2) \approx 0.977$

e

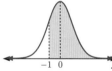

$P(-1 < Z) \approx 0.841$

f

$P(Z \geqslant 1) \approx 0.159$

2 a $a = -1$, $b = 2$ **b** $a = -0.5$, $b = 0$ **c** $a = 0$, $b = 3$

3 a

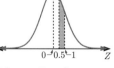

$P(0.5 \leqslant Z \leqslant 1) \approx 0.150$

b

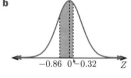

$P(-0.86 \leqslant Z \leqslant 0.32)$
≈ 0.431

c

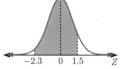

$P(-2.3 \leqslant Z \leqslant 1.5)$
≈ 0.922

d

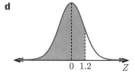

$P(Z \leqslant 1.2) \approx 0.885$

e

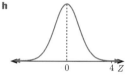

$P(Z \leqslant -0.53) \approx 0.298$

f

$P(Z \geqslant 1.3) \approx 0.0968$

g

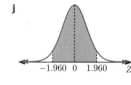

$P(Z \geqslant -1.4) \approx 0.919$

h

$P(Z > 4) \approx 3.17 \times 10^{-5}$

i

$P(-0.5 < Z < 0.5)$
≈ 0.383

j

$P(-1.960 \leqslant Z \leqslant 1.960)$
≈ 0.950

k

$P(-1.645 \leqslant Z \leqslant 1.645)$
≈ 0.900

l

$P(|Z| > 1.645)$
≈ 0.100

4 a ii ≈ 0.976

b i ≈ 0.910 **ii** ≈ 0.302

5 a i $z_1 \approx -0.859$, $z_2 \approx 1.18$ **ii** ≈ 0.687

EXERCISE 18D.1

1 a

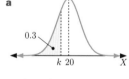

$k \approx 18.4$

b

$k \approx 23.8$

c

$k = 20$

d

$k \approx 22.5$

e

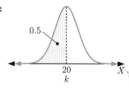

$k \approx 20.9$

f

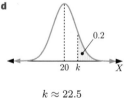

$k \approx 23.4$

2 a

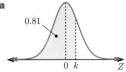

$k \approx 0.878$

b

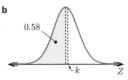

$k \approx 0.202$

c

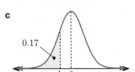

$k \approx -0.954$

d

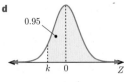

$k \approx -1.64$

e

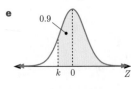

$k \approx -1.28$

f

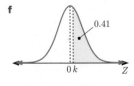

$k \approx 0.228$

3 a

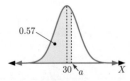

$\therefore \quad a > 30$

b $a \approx 30.9$

c **i** 0.43

    ii 0.07

4 a $k \approx 12.5$ **b** $k \approx 18.8$ **c** $k \approx 4.93$

5 a ≈ 0.212 **b** $k \approx 75.1$

6 a $a \approx 42.0$ **b** $a \approx 46.7$ **c** $a \approx 40.1$

7 ≈ 24.7 cm   **8** ≈ 75.2 mm

9 ≈ 501.8 mL to 504.0 mL   **10** $\approx 31.0°$C

EXERCISE 18D.2

1 a Greater. Data values less than 40 make up only 20% of all values.

   b $\mu \approx 45.0$

2 $\sigma \approx 3.90$ **3** ≈ 112 **4** ≈ 0.193 m

5 $\approx £96.48$ **6** $\approx 4{:}01{:}24$ pm

7 $\mu \approx 23.6$, $\sigma \approx 24.3$

8 a $\mu \approx 52.4$, $\sigma \approx 21.6$ **b** $\approx 54.3\%$

9 a $\mu \approx 4.00$ cm, $\sigma \approx 0.003\,53$ cm   **b** ≈ 0.603

10 a $\mu \approx 2.00$ cm, $\sigma \approx 0.0305$ cm   **b** ≈ 0.736

    c ≈ 0.153

REVIEW SET 18A

1 a The distribution of times taken for students to read a novel is likely to be positively skewed, and hence not normal.

   b The mean amount spent on groceries at a supermarket is likely to occur most often, with variations around the mean occurring symmetrically as a result of random variation in the prices of items bought and/or the quantities of items bought (for example weights of fruits and vegetables). So the distribution is likely to be normal.

2

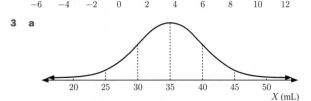

3 a

<image-curve>

| 20 | 25 | 30 | 35 | 40 | 45 | 50 |

X (mL)

b **i** $\approx 47.7\%$ **ii** $\approx 2.28\%$

4 a $\approx 2.28\%$ **b** $\approx 68.26\%$ **c** $\approx 95.44\%$

5 a $\approx 50.2\%$ **b** ≈ 7 oysters

6 a Harri's test score is 2 standard deviations below the mean.

   b $\approx 97.7\%$ **c** 7

7 a ≈ 0.364 **b** ≈ 0.356 **c** $k \approx 18.2$

8 a $\approx 6.68\%$ **b** ≈ 0.854

9 a ≈ 0.260 **b** ≈ 29.3 weeks

10 a $k \approx 28.1$ **b** $k \approx 26.5$ **c** $k \approx 25.0$

11 a $\mu = 29$, $\sigma \approx 10.7$ **b** **i** ≈ 0.713 **ii** ≈ 0.250

12 a **i** ≈ 0.0736 **ii** ≈ 0.0406 **b** ≈ 0.644

REVIEW SET 18B

1

<image-curve>

      cm

| 0 | 5 | 10 | 15 | 20 | 25 | 30 | 35 | 40 | 45 | 50 |

2 a mean $= 32$, standard deviation $= 5$

   b **i** $\approx 34.13\%$ **ii** $\approx 84.13\%$ **iii** $\approx 2.28\%$

3 a **i** $\approx 2.28\%$ **ii** $\approx 84.0\%$ **b** ≈ 0.3413

4 a **i** $\approx 76.1\%$ **ii** $\approx 96.0\%$ **b** ≈ 0.598

   c $x \approx 61.9$

5 $k \approx 1.96$

6 a ≈ 0.479 **b** ≈ 0.0766 **c** $k \approx 55.2$

7 ≈ 162 seconds

8 a $a \approx 9.05$ **b** $a \approx 13.7$ **c** $a \approx 10.4$ **9** 8.97

10 ≈ 0.0708 **11 a** $\approx 68.3\%$ **b** ≈ 0.0884

12 a **i** ≈ 0.722 **ii** ≈ 0.798 **b** ≈ 0.0563

EXERCISE 19A

1 a mean $= 70$, standard deviation $= 1.5$

   b Yes, 64 should be a sufficiently large sample for the Central Limit Theorem to apply.

2 a **i** 50 **ii** $\dfrac{9}{\sqrt{5}}$

   b, c Yes, as X is normally distributed.

3 a mean $= 80$, standard deviation $= 1.5$

   b Yes, 36 is a large enough sample.

c **i** ≈ 0.747 **ii** ≈ 0.0912 **iii** ≈ 0.252

d No, as we do not know if X is normally distributed.

4 **a** mean $= 3.6$, standard deviation ≈ 0.124

b Yes, 32 is a large enough sample.

c **i** ≈ 0.581 **ii** ≈ 0.210 **iii** ≈ 0.00767

5 **a** Yes, as X is normally distributed.

b X, as it has a larger standard deviation.

c **i** ≈ 0.309 **ii** ≈ 0.132

6 **a** Histogram B as it is closer to the normal distribution, and the values are spread out over a much smaller range.

b $\frac{64}{299} \approx 0.214$

c $\frac{205}{299} \approx 0.686$

For a normal distribution, the probability that a value is within one standard deviation of the mean is ≈ 0.6826

$\therefore$ our estimate is good.

7 **a** ≈ 0.369 **b** ≈ 0.0680

8 **a** ≈ 0.609 **b** ≈ 0.934 **9** ≈ 0.864

10 ≈ 0.908 **11** **a** ≈ 0.252 **b** ≈ 0.0228

12 **a** $\approx 14.5\%$

b **i** $\overline{X}_{64}$ is normally distributed with mean 267 days and standard deviation 1.875 days.

ii ≈ 0.0000945

c The answer to **a** may be affected but the answers to **b** remain the same.

EXERCISE 19B

1 **a** **i** ≈ -1.89 **ii** ≈ 0.0294

b As the p-value < 0.05, we reject H_0 in favour of H_1.

2 **a** H_0: $\mu = 80$ and H_1: $\mu > 80$

b $z \approx 3.40$, the null distribution is $Z \sim N(0, 1^2)$

c ≈ 0.000339 **d** As the p-value < 0.01, we reject H_0.

e We conclude that $\mu > 80$ at the 1% level of significance.

3 The customer's claim is valid at the 5% level of significance.

4 We conclude that the herd fineness has changed between 2013 and 2017, at the 5% level of significance.

5 It is not justified to adjust the machine.

6 There is insufficient evidence to support the underfilling claim at a 1% level of significance.

7 **a** Since the growth of carrots depends on many factors such as genetic makeup and environment, it is reasonable to assume the weight of carrots is normally distributed.

b The buyer will purchase the crop.

EXERCISE 19C

1 **a** In a Z-test, the null-distribution is $Z \sim N(0, 1^2)$, and does not depend on μ_0. For a given significance level α, the critical value, critical region, and acceptance region are the same for each set of hypotheses.

b Yes, the null distribution is still $Z \sim N(0, 1^2)$. The critical value, critical region, and acceptance region will be the same.

2 **a** H_0: $\mu = 1.6$ and H_1: $\mu < 1.6$

b **i** ≈ -1.96 **ii** $\{z : z < -1.96\}$

iii $\{z : z \geqslant -1.96\}$

c There is insufficient evidence to claim that the mean tread depth does not meet the legal requirement.

3 **a** **i** $\{z : z < -2.33 \text{ or } z > 2.33\}$

ii $\{z : -2.33 \leqslant z \leqslant 2.33\}$

b yes

4 H_0 will be rejected if $\overline{x} < -23.8$ or $\overline{x} > -22.2$.

5 The quality controller should not adjust the machine for $502 \text{ mL} \leqslant \overline{x} \leqslant 506 \text{ mL}$.

EXERCISE 19D

1 **a** H_0: $\rho = 0$ and H_1: $\rho > 0$

b test statistic ≈ 1.65, p-value ≈ 0.0804

c There is insufficient evidence to reject H_0 in favour of H_1 on a 5% significance level. The variables are not correlated.

2 **a** yes   **b** no   **3** yes

4 The variables are not linearly correlated at the 2% level of significance.

5 **a** ≈ 2.29 **b** $\alpha = 0.05$ and $\alpha = 0.1$

c p-value ≈ 0.0477, which is less than 0.05 and 0.1.

REVIEW SET 19A

1 ≈ 0.819

2 **a** mean $= 50$,   standard deviation $= \frac{6}{\sqrt{60}}$

b Yes, 60 is a large enough sample.   **c** ≈ 0.902

3 **a** H_0: $\mu = 90$ and H_1: $\mu < 90$

b Rosario's concerns are justified.

4 There is sufficient evidence to conclude that Yarni's pulse rate has decreased.

5 There is insufficient evidence, at a 5% level of significance, to suggest there has been a change in the colony's mean weight between 2011 and 2012.

6 $\mathcal{C} = \{z : z < -2.33 \text{ or } z > 2.33\}$

The test statistic $z \approx -0.685$ does not lie in the critical region. There is insufficient evidence to reject the manufacturer's claim at the 2% level of significance.

7 $£428000 \leqslant \overline{x} \leqslant £448000$

8 The variables are not linearly correlated at the 5% level of significance.

REVIEW SET 19B

1 **a** mean $= 500$ g,   standard deviation $= \frac{3.5}{\sqrt{20}}$ g

b Yes, as X is normally distributed.

2 **a** ≈ 0.296

b mean $= 63.4$ g,   standard deviation $= \frac{6.33}{\sqrt{10}}$ g

c ≈ 0.0447, this is the probability that the mean weight of 10 sausages will be 60 g or less.

3 No, the manufacturer's claim is not supported at the 5% level of significance.

4 There is insufficient evidence to reject the claim at the 10% level of significance.

5 **a** There is not enough evidence to say that the golfer has improved at the 5% level.

b Yes, there is now enough evidence to say that the golfer has improved at the 5% level.

6 $\mathcal{C} = \{z : z < -2.05\}$

The test statistic $z \approx -1.87$ does not lie in the critical region. At the 2% level of significance, there is insufficient evidence to conclude her form is below that of last year.

7 **a** $\overline{X}_{13} \sim N\left(\mu, \frac{3.71^2}{13}\right)$ **b** ≈ -1.94 and ≈ 1.94

c **Hint:** The significance level is the probability that the test statistic is in the critical region.

8 The variables are negatively correlated at the 1% level of significance.

INDEX